JOSEPH LOSEY
A Revenge on Life

by the same author

History

Fiction

JOSEPH LOSEY

A Revenge on Life

DAVID CAUTE

New York
OXFORD UNIVERSITY PRESS
1994

FOR MARTHA

Contents

PART XIII *The Films: A Viewing*

List of Illustrations

Introduction

'Cinema is the most interesting art form of this era of human history. But at the same time it's big business.'

LOSEY, Interview, *Film*, April 1973

'His life and art were inseparable but in a sense art came before life: that is, it took precedence where individuals were concerned. And what of the ethics of that: well, life lived and grew on his art . . . People got sacrificed sometimes.'

LOSEY ON BRECHT, '*L'œil du Maître*', 1960

'I have never made a film of which I was ashamed . . . although there have been many I would have *preferred* not to do or might have done better.'

LOSEY TO FRANCO SOLINAS, 18 April 1978

Our subject is both Joseph Losey and something wider: the motions of talent and ambition down a long gallery of characters, passions, conflicts, sets and locations. And down the subterranean passage where art is nourished by industry while savaged by its close cousin, commerce. Losey's life embraces a major crisis in political commitment (the 1930s) and public tolerance (the blacklist); his career, his *œuvre*, spans the most fundamental cultural confrontation of the century – between Marxism and Modernism, between progressive 'realism' and the avant-garde subversion of optimism.

His first film was shot in 1948; his last, by an Orwellian reversal, in 1984. Losey was both Hollywood Studio Man and European Nouvelle Vague Man; both a highly didactic messenger for good causes and a roaming lover of Old World ambiguities. His career carries us from expressionist agit-prop to a 'baroque' lavishness verging on decadence. He struggled in obscurity and exile; he became famous, lauded, honoured. His 'screen personality' – but not his character – underwent a dramatic change.

Losey's later years raise in acute form the question (loaded with contending critical vocabularies) as to what a good film really is. A striking paradox of the history of cinema is that the arrival of the 'talkies', clearly a technical and box-office advance, simultaneously threatened the film director's aspiration to fashion a new art form almost as exclusively visual as painting: *mise-en-scène*. Losey's early films, in Hollywood and Britain, were applauded by circles of dedicated film enthusiasts because they turned literary sows' ears into cinematic silk purses; because they demonstrated the triumph of *mise-en-scène*, 'the new art and science of cinema', over lousy scripts – the apotheosis of the new language.

The Paris 'cinephiles' of 1960 who likened Losey's early melodramas to the seminal works of the Renaissance and the Enlightenment were luminaries verging on high-altitude madness – though they (Bertrand Tavernier, for example) stubbornly deny the madness. To understand Losey's later work, the brilliant successes, the painful failures, we have to come down the mountain and situate ourselves within the wider art-house audience absolutely indispensable to the worldwide commercial success of a director like Losey. This audience consists of intelligent, well-educated people who don't necessarily read the *Cahiers du Cinéma*, *Positif*, or *Sight and Sound*. Appreciative of artistic innovation, this audience nevertheless remains attached to the literary-dramatic dimension of a film: well-turned characters, a plausible plot purged of melodrama, mature dialogue and motivation. This audience values quality but is not attending a film school; it does not separate décor, camera-movement, framing, editing and music from the whole; it does not possess that kind of analytical-deconstructional eye or ear.

The later Losey tended to disappoint an audience which wanted to be moved and involved as well. Experiments with time, disjuncture of sound and image, mirrors-within-mirrors – they do not banish the abiding demands of humanism or the cinema's inescapable dramatic heritage. Was Losey perhaps too hostile to the world to offer the magnanimity, the tenderness, essential to great art? Is his work a celebration of life or a form of revenge? Beyond dispute, he was a 'director's director', a pathfinder and pacemaker; but are we discussing one of the top ten film directors of his time, or was he a Yank at the Court of King Arthur (a phrase he humorously applied to himself) who had to settle for a lesser barony?

Losey's life and work may be seen as a succession of (perhaps familiar) dualities and contradictions in the successful socialist film director: disparaging of princes yet yearning for princely status, fiercely scornful of the patronage on which he depended, he wanted the money and he wanted – to use a Losey term – to fuck it. He well understood the pitfalls of the 'star system' but fell into the habit of fitting scripts rather than actors to parts. In Losey's case the romantic image of the Promethean artist fettered by prosaic businessmen should be adopted with caution.

This author never met Losey, who has been, since 22 June 1984, deaf to the host of questions one would like to put to him. But he did provide intensely personal responses to crucial questions about his life and work when interviewed by Tom Milne in 1966 and, a decade later, by Michel Ciment, the result being two book-length texts indispensable to the biographer – though by no means holy writ as regards accuracy.[1]

There are biographies of major film directors which tend to soar into a stratosphere of inspiration, sensibility and theory; the artist as Artist resides '*au dessus de la mêlée*'. In reality a film is both the finished product we see in the cinema and the outcome of a long process of collaboration and contingency – the writing, the financing, the available cast, the bad weather, the processing laboratories, the demands of the producers and distributors whom Losey loved to loathe. This book aims to depict the director as an artist requiring the skills of an industrial manager and a military commander – not to forget the perseverance of the devil. A film director is not a novelist; much depends on the collaborators he chooses (or who are thrust upon him). Losey did not, for example, write his own screenplays, and the casting of his films was invariably crucial to what is, after all, a performance art – a point evidently overlooked by the authors of a scholarly article, 'The Losey–Pinter Collaboration', which does not mention the contribution of a single actor to *The Servant*, *Accident* or *The Go-Between*.[2]

Biography seems to demand a strict observance of clock-time. Common sense links causation to sequence, to the order of events. However, Losey's biography, like his art, may kick against the tyranny of the calendar, the gloom of studying family trees before we discover why the boy born in La Crosse, Wisconsin, on 14 January 1909, is worth a book. Losey was no Welles (or Kazan). His theatre work in the 1930s and 1940s included two remarkable Living Newspaper productions and Brecht's *Galileo*, but had he died at any time before he made *The Servant* in 1963 he would have weighed lightly – a footnote – on the scales of potential biography. His major achievements (and failures) belong to the last two decades of his life. The seminal moment of his transformation was *Eve* (1962), but *Eve* was a self-indulgent apprenticeship for the masterworks which followed. I have therefore chosen, perhaps recklessly, to plunge in on Joseph Walton Losey III at the age of fifty-four, the moment when he became, with *The Servant*, a major director; when Joe Losey became 'Losey'.

So: as the credits come up against leafless winter trees, Barrett (Dirk Bogarde) is seen crossing the King's Road, Chelsea, rolled umbrella in hand, his pace leisurely but purposeful. As he enters Royal Avenue, Wren's Royal Hospital looms at an unknowable distance, against a grey sky, omniscient yet innocent. Prodding a half-open front door, Barrett moves with feline caution through the undecorated Chelsea house – then gazes down on his future master whom he discovers asleep, feet up on a chair.

PART I

Royal Avenue

I

The Servant: I

Harold Pinter* was thirty and living at 373 Chiswick High Road, London, when he first heard from Losey on 27 April 1960: 'I'd like you to know how impressed and moved I was by your play, *A Night Out*, this past Sunday. It has an intensity and inner truth both horrifying and purgative. There are few things, if any, I have seen on British TV that can compare with it. Congratulations.'

Thanking him, Pinter mentioned that Losey had spoken of the possibility of a film script when talking to Pinter's agent, Emmanuel (Jimmy) Wax: 'If this is a possibility, I'd certainly be interested to know what you have in mind.' Losey had nothing in mind. On 18 May he told Wax: 'I thought the play at the Arts [presumably *The Caretaker*] was beautifully played and staged, but I was not nearly so impressed with it as I was with the television.' Two and a half years elapsed before Pinter's collaboration with Losey began. Writing to Leslie Grade in May 1963, after shooting *The Servant*, Losey described Pinter as one of 'the half dozen top playwrights of this decade in the world, and a playwright who is in every way identified with what is contemporary, new and even fashionable'. Losey later told Michel Ciment: 'I don't remember one single production of any of Harold's plays in that period of time that I liked ...' Pinter's soaring international reputation – and

---

*Born in 1930, Harold Pinter had objected to National Service on grounds of conscience. After studying at the Royal Academy of Dramatic Art and the Central School of Speech and Drama, he worked as an actor, mostly in repertory under the stage name David Baron, supporting himself as a waiter in the National Liberal Club, as a dance-hall doorman, as a dish-washer, and as a door-to-door bookseller. His first (one-act) play, *The Room*, was followed by *The Birthday Party* and *The Dumb Waiter* – all in 1957. Almost unanimously abused by the critics, the first run of *The Birthday Party* closed in London after six nights. The play that made his name was *The Caretaker* (1959), which transferred from the Arts Theatre Club to the Duchess and ran until May 1961, winning the 1960 *Evening Standard* Drama Award. A successful Broadway production followed.

commercial success – as a playwright both pleased and unsettled Losey. Would this most prized of collaborators slide out of reach into his own slipstream?[1]

Losey first met Robin Maugham – the son of a Lord Chancellor and nephew of the writer W. Somerset Maugham – in the early 1950s. His interest in Maugham's novella *The Servant* was already apparent in August 1955, when he proposed a stage adaptation to Hugh Beaumont. The response was negative. In June 1956 another impresario, Stephen Mitchell, returned *The Servant* to him as 'too trivial'. By February 1961 Losey had heard that the director Michael Anderson had acquired the film rights and commissioned a screenplay by Pinter. In June Losey was in correspondence with Maugham about renting his Brighton flat at £10 a week. Maugham advised: 'I think you could persuade my housekeeper, Mrs —, to come in for two hours every morning to do your shopping – but I am afraid you will not be able to persuade her to do any cooking apart from breakfast . . .' She was the ailing vicar's wife and clearly no Barrett.

Later that year, Losey learned that Michael Anderson had not succeeded in raising funds and that the rights were buyable. While shooting *Eve* in Rome, Losey heard from Bogarde, who'd just read Pinter's screenplay in a hotel room overlooking the Bay of Cannes, the rattan shades slatting shadows across the pages. Bogarde was excited – Pinter wrote with 'the precision of a master jeweller . . . His pauses are merely the time-phases which he gives you so that you may develop the thought behind the line he has written.'[2]

Published in 1948, *The Servant* became something of a cult, though badly written and short of stamina. The setting is Chelsea in the immediate post-war years of austerity. Tony, the doomed hero, and Richard Merton, the narrator, have been regimental colleagues, two officers with a 'good war' behind them and very decent chaps.

Tony engages a new butler, Barrett, who 'knows the black market for miles around'. Describing the relationship between master and servant from the outside, through the narrator's third-hand report, Maugham provides very little dialogue – the heartland of any screenplay – between Tony and Barrett. Indirect narrative is alien to dramatic form; Pinter was obliged virtually to reinvent the master–servant relationship and to eliminate the narrator. With the narrator disappeared Maugham's archaic, leather-sofa mode: 'I interrupted. "We're old friends now," I said. "So do you mind if I ask a few questions?"
She smiled. "I don't promise to answer them all," she replied.'

Pinter updated the story from the post-war period to the early sixties. His Tony is a feckless young socialite (though not conspicuously social) listlessly dreaming of projects in Africa and Brazil while moving into a smart house in Chelsea and surrounding himself with family portraits and silverware. But the social scene of 1960 was no longer that of 1948. Rising standards of living had reduced the number of live-in domestic servants; for a young bachelor to employ a butler was now almost flamboyant, rather than the normal comfort of the officer class. Audiences viewing Losey's film in the 1960s were thus confronted by a domestic set-up which was exotic to the point of hedonism, or affectation, rather than normal. Pop stars had butlers.

Pinter salvaged Maugham's story from its blatant snobbery. Maugham's Barrett has a tendency not only to 'smirk' and 'snigger' but also to dress in poor taste, 'like an undertaker out on a spree'. His original sin is clearly poor breeding. Maugham's portrait of evil was also garnished with naïve sexual symbolism. His Barrett had a 'high-pitched querulous voice . . . He spoke in a prissy, affected voice, and the word "sir" sounded like "sahr" . . . in the middle of his sallow face were stuck a pair of rosebud lips, which gave him the look of a dissolute cherub. His lids were heavy and looked oily, I remember.' Not only Barrett's eyelids but also his voice was 'oily' – a typical public-school tag for Jews – and Maugham's Barrett is very much the insidious alien of the anti-Semitic tradition. Elsewhere the narrator describes him as 'a fish with painted lips', cold-blooded, not manly, faintly perverted, sycophantic. Pinter's screenplay rescues Barrett from the latent racism of the public-school subconscious while preserving what is most interesting in Maugham's novel: Barrett embodies the great fear of the English upper classes as to what will happen if ever their servants begin to use their brains ('cunning'). For Maugham's narrator, the ultimate horror is that his friend and the servant do crossword puzzles together.

Maugham's Vera was the fantasy-strumpet of the juvenile male imagination, a 'nymphomaniac' (another public-school term, suggesting both sex unlimited and a moral alibi for the honest male 'victim'). Maugham's narrator reports that Vera's 'particular weakness' is 'lust' and she ends up as a hungry streetwalker. Invited to her filthy little room off Shaftesbury Avenue, after a chance meeting on the pavement, the gentlemanly narrator is rising to leave when Vera begins 'breathing heavily'. It isn't the money she's after because he has already given her four pounds out of pity. 'She was panting now . . . All her body was trembling.' Pinter partially discarded these boyish fantasies, transforming Vera into a happy, amateur hooker of the permissive sixties.

Pinter's screenplay is unspoiled by the standard-formula grafts of melodrama, coincidence, and exposition which had marred Losey's earlier films. For the first time in thirty years Losey was offered the primacy of the implicit over the explicit, conflict percolating through the masking tape of received language, idiom and gesture.

What particularly attracted him to Pinter's adaptation were new scenes not found in the novel – from Barrett's rehiring to the final party. But Losey also wanted major changes: '*The Servant* was [Pinter's] first screenplay . . . and he was insecure. I confronted Harold who was – if anything – no less arrogant because he wasn't sure that he was Harold Pinter yet. And he said, "I'm not accustomed to writing from notes and I don't like this."' Pinter disputes Losey's recall beyond this point, but they are agreed that they both learned rapidly how to work harmoniously together. Clearly Pinter was amenable and receptive to Losey's editorial comments – provided Losey jettisoned the ingrained director's posture: 'Look, son, I know this business.'[3]

The successive scripts of *The Servant* offer a fascinating study of Losey's imaginative interaction with Pinter. The first screenplay (written for Michael Anderson) is dated 3 July 1961, and is henceforward referred to as *1961*. The copy studied carries Losey's annotations. Occasional carelessness in Pinter's *1961* script indicate something less than the full, imaginative commitment which emerged in 1963. Not only is Susan called Sally but Vera is usually referred to as Barrett's niece (as in Maugham) and sometimes as his sister. The revised script of January 1963 also carries Losey's handwritten annotations, and will be referred to as *1963*.

The Release Script reflects a number of Losey's annotations on *1963*, as well as other changes not previously signalled. Pinter's own printed version, not published until 1971, closely corresponds to the Release Script.

Losey's annotations not only reveal an acute ear and eye for British social detail, they also reflect his reading of individual characters and their potentialities. In *1961* Sally (Susan) and Tony have the following exchange about Barrett:

> SALLY: Does he give you breakfast in bed?
> TONY (*laughing*): No. I draw the line at that.

Losey intervened with a reversal, the *1963* version being:

> TONY: Of course.

In *1961* Tony berates Sally/Susan for criticizing his servant, Barrett – and therefore his own judgement.

> SALLY: All right. I'm sorry. I'm a fool.

Following a Losey suggestion, *1963* reads:

> TONY: I'm sorry. I'm a fool.
> SUSAN: You are.

In *1961*, Barrett tells Tony that Vera's skirts are too short.

> TONY: Rubbish. You're just an old stick in the mud . . .

Losey noticed the problem without offering a solution: 'Rewrite.' Pinter's *1963* text is better:

TONY: Well what the hell do you want me to do about it?

Losey admired Pinter for his lack of wordiness, but in *1961*, during the critical struggle between Barrett and Susan over flowers in the sick Tony's bedroom, Pinter's text reads:

BARRETT: I beg your pardon, Miss, but I'm afraid I really can't allow flowers in the patient's room without the doctor's permission. I'm sure you understand why. May I?

Losey crossed this speech out, substituting a stage direction: *He moves the flowers*. Pinter responded in *1963* with: *Barrett picks up the vase . . .*

It was Losey's idea that Barrett should wear white gloves when serving at table. He wanted Tony to comment on the gloves: 'Bit of Italian chi-chi.' Pinter (*1963*) came back with an improvement:

BARRETT: They're used in Italy.
SUSAN: Who by?

Some of Losey's major script cuts were no doubt dictated by a small production budget. He also edited twenty minutes out of the completed film, including a scene which showed Barrett living cheap by sleeping with his landlady and hurrying to Fuller's cake shop to satisfy her sweet tooth. Sellotaped to the wall of Barrett's room were pages from 'cheap sex magazines', 'pornographic calendars', 'nudes stuck in oil paintings'. This set-up was Pinter's, without debt to Maugham, and Losey cut it even though it was 'one of the best I've ever done in my life'. Losey emphasized Barrett's fascination with power, dominance, by wiping out indications of the man's vulgar libido. The emergent Barrett of the film can take sex or leave it; occasionally he has it off with Vera, but only to work up an appetite for a good smoke.[4]

On 14 March, while shooting, Losey turned to Pinter for an additional scene which would be highly stylized and set in a restaurant. Pinter asked for specifications. 'I recall that I told him "Vignettes: a young Bishop and curate, and two lesbians, a society girl and her escort."' By 19 March it was done. Pinter had broadly followed Losey's brief but (he explained to Losey): 'The priests suddenly became Irish. What do you think?' He added a personal word: 'Let me say here in black and white that it's been a great pleasure to work with you.'[5] It was mutual. Losey paid tribute to his new screenwriter: 'Pinter's words are few, economical, exact . . . Pinter also appreciated the usefulness of the overheard incidental line; of dialogue as sound effect . . . He understands how often the human creature uses words to block communication.'

Initially, Losey had no more luck than Michael Anderson with producers and distributors. 'The screenplay ... attracted, repelled, stunned and mystified many people. There were those who said it was an unshootable script, and to some extent they were right.' Leslie Grade was the saviour. On the look-out for a low-cost film of potential quality, he stipulated that if the budget could be held at £150,000, Elstree (a 50–50 combination of Leslie Grade and Associated British-Pathé (ABPC)) would put up 70 per cent. The National Film Finance Corporation (NFFC) completed the financing, with Pinter, Bogarde, Norman Priggen (co-producer) and Losey agreeing to deferments. The rights were acquired from Michael Anderson for £12,000.

The production budget of Losey's company, Springbok Films, itemized £14,500 for 'story and script'– £11,500 to Maugham, £3,000 to Pinter – a dramatic indication of the playwright's modest commercial status at this juncture. Losey (director and co-producer) and Priggen received £15,000 each. A mere £18,570 was allocated for the entire cast. Among the actors, Bogarde was paid £10,000 or, by his own recall, £7,000, Sarah Miles £4,000, with £2,300 for Wendy Craig as Susan, and only £1,000 for James Fox in the co-leading role, Tony. The studio rental fee at Shepperton was £7,500. The total projected budget was £141,725, but Losey later put the actual cost at £135,000. Was this a cheap film? Comparative budgets for other films in the same time-frame as *The Servant* are instructive. Jean-Luc Godard stated the budget for *Les Carabiniers* at $104,000 (£37,000); Joseph Strick directed *The Balcony* on a budget of $165,000 (£59,000); Roger Corman shot *The Intruder* for 'a little over $100,000 (about £36,000); Clive Donner made Pinter's *The Caretaker* for £30,000; Lina Wertmuller's *I basilischi* cost £28,000.[6]

As his co-producer Losey enlisted the services of Norman Priggen, known in the trade as 'Spike'. He was to serve in the same managerial capacity on seven of Losey's next eight films, their relationship and personalities not unlike that of colonel and sergeant-major. In the course of Losey's published conversations with Ciment, he mentions Priggen three times but says nothing substantive – on the top shelf of his consciousness he held little regard for the fetch-and-carry administrators on whose loyalty and efficiency he depended.

Born in 1924, Priggen had served as an RAF gunner during the war, then worked in Ealing Studios as third, second and first assistant director on such films as *Kind Hearts and Coronets*, *The Lavender Hill Mob* and *The Cruel Sea*. In December 1962 he received a rush of telephone messages from Losey who (he recalls) had a bad name in the industry: 'Production people said, "Stay away, he's a difficult man".' Determined to watch Fulham football club on the Saturday – he had a season ticket – Priggen called on Losey in Kensington before the match. 'Joe said this was a challenge, he gave me a script, and told me not to go to the match.' Priggen read Pinter's script while Losey plied him with Scotch. Over the weekend Priggen phoned Shepperton Studios and obtained a rental deal. By eight on Monday morning he and Losey had arrived at the NFFC with a budget. Bogarde regards Priggen as 'just what Joe needed' – efficient, technically competent, and willing to read the riot act. 'He was tough but without artistic pretensions.'[7]

Losey had first shown Maugham's novella to Bogarde – 'in lime-green covers', Bogarde remembers – after shooting *The Sleeping Tiger* in 1954. At that time Bogarde would have played the young master, Tony, but a decade later he was forty years old. Uncertain about playing the sinister butler, he suggested Ralph Richardson – but was told by Losey, 'He's too expensive, you'll have to do.'

There is no overt hint of homosexuality in Maugham's novella or Pinter's script. When Losey told a journalist, 'I don't want the film to be simply a study of a little homosexual affair,' he was being more provocative than serious. (In 1963 the film censorship code proscribed any direct, on-camera depiction of homosexual passion; eight years later, by which time consenting homosexual activity was no longer a criminal offence, *Sunday Bloody Sunday* showed a passionate kiss between two men, then took them into bed.) Nevertheless, Losey, Pinter and Bogarde are agreed that Tony Forward, Bogarde's companion and business manager, was nervous about the prospect of Bogarde acquiring a homosexual image in the cinema. As Bogarde puts it, Forward was afraid of 'two in a row', following *Victim*. (Bogarde had committed himself to the leading role of a bisexual lawyer in Basil Dearden's film *Victim*, which appeared in 1962, and which impressed Losey. Modest in its aspirations, *Victim* made an honest and highly committed plea for legal reform by depicting the existing law as a blackmailer's charter.) Bogarde remembers asking a rather inebriated Losey and Pinter whether *The Servant* was to be a homosexual picture – neither answered. 'I saw nothing homosexual in it,' Bogarde adds.[8]

Bogarde not so much grew into the part of Barrett, but grew the part. While making films, he told *Isis*, 'I nearly always do wear the characters a bit, but I wore Barrett really with a zip fastener down the back, it was extraordinary.' Losey won Bogarde's deep esteem as a director. Patient and considerate, he was immune to the constricting stop-watch philosophy then in the ascendant. 'No frantic cry from the Script Girl that a scene, a speech, a move even, had overrun its time.' Losey created spaces, physical and emotional, in which the actor's craft and intuition unwound. 'Like all the greatest directors, Losey never tells one what to do, or how to do it. Ever. Only what not to do.' There was no superfluous chatter, no in-depth discussions about motivation, soul, identity, truth. Bogarde also paid tribute to Losey's superb sense of the values of texture, of things, of wood, metal, glass, petals, playing-cards, snow. The actor was not placed 'in' a set, but made to feel an inseparable part of the tactile environment.[9]

Bogarde first spotted James Fox while watching a short television play, *The Door*, with Ann Todd and a handsome young actor calling himself 'Maurice Oliver'. Bogarde telephoned Losey in excitement, urging him to watch the rest of the young man's performance. 'Maurice Oliver' turned out to be one of the gifted sons of their mutual agent, Robin Fox. Soon afterwards Bogarde, Losey and Robin Fox talked to Sarah Miles at a film première supper. She, much in demand after a film with Olivier, *Term of Trial*, expressed a strong desire to play Vera in *The Servant*, and suggested her current boyfriend for the part of Tony – none other than James Fox (known by his real name, Willy). Losey screen-tested him before a silent camera, nothing spoken. Fox says it was done in 'someone's flat in Queen's Gate' (clearly Ruth Lipton's), at Losey's expense, 'crude and haphazard, no script'.[10]

According to the critic Paul Mayersberg, who observed the shooting of the film, James Fox's inexperience forced Losey to re-shoot scenes in take after take. Losey's skilful editing, in Mayersberg's view, was partially responsible for Fox's eventual glittering success on the screen. The film's editor, Reginald Mills, told *Isis* that Fox's performance was 'entirely Joe's performance rather than James Fox's. I think he was very malleable to the direction . . . a perfect piece of casting.' Fox himself recalls that Losey achieved his results by sheer conviction and commitment rather than by detailed tuition. Bogarde insists that Fox, although ignorant of the 'rules', displayed all the right instincts, knew exactly what he was doing, and caused Losey nothing but delight: 'Ripe as a peach.'[11] Fox himself commented to *Isis* that he found Tony 'terribly one-dimensional, I mean one does know stupid people, but not quite such stupid people . . .'

Tensions developed on the set where least expected. Sarah Miles and Willy Fox were already at odds when shooting began. (A self-confessed rebel-delinquent, at the age of twelve Miles had responded 'I hate it' when politely asked by the Queen Mother how she liked her posh school, Roedean. Expulsion followed.) The big, on-set, storm came in the scene where Vera is seducing Tony with the aid of her bare legs and a revolving leather chair. Losey had kitted out James Fox, as he recalled, in 'a kind of short karate dressing-gown, and fur boots. The fur boots were actually mine . . . the same fur boots that I used for Stanley [Baker] in *Eve*.' Sarah Miles arrived on stage. 'She was living with him, as everyone knew, and she said, "Oh you great stinking queer, you think I'm going to play the scene with you in that grotesque costume, you're crazy . . ."' Losey told her to go to her dressing-room and decide whether she wanted to be in the picture or not. 'She was an absolute bitch that day. I had to get her agent, who was also mine – Robin Fox – on the set . . .' Bogarde remembers Losey reading the riot act to Miles by showing her the 'Vera' scenes in the script that he was prepared to eliminate. In the end James Fox opted to play the scene barefoot, without Losey's fur boots.[12]

After shooting was completed Miles sent Losey a friendly note – she looked forward to her next picture with him 'more than anything in the world', adding elusively: 'You know how I feel about you – but this can't make you angry with me – I just have to grow up, that's all.' However, in July he wrote to her: 'I hear reports of your extreme dissatisfaction with me and the results of the picture,' adding: 'Your surface glitters, but the gold remains to be revealed, not only to audiences but yourself also.' He then softened: she should come to Venice.[13]

It was Bogarde who suggested Wendy Craig for the part of Susan. This turned out to be the least satisfactory role in the film, but the fault lay in the part rather than in Craig, who delivered a dedicated and professional perform-ance. Although he treated her well on-set, and she was moved by his gasping for breath (she knitted him a scarf while he was suffering from pneumonia), Losey was later inclined, during interviews, to slough the blame off on to the actress or her accent. He told Tom Milne: 'I asked her to try to assume an upper-class manner and accent; and it was utterly hopeless . . .' (In fact there is no fault in Craig's accent; Bogarde's theory that a northern Barrett would size up a northern Susan's social pretensions does not come across to the audience.) In 1969 Losey told a *Telegraph* journalist that Wendy Craig hadn't 'worked', drawing a pained letter from the actress: 'Knowing the back-stage story of *The Servant* and the dedication I felt for the role, which, let's face it wasn't the most rewarding in the world . . .' she couldn't believe that he could say she hadn't worked hard. Losey hastened to reassure her: 'Certainly nobody ever doubted the degree, intensity or quality of your work. Oh hell. Love.'[14]

One of the glories of *The Servant* was its overheard cameo roles, its passing pick-ups of idiosyncratic conversation and sealed-off obsessions. Bogarde told *Isis*: 'Then you'll find that our chums played all the tiny parts in the film because we couldn't afford to pay them . . .' (One of these 'chums' was Losey's estranged wife, Dorothy Bromiley, who played a girl waiting outside a telephone box in wind-driven rain while Barrett made a call.) Losey invited Patrick Magee, Alun Owen, Pinter himself, Doris (Knox) Nolan, and Ann Firbank to take part in the additional restaurant scene for a maximum fee of £25. Dorothy Bromiley's future husband, Brian Phelan, who'd played Pauley in *The Criminal*, obeyed a summons: 'I did it very Dublin. Just one take.'[15]

Shooting began on 28 January 1963 and continued into March. A week after filming started, in the coldest winter for years, Losey collapsed with pneumonia. Letters were exchanged between the producer, the doctors and the insurance company. X-rays and lab reports were submitted. Fearing that the producers would abandon the film and collect the insurance, Losey instructed Bogarde and Richard Macdonald to take over. Despite calls to the director's bedside for instructions, Bogarde found that Losey's meticulously annotated script 'covered every movement of camera and action, and every shade of dialogue'. When Losey returned to the set, gaunt, grey and weak, he was applauded. Frequently he lay down on an iron bed, swaddled in blankets and hot-water bottles, with a nurse in attendance. Everyone was asked not to smoke. Losey watched the footage that Bogarde had shot without comment, hoicking and spitting into a pile of paper handkerchiefs, then reshot most of the scenes.[16]

The Servant involved minimal location shooting. Royal Avenue was used only for exteriors; the home of Lord and Lady Mountset was shot at Chiswick House. As credited production designer on a feature film for the first time, Richard Macdonald exercised direct control of design on the set and built a spiral staircase (Losey recalled) 'right up to the top where it ends in an absolute blank trap in the room of Vera. And to the bottom where it ends in a trap that opens on to Royal Avenue in the snow. That was ... a masterful achievement ...' Losey constantly tortured the staircase into new perspectives, angles, shadows, vantage points, power points – a motionless object which was never at rest; the mute orchestrator of the human quartet, its banisters lovingly polished by Barrett, the coiling snail linking the separate rooms, each of them a place of hierarchy, privilege or sexual subversion.

Although a young man at home in 'swinging London', Tony's values (or his self-esteem) are restorative; his hesitant sense of the modern (an Elisabeth Frink over his bed) must be validated by borrowing the past, partly the Regency period, partly the style of the nineteenth-century club. Losey commented that Macdonald made the film look 'twice or three times as expensive as the money actually spent on it'.[17]

Losey's and Macdonald's purpose was to take the audience inside the house and keep them there. 'When you go in, stay in,' says Macdonald. 'Bergman burns out his windows.' The interior undergoes subtle transformations after the decisive rehiring of Barrett (Tony's Waterloo). Macdonald explained it as 'a creeping over the rest of the house of the disarray which started in the kitchen ... the curtains were hardly ever pulled back ... The flowers were replaced by dead twigs and pussy-willow or something ... we painted the niches and the ceiling-pieces black ...' The lighting cameraman, Douglas Slocombe, explains that the interior of the house passed through four phases: empty and cold; painted and beautiful; rotting; partially redecorated with a 'gaudy, meretricious look'. In Losey's words, 'a house airless, without composition, all smoothness and comfort gone, the gloss gone'.

Losey had wanted Christopher Challis, photographer of *Blind Date*, but Priggen notified him that Challis was not available. 'Whatever he's doing, get him,' Losey growled – the picture simply could not be made without Challis. Priggen then suggested Douglas Slocombe, with whom he had worked for fifteen years on Ealing Studios classics. Losey, Slocombe and his much-admired operator, Chic Waterson, embarked on a number of sustained shots running up to five minutes without intercutting. Slocombe comments that these long takes – particularly where 360 degree pans were involved – are difficult for the cameraman because he runs out of floor space to position lamps. Moreover, the sets for *The Servant* had ceilings, so no light could be found through the open top of the set. Keeping the camera and crew out of the curved mirrors was a further headache. Losey's masterly use of space and his exacting attention to detail impressed Slocombe and Waterson – stripping, repainting and polishing a banister; the careful arrangement of objects which the lighting cameraman intended to place in shadow; his habit of opening up scenes by dwelling momentarily on an inanimate object, usually an elaborately ornate one, a sculpture or a silver tray. 'If you finish one take on a face, don't cut to another face or people will think they're in the same scene,' Slocombe comments. Chic Waterson attests that *The Servant* stands out from the hundred movies he has made: 'Joe was the greatest as far as I'm concerned, a complex character.'[18]

Reginald Mills had been editing the film since the first rushes and produced a rough cut the day after shooting was completed. Mills regarded 'dissolves' or 'fade-outs' as 'very slothful devices', and reckoned he'd influenced Losey to eliminate opticals since they first worked together on *The Sleeping Tiger*. Losey's editing eye is reflected in his twenty-six Cutting Notes for Mills dated 6 April. They included two in which Losey mistook St Pancras Station for Euston:

6. High shot Euston not now in right place. Should go from LS to LS. Also high shot too long.
7. Shot of Barrett and Vera descending Euston steps. Eliminate pan from tower and cut much closer featuring candy bar.
16. Ball game. Start with ball into lens as Tony throws from landing. Back to LS as Tony throws from landing. Hold LS through breaking of vase if possible.
22. In final sequence, hold kiss through Barrett's snort.

Losey later regretted having edited out a shot in subtle slow motion when Vera brings Tony tea in bed; he watches her while pretending to be asleep.[19]

Interviewed by *Isis*, Mills, a veteran of the studios with a campily condescending attitude towards new talent, decided to condescend to Pinter, who, he remarked, 'should write the libretto for an opera. He repeats everything six times . . .' Mills also claimed that Pinter 'didn't really dare say anything to Joe . . . I cut out a lot of his lines and I'm sure he never knew; it just meant one repetition less.' This comment provokes speculation as to whether Mills influenced Losey's decision to cut the filmed early sequence involving Barrett and his landlady, where the repetitions, essential to Pinter's theatre writing, created a fourteen-line monologue for the landlady:

LANDLADY: I'm going to ask a couple of quid more for this room. It's worth a couple of quid more . . . Get a couple of ballet dancers in here . . . I can always get a couple of ballet dancers in here. There's any amount of ballet dancers . . . [and so on].

Mills's *Isis* interview then passed to comments about Pinter's clothes and accent. Mills received a letter written in cold fury, including the following remarks:

'I must apologize to you for my "Italian suit, old boy voice, and my incredibility". I have never considered myself incredible, but must clearly look again.

'Nor do I consider myself a dilletante [*sic*], an amateur or a cunt, all of which attributes are clearly implied in your comments. Nor am I "an awfully nice person". Nor do I wish to write a libretto. Thank you, however, for the suggestion.

'In plain terms, I, in turn, am "staggered" by your condescension and contempt.'

Mills was invited to launch a weekly gossip column or to 'go and fuck yourself, a function I am certain you will discharge with the utmost professionalism'.

The Servant, like *Eve*, celebrated Losey's love of jazz. He took pleasure in writing expert notes on London's various jazz clubs, his preference being for Ronnie Scott's in Gerrard Street – although no alcohol was served. John Dankworth and Cleo Laine brought a haunting quality of hedonism and lament to the film. Losey regarded Dankworth as one of the few first-rate musicians in England who took film music seriously. 'He is the only English jazz musician whose work as composer goes way beyond that of jazz.'[20]
Misha Donat noted that Dankworth's score divided into four thematic units: the opening theme (Tony and Barrett); the 'Vera theme', almost invariably on the tenor saxophone; the Cleo Laine song, occurring throughout the film; and the 'Mountset theme', heard only during the two Chiswick House sequences. The opening theme is sentimental, scored mainly for strings; as Barrett comes into the frame, saxophones intrude. Later, when Tony and Susan lie on newspapers in the unfurnished sitting-room, the jazz track is scored for saxophones and piano. The song theme ('Now while I love you alone') is associated with almost all the scenes in which Tony is alone with either Susan or Vera. The cross-cutting between the restaurant and Vera's arrival by train is matched by parallel musical patterning, and this

continues throughout the film, an ironic, mood-laden commentary not only on the present action but on the various dramatic variations. The hide-and-seek sequence on the stairs is all saxophones apart from a marimba or xylophone.[21]

The most haunting musical motif is Cleo Laine's song which Tony keeps permanently on his record player, 'Leave it alone, it's all gone', for which Pinter wrote the lyric: 'Can't love without you/Must love without you – alone . . .' According to Losey, 'John Dankworth, the composer, complained bitterly that Harold was no lyric writer.' Dankworth recalled doing some 'pilot lyrics' for Pinter to work on, 'to paraphrase just the actual number of beats – he'd never done any song writing before . . . But he didn't seem to be able to get on with this pattern . . .' The solution was for Pinter to write lyrics and Dankworth to compose a new tune on the basis of three different arrangements: simple, then more complex, then cacophonous in the final party scene.[22]

Losey had cultivated a good relationship with John Trevelyan, Secretary of the British Board of Film Censors, who had lunch with Losey and Mills on 21 May, before viewing a rough cut. Trevelyan raised a number of warnings which fell short of demands: '. . . it would be desirable for you to end the swivel chair scene with the chair swinging round so that its back blacks out the camera. I think that the final shots with [Vera's] legs at one side and [her] head on the other might well run into trouble.' (But this shot survived.) Losey had from the outset anticipated an X certificate, and was relieved that in the event Trevelyan imposed no cuts.[23]

2

The Servant: II

Wearing a pork-pie hat and carrying a rolled umbrella, Barrett (Dirk Bogarde) is seen crossing the King's Road, Chelsea, beneath the royal crest of Thomas Crapper's lavatory shop ('Sanitary Engineers, by Appointment'), and walking down Royal Avenue beneath leafless trees.

Finding the front door ajar, in squeaking shoes he cautiously penetrates the empty, uncarpeted corridors of a house stripped bare by sale, until he catches sight of the stretched-out legs of the sleeping owner, Tony (James Fox). Rich, indolent Tony – too much beer for lunch. Barrett coughs discreetly.

There are two chairs in the house: Tony forces the deferential applicant to sit rigid in one of them while circling round him with inquiries about his credentials and capacities as a cook. In future Tony will do the sitting and Barrett the standing. Tony – who has surrounded himself with family heirlooms, ancestral paintings, splendid silver services – takes it for granted that a bachelor in his position needs a 'manservant'.

The house is soon splendidly redecorated in cool colours, with Barrett threading his laden silver tray between workmen's ladders (his remarks and their glances indicating his allegiance to the master). Bogarde's Barrett dresses correctly – when he needs to. When he doesn't, an unfastened collar, or a tie-knot dragged down, signal the coming subversion.

Tony's relationship with his fiancée, Susan (Wendy Craig), is established as they dine out. She is well-groomed, practical, a potential wife and mother, by no means a romantic figure. Her hostility to Barrett, whose servility barely masks his insolence, is apparent from their first encounter; the uneasy trio are often glimpsed in a beautiful, oval, concave mirror placed over the fireplace. Tony himself seems to dwell in a time-warp. Out in restaurants, night-clubs, pubs, he is part of swinging Chelsea; at home he pursues pretensions of service and grandeur with the gravity of a clubman grown middle-aged before his time. Tony does not really know who he is or what part – in business, in love – he is cut out to play. Most of the time he solves the dilemmas by idleness. Barrett spots his prey and goes about his work, whistling Liszt's Hungarian Rhapsody No. 6 in D flat.

Trees are reflected in a pool of water on the pavement – a time lapse and also a delicate reprise of the bare branches seen against the sky in the title shot – then the camera pans up to the façade of Tony's house. In Tony's life there is no hurry.

In *Accident* and *The Go-Between*, the camera will adhere to a narrative

convention of singularity; in *The Servant*, of plurality. In *Accident* everything
is experienced by Stephen; in *The Go-Between*, by the boy Leo. Having
abolished Maugham's narrator, Pinter created a world where the only
'privileged' vantage point was the spectator's. The four main characters are
constantly paired off: Barrett and Tony; Susan and Tony; Barrett and Vera;
Vera and Tony; Barrett and Susan. The effect of this pairing is to emphasize
the exclusion of those not present – a succession of shifting conspiracies, with
Barrett alone omniscient even when not present. Gilles Jacob commented: 'In
The Servant the emphasis given to angles and staircases should be seen not so
much as a strategic advantage or disadvantage given to a particular character,
as the latent symbolism of the complex moral ascendencies and submissions
which work as interdependently as the water-levels in a chain of locks.'[1]

As for Barrett, is he *discovered* by the camera smirking and smoking down
in the kitchen? Or is it Barrett himself who *invites* us to spy on his game?
Bogarde's finely graded expressions – lips slightly parted to indicate secret
knowledge, eyes alight with furtive pleasure or solemn with deception, hastily
tossing a cigarette into the sink at his master's approach – perfectly excite the
spectator's desire to know what the game is. Pinter, Bogarde and Losey keep
us guessing; each hint is sly, elusive, terminal. Close-up of Tony and Susan
kissing; the camera pulls up to show them lying on newspapers in the
uncarpeted drawing room. Enter Barrett, without knocking. Susan is furious,
unforgiving. The battle is on. When she sends flowers to the sick Tony,
Barrett promptly moves them out of the bedroom. She brings them back in.
Cut to Barrett showing her out of the front door with a faint smirk:

BARRETT: I'm afraid it's not very encouraging, miss . . . the weather fore-
cast.

Barrett collects his 'sister' Vera (Sarah Miles) from the station, her pretty
legs trotting down the broad steps of St Pancras, he a confident gentleman's
gentleman, she a bubbling little provincial girl gawping at London from the
taxi while gobbling a chocolate bar. Arriving at Tony's house she is overawed
– never seen anything like it. But Vera knows her assignment. If not a
prostitute by trade, she's at home with the game.

But what is Barrett's game? In Pinter's view, Barrett initially phones Vera
because he wants to sleep with her himself. This interpretation, though perhaps
latent in the script, really depends on Bogarde's performance, which Pinter
liked because it wasn't 'completely Machiavellian . . . a certain dimension of
roundness in the character'. When asked whether Barrett entertained the idea
of dominating Tony right from the outset, Losey replied: 'Oh yes. He had the
idea of invading Tony, of bringing in his mistress, his friends, etc. Of taking
him over . . . I had some servants like this. They've tried it – it didn't work.'[2]

One doubts whether this was Pinter's position because in play after play he presents characters for whom 'intention' is a facet of response – they create themselves, step by step, by pressing against other people. Bogarde's acting strategy was to play up the 'menacing' quality in Barrett, and the imposture (confusing Viscount Barr with Lord Barr); by this interpretation Barrett never had been a gentleman's gentleman – though he'd observed the act from below stairs. 'I don't think he'd ever made a soufflé in his life,' Bogarde says.[3]

Naked in bed, Tony is a bit put out to have his breakfast brought by an attractive young woman – a perfect example of a Pinteresque 'intrusion', echoing Barrett's original discovery of Tony asleep in a chair. 'Where's Barrett?' Tony complains. Moments later we find Vera smoking in the kitchen, grinning, vampishly beckoning Barrett by pulling her short skirt up her legs. 'As he draws closer to her her head is suddenly flung back in a soundless sexual laugh, mouth open,' writes Pinter.

This is taxing on the actress, who is forced close to the extravagant. On the other hand, there is a 'Brechtian' aspect to the acting here: Bogarde and Miles are playing it up for the benefit of the audience. Their working-class lives are a series of expedients, up or down on your luck, living by the hour, like betting on horses (and Barrett will later pretend, to Tony, that Vera ran off with 'a bookie from Wandsworth' (Pinter) or 'a bookie from Wandsworth – Wandsworth!' – as performed by Bogarde). Barrett and Vera know that mischief can flower for a season; theirs is unplanned socialism.

Tony, meanwhile, constantly soothes himself with the richly sentimental voice of Cleo Laine. A spoiled, pampered, self-deluding fop, he can't even get himself a glass of water.

Frolic and riot follow: a sexy romp by Barrett and Vera in Tony's private bathroom, involving a sunray lamp and goggles, after the master, haughtily fussing about his cologne and the brushing of his suit, departs. Barrett opens an upstairs door in order to hear the front door close on Tony's departure; later Tony will do exactly the same to confirm Barrett's departure – such motifs and ironies belong to the reserved provinces of cinema, close-up and long shot; the middle ground belongs to theatre.

Barrett contrives to leave Tony alone with Vera in the house for a night. The maid seduces the master on the kitchen table under a huge 'operating theatre' lamp. The telephone rings; we know it's Susan; but Tony lets it ring and Vera smiles in triumph (but again Miles overdoes it). The motions and compositions of the seduction are perfect, the thrashing bodies reflected in dully gleaming pots. Seduced and besotted, Tony is now at Barrett's mercy, though the butler is in no hurry to play his full hand.

Tony creeps upstairs to tap on Vera's bedroom door (Losey has him reach through the banisters to tap). As Vera leaves the room we see Barrett lounging naked and smoking on her bed – another *coup de théâtre*. There follows the famous sex-in-a-swivel-chair scene, between Tony and Vera.

The increasingly ignored and angry Susan arrives when Tony is not at home, scatters new cushions about the furniture and outrageously humiliates Barrett. Light my cigarette. Come here. I want some lunch. It's enthralling stuff – Bogarde's performance is sublime – but there is an undercurrent of doubt. Would Susan bring herself to say to him, 'Do you use a deodorant . . . Do you think you go well with the colour scheme?' Would a 'lady' say it? Is she no 'lady'? And this raises the question of who, what, is Susan?

Pinter and Losey agreed that Pinter was never interested in her. She has no identity, family, job, or biography, and she is never more than an adjunct to the main action, an outraged spectator to the hothouse 'huis clos' of passion and perversity constituted by Tony, Barrett and Vera. Why should the pretentious Tony have engaged himself to such a woman when Chelsea abounded with posh dollies, the genuine article? Bogarde's answer is 'physical attraction'; but this does not wash; after all, it's the more obvious sex-appeal of the maid, Vera, which pitches Tony into social disgrace and psychological captivity. Physically, Wendy Craig was right for the part of Susan – provided she was endowed with a social background likely to persuade Tony that this was the kind of marriage one ought to make. The script provided her with no background at all.

We arrive at the scene of scenes. Arriving home early from a weekend in the country, Tony and Susan find a light burning in Tony's bedroom. At the foot of the staircase they hesitate – burglars, perhaps? Tony still believes that Vera is Barrett's sister, and Susan remains completely innocent of Tony's affair with the maid. Now they hear rising laughter, male and female, until the bedroom door is flung open and the shadow of Barrett, stark naked, leaning over the banisters in frontal silhouette, is cast in grotesque distortion on the staircase wall. Losey and Slocombe mounted an arc light on a rostrum above the stairwell; the sex of servants in the master's bed is filthy, defiling, unthinkable.

Summoned downstairs, Barrett arrives insolently attired in a dressing gown, a fag in his mouth, and resorts to sneering defiance, exposing Tony's affair with Vera in front of Susan. 'Get out!' yells Tony. He and Susan cower, horrified, stricken, in the drawing-room while Barrett and Vera pack upstairs, noisily, stoically, the landless peasants whose laughter carries the spirit of the Blitz. This packing is not in Pinter's screenplay; one detects Bogarde's generous acting instincts – no Bolshevik, he – in the display of working-class resilience, or fatalism, as Barrett and Vera make their singing exit while Tony and Susan choke in property-owning silence. We search for a moral centre of gravity but in Pinter's world there is none, only tremors and earthquakes.

Sinister Barrett may be in theatrical obsequiousness and in his eye for the main chance – but he is also an unpropertied man who has learned a trade to survive. He is youthful, spirited and easily wounded – but happy to enjoy a short season of luck. Pinter's approach to motivation is normally elliptical and laconic. In Pinter, the Sartrean definition – a man *is* what he does – encounters the absurdist objection: a man is what he would be if he weren't what he is – the lighted cigarette hastily drowned in dishwater at the master's footfall. Those critics who mistook the film for a fable about 'evil' – or about Faust – forgot that the Tonys of this world want a servant to remain one.

Left alone, Tony collapses into despair. Later he meets Barrett in a Chelsea pub; the butler talks his way back with some tall stories, the camera peeping past a pot of artificial flowers at the two men seated far apart at the bar. Losey then slams into a new mood and tempo, abruptly immersing us in the *menage à deux* that follows Barrett's return – a darkened, chaotic house in which the now tousle-haired, carelessly dressed servant shouts his complaints in a thicker, coarsened accent:

BARRETT: You still sitting there?

Losey's camera races from one altercation to the next.

TONY: You're a peasant.
BARRETT: I'll tell you what I am, I'll tell you what I am. I'm a gentleman's gentleman. And you're no bloody gentleman!

Yet, despite lapses into idleness and shared liqueurs (and a tendency to dope Tony from time to time), Barrett tries to keep working; he resents squalor as much as Tony does, and the two young men fall into a kind of camaraderie. When they play a rough ball game on the staircase, it is Barrett who gets hurt and howls the comic Pinter line, 'I'm off. I'm not staying in a place where they just chuck balls in your face!' (Slocombe set six lights on the staircase with overlaps to make them look like one.) A moment later, in a cathartic reversal, Barrett shouts, 'Well go and pour me a glass of brandy . . . Well don't just stand there! Go and do it!'

The staircase game, shot from above, with each banister shadowed on the wall, achieves an extraordinary, formal beauty. In *The Servant* the staircase is the living skin of the subverted household. So Barrett wishes to be master and autocrat? Not really. In this power-play both men are groping in the void, waiting for social reality (destiny) to reassert itself. Barrett is a cat pursuing a ball of wool wherever it takes him.

They play a new, ritual game of hide-and-seek. Tony hides, Barrett seeks. Tony cowers in the shower-room, his silhouette grotesquely distorted by the folds of the curtain, while Barrett advances up the stairs: 'You've got a guilty secret . . . I'm coming to get you, I'm creeping up on you.' The vibrating Tony stuffs a hand in his mouth. 'I can smell a rat . . .' warns the approaching Barrett. Tony's terror rises to a silent scream as he's discovered – but this terror is self-inflicted, a quirk of masochism in the master which Barrett services not only to achieve a sinister domination but to also earn his keep. In between such episodes Tony and Barrett sit down together, properly dressed, hair combed, to eat a meal cooked by Barrett – like 'old pals'.

Out of pouring rain comes a distraught Vera to beg Tony for money. She's destitute. Here Losey worked an interesting adjustment to Pinter's script. In Pinter, Barrett wants her out and off, and means it; Losey sticks to the lines but shows covert gestures and winks between Barrett and Vera suggesting it's all a charade to soften up Tony yet again. In the script, we never see Vera again after Barrett pushes her out of the door; this may be Maugham's legacy – Barrett as the ruthless egoist incapable of loyalty to his girlfriend. By bringing Vera back at the end of the film, Losey may be signalling the warmth which underpins working-class solidarity, but, as Gilles Jacob observed, 'the role of Barrett's pseudo-sister, planet of a subtly complex system, remains ambiguous to the end'.[4]

So to the final 'orgy' or 'dope scene' (as Losey's notes referred to it). This was Pinter's invention, with no provenance in Maugham. Pinter had originally placed it before the long scenes depicting Tony and Barrett living together as virtual equals. He had given the invited tarts some giggly lines. Losey wrote across the script: 'They don't speak.' The *1961* dope scene includes some crawling and whipping – but in the *1963* script and the film there is no crawling or whipping. The invited women laugh in a manner both mechanical and artificial, a small chorus of mocking 'furies'. Douglas Slocombe concedes that he wasn't sure what was going on – some kind of drugs, no doubt. Bogarde recalls being informed by Losey that everyone present was hallucinating from LSD. 'We'd never heard of LSD.'

It was at Losey's initiative that Susan reacts to her own misery by kissing Barrett, a baffling gesture – or surrender – since she detests the man. The idea came from the screenwriter of Losey's *Eve*, Evan Jones, to whom Losey had shown the script: 'I felt that [Susan] needed some expression of her despair, defeat, jealousy and disgust. Losey passed the suggestion to Pinter, who, to my surprise, accepted it.' Bogarde's view of Susan's motive is similar – 'a Judas kiss of loathing', a bitter acknowledgement of Barrett's final victory, laced with contempt. Losey himself came up with three reasons for the kiss (there may be three causes of the Protestant Reformation or the fall of the British Empire, but can there be three reasons for kissing someone?): '. . . she kisses him because she doesn't know what to do . . . probably a deliberate attempt to provoke a reaction out of [Tony] . . . also, though certainly not consciously, an attempt to maybe join him in the sense of degrading himself.'[5]

Judged as Bacchanalia or as 'political' revolt, the 'dope scene' is tame stuff. When the prostrate Tony yells 'Get out', Barrett, far from defying him, far from 'taking over', meekly obeys. Only Susan feels the full force of revolution:

ejected by Barrett along with all the strangers, she ends up clutching a tree in Royal Avenue, sobbing with shock and humiliation. Losey ties an extra knot by reuniting Barrett and Vera on the last shot, their laughter replete with life and survival. So ends the first of Losey's masterpieces.[6]

The producers were not convinced. Losey now embarked on a long struggle to promote the film to its backers. On 5 May 1963 he wrote a 3,000-word memo to Leslie Grade, reflecting his own mood of exaltation and his knowledge that he had made 'a remarkable film': 'I must thank you now and everlastingly, whatever follows, for the unique opportunity in my life up to this time of making a film as I and my group of enthusiastic collaborators conceived it, without any interference whatsoever. This is my fifteenth feature and my thirtieth year in theatre, radio, television and films ... And lastly my own name has come to have a strong promotional value on the Continent where the function of the director is better recognized and given far more credit than is usual in England. But, nonetheless, this reputation has been filtering back to England ...'

Long passages of his letter to Grade were devoted to strategies for promoting the film. Early screenings to highly select audiences – the 'word-of-mouth' factor – were vital. 'Fellini begins such private projections when his film is only in a state of rushes ... The chi-chi and prestige value has been incalculable ...' He offered Grade detailed plans on promotion, on the right style for posters and billboards, on the qualities required of the trailer which 'ought to be unlike any trailer ever seen ...' He was also keen to explain, in a heavy memorandum dated 2 July, what the film was 'about'. 'Thematically, *The Servant* is about the destructiveness of trying to live by obsolete and false standards ... a film about people for whom servility is a way of life ... If you like, it is also a kind of Dorian Gray tale, because it deals with the vain aspiration to eternal youth which leads instead only to age and death.' Pinter's approach to 'about' was briefer: as far as he was concerned *The Servant* was about the relationship of two men and what came to pass between them.[7] The probably apocryphal version runs, 'It's about a man and his servant.'

By April 1963 Losey was already jockeying for a showing at the Venice Festival. On 9 April Losey's future wife, Patricia Tolusso, whom he described as 'my assistant' or 'my secretary', cabled him at the Lord Byron Hotel in Paris, advising him that Professor Luigi Chiarini, chairman of the Venice jury, was 'reportedly left wing and appointed to his present job by socialists'. On 23 April Losey wrote to Chiarini supporting him against his enemies and reminding him that *Eva* had been withdrawn from Venice the previous year by its producers, the Hakims. Suffering from fractured ribs, Losey wrote twice to Pinter from the Lord Byron, urging his attendance at Venice. Pinter declined: he was due in rehearsal for a double bill of his plays. 'I know that you will consider that I have let you down, but let me just say this: I have to move very clearly and carefully these days, or else life will overwhelm me. Do you know what I mean?' On 25 August Losey wrote again: 'You can probably wait ten years now. I can't.' In August Losey was in Paris to supervise the Italian subtitling by LTC, horrified to discover as many as six joins per reel in the release print, some in the middle of a set-up.[8]

The British entries for the Venice Festival were *Tom Jones*, *Billy Liar* and *The Servant*. The Golden Lion went to Francesco Rosi's *Mani sulla città*. Losey was deeply disappointed by the poor reception of his film.* Even in Britain 'It was on the shelf. It was one of twelve films that had been considered to be unreleasable . . .' But then 'Arthur Abeles, who was booking for Warner's . . . chose it from among the twelve.' Patricia Losey believes that it was Richard Roud's invitation to show the film at the New York Film Festival in September that unlocked the situation. Losey complained that it was withheld from the London Film Festival by the producers, ABP, although the Festival asked for it.[9]

Opening night in London was 14 November, at the Warner, Leicester Square. Bogarde threw a party in the Connaught. Losey looked 'more like a weary don at Speech Day . . . head bowed to listen to constant praise, flushed with the pleasure he was so good at concealing'. The ongoing Profumo scandal, involving high-class prostitution and a disgraced minister of the Crown, gave the film a tangential boost. The critics responded with almost unanimous acclaim – though few of them were keen on the final 'orgy'. In the *Sunday Telegraph* (17 November), Philip Oakes hailed 'the best film of the best director now working in Britain . . .' *The Times* lauded direction, script, acting; in the *Manchester Guardian*, Richard Roud pointed to Losey's peculiarly American mastery of invisible direction, 'a degree of intensity that can seldom have been equalled'. Penelope Gilliatt's review in the *Observer*, headlined 'The Masterful Servant', called the film 'a triumph'; certain lines were used like sound effects, lines that floated up as overheards, the beginning of lost sentences employed like dissolves. Among critics Gilliatt was uniquely perceptive about the servant himself: 'Barrett has the enormous advantage of social resentment . . . Bogarde makes him seem like a man oppressed for too long, who simply has the strength and cunning of anyone struggling to be free.' Penelope Gilliatt promptly received from Losey an invitation to 'lunch or an evening drink, I should be delighted, any time that suits you'. More speculative lunches or drinks were offered to other friendly critics: 'Let us get

---

*However, in March 1966 Italian journalists awarded *Il servo* the Silver Ribbon (Nastro d'argento), as the best foreign picture released in Italy during 1965. In Spain the film won the Castillo de Plata Cine Club, Irun, awarded by II Journadas Internacionales de Cine.

together soon.' A batch of such letters was run off in a single day. Meanwhile, true to form, on 10 December Losey wrote to Robin Fox complaining about the quality of the trailer, about the selection of review quotes, the placing of quotes over pictures instead of on black frames, poor typography, bad soundtrack mixing, and a 'grey and gutless' print.

At the end of 1963 *The Servant* was listed as among the best ten films of the year by Penelope Gilliatt, Philip Oakes and Dilys Powell. Alexander Walker chose it as 'the film of the year', and the London Film Critics Guild named it the best picture of the year, with the best performance (Dirk Bogarde's). However, the 'Best Director' award went to Tony Richardson for *Tom Jones*. Bogarde received the Variety Club's award for 'Best Actor'.

The Servant received eight nominations by the Society of Film and Television Arts. The competition for 'Best Film' was stiff. Nominated were *The Servant*, *Tom Jones*, Lindsay Anderson's *This Sporting Life*, and John Schlesinger's *Billy Liar*. *Tom Jones* emerged the winner. ('Best Film' and 'Best Director' were not yet subject to separate awards.) Bogarde received the 'Best Actor' award; James Fox the 'Best Newcomer' award and Douglas Slocombe the 'Best Cinematography' award. Pinter's screenplay was nominated alongside David Storey's *This Sporting Life* and the eventual winner, John Osborne's *Tom Jones*.

Losey had arrived. 'Even if I make ten better pictures in my lifetime,' he told *Isis*, 'I don't suppose one could ever expect to have such unanimous appreciation and approval again.' But the supreme accolade arrived when Franklin, the popular *Daily Mirror* cartoonist, took up the theme of *The Servant*. The top box featured the Prime Minister, Sir Alec Douglas-Home, as an old retainer, bringing champagne, cigars, chicken and slippers to an ordinary bloke seated in an armchair. This is 'before election', but 'after election' the old servant – Sir Alec – sits in the chair with the cigar and champagne, his feet resting on the crouching figure of the ordinary bloke. (In the event Sir Alec lost the election.)

For Losey, the New York Film Festival was a remarkable occasion: for the first time in eleven years he set foot on American soil. Responsible for the invitation was the Festival organizer, Richard Roud, an American then working as programme director at the BFI in London, and a regular critic for the *Manchester Guardian* and *Cahiers du Cinéma*. Knowing that Losey feared trouble with US Immigration, Roud asked John Mazzola, Secretary of the Lincoln Center for the Performing Arts, to ensure that there were no incidents when Losey arrived. There was none. 'Joe seemed almost disappointed. . . . it was almost an affront,' Roud recalled.[10]

Following the excellent British notices, ABP sold the film to Ely Landau, of Landau Releasing Organization. In December Losey learned that his old friend James D. Proctor had been hired by Landau to head the 'special exploitation' of the film in America. On 5 February, Proctor cautioned Losey: 'An art film, as you know, has to make it very big first as an art film before it even has a chance to get circuit bookings, not to say major distribution.' *Tom Jones* and *La Dolce Vita* were, he warned, examples of 'commercial-art films . . . rated in another league entirely'. Even Bogarde was not nationally known in the USA. The *Tonight Show* and *What's My Line* didn't want him.

In March Losey returned to New York for the American première at the Little Carnegie on West 57th Street. The film also opened at Boston's Music Hall. The reviewers were ecstatic. For Brendan Gill of the *New Yorker* (21 March), the talents of Losey and Pinter were perfectly matched; the *Christian Science Monitor* lauded the 'exquisite . . . film craftsmanship'; both *Time* and *Newsweek* added their applause. In the *Herald Tribune*, Judith Crist greeted a film 'told in brilliant cinematic terms'. In the *Village Voice* (26 March), Andrew Sarris noted that this was the first film 'in Joseph Losey's tortured career to bear his personal signature from the first frame to the last'. Too many critics, Sarris wrote, had wrongly assumed that the motivation of the two main protagonists in the film conformed to a genre: Mephistopheles, Machiavelli, Faust, Dorian Gray. In Sarris's view, the servant takes over his master 'simply because someone has to take over someone else . . . the evocation of power as the dominant passion of a collapsing class society'. Sarris reminded his readers that Losey had been a Brechtian realist – a symbolic realist – before it was fashionable; and even if 'on the most sublime level of movie making . . . Losey lacks the unified vision of a Renoir in *The Rules of the Game*', he had made the most exciting movie of the year, a film which might come to be seen as 'a prophetic work marking the decline and fall of our last cherished illusions about ourselves and our alleged civilization'.

A half-page ad in the *New York Times* brimmed with impressive quotes. Eugene Archer described Losey in the *New York Times* (15 March) as an American who 'is just about the leading director in the British film industry'. Archer's plug for Losey raises an ethical question: that same spring he was urging Losey to set up and direct a film of Faulkner's *The Wild Palms*, for which Archer had purchased a $10,000 option. Requesting a script credit for himself, he proposed that he and Losey should co-produce the film on a '50– 50 deal'.[11] Archer happened, also, to be chairman of the New York Film Critics. In Judith Crist's top ten films of the year, *Dr Strangelove* headed the list, with *The Servant* at number six. In the *New York Times* Bosley Crowther placed Stanley Kubrick's film at the top, with *The Servant* in second place. In December the Independent Film Importers and Distributors of America chose Losey as the 'Best Foreign Director'. He was wryly grateful, though he thought he was an American.

By February 1965 – fifteen months after opening in London – ABP reported cash receipts of £174,868. Eleven years after its release in London, as at 31 December 1974, gross UK receipts were £238,893 and overseas revenue £150,383. After deduction of expenses, EMI's commission, and repayment of loans to National Westminster Bank and to the NFFC, the divisible balance of £141,631 was shared between the NFFC (33⅓ per cent), Elstree Distributors (30 per cent) and Springbok Films (36⅔ per cent). 'And it's one of the very few pictures on which I had a percentage and actually got some money,' Losey commented.[12]

But not from America. In January 1966 Leslie Grade informed Losey that US earnings had reached a modest £38–44,000. Losey convinced himself that Ely Landau (a) had not properly promoted the film and (b) was concealing his true profits. A New York law firm, Solomon and Finger, was prepared to investigate concealed profits for a fee of $3,000 plus a percentage of any profits revealed but the operation was too expensive, too uncertain, to pursue. In January 1968 Landau answered Losey's complaints: *The Servant* had unfortunately proved to be a most difficult picture to market for United States television. Censorship and prudery were two of the problems. No network would accept it. Even in the 'major urban and so called sophisticated markets' it was relegated to the late evening time slots. Losey was unconvinced. In October 1969 he suggested that the reported US TV sales gross of $50,000 was probably 'a diddle . . . I should have thought a figure of $500,000 would have been more like it.'[13]

3

Father to the Man

Losey's childhood, as he related it to Michel Ciment and others, owes much to the imagination. But it is precisely the selective imagination, and the psychology behind it, that illuminates the obsessive motifs of an artist. Whereas Ingmar Bergman as film-maker is both slave to and master of his childhood, Losey was a fugitive from the child who was father to the man. Despite periods of psychoanalysis, his self-alienation remained largely subconscious. His constant refuge was self-justification.

He was born on 14 January 1909 in La Crosse, Wisconsin, and christened (in the Episcopalian church) Joseph Walton Losey III, the elder of the two children of Joseph Walton Losey II and Ina Higbee Losey. Joe's sister Mary was born two years later. At the time of writing she – Mrs Mary Losey Field – is the only surviving relative from the inner family, the only witness to the full span of her brother's life. She didn't like Joe, much.

Joe's paternal grandfather had moved up the Mississippi from Pennsylvania to Wisconsin. The town of La Crosse took its name – Losey explained – from the fact that 'the Indians played lacrosse on a piece of island, flat land, at the juncture of the three rivers ...' La Crosse and its great river etched itself deeply on Joe's consciousness and in later years he frequently evoked its contours in romantic terms – the dances on steamboats, excursions, picnics, swimming and 'constant drownings'. The long winters brought ice skating, sledging, a landscape snowed under. La Crosse doubled its population, from 15,000 to 30,000, during his childhood: a small town in the mid-West with roots in Britain, Germany and Scandinavia.

Losey's scan of his family tree was mainly alert to two factors: wealth and culture. He told Michel Ciment that his paternal grandfather, Joseph Walton Losey I, 'became a tycoon, a vice-president of the Great Northern Railway ... and he had five [sic: three] daughters and one son, my father'. Mary Losey Field eliminates their grandfather's vice-presidency of the Great Northern: 'He was employed by J.J. Hill, founder of the Great Northern, to act on his behalf in bringing in the Burlington Railroad as a connecting line from Chicago to St Paul. Grandfather Losey was also, for some years, District Attorney for La Crosse County, but he was no "tycoon".[1]

Joe never knew either of his father's parents. Joseph Walton Losey I died in 1903, having given (according to Losey) 'an immense amount of land along the bluffs . . . which is now called Losey Boulevard'. But this, too, is a case of retrospective ancestral wish-fulfilment. 'Grandfather Losey,' Mary Losey Field explains, 'did not make a gift of extensive lands to La Crosse. He was largely responsible for the fact that the city set aside certain lands for parks, boulevards and other civic improvements which he had the foresight to plan and work for. He was a prominent and influential man but not a rich one . . . The Memorial Arch was given by friends and city officials as a tribute to him after his death.'

Joe was raised in a comfortable but far from palatial house at 1612 Ferry Street. He was prouder of his father's family than of his mother's, the Higbees. Wealth and culture are again the key factors. In the home of his father's sister, Mary Losey Easton, he found his model of a patrician lady: 'Aunt Mer', as she was known, was not only the first president of the La Crosse Home for Friendless Women and Children, and chairman of the American Red Cross during the First World War, she had also married the wealthy Lucian F. Easton. Aunt Mer represented culture – a gifted pianist but by no means the most musically accomplished of her sisters, one of whom (Losey recalled) was 'a concert violinist [with] a big career . . . really very well known'. This was Aunt Fanny, whose husband became a state Supreme Court judge and whose house 'I remember in great detail, a very beautiful house, lovely garden, on the crest of St Paul overlooking the city'. She had played the violin with the Budapest String Quartet and performed for President Theodore Roosevelt in the White House.

But it was Aunt Mer's house that Joe remembered – and reinvented – most vividly: 'I suppose there were thirty-five or forty rooms in the house, and a separate servants' house, and there were seven to ten servants, always. There was a vast carriage house and lots and lots of horses and carriages, and eventually automobiles. The first Cadillacs, the first Franklin in La Crosse . . .' Joe's sister Mary disposes of this film set as 'largely imaginary'. The Easton mansion had sixteen, not 'thirty-five or forty' rooms. There were 'three servants, not seven or ten'. Indeed, 'by the time Joe was on the scene there were no horses, only cars and they were not nearly as grand as he would claim' – despite the Franklin.[2]

Musicians and actors travelling on the concert and theatre circuits would visit La Crosse and be invited to stay with Aunt Mer: '. . . people like Josef Levine . . . Rachmaninov or Heifetz . . .' The house was full of Japanese art and fine porcelain, 'Dresden, Japan and Berchtesgaden, I remember.' Thanks to Aunt Mer's long cultural reach, Joe, an asthmatic 'from birth', was able to view at close quarters another distinguished sufferer, Ford Madox Ford. Aunt Mer also took an interest in Joe's education. 'When I was about twelve years old, I used to have to go and have dinner alone with Aunt Mary and read Racine and Corneille aloud after dinner. In French. Put me off a bit.' In 1970 he told Thames Television: 'By the time I was twelve I had read all of Dickens, all of George Eliot, all of Dumas – most of Scott . . .'[3]

Aunt Mer and her husband, Fred Easton, had four daughters and one son, Clark, who was some fifteen years older than Joe and who 'taught me how to canoe in the Mississippi when I was very young'. Losey described Clark Easton as a Greek and Latin scholar, a linguist and historian, who became 'the official translator for the King of Greece during the war and he was a captain in the army'. Joe had a craving for distinction in the blood and in the tree, but Mary Losey Field never heard of Clark Easton performing this romantic role in Greece, though she knew him well.[4] In March 1975 Losey again conveyed his romantic image of his primary film set, Aunt Mer's house, when writing to Clark Easton's widow, Joy B. Easton: 'I suppose you knew about, but never saw, the really quite fantastic house, full of the most beautiful and possessing objects and the most beautiful music and books and food. I remember Clark and . . . our adventures in his Oakland car and in the canoe . . . and . . . that exquisite house-boat and the (while it lasted) beautiful bowling alley and the lovely greenhouse and the fabulous workshop-cum-carriage house-cum-garage . . . I remember also [Clark's] Chicago flat and particularly his music . . . and his girls . . . he was a very male male and I was a very impressionable boychild.'

This revealing passage is packed with treasured themes from Losey's films (although the canoes did not surface again). Beautiful *objets d'art* are thickly spread around his film sets; a house-boat is found in *The Criminal*; an elegant bowling alley in *La Truite*; fine gardens in a number of films; a big greenhouse in *The Go-Between*; as for the Chicago flat, the music, the girls, they too are echoed from *Time Without Pity* to *Mr. Klein*. But Losey was sixty years old before he could raise the funds to bring the cherished splendour of Aunt Mer's house to the screen – although the architecture and décor were radically different: setting up *Secret Ceremony*, he invented a splendid family tree, heavy with the imagined fruit of European national and religious cross-fertilization.

Serving as an acolyte in La Crosse's Episcopal church – whose ceremonies, vestments and music impressed him – Joe absorbed religion as a touchstone of class. His parents and the Losey side of the family were Episcopalians, 'which is very snobbish, terribly snobbish. If you have pretensions to an upper-class background in the United States you're a [sic] Church of England or an Episcopalian . . . and I was very religious until I was about eighteen.'

Aunt Mer's impressive home, with its 'priceless collection of Whistler etchings', was in sharp contrast to the modest domestic arrangements of his mother's father and stepmother, the Higbees, where there was only one servant and where the big meal of the day was breakfast, 'which was served at six in the morning and consisted of meat and potatoes and eggs and gruel – a real farmer's breakfast . . . they were all in bed by nine . . .' 'My mother, who came from a relatively poor and uncultured background, was a terrible snob – very beautiful. And she, of course, had picked off the biggest catch in town, my father . . . Suddenly she found that she had married a pauper with no career.'[5]

Yet Joe's maternal grandfather, Judge Higbee, presiding judge of the Sixth Judicial Circuit, was neither 'poor' nor 'uncultured'. It was the legal and medical books in Grandfather Higbee's house which prompted Joe to consider medicine as a career. As Mary Losey Field puts it, Judge Higbee's library may not have contained French, German and English fiction, but rather Bertrand Russell, Havelock Ellis, and Ralph Waldo Emerson. 'The fact that he had supper at five and not dinner at eight is hardly a measure of "culture".'[6] Indeed not. But a younger cousin, Jean H. O'Neill, recalls that both Joe and Mary 'preferred to associate themselves' with the Losey/Eastons. The cultural difference between the Loseys and the Higbees was reflected in their choice of university: the Loseys headed East, for the Ivy League, whereas the University of Wisconsin was good enough for the Higbees. Receiving an honorary degree in Madison a year before he died, Losey felt compelled to remind his audience of his own Ivy League inheritance.

Joe attended a private 'normal school' where his two closest friends were Robert Lees and Jonathan Bunge, sons of partners in the law firm where Joe's father occupied 'a clerk's office'. 'We always had battles with the kids from the public schools who would ambush us. We always had a little more money than these kids; we had bicycles, sledges, toboggans . . . No mixing!' After 'normal school' Losey went on to Central High School, Cass Street, and 'loathed it. I was a terrible spoilt brat. Didn't like the way the kids smelled. I didn't want to have anything to do with them.' Unlike his 'protector' Jonathan Bunge, a football hero, Joe was never an athlete, a joiner, or a club boy. His health varied from delicate to terrible. 'Joe had every illness you could name,' his sister recalls: asthma, hay fever, heart trouble, scarlet fever, influenza (in 1918), violent appendicitis. Losey recalled suffering mumps twice as a child and again during his twenties.[7]

During his first year at high school he contracted scarlet fever with kidney complications and was confined to the house. A private tutor was brought in. This suited him fine: 'I could spend the whole day reading and writing and painting and then she would come after school and work with me from three till six.' He finished the four-year high-school course in three (which became two in recall) but clearly he exaggerated his own alienation: he played the saxophone and served on the editorial committee of the school annual, *The Booster*, his graduating epitaph being 'inebriated with the exuberance of his own verbosity'. *The Booster* annual for 1925 reveals his participation in the Falstaff Society's 'miniature musical comedy', *The Manicure Shop*; he played Prince Vladimir in a farce by Cosmo Hamilton, *The New Poor*, and took the part of Lord Andrew Gordon in a comedy called *Adam and Eva*. He was one of two boys chosen to represent Central in oratory. Today the new Central High School is located on Losey Boulevard South, named after Joe's grandfather.[8]

The young Joe was a film fan; he remembered seeing 'more films when I was a kid than I ever saw after . . . all the Fairbanks, all the Pickford, all the Buster Keaton, all of the Chaplin, all the Harold Lloyd, Pearl White, even Stroheim . . . When there was a Griffith film even my Aunt Mary went . . . I remember *Intolerance* and *Birth of a Nation*.[9]

Losey's drive in later life was to be both wealthy and progressive; the great crime was to be poor and reactionary. He described his father as 'very reactionary' – although Mary Losey Field recalls that their father 'once took me to see a cross being burned on the side of the bluff and explained to me that . . . the Ku Klux Klan . . . were evil and dangerous'. Joseph Walton Losey II, known as Walt, had begun his life on the higher branches of the tree of wealth and breeding. Educated at Lawrenceville and Princeton, he was a member (Joe proudly recalled) of the 'Tiger Inn at Princeton, the most snobbish of the clubs, and an all-American football player in the Princeton team of 1901. All his clothes were tailor-made . . .' Then catastrophe: Joe's 'wealthy' grandfather died, leaving no wealth. 'Suddenly [my father] found at the age of twenty-three that he had absolutely no money.' Unable to finish his studies at Princeton, he went to work as a claims agent for the Burlington Railroad, without legal qualifications. Clearly Joe admired, liked and indeed loved his father, craving his approval, but behind his admiringly masculine image of him lay a subverting sense of emasculation closely joined to his relative poverty. In Mary's opinion Joe was undermined – 'the damage was done' – by their mother Ina, a good-looking, vivacious, flirtatious woman who cosseted Joe in a possessive way. His sexual consciousness – sex as power – began with his mother's treatment of his father. Joe described her as 'very beautiful and . . . immensely ambitious. A little bit vulgar and terribly disappointed that my father wasn't a rich man . . . And my mother was one of those Ibsenian women who use sex as a kind of threat. I didn't understand it then, but the bedroom door would be locked . . . She was always threatening to leave if she didn't have a servant which she couldn't afford and she made him borrow money.' Mary agrees that their mother closed bedroom doors to make her point; she describes Ina as vain, always fussing about her appearance, 'too compulsively spotless', and inclined to threaten suicide: 'Maybe I won't be here when you come home from school.' Or, more bluntly, 'I'm going to kill myself.' Ina had 'eastern manners'. She was 'fast' and smoked in public. 'She wanted more money,' Joe recalled, 'she was always giving parties – and had male admirers who took her out in their cars. There were always big jealousy fights too, because Father got very jealous about that; she played the piano and she sang popular music and there were always people hanging around the piano while she played and sang. It was in the time of Prohibition. She drank a good deal of bootleg liquor. So I remember the smell of liquor and I disliked it very much.'[10]

In the mature Losey's artistic consciousness, wealth, sex and power are interwoven in a tapestry of constant torment and pain. In film after film the sexes butcher one another – but Losey avoided confronting his own childhood. Not only are mother–son tensions almost wholly absent from his films, one searches in vain for an echo of Ina's humiliation of Walt Losey. Joe's personal psychology was evasive in so far as it operated by refraction. Yet Monique Lange noted that at the end of his life he was still haunted by an unfavourable image of his mother. He told her about a letter his father had written him when he was 'nine', complaining about Ina. This incident parallels the storming row between the parents remembered by Mary, when Joe was thirteen. The cause was 'Uncle' Jimmy – James Hogan, Joe's godfather, a wealthy, *sportif* and rather decadent Irishman, a 'kind of a nut' who tried to write plays and complained that Eugene O'Neill had stolen his ideas. Uncle Jimmy and his wife Grace played poker and bridge with the Loseys and accompanied Joe's parents to dances. On that particular night Mary went to bed and heard a big fight downstairs. The following day an awful gloom prevailed – their father went away for a couple of days and Ina gave a big fat envelope to Joe to take to Uncle Jimmy. She also spoke to Mary of a probable separation.

Apparently it was his mother who inspired the lifelong hostility towards women which unrelentingly surfaces not only in Losey's films but in his private life: attraction, desire, offset by resentment, even contempt, and a probably subconscious impulse to exact retribution through humiliation. Here again relative poverty rears its head. Walt Losey was suddenly taken sick on 6 February 1925, the day after his forty-sixth birthday, with appendicitis; despite an operation peritonitis set in and he died on the 16th. Ina Losey, then aged forty-four, was left with little but a small insurance policy in trust for the two children. She sold 'hand-woven things from Kentucky and Cantagalli pottery' (her daughter recalls) before leaving La Crosse and placing Mary in the care of Aunt Mer. Joe, who was now without a home, felt robbed of his proper inheritance. 'The money was gone and the houses were gone . . .

The drive was gone . . . My father and my aunts and most of their children resulted in . . . nothing . . . not happiness, not production, not rebellion . . . not anything . . . So I am a rogue elephant . . .' (The dots are Losey's.) As for the admired paragon, Aunt Mer: 'the invulnerable dowager empress' went bankrupt. 'The Easton mansion was boarded up . . .'[11]

Mary Losey Field – whose own career was to be a distinguished one – acutely resents Joe's insistence on the singularity of his own achievement and – perhaps the fatal flaw in his later films – his lack of generosity towards others.[12]

4

Dartmouth

'I had a secret ambition to go to Harvard but I knew I couldn't pass the entrance exams, so I didn't even try. The principal of La Crosse High School was a Dartmouth alumnus so it was fairly logical to pick around among the smaller eastern colleges.' Mary Losey Field comments: 'In those days most Ivy League colleges required four years of Latin, which may have been one of the reasons he did not apply to either Harvard or Princeton.' Joe arrived at Dartmouth in 1925, at the age of sixteen, fatherless and poorer than suited him. 'I was put through college thanks to my godfather, who committed suicide in London the year before but had left me enough money to have a small income. I washed dishes for my meals and I worked in the library for spending money . . . and my Aunt Fanny used to send me $25 a month for extra money, for quite a long time.' (Joe's godfather, the notorious James Hogan, over whom Joe's parents had bitterly quarrelled, had got a local schoolteacher with child, causing a scandal in La Crosse. 'Uncle Jimmy' then went abroad and killed himself. Mary Losey Field guesses that his legacy to Joe was probably around $10,000 since it paid for three years at Dartmouth, a trip to Europe and a final year at Harvard.) Joe's younger cousin, Jean H. O'Neill, recalls him returning to La Crosse (where he now lodged with the Higbees) on vacation with a newly acquired eastern accent, an opera hat and a monocle: 'These pretensions did not endear him to the "locals" . . . Joe was a complicated and essentially morose person, from my point of view, and while he undoubtedly had a vision and achieved most of it, he didn't relate to ordinary people very well.'[1]

Joe had begun by studying medicine (pre-med) but was not particularly proficient in it. 'I had to take an exam in organic chemistry and I knew it well. But I sat there in a cold sweat. I knew the answers to all the questions but I simply couldn't get my hand to write.' He transferred to a liberal arts course and the crucial teaching relationships began to emerge. During his sophomore year he occupied a room in Sidney Cox's house: 'Sidney was the first person to suggest that my energies might be directed toward one of the creative arts . . .' Describing Cox as a great teacher who for years was passed over for tenure and promotion, he recalled long Sunday walks with Cox and Stearns Morse, 'another great and special teacher'. Cox evidently taught by provoking rage, passion, commitment, despair. 'He was cruel. He was kind. He was sadistic. He was masochistic.' After leaving Dartmouth Joe continued to correspond with him until a fatal encounter. 'I drove to [Dartmouth] in the

Thirties, probably about 1936, to talk to him . . . unburdened myself . . . in the deep winter and deep snow and took him to dinner in Windsor . . . Later I received a long hand-written letter from him "renouncing all the positions he had taken with me" . . .' Why? Apparently because they had led to Joe's present convictions (presumably Communist ones – but this is not clear), with which Cox entirely disagreed. 'I returned his own letter to him . . . a betraying sadistic shit . . .' In October 1952, blacklisted and facing exile, Losey drove to Dartmouth, 'and aimlessly tried to find [Cox] with a vague notion of making things up. He was dead (I believe).' But Stearns Morse was still alive when Losey returned to Dartmouth as a visiting professor in 1970 and 1975.[2]

From his freshman year Joe was active in the Dartmouth Players, serving as stage manager for Eugene O'Neill's *The Great God Brown*. By his own recall, he acted 'badly' in Pirandello's *Right You Are If You Think So*, and played in *Le Malade imaginaire*, performed in French. He was again stage manager for the official Dartmouth musical, *The Green Peach*, which went on tour. But Losey was loath to view any environment as supportive or congenial. 'At Dartmouth you were something only if you had either a lot of money or were a big athlete. And if you were interested in poetry or any kind of reading or films or theatre or music, it was just not on.' On the other hand: 'It was a strangely cultural place for a college that was so orientated towards masculinity and athletics.'

His first intimate introduction to professional theatre coincided with the arrival of a Theatre Guild road company production of Ferenc Molnar's *The Guardsman*. 'I watched Philip Moeller, the director, rehearse . . . and talked a lot; I helped Kate Lawson with the set dressing and the lighting over a period of six weeks or so.' From that time on Joe would visit New York during the vacations to watch Moeller rehearsing 'with the Lunts, Glenn Anders and many others — O'Neill to Turgenev!' He met O'Neill once at the Theatre Guild 'before he got Parkinson's disease . . . I think O'Neill was a chief influence on me.'[3]

Joe made his first visit to Europe in 1928 – Passport 577249 was issued to him on 29 May (the FBI gravely noted, twenty years later). During that summer he lived in Paris, first in the rue Gay-Lussac, then in the rue Bonaparte, 'and I went to the Alliance Française which didn't do me a bit of good but I met a lot of the Hemingway circle and a lot of newspaper men . . .' He also fell in with a prostitute, Tatiana Tykhorova, who lived at 50 rue Georges Sorel, Billancourt, and to whom he made love in the countryside while church bells rang. The price of the idyll was gonorrhoea. He related this story to his wife Patricia and Monique Lange in the early 1980s, later commenting to Lange, 'What I don't understand is why [Tatiana] hasn't tried to see me since I became famous in France.' (Fifty years had passed.) A casual acquaintance from that time was Ned Calmer, who later recalled Joe writing a play in the rue Gay-Lussac – one of the characters greeted everything with 'Fantastic!' 'There was a girl with a Russian name (Olga?) [presumably Tatiana] you used to bring to the rue de Bassano (*quartier des monuments*, remember?) and who used to say to me, "*J'encaisse, moi!*" ["I can't take it"].' There is one clue that Losey also visited Germany that summer. In 1975, while asleep and in the grip of a surreal dream, he informed the Administrator of the Paris Opera that he had visited Bayreuth 'with my Aunt Fanny in 1928 to hear *Parsifal*'. He reported the dream to his wife as if that had been the case. The memorable vacation of 1928 culminated in a two-week visit to England.[4]

During his last year at Dartmouth Losey was appointed student director, the top post in the Dartmouth Players. He also won the prize for the best one-act play in an annual contest – and chose to direct it. Written in blank verse, it was called *The Gods of the Mountain* and was 'very pretentious' and 'a horrible piece of romantic shit . . .' The theme was 'Weltschmerz, and the loneliness of the human soul'. It also demanded a complicated set, depicting mountains; during a rehearsal he fell nine feet through the hardboard heights, landing flat on his back with the faculty director of the Players on top of him. 'I broke my back and I just missed paralysis by a fraction of an inch . . . I was absolutely immobilized on a slab in the hospital. That was my senior year. I was one year in the hospital at Dartmouth . . .' His sister believes the accident happened in January or February of 1929; he spent two or three months in hospital, after which he wore a cast for some months but was up and about. 'I went to his graduation in June of 1929 and he was fine.'[5]

Joe belonged to the Delta Upsilon fraternity. Among his Class of 1929 acquaintances was Nelson Rockefeller, later Governor of New York and Vice-President of the USA. Periodically in later years Losey would attempt to exploit this connection to raise funds (for the newspaper *PM* in 1940, for the Proust film project in 1974), but without success – and without modifying his professed scorn for Rockefeller's politics.

In the summer of 1929 he quit Dartmouth with a '*cum laude*' arts degree to enrol with a small summer professional repertory theatre at Boothbay Harbor, Maine, a resort which he knew from holidays with his mother and sister. He was now taken on as actor and stage manager for Cocteau's *Orpheus* by 'some old ham . . . a horrible man . . . and I fell in love with the actress he was after which didn't help . . .' In the fall he entered the Harvard Graduate School of Arts and Sciences to study literature, taking a Shakespeare course with George Lyman Kittredge, a course on Victorian playwrights, and one on the Romantic poets. His main masculine friendship was with Henry Longfellow Dana, a man twenty-eight years his senior, the poet's grandson, a socialist and an enthusiast for Russian theatre. His main feminine friendship at Harvard was with the intense young Belgian-born actress and poet May Sarton, whose family lived in Cambridge. Only eighteen, she was already working in New York for Eva LeGallienne's Civic Repertory Company – her passion for theatre was to terminate abruptly four years later when Losey's production of Maxwell Anderson's *Gods of the Lightning* set the seal on a series of commercial disasters. In her memoirs she mentions him only in this context – nothing about his days at Harvard.[6]

He knew whom to befriend. The '47 Workshop', headed by George P. Baker, suffered a schism when Baker dramatized his grievances against Harvard by accepting an offer to set up the Yale School of Drama. As a result a rebel Workshop 47 emerged at Harvard with alumni support, and there Losey met the designers Lee Simonson and Robert Edmund Jones; the playwrights Sidney Howard, Philip Barry and Robert Sherwood; the critics John Mason-Brown, John Anderson and Brooks Atkinson. 'I got to know them and this of course was an immense help to me when I went to New York.'[7]

Joe's college years had further distanced him from his mother. Ina Losey had moved to a comfortable apartment in Washington, where she had cousins, and found a job (Mary recalls) in a small shop 'which dealt in elegant knick-knacks and the kind of stuff she had been selling in La Crosse . . . In the summers [of 1927 and 1928] she went to Saranac Lake, in the Adirondacks [where] she met a man who was a sort of "in house" stockbroker. His name was Mr Martin. I never knew anything about him except that he persuaded her to take what little capital she had and invest it in stocks "on margin". Of course when the crash came she was wiped out . . .' Joe's accounts of his mother's departure from La Crosse are replete with scorn and hostility: 'And then she got bored with La Crosse and they got bored with her because she was drinking excessively and she was a burden on everybody . . .' Fifty years later he told Françoise Sagan that a great sadness had been his mother's drinking after his father died and the 'family fortune' vanished. 'Neither the pastors nor the female neighbours could reason with her,' Sagan reported. Mary Losey Field comments: 'In my view she did not leave La Crosse because she was ostracized . . . she wanted to get away from a world of unhappy memories . . .'[8]

Moving to New York, Ina Losey found a job in Wanamaker's antique furniture department, near Washington Square, selling fine china and furniture. At some point Joe wrote her a devastating letter, reproaching her for drinking too much and for having lovers, among them 'Uncle Jimmy'. Losey told Monique Lange that he wrote this letter when he was 'twenty-two' – in which case he would have been living in New York; but he confided to Sagan that, full of the cruel intolerance of youth, he'd composed the letter when very pleased with himself after writing and directing a play – in which case he was still at Dartmouth. His mother did not reply. As reported by Lange, when she died in London in 1959, Losey found an envelope, 'For Joe', in which lay his own letter. As reported by Sagan, before dying his mother gave him a torn envelope – inside it lay his letter. In reality Joe's sister Mary found the letter in a trunk when visiting La Crosse in June 1958. 'And there was a large white envelope, marked "to be given to Joseph Losey upon my death".' Reaching London, where Ina was dying, Mary handed it to her brother, 'unopened and untorn'. Joe never told her what was in it.[9]

5

New York

Losey's portrait of his arrival in New York is a romantic one: 'I came into New York with nobody at all to help me and qualified for nothing ... I worked in the Actors' Diner Club to get food; I slept on people's couches or on the floor. I worked for Jed Harris ... and he didn't pay me anything ... He occasionally gave me $5 out of his pocket. I lived in his house, ate there and slept there – not very free and very lonely.'

Yet Joe had cultivated a network of theatre-contacts during his time at Dartmouth and Harvard, and he was soon writing reviews for *Theatre Arts Magazine*, the *New York Times* (under the patronage of Brooks Atkinson), the *Herald Tribune*, the *Saturday Review of Literature* (writing as Walton Losey), and *Stage Magazine*. The prevailing tone was of elegant condescension – on A.A. Milne, for example: 'Mr Milne joins the full ranks of those writers who, even at maturity, can't suppress an admiration for their juvenilia.' Reviewing Max Beerbohm's *Around Theaters*, he noted that Beerbohm 'is as amusing to read in 1930 as in 1900'. (How did Joe know?) His first (unsigned) review for the *New York Times* (19 April 1931) was summarily cut at its conclusion by an editor: '*Queen's Mate* is innocuous enough [and not so bad a tale as seriously to impugn the mental stature of those who may enjoy it.]' The same paper (24 May 1931) also cut his review of a novel by Arthur Meecher, Jr. Joe was parading style and his career as a reviewer was short-lived.

But his ambition was active involvement in the theatre. Hearing that extras were needed for Vicki Baum's *Grand Hotel*, he took himself to the Apollo Theatre and found applicants lined up in batches of twenty. Taken on by Herman Shumlin, Losey jumped from extra to assistant stage manager while understudying small parts. *Grand Hotel* opened at the National Theatre, New York, on 13 November 1930. Involved in the production for nine months, he was struck by a passionate attachment to the star, Eugenie Leontovitch, a glamorous White Russian with a jealous husband. Regularly he brought gardenias to her dressing-room and – he later reported – she took him to supper at her sister's 'every night, practically, and I was a very gallant escort for her. John Hammond lent me his family limousine at Christmas to take her out one evening to dinner ...'[1]

The crucial new friendship was with John Henry Hammond. A wealthy young aesthete, only nineteen, Hammond had dropped out of Yale on a $12,000-a-year income from Vanderbilt trust funds and various occasional family legacies. Passionately opposed to racial prejudice from an early age, Hammond celebrated his twenty-first birthday by moving out of his mother's 'marble mausoleum' on 91st Street into an apartment on Sullivan Street, removing his name from the Social Register on the way. Frequently he took Losey up to Harlem for night-long sessions at the Savoy Ballroom or at Smalls, or Covans, a speakeasy 'where all the great musicians played'. Through Hammond Losey met Benny Goodman and Count Basie. Hammond founded the Negro Theatre, on Lower 2nd Avenue. 'It wasn't successful. But while it lasted it was the heart of the revitalization of jazz.' Hammond led Losey to Monette Moore's Club on 133rd Street, where he had discovered the seventeen-year-old Billie Holiday, who'd already done time in jail for prostitution. Losey began to collect 78rpm records, mainly jazz. As with jazz, so with politics: 'John Hammond, although he remained essentially a liberal, rich, humanist patron, got me involved in the earliest stages of my left-wing politics.'[2]

In 1931 Hammond invited Losey to accompany him to Europe as his paid-for guest. They sailed, first class, on the *Homeric*. 'All very Scott Fitzgerald!' Losey later reflected. After their ways parted in Munich, Losey visited relatives near Hanover but, finding no theatre work, moved on to London, where he picked up a job through the impresario Gilbert Miller as stage manager of Jeffery Dell's *Payment Deferred*, starring Charles Laughton. Shortly before the play opened at the St James's Theatre on 4 May 1931, Laughton suffered a traumatic shock: he was taken to court on a charge of soliciting young men, and had to confess to his wife, Elsa Lanchester, that he was an active homosexual. *Payment Deferred* transferred to the Lyceum Theatre, New York, then moved to Chicago after three weeks. When an actor became ill, Losey found himself playing the prologue and epilogue with Laughton after only four hours of rehearsal.[3]

According to Losey, he introduced Laughton to Jed Harris, 'the wonder-boy director' of the 1920s who had made millions out of such successes as *Broadway*, *Front Page* and *Serena Blandish*. Harris produced Laughton's *Fatal Alibi*, from Agatha Christie, with Losey again stage manager. After touring New Haven, the play opened at the Booth Theatre, New York, on 8 February 1932. Laughton hated Harris, the play ran for only forty performances, and the actor became bearish towards Elsa Lanchester, who, still shattered by the revelation of her husband's vice, took refuge in the company of Losey. Together they visited exotic dives, searching out louche novelty, roaming Harlem in the small hours, hearing Louis Armstrong in afternoon vaudeville, dancing. One Sunday Losey and Hammond took her to a wealthy home in Mount Kisco for cocktails. Several of the guests had just returned from Russia, 'idle-rich types' who went to study the workers. Then

Losey fell badly ill with mumps, 'which I had caught from Laughton', and was still in bed when the play closed and the Laughtons left for England. 'If Joe hadn't got the mumps, we might have become involved,' wrote Lanchester, 'two lost, angry people floating around, overshadowed by Charles.' When she returned to America, 'I saw Joe for one last time. We met for an early dinner at his sister's flat. Joe cooked oysters, a quite elaborate dish served in shells on rock salt.' She disliked oysters. Yet forty years later Losey told John Loring that he and Lanchester had had the full affair. It may be so, but in July 1968 she invited Losey to contribute to a book about Laughton, on the basis of his memories from the time of *Payment Deferred*; the correspondence contains no hint of anything requiring discretion.[4]

Despite his slowly emerging political commitment, Losey was by temperament a loner who failed to attach himself to any of the city's avant-garde theatre companies. He preferred the catch-as-catch-can of the freelance market, the world of wealthy impresarios, to submersion in a collective. Now living at 110 East 64th Street, he made his first pitch at the film industry. On the recommendation of a Dartmouth professor, Arthur Mayer, he was interviewed by Walter Wanger, who presided over the Paramount Astoria Studios in Queens – but didn't get the job. Hired by Bill Stern at Radio City Music Hall, shortly before its opening, as assistant stage manager, he was fired after the big vaudeville show went on to 3.30 in the morning, by which time the entire gala-opening audience had gone home, including 'the Rockefellers in the royal box . . .'[5]

He shared two apartments with his sister Mary after she graduated from Wellesley in 1932, but he avoided contact with his mother, who lost her job at Wanamaker's in 1933 and returned to La Crosse, where she was set up in a 'carriage house' by Andrew Lees (father of Joe's boyhood friend Robert Lees), whose wife was in an institution for the insane and consequently could not be divorced. Joe used to take Mary to plays and, on occasion, to the Savoy Ballroom in Harlem. 'Joe couldn't dance for beans. Nor could John Hammond. I would dance with whatever came along.' In March 1933, tipped off that Roosevelt was about to close the banks, Mary withdrew her savings of $200 and entrusted them to Joe while she visited Princeton. Arriving back, she found that he'd spent most of it on hoarding food – particularly tins of baked beans they would never eat.[6]

Now came Joe Losey's first break as a theatre director, approaching John Hammond with a script he had read for Jed Harris, *Little ol' Boy*, by Albert Bein. Hammond was persuaded, bought the rights, and set up Henry Hammond, Inc. with Losey and Irving Jacoby. The Russian-born Bein was an unusual character who had done time in a reform school and in Missouri State Penitentiary, besides losing a leg while jumping a freight train. *Little ol' Boy*'s cast of thirty-five was headed by Burgess Meredith, John Drew Colt and Lionel Stander. The designer was Mordecai Gorelick. Losey raised sums of $1,000 from Robert Benchley, Lewis Milestone and others, the balance coming from Hammond (who later put his personal investment at $8,000 out of $10,000). Rehearsals began in March 1933 and the play opened at the Playhouse, 48th Street, on 24 April.[7]

'I discovered that I really could direct. We got immense critical acclaim, but the critics said, "It's propaganda and propaganda has no place in the theatre".' The play soon closed, 'a total failure and total write-off'. Writing in *New Republic* (21 June), Stark Young complained that the play had been driven off the boards as propaganda 'despite its many extraordinary theatre qualities'. John S. Cohen, of the *Sun* (30 June), also lamented that it had met an undeserved fate. It seems that a minority of the reviews alleging propaganda did the damage – for example, Robert Garland, of the *World Telegram* (25 April), found it 'more effective as protest than as a play', adding, in a phrase destructive to the box-office, 'another reform-school drama'. Yet many of the reviews were friendly. Quoting them in his CV, Losey later called *Little ol' Boy* 'my first real critical success'. However, of eleven New York theatre critics, only one included *Little ol' Boy* in his choice of the year's Ten Best Plays.

Losey continued to improvise, questing and lobbying in that remarkable spirit of ad hoc eclecticism endemic to the capitalist theatre. In the spring of 1934 he directed the annual production of the Harvard Dramatic (or Glee) Club, Dennis Johnstone's *Bride for the Unicorn*, at the Brattle Theatre, Cambridge. Two professional actors were employed, one of them Norman Lloyd, who later appeared in Losey's Living Newspaper productions. Virgil Thomson wrote an original score, mainly choral, for what Losey called 'a very simple but highly expressionist production . . . a combination of a Harvard-financed student group plus professional actors . . . We were put up in the Harvard dormitories . . .' Norman Lloyd describes the extravagantly gay Thomson, then teaching at Harvard, gazing out at the Charles River on a blustery day: 'I don't know whether to wear a vestie-westie with my coatsie-woatsie.' The production caused a small furore and the Boston press adjudged the parables and symbolism of *Bride for the Unicorn* to verge on the insane. Virgil Thomson – discovered as a wise and peppery gnome living in the Chelsea Hotel fifty-four years later – described Losey as a good director who knew what he didn't want. 'Taste is knowing what you don't like. The best will tell you what effect they want you to arrive at. The worst will tell you how to do it.'[8]

But to do what? Losey would have a shot at whatever came his way – in one sense a necessary condition of apprenticeship but, as his later film career confirms, an anticipation of a lifelong eclecticism. Turning again to the patronage of John Hammond, in 1934 he staged a heavy historical drama,

Jayhawker, co-written by the Nobel Prize-winning novelist Sinclair Lewis and the dramatic critic of the *Chicago News*, Lloyd Lewis (no relation), an expert on the Civil War. 'Joe Losey kept spurring me on to join him again in a Broadway theatrical production,' recalled Hammond. 'Joe's life was the theatre.' Losey got to know Sinclair Lewis well, a warm, affectionate, sad person, but an alcoholic. In Vermont the young Losey found himself wet-nursing the great writer: '. . . he could stay sober for six to seven months. Then he would have one drink and it was near death sometimes.' In his preface to the Doubleday edition of the text, Lewis thanked Losey for bringing a theatrical knowledge to the play which he himself lacked. Hammond used to visit 'Red' Lewis in Bronxville in the hope of getting him to rewrite the sagging third act, but the famous novelist was drinking himself into oblivion. (According to Losey, production costs had reached $35,000 by the opening night. With the eventual failure of *Jayhawker*, Henry Hammond Inc. was dissolved. Oddly, Hammond says he had 'no investment' in the production.)⁹

The play is set in the years 1861–4, the Abolitionist era, the action moving from Kansas to Washington. Jayhawkers were Kansas men who formed para-military bands designed to keep slavery out of the State. Ace Burdett, a fiery orator and fixer, rode the Methodist vote into the United States Senate at the outset of the Civil War. The play begins with a fifteen-minute speech by Burdett to a church meeting in the woods – and no one interrupts. The star actor was Fred Stone, normally found in musical comedy – 'Broadway song and dance king' – and now performing in his first straight play. Casting Stone and Walter C. Kelly (as a Confederate general) in tandem, Losey was unaware that they were deadly enemies. 'One or the other was always flat up against the back wall and doing everything that he could to destroy the other's performance.'

The play opened at the National Theatre, Washington, on 15 October 1934, moved to the Garrick, Philadelphia, on the 22nd, and reached the Cort Theatre, on West 48th Street, New York, on 5 November. 'Official Washington and society is expected to make tonight's première a real E Street occasion,' warned the *Washington Times*. The capital's press was respectful and sympathetic, with Fred Stone hailed as outstanding, but there was a general feeling that the play tailed off. As a result the authors worked on a new ending for the Philadelphia production. The Philadelphia *Inquirer* commented: 'No wonder Fred Stone floundered a bit, since, it is said, he had not seen the revised script until six o'clock that night.' The *Evening Ledger*'s Henry T. Murdock expressed greater displeasure: 'Unfortunately, *Jayhawker* . . . was permitted to make a public bow before it was ready . . . At its opening here it quickly stepped into one of the most completely disorganized and unrehearsed productions we have ever seen.'

Losey blamed Fred Stone for having failed to cope with revised lines. 'Nor did I realize that Fred Stone had memorized the whole original draft of the play, word for word, months and months before he came to rehearsal, so when I gave him changed lines in Philadelphia he went berserk because he could only remember the lines he had originally learned. One night he couldn't remember anything ... And the prompters were prompting from either side of the stage and he stopped, pointed and said, "Now wait a minute. Wait a minute. I can't hear both of you at the same time. *You* tell me." '[10]

And so to New York, where the Cort Theatre programme advertised Chevrolet's lowest-priced, six-cylinder, four-door sedan, at a price of $540. The Breakers Hotel, Atlantic City, offered full board and lodging for six dollars a day. Writing in the *New Yorker* (24 November), Robert Benchley regretted that *Jayhawker* went to pieces towards the end, but he nevertheless enjoyed it. Not a single review had mentioned Losey by name.

In the Fall of 1934, concurrent with *Jayhawker*, Losey directed a revised version of Maxwell Anderson's play about the Sacco and Vanzetti case, *Gods of the Lightning*, for the twenty-two-year-old May Sarton's company, Associated Actors' Theatre, at the Peabody Theatre, Boston. Due to the emotive nature of the subject – the execution of two anarchist aliens for the murder of a Massachusetts paymaster, despite thin and contradictory evidence – the project stuttered before it started. The play, Losey recalled, was 'perhaps a little crude but passionate and immediate'. Mordecai Gorelick again designed the sets, which Losey described as 'constructivist'. Norman Lloyd and Peter Kapell headed the cast. Opening on 25 November, the production closed after three weeks, losing Richard Cabot his $5,000 investment. 'We reached the end and I knew it,' wrote May Sarton. Her single, neutral reference to Losey indicates no lasting affection.[11]

According to Norman Lloyd, Losey now pointed him in the direction of 'the first attempt at a Communist collective theatre in New York'. The co-founders were Nicholas Ray (Ray Kienzle), who had attended the same La Crosse high school as Joe, and Elia Kazan. But the significant point is that Losey did not point himself in the same direction – his own links with the Communist Left at this juncture remained, so to speak, by private contract, shrouded in private patronage. He now headed for Russia via a Finnish landowner.[12]

6

To Russia and Back

Having borrowed $500 to buy himself 'a round-trip third-class ticket to Europe', Losey acquired a new passport on 19 December 1934, and left New York on 16 January 1935. 'I sailed steerage on the *Île de France* and I spent most of my time in first class with Gilbert Miller, who invited me up for drinks and such; then I retired to steerage where you had four or six people in the same state-room, no toilet facilities, dreadful food, a sort of community dining-room which stank, and the sea was rough and everybody was sick and it was horrible. The trip took me to the Soviet Union via Sweden and Finland, and then brought me back, via Finland to Germany and then to London and then to New York.'[1]

Ernestine Evans, editor and writer, had given Losey (twenty years her junior) an introduction to Hella Wuolijoki, a left-wing Finnish writer, Estonian by birth, and active in the timber business. Her sister was married to R. Palme Dutt, the leading ideologue in the British Communist Party. Losey was never to forget the midnight sun at Hella Wuolijoki's Marlebeck (Marlebaeckin) home, and three weeks of cutting trees in the snow, and falling for her daughter Vappu (Vappuli): 'One of the most fabulous months in my life, in deep, deep snow, and taking saunas . . . and jumping into the lake through holes cut into the ice that was anything from three to five feet thick. And in the middle afternoons of that winter I used, on skates, to push [Wuolijoki] around on the sledge chair on the lake.'[2]

Losey was also a guest at Wuolijoki's residence at 2 Jungfrustigen, Helsinki (Helsingfors), where she introduced him to theatre people including the architect Paul Blomstedt, the artist Juho Rissanen (whom Losey didn't like), Ronngren, the director of the Swedish Theatre of Helsinki, and the critic Hans Kutter. Losey was interviewed by *Helsingin Sanomat* (3 March), which described him as 'very young, very quiet and modest . . .' He had visited an amateur play in the Workers' Theatre in Sörnäinen but had found productions of foreign plays, including O'Neill, 'rather mediocre'. He also regretted that most Americans did not know where Finland was to be found.[3]

Losey was still at Marlebeck on 3 March, when he stated his intention to leave for Russia 'within a few days'. His first, self-confessedly 'dilatory', letter to Hella Wuolijoki from Moscow was dated 16 March. Losey gave his postal address as c/o Herman Habicht, Philiporsky Per.4, Apartment 6, Arbat – although he was writing from the New Moscow Hotel and was about to move into a room – found for him by the international revolutionary theatre group

MORT – where he expected to encounter 'my little animals' (he had a phobia about mice and insects). He subsequently stayed in Jay Leyda's room on the outskirts of the city, then moved to a cheap room in an engineer's house.

'I am doing very well, although to/day I nearly broke some dishes in the restaurant,' he wrote to Wuolijoki. 'Leningrad offered nothing excepting conventional ballet and opera, proficient but deadly dull.' Arriving in Moscow, he had soon seen Okhlopkov's production of Gorky's *Mother*. Nikolai Okhlopkov, he added, was the most interesting '*régisseur*' (director) in Russia. 'Incidentally, if you could whisper in the ear of Assmus that Okhlopkov couldn't have a more ardent disciple, and if he might hint as much to Okhlopkov – it might be of enormous advantage to me.'

His first letter also refers to the international network of pro-Soviet intellectuals including Ivar Lassy, a Finnish politician and writer, who had left for Russia in 1922 and was now employed in publishing: 'Lassy promises me some roubles editing in the Foreign Workers' Publishing House, but rather wanted some party recommendations. MORT is going to vouch for me . . . He [Lassy] is reading the play [Hella Wuolijoki's *1918*] . . . to advise on Soviet production . . .' Losey signed off by urging Hella Wuolijoki to send her daughter Vappu to Moscow without delay: 'I like her very much, [but] that shouldn't deter her . . .'

Forty years later Losey told Ciment: 'I was terribly disillusioned when I first arrived . . . I saw extreme poverty, dirt and discomfort, and I didn't see any of the positive things. But at that time the theatre was very alive and they were very effectively run by collectives and that was an entirely different atmosphere.' He attended rehearsals directed by Vachtangov, Meyerhold and Okhlopkov – whose staging techniques Losey later adapted for the Living Newspaper. 'Okhlopkov used a central stage with runways all around and stairs going up to second-level runways, but it was adjustable for each production.' Meyerhold – soon to perish in the purges, and whom Losey described as brilliant but faintly decadent – was rehearsing *Camille* and playing *The Inspector General*. Losey attended his classes.[4]

He recalled sitting in a 'box' near Lenin's tomb and watching for hours a great people's parade (presumably May Day). 'It had a tremendous emotion[al] impact. The old boy up there was Uncle Joe. It was impossible to think of him as other than warm, lovely . . .' The Chicago *Daily News*, he said, published his emotional report of it, but a search has failed to turn up any such report other than one from the Associated Press agency.

Losey's views on theatre architecture are of intrinsic interest. Having visited a new theatre in Kharkov and a children's theatre in Kiev, he reported to the Finnish architect Paul Blomstedt that, 'Both show progress architecturally, both are well equipped, but both are retrogressive as far as ideas are concerned – formal stage and proscenium, the yawning orchestra pit dividing audience and actor, boxes, bad sight lines, etc. It is discouraging.' Hella Wuolijoki had introduced him to Blomstedt and his futuristic design for a new theatre in Helsinki. On 19 May, writing from Moscow, he told Blomstedt: 'Two things continue to strike me, examining your sketches. One is the fatally low proscenium . . . Don't press down on the audience. Don't focus them too much on the stage behind the proscenium which a good *régisseur* will seldom use. And the other question is one of size . . .' He enclosed a page containing sketches of the layout for four of Okhlopkov's productions, called 'Start', 'Mother', [illegible] and 'Aristocrats'.

Losey's 'A New Theatre for Finland' appeared in the January 1936 edition of *Theatre Arts Monthly*. Heavily illustrated by drawings and models, the article described Blomstedt's plans for a new experimental Kansanteatteri, or People's Theatre, in the workers' quarter of Helsinki. A section of the orchestra floor could revolve through 180 degrees and create a central acting arena. 'The disk can be raised or lowered so that, if desired, the top tier may be on a level with the stage proper and the action may pour into the central arena over the stage and down the tiers.' Losey noted that Blomstedt's design incorporated some features from Walter Gropius's 1929 project of 'Total theater', *but he made no cross-reference to Erwin Piscator's theatre design* (which he regarded as similar yet superior to Blomstedt's). Blomstedt died on 3 November 1935, aged thirty-five – his theatre was never built.

Back to Russia: enter Sylvia Chen, a half-Chinese dancer known to Hella Wuolijoki. On 25 May Losey reported: 'Sylvia danced here last week and was dreadful. I felt most guilty and very angry with her for her presumption – and upset by her bad taste.' (Apparently Sylvia had rebuked him for having telephoned Moscow from Hella Wuolijoki's Finnish home; to Hella he expressed contrition: 'Will you chalk it up against me for payment in a better era when telephones cost nothing.') Twenty-five years later he recalled those whom he'd encountered in Moscow, among them Gordon Craig, with 'unlimited resources' to produce *Macbeth* at the Mali Theatre; Brecht, Hanns Eisler, Lotte Lenya, Joris Ivens; students from Vassar, and members of the Group Theatre including Harold Clurman and 'I believe even [Lee] Strasberg'. (Brecht travelled to Moscow on 13 May, at the invitation of Piscator and MORT.) 'It was like the Montparnasse of the 20s in Moscow then. Everybody was there.'[5]

Losey embarked on a speaking tour of the Ukraine, visiting Kharkov and Kiev, 'which must have been the ultimate presumption, and everything was translated'. On 26 June he reported to Hella Wuolijoki that he was making slow progress in placing her play, *1918*, with a Soviet theatre: 'Okhlopkov says he may do *Yellow Jack* in fall, but I have no faith in . . . his doing it . . . *Little Ol' Boy* seems fairly definite, but with uninteresting theatre. Piscator's plans for work with foreign *régisseurs* mature hopelessly slowly.'

Although he later described US Ambassador Bullitt as 'our protector', an odd feature of Losey's arrival in Moscow was his failure to register at the US Consulate until 8 July, four months later; he also gave his date of arrival in the USSR as 11 May, which was nonsense. (This comes to light in an FBI report of 28 November 1947 based on embassy records.) His decision to register may have been connected to a fall in his morale; by 4 July, his intention to remain in the USSR through September had shrunk to an intended departure 'around August 1'. *'Little Ol' Boy* now probably won't happen and I have no luck with *Yellow Jack* . . .' His visa would expire on 19 August unless he got work: 'I'd have to come back as an intourist, paying the fees all over again and with all the visa and room difficulties again . . . Meyerhold and all the rest are amiable, but they haven't any money and after all what good am I to them. Also they've gotten stuck with a couple of American phoneys. This is what really enrages me, that MORT should go on paying stipends to various eternal-nothings and so there is none for me.'[6]

Oddly, Losey's discovered letters to Wuolijoki make no mention of his production of Clifford Odets's *Waiting for Lefty* in Moscow with a cast of American expatriates ('an English-speaking group of actors, artists, writers and students'). Losey later called this production 'misguided'; the *Herald Tribune* journalist Thomas Quinn Curtiss, who saw it, believes it was sponsored by the Soviet cultural organization VOKS. 'The Losey version was a catch-as-catch-can affair, acted by semi-amateurs who spoke English (more or less) and looked like the soldiers in Cox's army instead of New York cab drivers planning a strike . . . It gave Losey a tough time and he spent the rest of [his] time in Russia this side of the footlights.'[*][7]

Losey later recalled having written a piece about the Moscow Circus which *Variety* published on its 'front page' (though a search has failed to show any such front-page article). An official called Oumanski called him in and said, 'We think you should leave the Soviet Union because we've read the *Variety* article and we don't like it.' Losey also recalled telling 'Kusinen' [Otto Kuusinen], the Finnish-born member of the Politburo, that he wished to stay in the USSR and maybe get a job lumbering in Karelia. He was advised not to be a fool: if he wanted to do something about changing the world, the United States was the place to do it.[8]

*Presumably Losey's Moscow production was inspired by news from New York of the tremendous impact of Group Theatre's production. When Losey directed *Lefty* at Dartmouth College in 1975 he wrote in the programme: 'I was sure I had seen one of the Broadway productions of *Waiting for Lefty* before I sailed for Europe . . . It seems not . . . So I must have read the published version of *Lefty* in *New Theatre* magazine and taken a copy with me to Moscow.' (The play was first published in the February 1935 issue; presumably Losey had obtained an advance copy before leaving New York on 16 January.) When interviewed by *Helsingin Sanomat* (3 March) before leaving Finland for Russia, he did not name Odets among his list of 'interesting' young American playwrights.

On 4 July he informed Wuolijoki: 'I'll be off via Germany where I must see my relatives and to England where if there's work I may stay a little . . .' His efforts on behalf of Wuolijoki's play, *1918*, had proved fruitless. By 11 August he was aboard the SS *Ariadne* sailing from Finland, and by the 19th he was writing to Wuolijoki from the Hotel Rubens in London. Evidently his visit to Germany was squeezed into the intervening week: he wished he had been able to spend a 'longer time in Germany'. He now planned to sail for New York on the 28th. This letter contains friendly references to John Hammond but begins, bafflingly: 'I'm not coming back to Moscow after all owing to the vagaries of rich youths who [?are?] alarmed by Mr Roosevelt's tax bills.' Was this Hammond? Who else? Had Hammond lent him the original $500 and subsidized the trip? He later recalled that Hammond had given him a huge collection of jazz records for the Russians. 'They were confiscated at the border, and later, in a great store, suddenly I hear coming over this loudspeaker some of my records. But I was never able to get them back, never.'[9]

A postscript on Hella Wuolijoki. Losey told Michel Ciment that she 'wrote a lot of plays . . . She had known John Reed and people like that . . . I was instrumental in sending Brecht to her and Brecht also stayed with her [and] used one of her plays and collaborated with her on it to write *Herr Puntila*.' Her daughter Vappu Tuomioja writes: 'She had strong leftist sympathies, but she never was a member of the Communist Party . . .' In 1942 a female parachutist 'dropped out from the sky and spent a night under her roof in the country . . . So Hella was sentenced to prison [for harbouring a Soviet spy] and freed when the peace came in 1944. After that she was nominated director of the Finnish radio. Her plays are still performed all over Finland . . .' Wuolijoki's grandson, Erkki Tuomioja, describes her as 'very much a fellow-traveller' who became, after the war, a member of parliament for a Communist front, the People's Democratic League. She died in 1954. Despite post-war correspondence Losey never met her again. From the late 1930s he began to mis-spell her name as 'Wou' rather than 'Wuo', an error which stuck.[10]

The impact of the trip to Russia on Losey's politics is hard to assess. He later set out his radical graduation as follows: 'When I went to college I didn't want to be politically committed. I wanted to lead my own intellectual life, making my own comments, as though I were a privileged member of the Bloomsbury set at the end of the nineteenth century . . . I was indifferent to social problems, up to the moment when, brutally, I was precipitated into . . . the great economic crisis of 1929 . . . The Depression swept away all our preconceived ideas on life and our place in society.' By his own account he took part in a protest march in Washington against the destruction of the veterans' camp – First World War 'Bonus Marchers' were living in tents – by General Douglas MacArthur, acting on Hoover's orders. This occurred in 1931.[11]

Losey's involvement with the far Left probably began in 1932, when John Hammond put money into the Theatre Union (at the Civic Repertory Theatre on 23rd Street), where Losey met Charles Walker, a Trotskyist 'who gave me most of my early political reading . . . I read about William Foster, the IWW movement, I read Emma Goldman's autobiography, *The Anarchist*, Prince Kropotkin and so on.'[12]

In retrospect Losey tended to present himself as directly responsive to world events: I was There. Here, for instance, he describes the outbreak of the Spanish Civil War in July 1936: 'I was driving with my wife [*sic*] to our summer place in Vermont. It was a warm and beautiful day. The top of the convertible was down. The birds were singing. The radio announced the horrible event. Both my wife and I realized that for us this was the end of a post-war 1918 period and the beginning of the pre-war 1939 period that would change all our lives.' (Few people realized in 1936 that they were entering 'the pre-1939 period' – the point being Losey's inclination to 'realize', instantly, what everything meant and presaged.) A certain arrogance accompanied induction into the all-knowing universe of the Party line. Interviewed in 1988, in the Chelsea Hotel, New York, a very old, deaf but sharp Virgil Thomson recalled: 'He [Losey] was a sour puss like a great many of those Communist boys. He disapproved of everyone, particularly those who were not Communists. He spoke the lingo. We used to tell Communists by their vocabulary . . . The Communist agitation was always to make you feel inferior.'[13]

Losey's sister Mary recollects that she, Joe and Joris Ivens visited the White House to persuade Eleanor Roosevelt to show Ivens's *The Spanish Earth* to FDR. In his memoirs, Ivens describes in detail the evening when he and Hemingway showed the film to the Roosevelts, but does not mention either Joe or his sister. But Losey later made the most of his First Lady connection. *New Masses* (1 February 1938), a weekly journal effectively published by the CP, reported that Losey had taken part in a benefit on its behalf. This was one of the very few entries from the 1930s in Losey's FBI file. He always denied having been committed to any political group in America during the 1930s. 'But there was no question of organizational commitment. I was not a member of the Communist Party at that time.' His FBI file offers no indication that he was suspected of Party membership in the 1930s. However . . .[14]

Take the case of a woman who knew Losey well, and who recalls an occasion 'about 1936', in Jack Dempsey's, when Losey explained to her his involvement 'in the CP' and urged her to do likewise. 'He wanted me to attend CP Marxism classes.' She declined. He never solicited her again. But – in her view – it meant that in later years she was lying when she denied any knowledge of Losey's Communist connections to the loyalty – security board which introduced his name while interrogating her. She adds that the prominent Soviet intellectual Karl Radek became one of Losey's contacts. (Radek was put on trial in Moscow in January 1937.)

Many years later Losey confided both to this woman and to his wife, Patricia, that he had been a courier for the Communist underground in the 1930s, carrying letters whose contents were not known to him; he was more useful outside the Party. One significant line of indirect evidence confirms his involvement with the Communist underground – his relationship with the Hungarian-American Communist József Péter (known in the USA as J. Peters), whom Losey's New York doctor, Lewis Fraad, himself a member of the Communist Party at that time, calls 'Joe's mentor'. According to Irma Fraad, 'J. Peters was in charge of the Arts Project which included the theatre. Jim Proctor, a theatre publicity agent, was in Peters's unit along with Joe Losey.' Fraad had been Proctor's room-mate at college. 'It seems,' his daughter writes, 'the Party told him to go to medical school as a cover for Comintern work in Vienna.'[15]

Hungarian by birth, J. Peters was fourteen years Losey's senior. As the Cold War gained momentum and anti-Communist hysteria mounted, Peters assumed a demonic profile, described by the most famous of ex-Communist witnesses, Whittaker Chambers, as 'the head of the whole [Communist] underground' who had allegedly introduced him to Alger Hiss in 1934 and thereafter acted as minder to Chambers, Hiss, and the Communist underground cells in both Washington and Hollywood; a purveyor of false birth certificates, a forger of passports, the key liaison with Soviet agents like Chambers's 'Colonel Bykov'. In his book *Witness*, Chambers claimed that Peters had been, as 'Alexander Goldberger', a minor member of Béla Kun's Soviet regime which ruled Hungary in 1919 – 'a short, dark, friendly Hungarian'. In 1947, Louis J. Russell, a former FBI agent now acting as an investigator for HUAC, testified to the Committee that on 3 May 1942, 'Alexander Stevens, also known as J. Peters, whose real name was R. Goldberg' had visited Los Angeles and made contact with a number of Hollywood Communists, including Herbert Biberman, Waldo Salt, John Howard Lawson, Lester Cole and others. (Losey was still working in New York, but these men were, or became, his friends and comrades.) When Russell testified, HUAC had already issued a subpoena to Peters but its agents were unable to serve it due to lack of cooperation from the Justice Department's Immigration division, then holding him on deportation charges. Chambers turned up at the public deportation hearing at the Federal Building, New York, to identify J. Peters.

Of contact between Losey and Jószef Péter during the first two decades of the Cold War there is no found evidence. The Fraad family visited József and Ana Péter in Budapest in 1964. At the turn of 1971–2 Losey was editing his film *The Assassination of Trotsky* in Rome. With a wry smile and a laugh he told his screenwriter, Nicholas Mosley, that he had received a call from his old Communist contact in New York, now living in Eastern Europe, expressing interest in the film. Losey did not name the caller but it was clearly Jószef Péter (indeed, Lewis Fraad heard in Budapest, from 'a third party', that Losey had invited Péter to serve as a consultant). From Mosley's memory the exact content of the call is not clear – but Losey somewhat teasingly conveyed the implication that his own progressive credentials were under scrutiny.[16]

The first discovered letter from Losey to Jószef Péter is dated 8 July 1971. 'Perhaps,' Losey wrote, 'you know that I tried to reach you in Budapest from Vienna where I spent Christmas about three years ago [December 1967]. The address I had then came via Jim [Proctor] and I got no response so I have had no way of communicating with you.' (Jim Proctor had acted as New York film publicist for *The Servant*.) Losey's letters to the Péters were discreet and carefully voided of content. On 9 February 1972 he reported an invitation to Elizabeth Taylor's birthday party in Budapest. Would the Péters be able to see him 'without any embarrassment to yourselves?' Losey requested a note saying merely 'yes' or 'no' – 'otherwise Patricia will ring you and say she is Patricia, and you can say yes or no to her, and where and when if'. The Loseys attended Taylor's birthday party on 27 February, and visited the Péters' apartment for tea. A wood stove was burning and the time was spent in affectionate and discreet small-talk. Patricia does not remember any discussion of what was most on Losey's mind, *The Assassination of Trotsky*, then in its final, tortuous, phase of editing. Later that year Losey reported his visit to Budapest to his old friend of the 1930s, the Red millionaire F.V. Field, with a circumspection he clearly enjoyed: 'Our friends are well . . . old . . . and a little sad . . . not so much changed as one might have imagined.' The dots are Losey's. Questioned about this, Field is 'evasive' (the word is his).[17]

On 15 August 1975, Ana Péter wrote from Budapest, thanking Losey for sending a book: 'I'm sure that you'll be interested to know that Pete [J. Péter's nickname] has just celebrated his 80th birthday and was given many honors here including one of the highest decorations by the [Kàdàr] government.' On 20 December 1977, Losey wrote from Paris mentioning several mutual friends: 'I see a good deal of Joris [Ivens] who has completely transferred his confidence to China . . . As for J. Proctor, I have written him off. But not Lou Fraad.' Quarrelling with Proctor in a letter dated 14 January 1975, Losey had compared him unfavourably to Jószef Péter: 'If such are my friends, "who needs enemies", as they say in Pete's country.' Losey's lifelong loyalty to Stalinism was essentially a loyalty to his own youthful commitment.

How did Losey regard events in the USSR in the era of trials and purges? 'We all believed that these people were – must have been – traitors. It's extraordinary. It was very well managed by the Stalinists and by the Russians.' (In the late 1950s Losey lived next door to Arthur Koestler in London, but does not appear to have talked politics to the author of *Darkness at Noon*.)[18]

7

The Living Newspaper

By September 1935, Losey was back in New York after some eight months abroad in Europe and the USSR. On 1 October he wrote to Hella Wuolijoki referring to 'the tremendous new Federal Theatre project . . . it is sure to run amok [*sic*] of labor difficulties, etc. etc. . . . My life at the moment is too routine, and dull routine at that, to write about. I look at radio, at Hollywood, at bad plays, and am fundamentally idle.'[1]

But never for long. In January 1936 he directed Paul Green's one-act play, *Hymn to the Rising Sun*, for the Actors' Repertory Company at the Civic Repertory Theatre on 23rd Street. It was performed on a bare stage with no scenery. No one was paid. The captain of a chain gang addresses his convicts early in the morning of Independence Day. Brooks Atkinson reported in the *New York Times*: 'Partly as a result of Charles Dingle's masterly playing, Mr Green's *Hymn to the Rising Sun* is an overwhelming piece of work.'

As assistant to John Houseman, Losey was now involved with the Federal Theatre Project's Negro Theatre at the Lafayette Theatre, 'where John [Hammond] and I had spent all of our nights so many years before'. Here he directed *Conjure Man Dies*, a comedy in three acts and fourteen scenes written in 1933 by Rudolph (Bud) Fisher, a Harlem physician and novelist. Losey described it as a conventional 'fourth-wall realistic production with realistic sets'. A fast-moving, big-city entertainment crammed with inside allusions and bitter minority jokes, it opened on 11 March 1936 and was a smash. Clearly Losey was at home in jazz-theatrical Harlem; the play itself is so funny (though chaotic) that it required only theatrical discipline and know-how to bring the roof down.

We begin in Crouch's undertaking parlour, 130th Street, Harlem. A table bears a sheeted corpse. The next body to arrive is supposedly that of N'Cana Frimbo, the conjure-man, circus-man, prophet and caster of spells. Enter the eager Detective-Sergeant Dart (bound to make a fool of himself sooner or later, probably sooner). Jokes on Harlem dialect abound: 'BC' means 'Befo' Circuses'. The super-religious Aramintha Snead is asked whether she's an American: 'I is now. But originally come from Savannah, Georgia.' Act 2 moves us to the Forty Club Cabaret in the Cellar Honky-Tonk. It emerges that Frimbo had been in the numbers racket. His corpse gets stolen. And who owned the handkerchief found in Frimbo's throat? Detective Dart accuses Webb, a black, of 'Carrying a rod, running numbers, bribing cops, muscling in on half the cigar stores in the avenue, running a blackjack and dice joint . . .'

Webb nods. 'All right – what's the charge?'

Inevitably Frimbo comes to life and claims he was murdered. After a prayer meeting, with hymns, and some shoot-outs, he is suspected of . . . (too complicated to explain). This provokes from Frimbo a long speech about his childhood in East Africa, involving feast ceremonies and rituals in the state of (you guessed) Buwongo. Finally all the suspects are assembled in a darkened room and Frimbo is shot dead by someone disguised as someone else. Sexual jealousy may be behind the whole thing. Terrific stuff, but the downtown critics emerged from the Lafayette long-faced. The *New York Times*'s review was headed 'Harlem Mumbo Jumbo': 'To a paleface, fresh from Broadway, the new play seemed like a verbose and amateur charade, none too clearly written and soggily acted. But the Lafayette Theatre was bulging with family parties last night who roared at the obese comedian and howled over the West Indian accent of a smart Harlem landlady. This column doesn't know everything after all.'[2]

The big production which held the key to the Negro Theatre's future was *Black Macbeth* (1936). Enter another boy from Wisconsin, the flamboyantly gifted Orson Welles, then only twenty years old and already a highly paid star of radio. Backstage at the Lafayette, to make room for huge slabs of scenery and painted backdrops, *Conjure Man Dies* was 'gradually being edged down towards the footlights, to the full fury of Joe Losey, its director' (recalled John Houseman, who intensely disliked Losey, regarding him as an envious opportunist with an appetite for setting up elaborate feuds). Welles pulled off a smashing success, the 1,233-seat Lafayette Theatre overflowing during a nine-week run which earned such esteem from the FTP's director, Hallie Flanagan, that she allocated Welles his own theatre and an assignment to launch the FTP's classical unit, with Houseman in charge of administration. This episode precipitated Losey's lifelong sourness towards Welles. 'I saw Welles once or twice a week . . . I don't think he liked me and I didn't like him . . . He did a production with Paul Bowles's music of *An Italian Straw Hat* . . . I went home and sent him a three-page wire . . . I thought it was the best theatre production I had ever seen in the United States.' Losey got no response; years later he was told by Houseman that Welles had assumed Losey was pulling his leg.[3]

The Theatre Union production of Brecht's *The Mother* had opened on 19 November 1935. Brecht made his first trip to America and hated the production. Losey recalled him rising up in the middle of a rehearsal to exclaim '*Drech!*', then staging a furious exit. 'And he was right. The production had no style. It had no passion. It was just the shell and it was heavily sentimental.' In Moscow, Losey had met Brecht's esteemed colleague Erwin Piscator, the moving spirit behind the international theatre conference held in

April–May 1935. In 1964 Losey made an odd comment: 'The more I got to know Piscator, the less interested I became . . . In Moscow he was already half lost, in New York completely so.' The heart of the matter is that Piscator worked in New York with the Group Theatre (Harold Clurman, Lee Strasberg), with which Losey was never actively associated.[4]

Losey introduced John Houseman to Brecht and suggested that Brecht's *Round Heads and Pointed Heads* might be suitable for the WPA Negro Theatre, but, according to Losey, 'it was stopped by Hallie Flanagan as being too political'. Losey himself moved on to direct two productions for the Living Newspaper Unit of the Federal Theatre Project, which, as he said, changed 'the style and direction' of radical theatre. Brecht – he reported – adored the Living Newspaper. Despite his own austerity regarding colour, his white lights and neutral hues, he particularly liked a sequence in which Losey dressed his proletarians in fuchsia and pink. 'I remember Brecht cross-legged on the floor of the drawing-room, cap and cigar and denims against the satins and Japanese mats and Naguchis and the Calders, Mirós, Lurçats that were the background of my then wife . . .' Losey wasn't yet married to Elizabeth Hawes, who designed costumes for the Living Newspaper, but Losey's *coup d'oeil* marks a marriage of Marxism and Modernism.[5]

The New York Living Newspaper Unit was established late in 1935. By an agreement with the Newspaper Guild, unemployed journalists became the research force for a documentary theatre dramatizing social problems and class conflict from a radical, anti-capitalist perspective; its agit-prop techniques owed much to the proletarian theatre movements in Germany and Russia. Losey himself stressed his own debt to Meyerhold's protégé Nikolai Okhlopkov, whose work he had observed in the Krassnya Presnya district of Moscow: 'Okhlopkov was breaking down the proscenium and presenting theatre in the round and the rectangle and the hexagonal as it has never been dreamed of before or approached since.'[6]

Morris Watson (a Newspaper Guild official who'd been fired by United Press) and his 'managing editor', Arthur Arent, approached Losey when he was working as assistant to John Houseman in the Harlem FTP company. Losey's first Living Newspaper production was *Triple-A Plowed Under*, which opened on 14 March 1936 at the Biltmore Theatre and ran for eighty-five performances. Staged in twenty tableaux, the production took American agricultural policy as its agitational theme: War and Inflation; The Price of Milk; Farmers Organize; Milk Strike; The Harvests Burn; Drought; The Sherwood Affair; The Supreme Court. The settings were by Hjalmer Hermanson, with a cast of one hundred.

'In *Triple-A Plowed Under* there were a lot of little vignettes,' Losey recalled. 'It was approaching a movie technique: parts of the stage on different levels were picked up by spots – like film cuts . . . The music consisted of a large orchestra with only trombones and percussion; the rest were all fire sirens and sounds . . . from the orchestra pit. It was a plastic set entirely moulded by light . . . We had a lot of designers and it was a kind of collective. The costumes I did with my then wife, Elizabeth Hawes' (they were not married until 1937). Everyone was paid $30 a week, from director to stage hand. 'If anybody got a job in the commercial theatre, he immediately wanted to leave.'[7]

Losey's production used a proscenium-arch stage but (comments Arnold Goldman) 'made effective steps towards "unity of design" in its use of atmosphere, episodic treatment, mime and de-mystifying didacticism, and the thematic relationship of the involved groupings (farmers, consumers, middle-men, politicians) mediated by the stage company . . . Patient, unembroidered explanations are required to show to farmers how the prices they receive for their produce are inflated by middlemen and passed on to the consumers . . . Government intervention, in the form of the New Deal Agricultural Adjust-ment Act (AAA or Triple-A) is soon exposed as vulnerable to manipulation . . . The play's agit-prop ending, calling for and heralding the beginnings of a "farmer-worker alliance", is in fact prepared for and inherent in the whole play.'[8]

On 20 March 1936 Losey reported to Hella Wuolijoki: 'I have just opened a play in the new Federal Theatre set-up which has caused riots and much excitement. The government is beautifully scared and wondering how it all happened.' The FTP's national director had condemned the production as 'outright Communist propaganda . . . She [Hallie Flanagan] and Philip Barber tried to stop it. Morris Watson, Arthur Arent and I said, "We either open it in three days as planned and as announced or all three of us will resign . . ." So they let us open and it was a tremendous night . . . attended, of course, by Eleanor Roosevelt, Mayor LaGuardia, by [Congressman Vito] Marcantonio, by Earl Browder, the head of the Communist Party. The place was swarming with secret servicemen, and there was a meeting that afternoon of the American Legion Chapter. The show was marched on by the veterans, repelled by the police because of the kind of people who were inside . . . Later on in the production, they destroyed the switchboard which was a very expensive one and very elaborate. So it materialized into an actual physical conflict.' Elsewhere Losey wrote: '. . . the protest . . . [was] headed by the stage-manager of the production . . . he addressed a rabble-rousing speech to a veterans' meeting on the afternoon of our opening, organizing a march for that evening . . . the going was exceedingly tough and not unrelated to the kind of struggle which was a life and death matter in Detroit in 1936 and after, where I was also present.'[9]

On 18 March the New York Times's intrepid critic, 'B.C.', reported 'an atmosphere bristling with tension, uncertainty and blue-coated guardians of

the law . . . Up to the moment that the curtain rose there was speculation in
the audience as to whether [the] event would actually occur . . .' The reviews
were friendlier than might have been anticipated. The *Times*'s 'B.C.' reckoned
that *Triple-A*, 'a hard-biting, necessarily sketchy but frequently brilliant
review of the American farmer's plight, violates one rule of a good newspaper
story. It waxes editorial. It takes sides.' As for the theatrical aspect, 'The
staging . . . is economical, fast and occasionally confused, with an amplified
voice prefacing each scene . . . It is definitely the method of the newsreel,
with the stereopticon, indeed, used in several instances . . . The Living
Newspaper must be credited with a pretty sensational beat.'

Triple-A closed on 2 May, though it was briefly revived to fill the gap until
Losey's next production opened on 24 July. This was *Injunction Granted* –
the title refers to the phrase repeated by judge after judge against combinations
of workers attempting to defend themselves by creating labor unions. The
production – again a collaboration between Losey, Arent and Watson –
followed a chronological narrative of labor history from indenture to the
formation of the Congress of Industrial Organizations under John L. Lewis.

The production was designed as a single-unit set with ramps leading to five
or six levels. Goldman writes: 'In *Injunction Granted* Losey was able to devise
an adaptation of Okhlopkovian staging within the restrictions of a proscenium
arch . . . A system of runways, platforms and hatches provided the planes and
areas which could be selected by complicated lights and which could overflow
into the audience . . . scenes were "spotted" at various areas . . . the rest of
the stage remained dark. Only the last scene, as a summation of all that had
gone before, used the entire stage.' Projected headlines and cardboard signs
provided both pace-quickening exposition and a way of presenting history
through newspaper techniques, particularly headline and cartoon. A clown-
figure was used to parody the callous antics of the judiciary and to convey the
obsequious folly of the naïve and gullible. Norman Lloyd played the Clown
in tennis shoes, a wide-sleeved jacket, patched trousers and a Harpo Marx
hairdo. Goldman contrasts the parody and farce used in *Injunction Granted*
with 'the grave decorum and even pompousness of *Triple-A*'.[10]

In Losey's view, *Injunction Granted* was 'much more advanced theatre'
than *Triple-A*, the latter being 'more successful because it was more conventional, and also because it didn't tax the audience at all . . . they didn't have to
cope with Virgil Thomson's score, and with the mime either'. Thomson, who
used percussion and drums to punctuate the text, recalls a score with several
hundred cues, using sixteen snare drums and sixteen base drums. Losey
would suggest points in the script suitable for some topical musical reference.
'I would say "TTT – tunes take time, that will slow it up".' Thomson
chuckles: 'It scared hell out of the US Government'.[11] What caused the
gravest offence were the post-intermission scenes, covering the years 1929 to

1936. No one was too worried about the finer details of seventeenth-century history (although dramatic criticism obviously extended to the whole production).

In Scene 18, a Demagogue preaches the unity of capital and labor, denouncing Bolshevism, Fascism, Socialism and Un-Americanism while workers and capitalists applaud the parts of his speech which appeal to them. John D. Rockefeller and Howard Heinz put in an appearance to reinforce the message: capital and labor must work together in harmonious partnership.

Scene 19 depicts the Gastonia, South Carolina, strike of 1929. We meet a heroic politician, LaGuardia, seeking to restrain judicial interference in labor disputes and to defend the workers' freedom of association. The Norris–LaGuardia Act is passed: a major step forward. The employers cleverly sidestep the Federal courts and appeal to the State courts. The injunctions are again granted.

By Scene 23 we are at 1933 and attempts to unionize journalism. Arthur Hays Sulzberger, proprietor of the *New York Times*, says: 'I think reporters have the right to organize . . . but I think it is a right they ought to forgo!'

The San Francisco general strike of 1934 arrives with Scene 24. Enter here the representative of the New Deal, General Hugh S. Johnson, 'wearing frock coat with gold epaulettes'. First he announces that 'the right to strike is inviolate, like the law of self-defense!' Then he denounces the 1934 general strike as 'a threat to the community . . . a menace to government . . . It is civil war!'

Scene 25 involves the conflict of the Newspaper Guild with William Randolph Hearst, using the idiom of a boxing match. The Referee is on Hearst's side and finally proclaims him the winner, although he has been knocked unconscious. In Scene 26 the Liberty League is attacked and its wealthy patrons – particularly the Du Ponts – listed.

Scene 28 is 'Labor – 1936 – Finale'.

(*Light up center. Line of women come in, and immediately begin a pantomime of ironing. They stand in a straight line facing the Audience. The Orchestra accompanies the ironing with sound effects.*)

At this juncture the judges (i.e. the Supreme Court) are ruling out a lot of pro-labor New Deal legislation as unconstitutional.

VOICE OF LIVING NEWSPAPER: . . . the CIO . . . headed by John L. Lewis is formed by twelve unions of the American Federation of Labor. The CIO meets its first test. The drive to organize the steel industry.

Large numbers of workers, carrying their (mainly pro-Communist) union banners, arrange themselves behind the figure of Lewis – who has the final word of the production. Expressing 'the sentiments, hopes and aspirations' of millions, he 'accepts the challenge of the overlords of steel'.

(*At the word 'challenge' all signs are lifted up. Curtain.*)

The critical impact was explosive. Richard Watts, of the *Herald Tribune* (25 July), reported that Losey displayed a genuine gift for stage movement and for stylized simplicity, but the later episodes were vitiated by 'a sort of hysterical caricature that is both ineffective and anti-climactic'. Too much ground had been covered, the result being 'chaos'. 'Mr Losey is a director to be watched, but the show at the Biltmore is not a triumphant editorial.'

Indeed there were misgivings at the highest level of the Federal Theatre Project. Both Hallie Flanagan and her deputy, William Farnsworth, raised objections after a rehearsal, Mrs Flanagan asking Losey and Morris Watson to 'clean up the script and make it more objective'. They didn't. After the opening night on 24 July she wrote them a long memorandum:

'1. The production seems to me special pleading, biased, an editorial, not a news issue . . . Whatever my personal sympathies are I cannot as custodian of federal funds, have such funds used as a party tool. That goes for the communist party as well as the democratic party. To show the history of labor in the court is appropriate; to load that document at every turn with insinuation is not appropriate.

2. The production, in my opinion, lacks a proper climax, falling back on the old cliché of calling labor to unite in the approved agit-prop manner.

3. . . . [The] production uses too many devices, too much hysteria of acting . . .'

Watson replied that the play was drawing crowds – and made no changes. On 20 August, Flanagan expressed her exasperation: 'Everyone knows that those crowds are being sent by their Unions . . . I will not have the Federal Theatre Project used politically . . . I took your word and Joe Losey's word . . . and I think you both let me down . . .the avalanche of unfavorable publicity . . . is all ammunition against the project . . .'

Some changes were in fact made to the production, although they did not satisfy Hallie Flanagan. On 22 September, the *Daily Worker* reported: 'WPA Theatre Revamps Shows. Changes made in line with criticisms offered by reviewers.' Instead of the confusion prevailing at the finale, brief staccato scenes had been substituted to serve as a summation. But clearly Hallie Flanagan was not satisfied and the play was closed in October, well before the next production was ready. *Triple-A* was brought back to fill the gap. According to Losey, it was 'grotesque and ridiculous' of Hallie Flanagan to claim that the Unit was being used as 'a party tool' since 'I was not a member of the Communist Party at that time – nor, do I believe, were the others.' Losey handed in his resignation.

Injunction Granted represented the apex of the Living Newspaper's radicalism; in retrospect it can be seen as the beginning of the end of the FTP. After long delays, Mrs Flanagan's request to defend herself was granted by the House Special Committee on Un-American Activities (the Dies Committee). She took the witness stand on 6 December 1938. In 1973 Losey wrote to John Houseman, expressing surprise that his recent book had 'whitewashed Flanagan, considering . . . the fights that both you and I had to maintain our projects'.[12]

8

Love and Oil

By September 1936 Losey was back in action with *Who Fights This Battle?*, a play written early in the Spanish Civil War by Kenneth White, with a musical score by Paul Bowles. Losey staged two special performances in the ballroom of the Delanee Hotel. Funds were raised by collection, everyone worked free, profits went to Loyalist Spain, and the *Daily Worker* praised Losey's direction. A Ralph Steiner photo-portrait of him at this juncture shows arms resolutely crossed, mouth heavy, eyes faintly resentful, expression resolute and sullen.[1]

He began planning Social Circus. The first programme, *Convention*, was going to be 'an open theatre using acrobats, dancers, singers and actors, who would dramatize the horrors of the American political convention system'. But, despite Elia Kazan's involvement, Social Circus never got off the ground. Casually Losey added: 'I directed Kazan in a thing called *Newsboy* some time around 1936, which was a highly left-wing agit-prop but very successful. I never liked him much.'[2]

In 1937 he staged a number of Political Cabaret shows, sponsored by Theatre Arts Committee, which took place (according to Losey) in the crypt of a deconsecrated church on East 54th Street, between 3rd Avenue and Lexington, on Sunday afternoons, 'cocktail hour'. He described the ambience as 'anti-Nazi, pro-trade union, anti-capitalist ... We served drinks and sandwiches and the audience sat at tables', but the New York version lacked, he admitted, the 'sophistication and experience' of the German political cabaret of the late 1920s. 'I guess it lasted a year, maybe two, quite a long time.'[3]

Collective enterprise did not soften personal ambition. In March 1937, while living at 21 East 67th Street, Losey wrote a note to the curator of the New York Public Library's Theatre Collection. 'Curiosity and vanity' had impelled him to look himself up in the catalogue and he wished to set the record straight. 'Let's have lunch some time.'[4]

On 23 July 1937, Elizabeth Hawes Jester, 'internationally famous designer of women's clothes', and Joseph W. Losey were married at Hoosick Falls, NY, by Justice of the Peace Theodore Wucks. At thirty-three, the bride was five years older than the groom. The *Herald Tribune* reported that Elizabeth Hawes, 'Stylist' and 'Creator of Fashion Shows Causing Sensations Abroad', had wed Joe Losey. The picture was of a smiling Hawes – she was the news.

Indeed, Elizabeth Hawes entered *Contemporary Biography* in 1940 – whereas Losey did not surface in its pages until 1969.

Descended from a New England family, with maternal and paternal American lineages extending back to the 1630s, Lizzie Hawes was a small, dark, effervescent woman of remarkable energy, talent and imagination. After graduating from Vassar in 1925 she had set out for Paris, where she worked as a fashion journalist, produced sketches for American manufacturers, and took a job with the *couturière* Nicole Groult. The firm of Hawes–Harden opened on East 56th Street on Hawes's twenty-fifth birthday. In July 1931 she exhibited her own, distinctively American, fashions in Paris. After hiring her services to a 7th Avenue dress manufacturer, she was able to re-establish her own business on East 67th Street. In December 1930 she married Ralph Jester, a designer–artist who introduced her to Alexander Calder, Joan Miró and Jean Cocteau, but their lives drifted apart and the marriage ended in 1934, soon after she had begun a passionate affair with the artist and tapestry designer Jean Lurçat (who was happily married in France). 'He was the first man I ever slept with who had no inhibitions and was also very potent.' But – we are told – she needed him too much and broke it off.[5]

According to Losey, he met Hawes in 1933 – but he also said 1934. Their paths converged at the Askew Salon. 'We met at various salons of the early thirties art coteries . . . people like Kirk Askew, Alfred Barr, Sandy Calder, Julian Levy, Virgil Thomson, Lurçat.' Evidently their relationship began at the time of his production of Sinclair Lewis's *Jayhawker*. She reached Moscow three months after Losey, having set up an official visit as a dress designer through the Soviet consul in New York. Losey met her on arrival in Leningrad, but at this stage his affections had been diverted by a brief meeting with Hella Wuolijoki's daughter Vappu in Finland. She – now Mrs Vappu Tuomioja – writes: 'Lizzie Hawes: according to Joe her visit did not turn out very well.' Hawes and Losey made their way home to America separately. 'I met Joe again in Finland in August. His stay was very brief, but long enough to start a friendship that lasted through the seventeen years when we only communicated through letters.' To Vappu's mother he reported: 'Vappuli is swell . . . We had an excellent time together of which she will tell you as much as [she] chooses. I hope deeply that she will come to America . . .' His hopes were dashed; Vappu fell silent. On 6 December he expressed sadness to Hella that he'd heard nothing from Vappu; on 20 March 1936 he wrote again: 'And, incidentally, tell Vappuli that I am sorely disappointed that she never pursued the opportunities at Vassar . . . I am quite lonely in fact . . .' On 17 April: 'I am quite lonesome to hear from you and Vappuli . . .' Might Vappu not, after all, reconsider about applying to Vassar? Although he was now living with Lizzie Hawes he clearly remained under Vappu's spell. On 20 June 1939, he wrote to Hella asking for Vappu's married name and address.[6]

In later times, when discussing the year 1936, Losey frequently referred, incorrectly, to Lizzie as 'my wife'. Lizzie told her son Gavrik that she married Joe for the only good reason to get married – she was pregnant. However, Gavrik was born a full year after the marriage. Joe's sister describes Lizzie as bright, nervy, brilliant, opinionated – she and Joe kept up on the social scene. Elizabeth Hawes achieved fame largely as a result of her writing, journalism and broadcasting. Her debunking book, *Fashion Is Spinach*, published by Random House in 1938, argued that American women were the victims of a 'deformed thief', fashion. Style, yes, fashion no. Style was functional, whereas fashion responded to the whim of designers and manufacturers. A dress in good style could be worn for three years.* But this enlightened attitude was offset by the crassest reduction of woman to an appendage of man. She told the New York *Daily Mirror* (5 February 1938) that her cheapest creation cost $135. 'When I dress a woman, I dress her for a man. That's the whole story. For a man ... So, to do the job well, I have to know the man ... what he likes, what he hates, what his friends like ... What the woman who wears the dress likes is more or less incidental.'

Hawes also wrote about men's clothes, bringing the reporters hurrying to interview Joe (who didn't mind at all). The Richmond (Va) *News Leader* of 16 August 1939 carried the headline 'Joseph Losey and Elizabeth Hawes Don't Agree on Trousers – For Her.' She had crusaded for comfortable clothes for men, but Losey preferred not to be conspicuous: he favoured dark-green plaid jackets, grey trousers, blue shirt, socks and tie – and did not intend 'to walk the city streets in a slack suit in the near future'. After two years of marriage to Lizzie, Joe told a newspaper, 'We like the same books – anything that isn't a novel. Same kind of food – very expensive. We both like to travel. To dance. Hate formal occasions ...' According to Virgil Thomson, who spent a weekend at their rented house in Vermont, 'Both she and Joe had enormous reserves of disapproval – the lemon pair, always complaining.' Joe recalled: 'We had a house in Vermont. We went up to it at Christmas time and the snow was right up to the door when we woke up in the morning ... we took our night clothes off and ... and washed ourselves in the snow, and then went back in front of the fire and had breakfast which was lovely.' Lizzie owned a pair of Afghan hounds.[7]

Joe's sister Mary remembers an occasion when Lizzie and Joe were invited to the salon of Muriel Draper, mother of the dancer Paul Draper, for a New Year's Eve party in 1937 or 1938. Mary and her husband had invited some documentary film people to a party of their own. 'Joe dropped in and said pompously, "I'd be glad to take you all to Muriel Draper's but I can't take

---

*Later books by Elizabeth Hawes included *Men Can Take It* (1939), *Why Is a Dress?* and *Good Grooming* (both 1942).

that man because he's drunk."' He indicated a German who never drank but was playing the clown. 'Joe basked in glitter and chic,' adds his sister. Lewis and Irma Fraad remember, when they first met Joe, a kind of 'hayseed quality, a kind of innocence'. But this evaporated. Irma Fraad introduced Joe to an eccentric white doctor practising in Harlem; afterwards Joe commented that the man wore an impossibly cheap suit. 'Joe couldn't see further than that. He admired class.' Joe and Lizzie led a smart and elegant existence in a brownstone on East 49th Street, frequenting Café Society, a Village nightclub: during the war Joe shifted his patronage to the more elegant uptown Café Society. The Loseys' friends (Ralph Ingersoll, Frederick V. Field) tended to be both left-wing and wealthy. Field recalls that his friendship with Losey 'extended over a good many years . . . in which I was actively associated with the CP', but he prefers to remain 'evasive' (as he puts it) regarding the detailed questions put to him about Joe.[8]

According to Gavrik, who was born in July 1938, Lizzie took off for Peru, leaving Joe holding the baby. Gavrik promptly fell down the stairs and ruptured his appendix. Joe's sister, now married as Mary Losey Mapes, was 'totally outfitted' by Lizzie for four years: 'I knew she neglected Gavrik. She had an authoritarian quality and said children shouldn't be dependent.' Who wore the trousers? A crucial point to Gavrik is that Lizzie was earning (he says) 1,000 dollars a week, whereas Joe was working as an itinerant off-Broadway director 'at twenty-five dollars a week'. (Both figures require a large pinch of salt.) Gavrik believes that his father could never come to terms with being supported by a woman. Losey himself later recalled: 'So with the exception of the Rockefeller films and the *Pete Roleum* film and *Sunup to Sundown*, practically all the work I did from 1936 to 1943 I got paid nothing for. Barely enough to eat. Fortunately my wife at that time was making fairly reasonable money so it wasn't a big crisis for me.' (The $10,000 he got paid for *Pete Roleum* was at that time sufficient to purchase nine or ten six-cylinder Chevrolet saloons. A first-class hotel room cost five dollars.) When Lizzie relinquished her exclusive dressmaking establishment in January 1940 it was bringing her a personal income of $18,000 a year; the business was taken over by her former employees, as Hawes Customers, Inc., for which Lizzie did occasional designs.[9]

On 1 February 1938, Losey's production of *Sunup to Sundown*, by Francis Faragoh, opened at the Hudson Theatre on 44th Street, financed by D.A. Doran, who later became head of Paramount. Set on a tobacco plantation, the play had a cast of twenty-two, mainly youngsters, and was designed by Howard Bay (who didn't think much of Losey as a director, according to Lewis and Irma Fraad). Sidney Lumet, then thirteen, played the juvenile lead – Losey possessed a photograph of himself in rehearsal with Lumet, the director's nose in a splint and two black eyes. The cause of injury has not been unearthed and Lumet does not recall.[10]

Despite universal admiration for Howard Bay's tobacco-loft set, the critics were of one mind: lack of vigour and vitality in the play itself. Writing in the *Journal American*, John Anderson complained that the author did not hit out hard enough, and the story came disastrously close to being not so much about exploited children as about a fifteen-year-old Mexican girl, old enough to be pregnant, too young to be married. But the production was brilliant. 'Mr Losey, whose direction of *Little Ol' Boy* indicated that he had a way with younger actors . . . has . . . coaxed from them a general performance that is extraordinary.'

Despite an award from the International Ladies Garment Union and an offer from the National Child Labor Committee to enclose fliers for the play in its mailings, *Sunup to Sundown* was doomed to run for only seven performances. The young actors were distressed; on a sheet of lined white paper sixteen of them, including Sidney Lumet, declared – almost in blood – that 'the kids' in the cast 'have joined together in a plan which they hope will help prevent this mishap [i.e. closure]'. The plan was to turn back their salaries 'to relieve the strain from Mr Doran's shoulders. This is only a gamble but we can all afford to gamble as did Mr Losey and Mr Farrogoh [*sic*] who staked their names and reputations for us.' But to no avail. A final critical cut came two weeks after the play closed: writing in *New Masses* (19 February), Nathaniel Buchwald accused both the playwright and the director of burying the social message. 'Joseph Losey did all he could to put the play into the morass of futile sentimentality.'

Ernest Hemingway's play *The Fifth Column* came within Losey's grasp during the summer of 1938. Following disagreements with the impresario Jed Harris about rewrites, Hemingway came north from Key West and sold the play to Austin Parker, then left for Spain. Parker died a few days later. Losey picked up the option for $1,000 while Hemingway was in Barcelona. On 20 April both the *New York Times* and the *Herald-Tribune* reported that 'Contracts have been signed and there is talk in the Losey office of persuading Gary Cooper to play the leading role . . .' Ralph McAllister Ingersoll was named as the principal backer, with production costs estimated at $30,000.

Set in five scenes in various parts of Madrid, it was Hemingway's first play, and exceedingly long and verbose. Act 1 is set in the Hotel Florida, Madrid, the main characters being Dorothy, a nice, cool, affected, stupid, cultivated Vassar girl; and Philip Rawlings, an impressive drunk dedicated to the Republic. Much stage-smiling is wrung out of the tendency of Spaniards to speak wrong English. Hemingway's unmistakable style soon surfaces:

PHILIP: I am thinking, you see, what the man is going to do that you
   let get away, and how I'm going to get him in a nice fine place like
   that again before he kills somebody.

Dorothy and Philip settle down to discussing Life. The occasional assassination doesn't relieve the tedium. Better is promised when Act 2 moves to a
room in Seguridad Headquarters, with Philip talking to the first serious
Spaniard to surface in the play, Antonio. But then Philip tells Antonio that
he'd like to marry Dorothy 'because she's got the longest, smoothest, straight-
est legs in the world, and I don't have to listen to her when she talks if she
doesn't make too good sense'.

After more tedium, action follows, with the emergence of the Franco fifth-
columnists snarling, 'We'll kill plenty of them tonight. The Marxist bastards.'
Philip apparently arrests 300 of them, more or less single-handed. Executions
follow, off-stage. Philip is every bit a Hemingway hero: idealistic and cynical,
lustful and world-weary, writer and traveller. 'I've been to all these places
and I've left them behind,' he remarks. He's also very rude to women and
they love it.

On 16 June the *World-Telegram* announced: 'Ernest Hemingway is due in
town this week to consult with Joe Losey . . .' The services of Robert
Montgomery were being sought for the main part. D.A. Doran would co-
sponsor the production. Losey's notes prior to a meeting with Hemingway,
dated 2 June, are more concerned with raising money than with silencing the
play's hero. 'Wednesday. Wire D.A. [Doran] about Oscar Hammerstein. Next
week. Go over entire money situation of play. Get introduction to Mrs Cyrus
McCormick of Chicago.' A list of relevant names includes: [Ralph] Ingersoll,
Liebman, [Lillian] Hellman, Pratt (?), [Robert] Reilly, Losey, Hemingway (?).
The queries are Losey's.

According to Carlos Baker's biography of Hemingway, the writer came to
New York on 22 June principally for the Joe Louis–Max Schmeling
heavyweight title fight, but in fact Hemingway also dined with the Loseys –
or one of them. Revolted by Lizzie's pregnancy (Gavrik was born the
following month), he refused to sit down to eat with her. Lizzie relayed this
to Lisa Giorgi, who in turn passed it on to Hawes's biographer, Bettina
Berch. The *New York Times* now reported that Fredric March had been
approached for the leading role but demanded changes that Hemingway was
not willing to make. The writer was by now heartily sick of the play and on
24 August the press reported that Losey had abandoned his plans due to 'a
combination of casting and financial difficulties'.[11]

In November 1937 the *New York Times* had reported: 'Joe Losey, stage and
radio director, is the new production supervisor of the Experimental Film

Project of the Commission of Human Relations, a branch of the Progressive Education Association.' Some twenty-six shorts were made in 16mm film, mainly for progressive schools, and subsidized by the Rockefeller Foundation. In 1938, after *Sunup to Sundown*, Losey supervised the production of some sixty educational montage films. No original footage was shot. With Helen Van Dongen serving as chief editor, Losey carved shorts out of Hollywood features.[12]

And now, in an extraordinary *démarche* worth an exclamation mark, Joe offered his propaganda skills to the filthy capitalists. First shown at the World's Fair on 22 May 1939, *Pete Roleum and His Cousins* sought to mobilize public opinion against the fiscal burden levied on oil sales – 35 cents in every dollar. As the *Brooklyn Eagle* (28 December 1938) explained, the oil industry gave Losey 'carte blanche – except for seventeen points which he was to be sure to cover'. Top of the list was the petroleum industry's claim to be making only one cent profit per gallon.

Pete Roleum was an animated puppet cartoon in Technicolor – Losey's first screen venture – in which all the characters were whimsically moulded as oil droplets. The plan was to show the twenty-five-minute film continuously, once an hour, seven days a week, for six months, in the triangular Petroleum Building at the Fair. The shooting script, by Kenneth White and Losey, is dated 27 July 1938. Pete Roleum, representing Oil, heads the cast of characters, along with such cousins as Stinky Lube, Grease Boy, Gassy, Miss Polish, and Hi Test. History being important, the first scene is set in 1859 – *Injunction Granted*, so to speak, in ideological reverse.

A leading voice belongs to the sceptical Heckler whom Pete R. ironically addresses as 'Connie' – short for 'Conscience'. In 1981 Losey recalled: '. . . and a "heckler" at the back of the theatre [was] recorded in sync with the picture and projected from loud speakers at the back of the hall. This never worked.' But, evidently worried that Connie might get the best of the argument, Losey blue-pencilled out the following lines: 'Well, you take all the gas that's sold in this country in a year and multiply it by one cent [profit per gallon]. And you take all the other things the oil industry makes a profit on . . . profit, profit, that's all they make.' Black and white drawings illustrated Pete R.'s explication of the industry's operating costs: an oil derrick, a refinery, pipelines, workers, a service station. During a long exchange about oil costs, taxes and profits, *'Uncle Sam's arm reaches down and takes up a quarter and a dime. They disappear from the pile of coins, leaving only one penny.'*

Louis Bunin and his brother were commissioned to make the puppets (composed of flexible rubber skins); animation was by Charles Bowers, photography by Harold Muller, president of the East Coast cameramen's union. The editor was Helen Van Dongen, wife and collaborator of Joris Ivens. Some seventy-five figurines were made, each about eight inches high, for which Howard Bay designed some fifty three-dimensional sets. Losey's 'laboratory' was an animating studio on Broadway at Times Square. Hanns Eisler wrote the music. 'And I remember one song,' Losey recalled. 'It starts on a bleak world, completely destroyed by war, and also destroyed by the removal of oil . . . Then slowly the [oil drops] came up over the bleakness of the destroyed world singing a song, "We are coming back *if* – you want us to come back!"' [13]

The picture took a year to complete. Losey's fee as director/administrator was $10,000; Howard Bay was paid $4,000, Kenneth White $2,000, and Charles Bowers $8,500. Equipment cost $13,302, marionettes $23,200, film stock $33,356. The project was financed by all of the major petroleum companies operating in New York and cost $115,546.

In April 1939, when Losey arrived in Hollywood for final editing at the Technicolor Motion Picture Laboratories, *The Hollywood Reporter* predicted that 'Mr Disney and other big boys . . . have a treat in store.' They might even persuade Joe Losey to stay on the West Coast. A few days later 'more than fifty production departments' representatives attended a special showing of Joseph Losey's new cartoon process at RCA projection room yesterday. Inventor planed to New York with film.' (Losey, of course, was not the 'inventor'.) On 17 May the *Journal-American* reported: 'Losey's dream-child, which points the way for a new phase of screen technique . . . is currently being considered as a regular production feature in major studios on the Coast . . . The youthful Mr Losey . . . not long ago . . . aroused a storm of praise for originating the Living Newspaper technique for the Federal Theatre . . .' *The Hollywood Reporter* heard that Losey was engaged in negotiations with Michael Myerberg, of Universal, for a full-length puppet feature.

So why did nothing come of it? We don't know. Almost certainly Losey at that time used the same promotional technique when talking to the press as he did in the 1960s and 1970s. In masking the self-promoting source of its reports, the press tended to blur the line between wish and reality. Nevertheless, universal acclaim greeted *Pete Roleum and His Cousins*:

'. . . a SENSATIONAL short – a forerunner of bigger and better things in a field destined to become increasingly important' (Irving Hoffman, *The Hollywood Reporter*).

'Here's a real innovation' (*Hollywood Film Daily*, 5 May).

'The fairest screen fare at the Fair' (Walter Winchell, *Sunday Mirror*).

'Colorful, imaginative, educational and tuneful (with specially composed music by Refugee Hanns Eisler and Pianist Oscar Levant)' (*Brooklyn Eagle*).

'Perhaps the most ambitious puppet film ever made in America' (*Newsweek*, 29 May).

'One of the best animated Technicolor films ever seen' (*Cue*, 20 May).

The technique was also greeted as a long-term innovation by the *New York Times* (5 and 23 May). An undated, typed 'Discussion of Three-Dimensional Animations', written by Losey, held by the Museum of Modern Art, offers a history of the technique and its practitioners, starting with the work of Méliès at the end of the nineteenth century. Losey called *Pete Roleum* 'the most ambitious experiment of its kind to date' and congratulated the petroleum industry for its boldness and imagination: 360,000 people had already seen *Pete Roleum* and the advantages of 'three-dimensional animation', involving a Latex skin wrapped round an armature, over cartoon films, including Disney's, were now clear. *The Wizard of Oz* would have been 'ten times more effective, and much less expensive, if produced in our medium . . .' In August 1939 he made a plea to John Marshall, of the General Education Board of the Rockefeller Foundation, stressing the potential of the three-dimensional puppet for satire, caricature, advertisements, PR, or propaganda. He therefore requested funds to make a feature, *The Tale of Reynard, the Fox*, again written by Kenneth White, which he budgeted at $311,478, of which $15,000 would be allocated to 'direction'. No reply is found.[14]

In 1939 Losey directed a jazz concert set up by John Hammond and sponsored by the Theatre Arts Committee, at Carnegie Hall. Among the star performers were Benny Goodman's Sextet and the Kansas City Six, with Buck Clayton and Lester Young. According to Gavrik Losey, no doubt echoing his father, this was the first occasion that jazz, with its strong black associations, had gained social respectability in America. John Hammond, who set up the show, later commented that this Spirituals to Swing concert was less of a success than its predecessor. 'It was staged by Joe Losey and in some ways better performed, but it lacked the audience–performer rapport of the earlier concert.'[15]

Losey told Ciment that his close friendship with Hammond came to an end because Hammond's progressive politics and philanthropy 'always stopped short of any organizational commitment'. This is misleading. Hammond held in suspicion the Communist Party and its front organizations, but he had been involved with the National Association for the Advancement of Colored People from an early age. His memoirs, published in 1977, brim with enthusiasm for those he admired and loved, but indicate little tenderness for Losey – 'a persuasive young man with ambitions to become a theatre director' who later 'went on to become a successful producer [*sic*] of B [*sic*] pictures in Hollywood from about 1937 to the early 1940s [*sic*]'. These errors betoken estrangement as much as carelessness; indeed, Losey never met Hammond's wife, Jemison, whom he married in 1941, although both men were living in

New York. Forty years after the 1939 concert Hammond, Director of Talent Acquisition for Columbia Records, sent Losey a polite, but guarded, appreciation of *Don Giovanni*.[16]

9

Whose War?

Losey directed two short documentary films which he later placed in 1939, but it seems that they were shot and released in 1940. Both are highly didactic; the second ranks as outright propaganda on behalf of the New Deal. Propaganda had become, *faute de mieux*, Losey's speciality. He brought vitality, immense competence and technical craft to everything he touched – but what his younger Wisconsin contemporary, Orson Welles, touched was *Citizen Kane* and *The Magnificent Ambersons*.

In setting up *A Child Went Forth* for the National Association of Nursery Educators, Losey received help from Frontier Films, a left-wing production company. The title was borrowed from Walt Whitman and Losey later claimed the script as his own, but on the film print itself Munro Leaf is credited as the writer and Lloyd Gough as the narrator (though most filmographies award the narration to Leaf). The music was written by Hanns Eisler (supported by a Rockefeller Foundation grant) and the photography was by John Ferno (cameraman to Joris Ivens in Spain).[1]

Having borrowed $500 – with $2,000 credit from the lab – Losey sold the film to the State Department for 'a rather considerable profit'. With war looming, the Government was interested in a model solution for the evacuation of children. Distributed by the Rockefeller Committee on South American Relations and the Office of War Information, *A Child Went Forth* had been dubbed in twenty-four languages by October 1945.

Eighteen minutes long, the film shows children between the ages of two and seven enjoying life at the Nell Goldsmith nursery camp, Woodlea. Walking them through an idyllic, flower-carpeted meadow, the narrator invokes Whitman: 'And the first object he looked upon, that object he became. That object became part of him for the day or a certain part of the day or for many years or stretching cycles of years – so spoke Walt Whitman.'

'Children are people and how they play, and stretch those muscles and minds of theirs is of real importance.' Most of the kids are six or seven, but a few are down at three and two (Gavrik Losey, for example). They hammer nails, dig their hands in clay, get thoroughly dirty in the company of mud pies: 'Did you ever find good clay or mud without someone shouting, "Don't get your clothes dirty! No, no, don't do that!"'

Learn to live together. Nature studies – playing with mice: 'No silly screaming or jumping on chairs when these children grow up!' Pig and piglets; clipping wings of ducks; swimming in rubber rings, swinging from ropes, fishing nets, frogs – here Losey achieves beautiful close-ups of activity on the water. Home to the farm they all come in a station wagon. 'Another big day.'

The commentary then makes its sales pitch: if mothers have to be called away into war industries, this may be a solution, or 'food for serious thought'. The voice is didactic but pleasant: 'Or should we have to face the stark necessity of evacuation of our cities . . .'

Youth Gets a Break followed hot on the heels of *A Child Went Forth* and was commissioned by the National Youth Administration, headed at that time by Lyndon Johnson, who attended the first screening in Washington, along with 6–700 others, 'a fairly impressive occasion', Losey recalled. The film was shot mainly in New Jersey during 1940 and reached New York in October. Eleanor Roosevelt went to see it. 'I hope it will be run by every motion picture theatre in this country, for it has a deeply moving story to tell,' wrote the President's wife in her regular column.[2]

John Ferno was again on camera, although this twenty-minute film was a big enough undertaking to require three crews and cameramen, the others being Willard Van Dyke, later head of the Museum of Modern Art archives, and Ralph Steiner. The orchestra and choral group was from the New York City NYA Radio Workshop.

The establishing shot immediately presents two economic nations: a car passes young men hitching a lift. The driver, in a white hat and glasses, stops at a diner and guzzles a hamburger, watched by the unemployed. Chirpy jazz music in the background. Cut. Now we're in the city. A man with a sandwich board: 'Please, please – give me a job. I will do anything.' He gazes into a grocery store laden with food.

Cut to the countryside: unemployed men watching a tractor at work in a cornfield: watching and wanting. A man, then a woman, walking sadly and slowly: wistful clarinets . . .

Cut to a pickaxe biting into the ground, brawny shoulders, dynamite, ditch digging. Against this, a didactic conversation between a young man working and another watching, about how many are employed by the National Youth Administration.

Cut: young people learning how to make radios and become radio operators: 'It's a simple thing, when you know how.' Woodwind represents hope; the musical background is incessant as we pass from one industrial process to the next. Women in a millinery factory: 'I was the girl who thought young people couldn't get work in this town.'

Harvesting corn, milking cattle in spotless white overalls, the milkmaids as glamorous as actresses. Male voice: 'Making plants grow, nourishing them, growing, growing, thriving . . .'

Cut: a bugler. Lads running from their tents and leaping into the waves. Cut to an aerial view of a building complex: 'An outstanding example of a project of the NYA. Maintaining the grounds and buildings of the Federal Government.' Cut to the showers after work; medical checkups, eye tests.

Young Pioneers doing exercises – very Soviet in mood – each leaping up as the name of his State is called. Now collective virtue in a general assembly, everyone democratically conferring about cases of negligence, about dirt and garbage disposal. 'Report of the Activities Committee.' Cut to noisy discussions, shouting, elections, cheers: Democracy. A torchlight procession at night with a big band, heroic music. Then be-bopping to a jazz band, the boys in double-breasted suits. Finally, a succession of youngsters, black and white, testifying what the NYA has done for them, each looking up from his or her work bench to bear witness.

1 September 1939: 'I was dining on a summer evening with my first wife in a house on an inlet of Long Island which we had rented from John Howard Lawson [the doyen of Communist playwrights and film writers] ... The news of Hitler's invasion and his speech came over the air.' Ten days had passed since the Nazi–Soviet Pact; where he was dining on that occasion Losey does not record. In the 1940 presidential election he voted for the Republican candidate, Wendell Willkie, and not for Roosevelt – an expression of Communist opposition to American intervention in the war. 'On the left we all believed that the war was a phoney war, that Germany would never attack England, and so everybody stayed removed from the war until the actual attack on Russia when it then clearly had to be an anti-fascist war.' Losey made these comments in 1976, although four years earlier, interviewed by *l'Express* (2–7 May 1972), he had admitted that the Gestapo and the GPU had worked in collaboration after the Nazi–Soviet Pact was signed. Not everyone on 'the left' was of Losey's persuasion; his sister, Mary Losey Mapes, was deeply disturbed by the fall of France and the aerial blitz on Britain.[3]

After Hitler's invasion of the USSR in June 1941, Losey was appointed head of the production department for Russian War Relief (RWR), staging mass meetings in Madison Square Garden. According to the Fraads, 'He had an imperious quality and was jealous of his position.' Losey noticed that a junior in his department, Florence Watt, had a larger, handsomer desk than his own. 'Three days later the desk was in his own office.' Letters of reference on behalf of his application for a commission in the armed forces throw light on his work for RWR. Edward C. Carter, President of Russian War Relief, Inc., praised the 'exceptional dramatic quality' of the meeting staged by Losey at Madison Square Garden on 27 October 1941: 'a man of great organizational capacity.' Serge Semenenko, Vice-President of the First

National Bank of Boston, and Treasurer of the Massachusetts Committee of RWR, lauded Losey's organization of a mass meeting held in the Boston Arena on 14 December 1941 (admission prices varied from 55 cents to $5.50 for box seats). Losey later described Semenenko as 'a White Russian who was very patriotic about Russia . . . So I only had to say . . . I want Tyrone Power or I want Eleanor Powell . . . and he produced these entertainers.' Performers included Paul Robeson, Paul Draper and Larry Adler, while Kenneth White did most of the writing, Martin Gabel most of the narration. Many of these artistes were later blacklisted. 'I had a technique', Losey recalled, 'of mime with voices being spoken over a microphone from a stage box.' The 28 October 1941 edition of *PM* displays a huge aerial photograph of the RWR meeting held at Madison Square Garden. Powerful lights from all sides beam on to a central podium the size of a boxing ring. Allied flags are suspended across the hall. The sum of $175,000 was collected before and during the event. Of the guest speakers, only the British ambassador, Lord Halifax, was booed with shouts of 'Open up the Western front.' Losey believed that Churchill's non-invasion strategy was 'highly dubious' and that the British plan was for the Nazis and Russians to decimate one another or 'that the Russians would be defeated by the Germans . . .' Full page ads in the Press declared, 'Russia's valiant armies are now fighting our battle on the Moscow, Leningrad and Rostov fronts . . .' On the 'Society' page of the *Washington Post* (5 April 1942), devoted to 'Fashion – Clubs – Gentler Sex', one reads: 'Official Washington turned out en masse to hear Serge Koussevitsky conduct the Boston Symphony Orchestra at the Russian benefit concert in Constitution Hall.' Pictures showed Ambassador and Madame Litvinoff; Mr and Mrs Joseph E. Davies; Mrs Woodrow Wilson; Mr and Mrs Nelson Rockefeller. 'Capital society were on hand to applaud the Star Spangled Banner and the Internationale alike.' Losey stage-managed the show, but got paid only $30 or $35 a week for his work.[4]

Losey made efforts to enlist as an officer in the armed forces – but on his own terms and at his own ranking. 'I volunteered to head a camera combat unit in the Air Corps.' A medical examiner told him he had varicose veins and asthma but promised, 'I will waive it in terms of okaying your commission.' Nothing happened. He tried the Navy and (he claimed), 'was commissioned a lieutenant, had my medical examination waived . . .' Again nothing happened. He pulled strings, advising former Ambassador to the USSR Joseph Davies that his conscription difficulties must be political. By Losey's account, after some weeks Davies, author of *Mission to Moscow*, called him back to say: 'There is a dossier, a rather long dossier on you. I'm not at liberty to tell you what's in it . . .' Davies advised him to write a complete report on his past political activities. Losey did so 'and I never heard from him again or from anybody'. For all of this there is no substantiating documentary evidence. What the files do show is that Losey's application for a commission in the Army Air Corps was turned down by the Surgeon-General's office on account of a history of allergies. Almost certainly his application for a commission in the 'USNR' was rejected for the same reason. Dr Lewis Fraad treated his asthma and it was this affliction, in Fraad's view, that prevented Losey from being accepted as an officer.[5]

Paradoxically, Joe's sister Mary suffered persecution many years before he did. She had been working for the British Information Services in New York and Washington when 'a little jerk from the FBI' arrived at her apartment on 58th Street to ask questions. When working for the Office for War Information 'in 1943 or 1944', she was denied a passport to travel to England. On 6 May 1944, having applied to be transferred to the OWI's Motion Picture Bureau on a permanent basis, she was called to a special hearing by the Investigation Division of the US Civil Service Commission and invited to explain her past membership of the American Russian Institute, and her role as founder-secretary of the Association of Documentary Film Producers. (Losey himself was a member, as was Joris Ivens, but he was never mentioned during Mary's wartime interrogations.)

Losey's career in radio drama had begun in 1937 with a Columbia Workshop Production of Albert Maltz's *Redhead Baker*, its theme being prejudices about education and the young, its tone highly didactic. Ray Baker plays truant from school, gets into fights, receives bad reports, lies to his despairing parents, and is threatened with Reform School. His teacher, Miss Smiley, is unsympathetic and acidic. The enlightened teacher Collins visits the Baker home and explains Miss Smiley's attitude: '. . . she's got five times as many kids as she should have – they're sitting two to a seat'. Collins then suggests a 'progressive school' for Ray, adding: 'When enough people learn about it and when the country wakes up to its possibilities, it'll become the form of education that all children get.' Collins's speeches get longer; the two Baker parents remain admiring foils for his wisdom. When he tells Ray about all the things he need not do at a progressive school, and all the sports available, Ray is amazed.

RAY: Gee, I think I'm gonna like this school.

Maltz wraps his didacticism in the courtroom genre – the play begins and ends in the juvenile courts. Here, at the finish, the Doctor is explaining to his Friend: 'Now Ray's on his way to college to study electrical engineering . . . He's getting ready to take a useful place in society.' (*Doctor laughs.*) 'Oh, and I forgot – he plays the violin now, too.'[6]

Variety (30 June 1937) called it a 'gripping dramatic show,' marking 'the

début as a radio director of Joseph Losey, stager of legit. productions . . .
Transitions were smoothly executed and logically conceived . . . Losey's
direction and the acting . . . carried a wallop.' *Radio Daily* (22 June) praised
Losey as guest director and described the production as 'engrossing'. If one
interprets Losey's retrospective remarks correctly, *Redhead Baker* had
originally been conceived as 'an experimental television show – but it didn't
go to the public because there wasn't at that time public television so it was
only on closed circuit . . . at CBS'. The Maltz script on file is definitely for
radio, not television.[7]

From 1939 to 1943 Losey directed and produced occasional commercial
radio programmes for NBC, where he worked full-time for nine months in
1943, producing and directing a succession of sharply contemporary, always
didactic, dramas, notably the *Day of Reckoning* series, sponsored by the
Council for Democracy, *Worlds at War*, dramatizing important war books,
Red Cross Show, Inter-American University of the Air, and 'some Treasury
performances'. The actors included Peter Lorre, Raymond Massey and Eva
LeGallienne.

In Elmer Rice's *The People vs Pierre Laval*, we begin at the gate of Heaven
– the Messenger announcing, 'But straight [*sic!* strait] is the gate and narrow
is the way'. The Messenger then descends on Vichy France and kidnaps
Pierre Laval (wrongly described as the French Head of State) for an interview
with the Recording Angel. Summoned to testify from the eighteenth century,
Lafayette blasts Laval for betraying France, then takes a swipe at Mussolini,
whose imperialism is also denounced by an Ethiopian woman. 'But a madman
in Rome cast his frenzied eye on our land . . . he craved empire – empire to
bolster his pride and glut his vanity.'

General Benedict Arnold is wheeled in to explain why men are weak and
corrupt. Enter a German worker to testify against Hitler. A French woman
declares: 'That's why my son had to die in 1940 at Sedan – a victim to those
Lavals, whom we trusted, and who neglected to arm us, while they carried on
their slimy intrigues with the moguls of the Bank of France, class against
class, gentile against Jew . . .'

Enter Joan of Arc. Laval quakes: 'Yes, Heavenly maid, I hear you.'

JOAN: Down on your knees, then, and listen to what I have to say, for I
   speak with the voice of saints and archangels.

Joan describes how she saved France from foreign foes. 'I drove the
usurper from our soil.' (But England, the usurper, is tactfully not mentioned.)

Norman Rosten's *The People vs The Unholy Three* was broadcast on
Saturday 19 April 1943. The theme music comes up behind the initial
announcement: 'The National Broadcasting Company, in cooperation with
the Council for Democracy, brings you . . .' Rosten's script follows the same

other-worldly formula: Goebbels, Goering and Himmler are put on the rack
by an all-powerful, all-wise Narrator and a woman called Anna.

> ANNA: Are you afraid, Herr Goebbels? You have no guards now.
> You're a coward, a shivering mouse.
> GÖEBBELS: I made Germany great, remember that! Stay away from me.
> Help!
> NARRATOR: People of the earth, you are the court! You are the tongue
> of justice. Speak! (*Later.*) Do not forgive them, O Lord/ For they
> know what they did! (*Volley of rifles from firing squad.*) The sentence
> of the people is carried out. We turn now to the rebuilding of our
> world.

Broadcast on 8 May 1943, *They Burned the Books*, by the late Vincent
Benét, was a play written to mark the tenth anniversary of the Berlin book-
burning of 10 May 1933. Heard are the Narrator, the Voice of Schiller, Nazi
Voice, the Voice of Heinrich Heine. Blank verse is the formula.

> 'Excuse me, sir, my name is Friedrich Schiller,
> A name once not unknown in Germany.' (*Sound of flames periodically.*)

The Nazi Voice keeps cutting in on Heine: 'To the flames with him! How
dare you speak, exile and Jew?' The Nazi confesses that his conquests will not
be secure while there remains alive a single man who remembers 'the views of
freedom':

> And till we've taken an electric wire
> And burned the brain cells from his very brain
> So he will be a dumb and gaping slave . . .

Losey later recalled having staged two Norman Corwin plays, 'but they
were not for NBC. They were done in peripheral theatre.' This is difficult to
explain; on 27 May 1943, NBC broadcast Losey's production of Corwin's
The Long Name None Could Spell – which meant Czechoslovakia. Confined to
two voices, plus an orchestra, this drama was written in a version of blank
verse:

> The bartered peace has since been renegotiated by the millions who shall
> walk no more:
> Come with your guns, Allies, and with your fresh young armies . . .

The accompanying music was the Czechoslovak national anthem, plus bells of
jubilation. Most of these radio plays made a blatantly sententious pitch for
the religious sentiments of their audience, employing holy choirs, awesome
glimpses of Heaven's judgement. Apparently they were broadcast live.

While at NBC Losey received a summons from Louis. B. Mayer to the Warwick Hotel. 'Hollywood had never shown the slightest interest in me when I was in the theatre. There was never any offer.' He found Mayer stretched out in the barber's chair in his suite, 'with someone giving him a pedicure, another one giving him a manicure, and another one shaving him . . . He said, "I heard some of your radio shows. They're damn good. You want to come to Hollywood?"' Mayer proposed a seven-year contract beginning at $350 a week. Years later Losey recalled: 'I refused. I didn't want a slave contract.' He turned to 'a big agent', Arthur Lyons, to negotiate with Mayer. What came of this is not clear – except that Losey flew to Hollywood in December 1943. 'Louis B. Mayer kept me standing for five minutes while he was on the phone . . . The office was like a small chapel with a vaulted ceiling, and his desk was in the middle of the room.' Mayer told him to view pictures, visit sets, meet producers, find a story. In a CV he wrote just before military service, Losey described his first assignment as associate producer of *Mrs Parkington*, starring Greer Garson. But only one month after joining MGM he was drafted into the army.[8]

To Losey's disgust, America wanted him as a private soldier. On 11 December 1943, nine days after arriving in California, he received a telegram from the New York City Draft Board ordering him to report for a physical examination. According to the FBI, on 14 January he told Hanns Eisler's wife that the Walter Reed Hospital had sent a complete medical report to the Draft Board 'and he thinks this will work'. But the following day he told her that the local Draft Board had refused to consider the report from the Walter Reed. According to his military discharge certificate, he was inducted on 19 January 1944 in Los Angeles; according to another document, his service began on 2 February at Fort MacArthur, San Pedro, California.

Ruefully he recalled: 'And they didn't care about my asthma or my varicose veins . . . So, knowing that cats gave me asthma I took two kittens to bed with me before I had my medical examination. They had absolutely no effect on me. In the morning when I was examined, I was passed! Three or four days later I was on a troop train . . . for three days, all blacked out, and we finally wound up in North Carolina.' The air and food were good. He went through sixteen weeks of basic training, 'forced marches with forty pound packs, night reconaissance'. On 4 May he qualified as a 'sharpshooter' on the .30 Cal. M-1917 Course C. Although he had never before held a weapon, he became a member of 'Five-Forty-Five Club', requiring five bulls-eyes in a row with a revolver (as he related to Tennessee Williams in 1967, offering the playwright physical protection against enemies unidentified). He felt terrific, '146lbs of sheer muscle'.[9]

His military service was brief – less than ten months. After basic training he was transferred to the propaganda film unit at Astoria, Long Island, where he made two shorts for US Production Army Pictorial at the Signal Corps Photographic Center's studios. His status was 39724734, Private, Unit B. A lot of Hollywood people were there, most of them officers. 'My cameraman was a captain . . .' The atmosphere was 'ugly and horrible' – even though he was permitted to spend nights in New York with his new girlfriend, Louise Moss, reporting back to camp at 6.30 a.m. Joe and Louise were living at 15 Charleston Street, in Greenwich Village. The Fraads remember Losey with Louise in a small apartment – a tall, willowy, pretty blonde with an oval face, beautifully dressed 'in rags' and standing at the ironing board 'like Russian royalty deposed'. Gavrik describes her as 'a very tall, statuesque blonde woman, loving and warm and friendly to me'. Prominent in Gavrik's memory – he was six – is Louise's fabulous 1940 Lincoln Continental drophead automobile. They used to collect Joe from Long Island: 'He had to line up in his uniform and be dismissed.'

While in the army Losey was able to stage 'a couple of voters' rallies for Roosevelt and Truman' and 'had to listen to [Truman] play the piano'. His asthma and leg trouble having resumed (he'd suffered not at all during the tough basic training course), he was sent to hospitals on Staten Island and Governor Island before being honorably discharged on 14 November. He was entitled to wear the Good Conduct Medal and travel pay was authorized to Santa Monica, California.[10]

Losey's military service provides an interesting parallel – and contrast – with that of his friend, John Hammond. Hammond was also inducted as a private and found himself at Fort Dix and Fort Belvedere – but in his case braving one confrontation after another with sergeants and officers over the illegal racial segregation imposed by local army commanders. Losey's training in North Carolina undoubtedly involved the same racial segregation – but he nowhere mentions it. For Hammond anti-racism was a passionate instinct; for Losey, an issue which the CPUSA had buried in the interests of national unity. When the Communists revived the race agenda after the war, Losey began campaigning against the Ku Klux Klan.[11]

Joe's marriage was heading for the rocks.

In May 1940 Elizabeth Hawes became editor of the 'News for Living Department' of New York's new, progressive, tabloid newspaper, *PM*. In a later job application, she indicated that she worked for *PM* until February 1942; elsewhere she described herself as having been 'fired from *PM*', adding that her membership of a union assured her a pay-off. According to Losey, she was fired after a political disagreement with the founding editor, Ralph Ingersoll. The Loseys were probably unaware that working for *PM* had generated an FBI file on Hawes – a file which managed to confuse her life with that of another woman, Elizabeth Day Hawes, the result being a tangled skein of misinformation. The FBI categorized Lizzie as an 'S' (Communist Sympathizer) member of *PM*'s staff. Also noted was her membership of the League of Women Shoppers, described as 'a standard Stalinist front'. The Director of the FBI, J. Edgar Hoover, recommended that she be considered for 'custodial detention in the event of a national emergency'; in 1942 the War Policies Unit of the Justice Department agreed to classify her.[12]

From October 1940 the Loseys were living at 217 East 48th Street; in September 1942 they moved to 52 Jane Street. In March 1943 Lizzie went to work for Wright Aeronautical, in Paterson, New Jersey. Graduating from the trade school, she got her first job as a 'grinder', making screws for sixty cents an hour. Later she worked night shifts, leaving home at 10.30 p.m. and returning by 8.30 a.m.; her waking 'day' began at five in the afternoon. Gavrik pined for his mother's kisses but she didn't believe in it. In *Why Women Cry: Wenches with Wrenches* (1943), her character Amanda says: 'I love children ... but I don't want to be with them *all* day ... But the nurse in your housing project will love babies and she'll be a good cuddler. I wouldn't have to do all the cuddling.' Hawes quit Wright's in May 1943, after only three months, then went up to Vermont to finish *Why Women Cry*. According to the Fraads, Joe seemed happy in his role of caring nurturer for Gavrik, looking after the boy with keen and loving interest, but Joe's role as full-time father can be exaggerated; he kept himself busy in the world outside with War Relief rallies and radio drama; the existence of a 'nurse' at home was taken for granted. Fifty years later Gavrik recalls his mother as a warm and physically demonstrative person – unlike his father. 'He was never an open physical person.' Irma Fraad remembers Gavrik as a boy craving for attention.[13]

Joe dated his separation from Lizzie as November 1943, when he met Louise Moss: 'Then my first marriage busted up amicably, but none the less irrevocably ...' Joe's sister, Mary, remembers seeing Lizzie 'the Christmas after he left' (presumably 1943) and finding her distraught. 'She told me over and over how afraid she was of him and his criticism, and how hard-boiled he was in his demands.' The divorce took place on 7 August 1944, in the Court of First Instance, Provost District, Chihuahua, Mexico. Neither Joe nor Lizzie bothered to inform Gavrik, who was now at Poughkeepsie Day School; he heard about it from another boy. He was now semi-abandoned by both parents. Hawes moved to Detroit, to work for the United Auto Workers.[14]

What was Lizzie's afterview, or overview, of Joe? In 1983 Patrick Mahoney sent Losey a strange typescript, apparently written by Lizzie in the early 1950s and given to a friend, Mary Bancroft, as part of an intended book. It is written in the fictional form of a letter 'to Harriet'. The fiction has to be interpreted, but the references are clear. Jean Lurçat, for example: 'Do you remember my Frenchman? The one I was pursuing in 1935?' In the passage that follows, Joe is called 'Sam', a 'playwrite' [*sic*] and Gavrik is 'Sammy'.

'When I married him he'd just had his first success. Then he had two years of horror during which I paid for everything including Sammy's being born. It never seemed to matter, although in some ways I was too busy earning money to notice perhaps. Finally Sam had a mild success upon which he came round and said he thought he had fallen in love with an actress he'd been working with. Well, I was very tired at the time. The war was going and the servant problem fierce and Sammy only about two years old [*sic*: five] and I was breaking myself to pieces trying to help the war effort . . . So he got to work seriously to leave home . . . I got it straight in the face. I had been, in his opinion, a blasted dictator who was ruining his life all the time I was supporting him . . .' She concluded: 'Both my marriages broke up because I was supporting my husbands. Of course I now recognize the fact that I was plenty dictatorial.'

In the second chapter, Lizzie went on to discuss sex. 'I don't remember how many men I have slept with. Every time I try to remember I later recall that I left out several. I think I must have an overall count of about 25 or thirty men.' And Joe? 'I don't think I ever gave Sam a chance to be a real lover. I married him on the rebound from two real ones taken in close succession . . . a man one starts sleeping with under such circumstances is also behind the eight ball as far as developing a real love affair is concerned.' When this was written Hawes was deeply unwell and drinking heavily. Her final published book was already behind her. Habits of popular journalism contributed to the slapdash sleaze. Confronted with this manuscript forty years after his marriage to Lizzie ended, Losey responded with touching dignity and generosity: '. . . she was older than I . . . and frankly sometimes overwhelmed by both her personality and success.' He denied that she had been dictatorial. 'I should like to add simply that in our period together we had many good mutual experiences, travelling, talking, sharing and sometimes enjoying the sexual relationship. I learned a great deal from her . . .'[15]

PART III

The Film Director

10

Hollywood

When Losey moved to Hollywood in December 1943, the FBI stated his address as c/o Hans [Hanns] Eisler, Pacific Palisades. The beautiful Louise Moss (born Louisa Stuart) was two years younger than Joe and married to the producer Jack Moss (having changed her name to Louise when under contract to Paramount). Her acting career had evidently been short-lived. What attracted her to Joe? 'He looked like every college professor I'd ever had a crush on.' By her own account her husband was 'a darling funny man' but the sexual relationship was nil. At a meeting Joe heard her challenge the producer Dore Schary and (she says) was impressed. 'And I knew everybody in Hollywood – he saw me as an entrance.' With Joe came sex but (she says) 'he needed the lights on low, big mirrors, so that he could watch himself. I hated that.' She first met his mother when Ina visited them in New York during 1944. '"Have you made any nice chums in the army?" she asked Joe. He couldn't bear her. When she came out to Hollywood and had 'flu, he just wanted her to die.'[1]

According to their separate recalls they were married in a Methodist church on 19 October 1944, in Bucks County, Pennsylvania. Oddly, their Mexican divorce certificate, dated 14 March 1953, gives the place of marriage as Frenchtown, New Jersey. Talking to Michel Ciment in 1976, Losey did not deign to mention her by name. 'And the woman and I were married in Bucks County with John Wexley as best man.' (Wexley was a Communist author and scriptwriter.) When she read this in the 1980s, Louise suffered the same shocked reaction as Joe's third wife, Dorothy, who was also 'wiped out' in Losey's conversations with Ciment: 'I'm described as "*a woman* from Hollywood with a big Lincoln convertible . . . the *woman* and I were married". The best man has a name though: John Wexley!.'[2]

After leaving the army and marrying Louise, Losey resumed his interrupted career with MGM. He never saw Louis B. Mayer again, 'except at a distance'. Resuming the agenda that Mayer had set for him, he viewed two or three films a day, 'and saw more films than I ever had in my life. I saw everything, from any studio. I just ordered what I wanted and I had it . . . I hated Hollywood because there was no life . . . excepting a sort of pseudo-political life.'

Losey's notes on possible MGM projects and properties indicate that he was using Louise as a reader: 'Louise read. Says phoney English. Unbelievable. Terribly cute. Almost anti-British in effect.' Again: 'Louise read, says cliché corn.' Among Losey's 'synopses' we find: '*Oh, Bury Me Not*, by Patricia Colemna. Play bought for $50,000 at insistence of Hepburn . . . Louise read play – says, naïvely old hat excepting for occasional hard-boiled humor. Anti-semitic, etc.' In January 1945, Losey wrote a long memo recommending *London Bridge Is Burning Down*, by Madge MacDonald: 'A refugee (British) child story will cash in on the British–American differences, complexities, similarities . . .' In March he sent a memo to Robert Sisk: 'You asked about stories for Elizabeth Taylor. Do you know *The Little Princess* by Frances Hodgson Burnett? She wrote *Little Lord Fauntleroy*, of course, and this has all the sure fire sentimental elements of that piece. Sure it's dated as to emotion, but it's the old Cinderella yarn and could be a wonderful vehicle for Taylor . . .' (Many years later Losey remarked that he'd auditioned Elizabeth Taylor in Hollywood when she was 'ten', but she was fourteen in 1945.)

MGM insisted that he make a short, despite his protests that he was under contract to direct feature films. The upshot was *A Gun in His Hand* for the 'Crime Doesn't Pay' series. The police arrest a young detective who turns out to be a gang leader. 'It was a week of shooting with a very bad script, and a very bad cameraman . . . I had shot for three days when the first big strike began in Hollywood with picket lines and fire hoses.' He was made to shoot the night scenes by day. This was 'absolute hell . . . I hated it.' Then the head of the shorts department, Jerry Bressler, 'an awful, vulgar, tasteless man', said 'You don't know anything about films – this stuff is uncuttable.' Losey became so emotional that he burst into tears. As a result he was allowed to cut the film himself 'and it won a nomination for an Academy Award'.

He wrote a bitter memo to himself, dated 26 April: 'All this proves the smartness of Kazan and others who wouldn't come out [to Hollywood] without a specific assignment and set-up . . .' He felt acutely that Metro had not permitted him to direct his actors. 'I have taken such a kicking around as never before in my life, the lesson will be indelible!' Lessons for the future included: 'NEVER BE NICE to anyone or about anything, if it gets in the way of your job. NEVER AT ANY PRICE permit personal indignity or lack of respect to you as director from the producer on down . . . Tell this kind of producer to go to hell . . . BE TOUGH.' At the end of 1945 MGM took up their option on his services, 'So I was stuck for another year, during which I did absolutely nothing.' In 1946 and 1947 he staged the Academy Awards Presentation at Grauman's Chinese Theatre for Dore Schary, but he remained in deep freeze at MGM and was granted leave of absence by the studio in September 1946 to pursue his collaboration with Brecht on *Galileo*.[3]

A letter to Hella Wuolijoki, written in January 1947, indicates a still-happy marriage: 'I was re-married to a most fine girl, named Louise, whom you would love and she you . . . Soon we hope to have our own [child]. The frightening world permitting.' Gavrik remembers visiting Brecht's New York apartment while they were working on *Galileo* with Charles Laughton. 'Laughton was appalling to me. He called me a monstrous child.' But Brecht played chess with the boy on a huge board. The presence of Louise's son Mike, six years his senior, further disturbed the already disturbed Gavrik. Forty years later he still believed that Mike's father was Louise's previous husband, the film producer Jack Moss, but this is not so. In retrospect Louise is bitter: 'As for Michael, he was *forced* to take Joe's name – after Joe had "explained" to him that he was only adopting him for tax purposes . . . But the one he loved and who loved [Mike] was [Jack] Moss.' In general, 'Joe was a disaster to Mike and me as well. Perhaps if I'd been more understanding about his bisexuality . . .' The dots are hers. Joe later told his fourth wife, Patricia, that Louise had persuaded him to adopt Mike against his better judgement.[4]

On 7 April 1947, days before he left MGM for RKO, Losey wrote to Hella Wuolijoki, describing Metro as 'a great factory operating in a vacuum. I stay alive inside by working on elections and on an atomic energy radio series.' Dore Schary, now in charge of production at RKO, signed him up on a seven-year contract – long contracts being entirely to the studio's advantage, since they bonded only the employee – which set his salary at $500 a week. If the studio took up the option to employ him as director of *The Boy with Green Hair*, his salary would be $600 a week for one year following the exercise of the option. Salary escalator clauses covered subsequent renewal of the option by RKO, which retained the right to 'lay the Director off without compensation' for not more than twelve weeks within any one-year option period. During a period of suspension he would be free to pursue other work (theatre, radio, television) but not a motion picture. Artistic control belonged to the studio; one of the director's duties was to '*assist* in cutting and editing the Pictures' (emphasis added). The contract made no reference to political affiliation but contained a standard clause which provided the studios with a legal pretext to impose wide-ranging political sanctions: '. . . the Director . . . will not do anything which will tend to degrade him in society or bring him into public disrepute, contempt . . . or that will tend to shock, insult or offend the community . . . or prejudice the Corporation . . .'[5]

In fact Losey spent the first eight months of his time at RKO on continued leave of absence, setting up the Hollywood and New York productions of Brecht's *Galileo*. On 30 March 1947, while he was still under contract to MGM, his production of Arnold Sundgaard's *The Great Campaign* opened in New York for a five-night run. Set in rural America – Minnesota, Illinois and Ohio – with a large cast, the play took as its theme populist revivalism, with the Lord's Day of Judgement on the horizon for hayseeds capable of both credulity and devotion. The text is slack and verbose.

At RKO Dore Schary planned to launch a series of low-budget films on ambitious social themes. Adrian Scott had adopted a war orphan who remembered bombardments; from this was born the idea of *The Boy with Green Hair*. (But Scott, producer of the money-spinning social drama *Crossfire*, was forced out of RKO in November 1947, after being subpoenaed by HUAC.) 'I am glad that after all these years you have finally got your hands on the bat,' Schary wrote to Losey in February 1948. A photograph shows a smiling cast assembled in Schary's luxurious office at RKO: the young Dean Stockwell, Barbara Hale, Robert Ryan, and Losey wearing a double-breasted suit with white shirt and dark tie – at this period he normally wore spectacles and cut his hair short; it didn't begin to flow over his neck until he boomed with the Burtons in the late 1960s.

In mid-1948 Howard Hughes took over RKO and Dore Schary was forced out. According to the FBI, early in 1949 Losey was still employed by RKO 'as a writer' at a salary of $600 per month. At the tail end of his time with RKO he was offered a film called *I Married a Communist*, 'which I turned down categorically – I was the first. I later learned that it was a touchstone for establishing who was not a "red".'[6]

The Loseys were now living at 2212 Holly Drive, Hollywood, telephone Hillside 6438. Their lifestyle is described by Jean Butler: 'Comfort, simplicity: the fruit in the wooden fruitbowl, pillows if you felt like sitting on the floor. Louise was a good cook . . .' She also had beautiful feet and glided around barefoot. Gavrik, meanwhile, had been holed up in a boarding school in the East, but now Joe found a school in Westwood dedicated to children with 'learning disabilities', attached to the University of California. Gavrik recalls: 'I was whisked out in the middle of the Poughkeepsie term and thrown into this school the following day.' A small part in *The Boy with Green Hair* – as one of a mob of boys chasing Dean Stockwell through the woods – offered a minor compensation for a daily three-hour round trip to school by bus and trolley, but finally the pressure yielded its dividend – he could read and write. At some point after May 1949 Joe and Louise moved from Holly Drive to 12045 Maxwellton Road, 'the other side of the valley'.

Schary returned to MGM; Losey followed – but only as a screenwriter, a second-best role for a director, marking time. He worked uncredited on *Mystery Street* (directed by John Sturges) and unrewarded on *Man on the*

Train (inspired by the Baltimore plot to assassinate Lincoln before his first inauguration), which he wrote with George Worthing Yates and Daniel Mainwaring. Losey was keen for the historian Carl Sandburg to introduce the picture – and equally keen for Lena Horne, then under contract to MGM as a singer, to play the black maid. This project was eventually filmed by Anthony Mann as *The Tall Target*. Losey later described Lena Horne as 'a close friend of mine' – evidence of his professional interest in her continues into the 1950s. It was a frustrating, piecemeal existence. The young writer Roland Kibbee 'wrote a very funny script for me about how a dog destroyed the marital life of a young couple ... I spent a lot of time with him on Malibu Beach where I had a house.' Schary set him to work on a project called *Murder at Harvard*, written by Richard Brooks; Losey made a research trip to Boston but nothing came of it. He was never again on the payroll of a major studio as a director; his four subsequent Hollywood pictures were made for independent producers with major-studio backing. On 13 July 1950, the Los Angeles *Daily News* reported: 'Joe Losey, one of Hollywood's busiest young directors ... has plans to make *The Woman of the Rock* as an independent venture with Evelyn Keyes in the lead.'[7]

Losey struck up an association with Irving Allen Enterprises. Endorsing Losey's 1950 passport application, Allen stated that he was employing Losey as the director of *The Man Who Watched the Trains Go By*, to be made in Paris and London. Nothing came of this. In September the FBI reported that Losey was employed as an independent director by Motion Picture Center, Los Angeles. On 18 November, the Los Angeles *Examiner* announced that Losey would direct, and Irving Allen produce, Jan De Hartzog's London hit-play *The Four Poster*. Losey, Allen, Evelyn Keyes and Hugo Butler owned equal shares in the project. Forgoing salaries, they would take their returns out of the gross on this art-house project; in April 1951 it was reported that Losey would direct the same film for Stanley Kramer. Whether these press items reflected anything more than Losey's own hopes is an open question – but the blacklist descended, Losey fled from a subpoena, and the question became academic.

11

Sending a Message,
Signs of a Signature

The Boy with Green Hair

'You want to send a message? Send it by Western Union.'

This joke is recalled by Ben Barzman, co-screenwriter, with Alfred Lewis Levitt, of *The Boy with Green Hair*, adapted (heavily) from an original magazine short story by Betsy Beaton. Their message-loaded screenplay offers us a plucky little American boy, recently orphaned and living in a small town, who observes his mop of hair turn bright green after he becomes faintly aware of the evils of war. Persuaded by dead orphans that his new destiny is to carry their message, he is confronted by the resentment and conformism of the townspeople. This simplistic storyline carried music to match – an elaboration on a King Cole hit-song, 'Nature Boy', sung by Eden Ahbez, for which the studio paid $10,000, at that time the highest-ever fee for movie rights in a song.

Losey's first feature film was made in Technicolor on a large budget of $900,000, but he was confined to a stock studio set and allowed no opportunity for location shooting. Under Dore Schary working conditions were excellent. From the producer, Stephen Ames, 'a major investor in Technicolor', Losey received only 'encouragement and kindness'. George Barnes was the cameraman. But Losey's desire to shoot on location was resisted by RKO, which already possessed an expensive small-town set. Remembering La Crosse, Losey announced that the most prejudiced, bigoted, racist people 'are always the people from the small towns'.[1]

The key to the film lay in the hands of Dean Stockwell, whose career had begun at the age of eight on the Broadway stage. Many years later, in 1976, Losey described the young actor as 'a sensitive, unhappy, suspicious child . . . he was suffering, not only from conflicts with his family, but also from the familiar traumas of a successful child actor . . .' Losey frequently had lunch with him, talking through the scenes coming up. 'In fact the most awful days on that picture were the days in which I had to make [Dean] Stockwell cry. And I didn't ever put anything into his eyes. So I had to make him cry by cruelty – of a sort.' His twelfth birthday party was held on the set – Losey gave him a dachshund puppy. After making the film, Losey wrote a publicity-associated article, 'So You Want Your Child to be an Actor!':

'There is one cardinal offence in discussing the values of a scene with a child and that is to talk down to him ... it is far better for the child not to understand every word you use than to have him feel condescension ...'[2]

Shooting began in February 1948 and continued into April: the schedule provided for thirty-two days of filming. Howard Hughes took control of RKO before *The Boy* was released. Dore Schary, Losey's patron and protector, departed. The film was now the focus for an ideological struggle. Peter Rathvon, 'a very nice man' with a 'vast, rather Mussolini office', called Losey in and handed him scribbled notes on yellow scrap paper which, in Losey's view, 'came from Hughes's office'. By Ben Barzman's recall, Hughes, a major munitions manufacturer, summoned young Dean Stockwell and Pat O'Brien. Hughes wanted Stockwell to supplement his film line, 'War is harmful to children and to all living beings', with an additional line: 'And that's why we must have the greatest army, the greatest navy and the greatest air force in the world.' Dean Stockwell thought about it but refused.[3]

On 15 August 1948, the New York *Star* reported that the film's message would be doctored or removed by Hughes. On 6 December, *Life* devoted three pages, mainly pictures, to the film: 'Green Hair Trouble – Hollywood titans wage a battle over a modest little movie with a message. Howard Hughes hates messages ... A crew of skilled technicians was turned loose with orders to blast the message out ...' But failed. *Life* found two ideas in the film: that physical discrimination is bad; and that wars create suffering. 'These explosive ideas are so bathed in gentle humour and fantasy and pretty Technicolor and cloying music, and the story moves around so haphazardly that no one should be violently affected.' The December 1948 issue of *Ebony* – supporting the film's condemnation of racial prejudice, for which green hair was a metaphor – claimed that Hughes spent $200,000 trying to change the film's content – but without success.

Documentary evidence shows that on 26 June Floyd B. Odlum, of RKO, advised Rathvon that four speeches in the grocery store scene were liable to be misinterpreted and 'should be tempered somewhat'. The scene involved a clumsily contrived dialogue between two town ladies representing the progressive and simple-minded attitudes towards war. On 9 July Rathvon proposed some revisions:

MRS A: Right now let's talk about being prepared and give the world time to talk about peace based on understanding. If we don't –
MRS B: If we don't we'll have another war just in time to get the next crop of youngsters.

Rathvon also wrote, in pencil: 'We were not ready for the fight ... People must be ready to fight for what they believe in.'[4] The conflict here is not fundamental: in 1942 Barzman and Losey would have endorsed Rathvon's lines. Barzman's script reflected the shift in the CP line as the 'patriotic war' of the anti-Nazi alliance yielded to the Cold War.

Following a Hollywood preview on 19 November, general release followed on 8 January, 1949. RKO put out an advertisement filled with Dean Stockwell's puzzled, angry face under a mop of green hair. 'PLEASE DON'T TELL WHY HIS HAIR TURNED GREEN!' The critics were friendly, despite misgivings. Writing in the *New York Times* (13 January), Bosley Crowther praised this 'novel and noble endeavor to say something withering against war on behalf of the world's un-numbered children who are the most piteous victims thereof', but concluded that 'the gesture falls short of its aim'. Despite Dean Stockwell's splendid performance, the 'green hair' idea was 'weakly motivated' and 'banal'. Regarding the anti-war and anti-racist themes, Crowther held the scriptwriters and Losey responsible for confusion. In February 1949 *The Boy* received the National Screen Council's 'Blue Ribbon' Award for 'The Best Picture of the Month for the Whole Family'. When the film reached La Crosse's Rivoli Cinema, in July, the top of the ad blazoned, 'Directed by La Crosse's own Joseph Losey'. The local *Tribune* carried a feature.

The Boy did not surface in England until June 1950, when RKO-Radio released it with a U certificate. It was ignored by *The Times*, the *New Statesman* and the *Spectator* – a crushing verdict-by-silence. Losey himself concluded in retrospect that the weakest scenes were the symbolic ones: the posters of orphans in the gymnasium, the meeting with the orphans in the glade, the breaking of the milk bottle in the grocery store. 'There are certain things in the film that embarass me terribly because they are so blatantly sentimental.'[5]

The Lawless (*The Dividing Line*, UK)

The writer Daniel Mainwaring proposed Losey, now free of RKO, as director of *The Lawless* to the notorious tandem, William Pine and William Thomas, successful producers of some sixty B-movies and known in the trade as the 'Dollar Bills'. Writing under the name Geoffrey Homes, Mainwaring laid the story in farm country north of Sacramento where migratory workers of Mexican origin were employed on the fruit plantations. The film was made on location in Marysville and Grass Valley for a modest $407,000. Working from chosen locations, Losey and his uncredited production designer, John Hubley, drew detailed pre-designs of the main sequences.

Some fifty locations were needed, including an unpicked tomato field, a newspaper office whose owner wouldn't mind having it wrecked, and – most difficult – acres of dredged-over land (owners of gold dredges being sensitive to what they were doing along the rivers of northern California). For Losey location filming was fulfilment. He sought natural expressions impossible to obtain from stock actors and Hollywood professionals. The citizens of Marysville and Grass Valley were persuaded to riot with zest, hurling stones and overturning cars – but they were not told what the film was about. (Hedda Hopper reported that the studio spent $100,000 in Marysville, paying the Mayor $55 to deliver one line.) For the mob scenes, Losey had studied news pictures of the recent Peekskill riots in New York State, when right-wing Legionnaires and vigilantes wrecked two Paul Robeson concerts on 27 August and 4 September 1949. He had also viewed and admired the lynch-mob scenes in Fritz Lang's film, *Fury*, while idling at MGM.

Shooting began in October 1949 and Losey finished the film in twenty-one days, assisted by a brilliant cameraman, Roy Hunt, who had refined the Mitchell camera. 'He was the one above all others who taught me how to work intensely and well and still fast. We'd hardly get finished with a shot and he'd have the camera on his back, still on the tripod, and run to the next set-up ... He was marvellous.' Pressed for time during the cross-country pursuit sequence, Losey and Hunt left the heavy equipment behind, using a light French camera and hand-held mikes.

Pine and Thomas were 'monsters, absolute monsters. And they interfered in the worst possible way at all points,' imposing a musical score that Losey detested: 'It made it cheaper and more melodramatic and it slowed its tempo.' Script conferences were held while Thomas sat on his lavatory with the door open. 'On location in Marysville, California, I threw the script at him one night at three o'clock in the morning and said, "Direct your own fucking picture!" and went upstairs to bed.' When Losey refused to apologize, Thomas said, 'Well, nobody saw it but Dan [Mainwaring] and you, so as long as you don't tell anybody we'll let you finish the picture.' On another occasion Losey had a fight with the production manager, Doc Murman. 'I jumped at him with a kind of intention to kill him but with some Freudian repressions or inhibitions.' Murman's bodyguards moved on Losey and Freud may not, strictly, have been involved.[6]

MacDonald Carey played the lead role, a highly professional actor whom Losey liked (and used again, twelve years later, in *The Damned*). Losey discovered the two young Mexican–Americans, Lalo Rios, a carpenter in Los Angeles, and Mauricio Jara, an actor working in the Mexican Theatre in Padua Hills. Rios was a 'natural' whose performance was universally praised. The problems arose with the female lead, Gail Russell: 'I had a tragic time with her. I think she had the most beautiful eyes I have ever seen ...' The first shot with her was a long, night-tracking shot, with three or four pages of dialogue – she couldn't remember a line. After patient coaching Losey displayed exasperation. 'And she grabbed me, her hands were icy cold, she was absolutely rigid ...' She'd never wanted to be an actress, she'd never been faced by such a long scene before – and she needed a drink. Despite Paramount's injunction, he got her a drink and she did the scene. (The spectator, however, would guess none of this.)[7]

The Lawless is swift, vigorous, clean-cut, and graced by brilliant pre-design, fine photography, superb sound effects, and the emergent eye of a major director. On the debit side, the characters are stereotypes, the music banal, and dialogue is planted to hammer home social or biographical points. It seems that the original script may have been even more didactic than the final version, since Daniel Mainwaring conceded that it had benefited from changes demanded by Joseph I. Breen, chief of the Production Code Administration: for example, the insertion of some decent solid citizens alongside all the bigoted ones. As Foster Hirsch notes, Losey was clearly more comfortable with the real locations and socialist-realist mode of *The Lawless* than with the studio sets and coy '*fabliau*' quality of *The Boy*. In 1957 Losey commented: 'By its content, it's closest to the kind of film I'd have chosen to make.' Apparently it certainly wasn't the kind of film that Paramount's illiberal boss, Y. Frank Freeman, wanted to distribute.[8]

The film was first released in July 1950. The racial dimension drew warm support from *Ebony*, alongside advertisements for Negro hair-styling: 'Page Boy', 'Cluster Curls', 'Half Glamor'. *B'nai B'rith Messenger* (16 June) came out with some dubious biographical notes on the director: 'When Losey, in 1945, came back from the war . . . he was shocked to learn that even . . . his home town, La Crosse, Wisconsin, was infested with the poison of race-hatred, anti-Semitism, anti-Negro and anti-Roosevelt sentiments.' *Time* (3 July) greeted the film warmly: '*The Lawless* is not only a good movie but, considering its makers, it is also as unexpected as a slum documentary by Cecil B. DeMille . . . Using unvarnished photography on the streets . . . and people of real California towns, Director Joseph Losey has given the picture a startling look of reality . . . His mob scenes crackle with a spontaneous movement and raw vitality usually found only in bag-up newsreel footage.' A quarter-page display ad headed 'Paramount on Broadway!' showed the Astor Theatre decked out for *The Lawless*. At the bottom of the ad was a different message: 'There's still a good chance to kill that 20% tax! Act fast! Write, wire, talk to your senators. Do it now!'

Released in the UK as *The Dividing Line*, the film opened at the Carlton Cinema in May 1950 – two months before its American release – and was far better received by British critics than any of Losey's other Hollywood films. *The Times* praised its strong visual values: '. . . this is a director's film, and the director is Mr Joseph Losey'. In the *New Statesman* (13 May), William Whitebait was equally enthusiastic, putting the film on the same level as *Intruder in the Dust*. Whereas *The Boy* is an artificial fable tangled in two messages and reeking of studio-set, by contrast *The Lawless* carries an authentic, naturalistic energy and a savage beauty which combine to transcend the plot contrivances.

The Prowler

Of his five Hollywood films, Losey was proudest of *The Prowler*. When exiled and seeking work, this was the one he most commonly displayed to British producers and television executives. Paradoxically, *The Prowler* displays Losey's innate cinematic sense precisely because of its intimate, small-scale theatricality; visual signs codify the claustrophobic relationship between a cynical policeman and an unhappy wife more effectively than the often cluttered dialogue.

Losey had seen Billy Wilder's *Double Indemnity* 'many, many times . . . one of the best films of the sort ever made'. In both films a pillar of society steps into an expensive West Coast imitation-Spanish residence and is induced by greed to take another man's frustrated wife and murder the husband, covering his tracks by exploiting the professional knowledge he has betrayed. However, the complexity of *Double Indemnity* resides strictly in its Raymond Chandler plot – *The Prowler*, by contrast, offered psychological ambiguities and a social-class dimension. That Losey borrowed from Wilder cannot be doubted – in particular the tendency to shoot at night, with strong, *film-noir* lights and shadows.

An Eagle–Horizon production for United Artists, this was to be Losey's only completed collaboration with Sam Spiegel ('S.P. Eagle'). Losey's recall of this experience is contradictory: on the one hand, Spiegel had provided the best possible technicians and 'I had no battles with Spiegel on *The Prowler* at all'; on the other hand, Spiegel had started looking at the rushes before Losey did 'and then coming on the set and telling me what was good and bad . . . I did throw him off the set. In my early days I was pretty intense and nervous . . . I was given to uncontrolled rages.' By good fortune the co-producer was John Huston, then working for Spiegel, whom Losey admired and liked.[9]

The original story and script, written by Robert Thoeren and Hans Wilhelm, was 'pretty awful' Losey recalled. Spiegel called in Dalton Trumbo, a blacklisted member of the Hollywood Ten, who (Losey observed) 'lived incredibly in a vast, incongruous place somewhere in the mountains outside Hollywood where he'd built his own trout lake and imported Italian marble and French antique furnishings'. The incomplete script in the Losey archive contains annotations, sketches, detailed camera instructions, but also numerous page revisions, deletions and notes, indicating massive uncertainty concerning the second half of the film. According to Trumbo, he wrote two drafts for Losey, for which he received $7,500, the final third of which he collected from Spiegel five years later in a legal action. It took Losey two years to extract the $10,000 Spiegel still owed him. Trumbo was paid a princely $35 for playing, uncredited, the radio voice of the cuckolded husband – a fine gesture of ironic bravado.[10]

Losey and his assistant director Robert Aldrich visited Ryolite in mid-March 1950, to search for locations. After a two-week rehearsal period he then shot the film in only nineteen days.

John Hubley designed the three main locations with impressive accuracy: Susan's luxurious, Spanish-style house; the utilitarian motel; and a crumbling wood cabin in the desert. The Spanish house was built as a composite set, with arched doorways, to convey emotional emptiness within material comfort. Paintings, tapestries and expensive furniture don't make the house lived in: it lacks and awaits a child. The same emotional vacuum extends to the motel, with its plain white walls. As Hirsch notes, 'A flashing sign of "vacancy" in front of the motel is an appropriate indicator of the characters' own vacancy.' The desolation of the desert shack is reinforced by piles of packing crates, a lonely phonograph, a travelling first aid box. Paul Mayersberg later commented: 'In *The Prowler* the human development is reflected in the décor . . . In his analysis of human behaviour and in the scale of his themes, Losey is already the von Stroheim of the post-war cinema.'[11]

The Prowler was released in the USA by United Artists on 25 May 1951, and on 2 November in the UK, where it opened at the London Pavilion. The reviews did not anticipate the high regard in which this film later came to be held. In the *New York Times* (2 July), Bosley Crowther noted that, sordid tale though it was, 'it is a story spun with conviction and mounting force'. In Foster Hirsch's view, *The Prowler* is 'the most technically polished of Losey's early pictures' and 'strongly reminiscent of *Double Indemnity*, the ultimate *film noir*. . . Losey's signature is evident in the absolute clarity and detachment with which characters are observed . . .' As Hirsch points out, this is a first essay at Losey's famous 'intruder' theme. Both *The Prowler* and *The Third Man* are morality plays subscribing to the storyline 'Nice Woman Loves Bad Guy Who Bites Dust', but there the resemblance ends. In Graham Greene's script the tragic logic of events is written deep into profound characterization, superb individualization, and the specific tensions of post-war Vienna. Losey and Trumbo intended an exposure of American petty-bourgeois materialism but lost their way in melodrama.[12]

M

Although neither Losey nor the critics thought so, *M* is the pick of his Hollywood films and his best prior to *The Criminal* (1960): a film of force and freedom. But hanging over the enterprise was the stigma of a 're-make'. Fritz Lang's *M* (1931) had become a cult classic; Losey had seen it in Munich in the year of its release and viewed it again before shooting his own version nineteen years later. Seymour Nebenzal, the producer of Lang's *M*, proposed a version closely modelled on the original script (by Thea von Harbou and Lang) but set in contemporary Los Angeles. Losey was reluctant: 'I had twice refused to direct it but finally my financial situation dictated my acceptance of the project.' The contract is not found but legal correspondence four years later indicates that it involved a $10,000 deferment, once the bank loan had been paid off and a number of other charges met. By March 1954 Losey had received no statement of account from Nebenzal or Columbia. According to Thomas Elsaesser, Lang never forgave Nebenzal 'and was not prepared to acknowledge the director [Losey] as a member of his profession'. (Yet Lang himself was not above directing remakes.)[13]

The new screenplay, by Norman Reilly Raine and Leo Katcher, was graced by additional dialogue from Waldo Salt. Losey was anxious to update the psychology of the fictional child-murderer. The production file notes on 'Martin Harrow', the killer, dated 31 May 1950, are based on a study by Wertham and Menninger of two actual psychotic killers: 'Harrow was isolated in his youth by religion and by poverty. He is suffering from hyper sensitivity. He was sexually attached to his mother. This resulted in frustration, hatred of father.' No less simplistic is the explanation for the killer's habit of collecting his small victims' shoes – not found in Lang's version: '. . . the shoe and foot as sexual symbol – contact with earth, fecundity'. Losey also contributed his own, abiding fixations (as an artist should): 'And I wanted to present him as a product of a mother-dominated and materialistic society of lower middle-class America, where everybody had to be big he-men otherwise they were sissies . . . this man undoubtedly was a concealed homosexual, totally in conflict with everything including his own mother whom he adored and hated.' This approach was in strong contrast to Fritz Lang's, with its debt to *The Cabinet of Dr Caligari* and the German expressionist fascination with the demonic loner, the ruthless and anarchic individual ego.

A first-class crew and cast was assembled, with Robert Aldrich the assistant director, John Hubley the production designer, and Michel Michelet composing the music. The director of photography was Ernest Lazlo, whose talents were ideally responsive to Losey's.

Peter Lorre, who had played the psychopathic killer in Lang's film, later joined the German refugee colony in America; Losey had employed him in *The Day of Reckoning* radio series for NBC, but, writing to Brecht in May 1948, he dismissed Lorre as unacceptable for a remake of *M* because 'he is now regarded by the American movie public as a clown'. To play the child-killer, Losey chose David Wayne, an actor mainly associated with high comedy parts – Losey had seen him on the stage in *Finian's Rainbow*. A fine supporting cast (many of them doomed by the descending blacklist) included Luther Adler, Howard da Silva, Martin Gabel and Karen Morley.[14]

The splendour of Losey's *M* is its urban locations, set mainly in the dilapidated downtown Bunker Hill area of Los Angeles. The final manhunt was set in the Bradbury building, at the corner of Third and Broadway, a vast warehouse rising tier upon tier, its wrought-iron staircases echoing with angry feet, a prelude to the prison in *The Criminal*. Losey also introduced echoes of San Francisco: precipitous streets, cable cars. The film was shot in only twenty days.[15]

Substantial themes were cut from Lang's final scenario, including the gangster leader's denunciation of liberal law courts and soft justice. Lang was staring at fascism; Losey's chief racketeer seeks to exploit public gratitude and so get a grand jury off his back. Lang was trading in metaphysics, Losey in urban pragmatism. Whereas Peter Lorre's pop-eyed, demented killer breaks off from his compulsions to chuckle villainously – hee hee hee – over a whole city's fruitless search for him, David Wayne steps over the screaming headlines like a stranger to himself. Wayne has a psychiatric history, Lorre was born to the devil.

M was released by Columbia in March 1951. Trade press reviews were mixed, agreeing only that it was a 'gruesome' film (*Variety*) and 'morbid and horrible' (*Harrison's Report*). According to Losey, the film was censored and 'totally banned in eight states . . . So it's a kind of a dead picture . . .' Heavily cut, *M* opened at the London Pavilion on 14 December 1951, only six weeks after *The Prowler*. In the *New Statesman* (29 December), William Whitebait gave it five lines, describing the film as 'a remake or crib from Lang's atmospheric old shocker, which succeeds in worsening its original at every point; some attempt is made at psychological understanding, but this quickly vanishes in a welter of cops and crooks'. In the BFI's *Monthly Film Bulletin* (August 1951), 'K.R.' found it inferior to Lang's original in every respect.

Losey himself was unnecessarily apologetic about *M*: 'I couldn't believe myself in the idea of the whole underworld ganging up against the killer . . . It made a kind of mélange of contemporary Los Angeles, 1920s *Beggar's Opera*, middle Europe which just couldn't mix.' David Thompson rightly called *M* 'a marvellous, frightening film, years ahead of his [Losey's] time and typical of his American virtues'. In Foster Hirsch's view, 'Lang's creation of atmosphere, his use of closed frames, his famous counterpoint between sound and image, are unsurpassed . . . Losey simply did not have the experience or the artistic freedom to compete with Lang.' This is debatable. Lang's gangsters and policemen, in their smoke-filled rooms, are tediously garrulous, as if the director was revelling in the recent liberation from silent cinema. In addition, Lang's film reeks of 'studio' and painted sets, whereas Losey achieved the liberation of location shooting, capturing what Hirsch aptly describes as 'the almost surreal desolation of . . . the paintings of Edward Hopper'.[16]

The Big Night

Losey directed his last Hollywood film, *The Big Night*, for Philip A. Waxman Productions during the early summer of 1951. Based on Stanley Ellin's novel, *Dreadful Summit*, the screenplay was credited to Losey and Ellin but was mainly the work of Hugo Butler, until he was blacklisted, and then of Ring Lardner, Jr, one of the Hollywood Ten and only recently out of jail. The budget was only $300,000, which (Losey recalled) required 'a lot of faking, which partly worked and partly didn't'. For example, the Madison Square Garden sequence was 'just cardboard and it shows a bit'. Nicholas Remisoff was the art director, Hal Mohr the photographer.[17] *The Big Night* contains its 'Losey moments' but is inferior in almost every respect to his previous Hollywood films. This is voice-over narrative flashback at its worst. The story is told by the seventeen-year-old hero, George La Main (John Barrymore, Jr), whose performance invites unfavourable comparisons with James Dean – his tortured monologues are clogged by exposition and cliché. George proceeds from one studio set to another during a bizarre night out, seeking revenge for the inexplicable beating of his father in his own bar. The dialogue is third-rate: 'You're lying to me . . . lying . . . lying . . .' A revolver keeps getting into the wrong hands. The music (by Lyn Murray) is bad to terrible and the climax is slop: 'Georgie – you matter to me. I should have tried to tell you.' Among the performers, Philip Bourneuf's wit stands out. The women (Dorothy Comingore, once the wife of Citizen Kane, and Joan Lorring) are as good as their lines allow.

Occasionally there are signs of Losey's class: the arrival at the bar of Al Judge (Howard St John) and his two thugs is initially seen as a close-up of a cane and two broad backs. Losey's eye keeps surfacing above the hack script: the extinct candles of a birthday cake in an empty bar; opticals showing a boxing match advertisement over the men in trilbies arriving at the arena; George inadvertently insulting a black night-club singer he admires; a sinister shadow on a staircase.

In July 1951, Losey left hurriedly for Italy and the film was re-edited in his absence. The producer, Philip Waxman, decided that the 'one big flashback' should be changed to strictly chronological order, 'and I think it was very damaging to the film'.[18] *The Big Night* was released by United Artists on the West Coast in December 1951, reaching Loew's neighbourhood theatres in New York in March 1952. The reviews were terrible: in the *New York Times* (20 March), Bosley Crowther complained of 'the dull and dismal toils of a bleakly pretentious melodrama . . . Not only is the story presumptuous and contrived, without any clarification of character or theme, but it is directed by Joseph Losey in a provokingly ostentatious style, and it is played by a cast of professionals as though it were an exercise at dramatic school . . .'

Two years later, and living in England, Losey wrote asking his lawyer, Sidney Cohn, to inquire about his share of the profits. Cohn replied that the film had been a total loss, with United Artists over $100,000 'in the hole'.

A decade later, the devoted Raymond Durgnat discovered exceptional moral virtues in the film; Gilles Jacob noticed how the camera's constant, hesitant reframing suggested the clumsy excitement of first love between George and Marion. He admired the interiors 'flaunting their bronzes, ornaments, miniatures, chandeliers, in execrable taste but loudly signalling the presence of a woman . . .' Stressing Losey's 'ability to recycle pulp', Foster Hirsch admires the smoky, claustrophobic, darkly lit, crowded scenes in the prizefight arena and the nightclub. Losey's own, retrospective, view of the film focused on the pain of a bewildered youth striving for self-respect in the face of his father's humiliation. 'And I've experienced it in my life with my sons and with my father . . .'[19]

Stranger on the Prowl

This ill-fated film, shot in Italy, was the first that Losey made outside the United States and the first to be released under a pseudonym, 'Andrea Forzano'. The producer was John Weber, and the intended director Bernard Vorhaus, until he was blacklisted. Losey's friend Ben Barzman had written the script, based on a novel by Noël Calef, *A Bottle of Milk*. Calef didn't like Barzman's screenplay and Losey later reflected that 'in many ways he was right . . . Barzman was still a Hollywood-oriented writer'. Losey developed an affection for the generous, sensitive Calef, frequently visiting him and his wife in Paris.

The film was shot mainly in Tuscany, and partly in Rome. 'The locations [at Taranto, Livorno and Pisa] were fantastic.' Losey was 'overwhelmed' by Tuscany and Venice, and by the 'impact of the Renaissance'. Shooting began at Tirrenia Studios near Livorno on 24 September 1951, with an eight-week schedule which finally extended to eighty-six days. A catalogue of disasters ensued. The changes Losey wanted in the script didn't happen; he was ill for a period and had to leave the studio; the working conditions were the worst he'd known. 'I found that my American left-wing Communist friends [presumably Weber and Vorhaus] had made a deal in Italy with the Forzanos, who had the new studios near Pisa.' The elder Forzano had been a lyricist for Puccini and later 'a kind of minister of culture for Mussolini. The backing came from a man who had owned mercury mines and had been condemned to death as a war criminal' – in short a Communist–Fascist co-production. The Italian actors in the cast had to learn their lines phonetically. Scenes which ought to have been shot on location were filmed in the studio and vice versa; sets constructed on too small a stage presented the photographer with insoluble problems. Then payments were delayed, including Losey's, 'so we held a meeting in which we all agreed to strike against the production, which was terrible because we were all sharing beliefs to one degree or another ... very ugly'. The British producer Leon Clore later heard from both Losey and Bernard Vorhaus that they had had a falling-out; Vorhaus, who claims to have had no part in the film, and who was expelled from Italy while working on another, will not discuss his quarrel with Losey. Both later worked at Merton Park Studios under pseudonyms.[20]

Shooting in Italy had its compensations, including a role for Losey's mistress, Joan Lorring, and a marvellous child actor, Vittorio Manunta. (On 10 October, *Variety* reported that Lorring had reached Italy a week earlier to fill a role previously assigned to Lea Padovani.) The leading actor, Paul Muni, did not figure among the compensations. According to Losey, 'Muni was the quintessence of everything that's bad about professionalism. He used to rehearse all day and night with a tape-recorder ... working in front of a mirror ... he deeply resented any attempt to get into his insides ... He was absolutely petrified that he might be associated with a Communist or somebody who was going to be blacklisted.'[21]

This was Losey's first collaboration with the great photographer Henri Alekan* (who was irritated by the permissive attitude of a man called 'John' – presumably Weber – to his spoilt brat: very American, Alekan thought). Alekan worked with Henri Triquet as his '*cadreur*' (operator) on every film; Triquet spoke English fluently and Alekan let him discuss details with Losey while – as he recalled – feigning to understand by 'nodding my head approvingly like a perfect imbecile, though Losey never said so'. Each day Alekan drove Losey to the studios at the great speed he imagined an American demanded. Not grasping the American accent, the swallowed words, he had little chance of understanding him unless the car windows were closed. Alekan was struck by the extreme care with which Losey selected his settings, whether in the port of Livorno or in Tirrenia Studios. 'Collaboration was total and unreserved.' Losey was so fearful of contracting diseases from the war-ruins that an assistant was instructed to precede them with a big can of DDT. When Losey viewed the rushes he spotted every weakness, but blamed himself.[22]

Stranger on the Prowl is the only Losey film for which no material of any description exists in the BFI's Losey archives. The story was a variation on

---

*Born one month after Losey, Alekan had worked as *chef opérateur/directeur de la photographie* for Max Ophuls, Marc Allegret, Marcel Carné and Abel Gance. (Alekan's exploits under the German Occupation had won him the *medaille des Combattants volontaires de la Résistance*.) In sharply contrasting style – a measure of his versatility – he photographed Réné Clément's *La Bataille du rail*, a drama of class struggle, Jean Cocteau's surreal *La Belle et la bête*, and Julian Duvivier's *Anna Karenine*, starring Vivien Leigh.

the 'impossible flight' theme that Losey constantly explored: the physical flight of *The Lawless*, *The Prowler*, *M*, *The Sleeping Tiger*, *The Criminal* and *Figures in a Landscape*; the psychological flight of *The Intimate Stranger*, *The Romantic Englishwoman* and *Mr. Klein*. The action takes place in a poverty-stricken, half-destroyed Italian port which Losey photographed under the spell of 'neo-realism': ruins, desolation. A hungry, disillusioned and desperate vagrant (Paul Muni) seeks to peddle his revolver in order to get the money for an illicit passage aboard an outgoing freighter. After several rebuffs, the now famished wanderer snatches a cheese from a dairy shop and, in attempting to stifle the owner's outcries, he unwittingly strangles her. A boy (Vittorio Manunta) who has stolen a bottle of milk from the same shop joins the man in flight. Later the fugitives hide in the room of a serving maid (Joan Lorring). This results in a romantic interlude between Muni and Lorring. Finally Muni, cornered by the police on a rooftop, is shot dead. This was the first film in which Losey displayed his nascent passion for the Mediterranean, for sleepy, sun-drenched landscapes, wild rocks, the sea beating on the shore.

The film was first shown in Italy in May 1952 under the title *Imbarco a mezzanotte*. A score was imposed 'absolutely wrong for the film. It was heavy, it was symphonic, it was all wrong,' Losey complained. Not until 2 November 1953 did United Artists release it at the World Theatre in New York – without the names of Losey and other blacklistees. The screenplay and direction were credited to Andrea Forzano, a son of the owner of the studio. The *New York Times* (10 November) praised the film's pace but found nothing beyond 'surface drama'. Muni showed 'strength if not stature' as an unlettered man with a sense of justice, despite a hampering script.*[23]

'I gather it was wildly re-edited and shortened,' Losey recalled. 'I never saw it, I don't know.'

---

*After Losey's death the Kinoteka of Ljubliana held a Losey retrospective, including a seventy-eight-minute version of *Stranger on the Prowl* in English with Serbian subtitles. This print belongs to the Jugoslovanska Kinoteka of Belgrade. Another print is held by the Library of Congress.

12

Nightmare:
Surveillance, Subpoena, Divorce

Losey never applied, under the Freedom of Information Act, to be shown his own FBI file. He would probably have been surprised by the sheer scale of surveillance to which he was subjected from 1943–4, and the number of regular FBI informers within the Hollywood studios. But he would have had to guess their identities from fragments of circumstantial evidence. The FBI file consists of 750 pages, of which 518 pages have been released; of these large sections have been blacked out to protect third parties, informants, and certain surveillance methods. In the précis that follows, FBI descriptions of individuals should not be taken as necessarily accurate.

Losey's file seems to have been activated by his friendship with Hanns Eisler, the German Communist composer with whom he stayed when he temporarily moved to Hollywood in December 1943 (although he also seems to have stayed with Jack and Louise Moss, precipitating the end of their marriage). He had first met Eisler in 1935; subsequently Eisler wrote the scores for *Pete Roleum* and *A Child Went Forth* – and dedicated a movement of his 'German Symphony' to Losey. In January 1940 Losey had written to the American Consul, in Havana, Cuba, on behalf of Eisler's application to be admitted as a resident of the USA.[1]

Eisler's home in Santa Monica was under heavy surveillance, his telephone tapped. On 18 November 1943, Mrs Eisler called Robert Thoeren, 'an active Communist in the Los Angeles area' according to the FBI, advising him that Losey was coming to the coast to work for MGM. (Thoeren later contributed to the screenplay of *The Prowler*.) On 16 December Losey talked to an unidentified woman 'and discussed the Lanham Act vs the Thomas Bill', suggesting that 'somebody must go to Washington to work on the educators and smooth out the situations [*sic*] with the CIO'. On 29 December, Losey was in contact with Yip Harburg, 'well known Communist in the movie industry' and with George Wilner, 'known Communist'.

Following military service, and married to Louise, Losey returned to MGM and was reported by the FBI to be staying with the Eislers. The day that Roosevelt died, he was 'driving along the Roosevelt Highway to my home in Malibu . . . For me this was the blackest day and blackest event of my life . . .' (He had voted against Roosevelt in 1940.) In April he staged a Roosevelt Memorial in the Hollywood Bowl, with Dore Schary acting as producer. Then came 'the even more terrible event which was to occur shortly after, namely, Truman's dropping of the atomic bomb on Nagasaki and Hiroshima'. (In fact, American Communists raised no protest at the time, because Stalin didn't. Losey's instant horror was reconstructed, *post hoc*, after the Cold War began.)

In July the Bureau noted that he had been reported as an agent of the NKVD or OGPU. Mail cover was authorized. The Loseys moved to 3827 Hollyline Avenue, Van Nuys, in the Santa Monica foothills, described by the FBI as a rather isolated area. In September Losey received a letter from the Southern Conference for Human Welfare, New York, regarded as a Communist front. On 9 October, a sixty-day mail cover was renewed.[2]

'Obviously I got to know the left-wing people – Dalton Trumbo, Adrian Scott – I already knew John Howard Lawson, Sidney Buchman, Francis Faragoh.' According to Losey, he joined the CPUSA 'after the war', more specifically in 1946. 'We believed in some kind of theoretical revolution, which obviously was fatuous. I was full of certainty. I'm not full of certainty any more.' Losey's wife of that time, Louise, believes that he was already a Party member when he arrived in Hollywood, but she may be confusing commitment with membership. 'It was a kind of Hollywood guilt that let me into that kind of commitment,' he recalled. 'And I think that the work I did on a much freer, more personal and independent basis for the political left in New York, before going to Hollywood, was much more valuable.' In Hollywood it became 'a lot of meaningless so-called Marxist classes which were a bore and which never had any practical result either in terms of the films . . .'[3]

On 14 January 1946, J. Edgar Hoover signed a memo to the Attorney-General, describing Losey as a contact for various Soviet espionage agents and recommending a 'technical surveillance'. On 11 March permission was granted to renew a sixty-day mail cover on the Loseys, now living at 4248 Arch Drive, North Hollywood, and to station agents on spot surveillance. On 14 April the FBI, Los Angeles, reported that Losey and his wife were active in the Hollywood Independent Citizens Committee of the Arts, Sciences and Professions (HICCASP), and in the 'Win the Peace' movement. On the 16th, Losey contacted [name blanked out], 'a Communist', and advised him that he intended to buy into a housing project, Community Homes, Inc., at Encino, which did not practise racial restrictions. On the 21st Losey sent a telegram to the reactionary Republican Senator, William Knowland: 'I am a veteran and will make it clear in the fall we did not fight for the kind of life the House OPA bill makes certain.' Also in April, Losey was bugged while telephoning an unidentified man at MGM, Culver City, where he worked. Joe said he had some 'stuff'. The man asked 'if it was cards'. Joe said no, 'some stuff on the atomic'. The FBI concluded that this was a script about the atomic bomb.

Added to his file was his employment, in 1941–2, by Documentary Films Production, Inc., New York, which was allegedly under Communist influence. This had first been noted by the FBI, Washington, on 29 May 1941, but two days later a memo found nothing 'derogatory' on the film company and no link between Losey and people listed by the FBI.

On 24 May 1946, Losey contacted an unidentified woman and discussed Truman's anti-strike legislation, 'fantastic, incredible, a step towards Fascism and would lead to revolution'. Two days later he telegrammed Senator Downey, protesting the President's 'strike-breaking speech'. On 3 July the FBI, Los Angeles, reported: 'Subject is attempting to organize a mobile motion picture unit to fight the Ku Klux Klan activities locally. Contacts of Subject with Communists locally set out.' The American Veterans Committee and HICCASP had agreed to sponsor the project; Losey's idea, evidently, was a mobile theatre ready to move to the scene of any KKK outbreak.

An FBI, Los Angeles, memo covering the year 1946 reported Losey as active in Communist fronts and observed at many CP meetings, 'several such meetings having been held at his home', attended by writers such as John Howard Lawson and Waldo Salt. At this juncture the FBI became obsessively interested in Losey's plan to direct 'Galleleo [sic] by Bertoldt [sic] Brecht', which might be 'a Communist propaganda medium'. The Bureau had informers in the studios and knew when MGM released him from his contract at his own request. Losey, the FBI accurately noted, felt that he was not being given important enough work at MGM.

Arriving in New York on 17 September 1946, Losey was under close surveillance from the moment he reached LaGuardia airport. The files allege that he made numerous contacts with Communists and sympathizers; for example, he contacted [name blanked out], 'a CP member and writer for the CP newspaper', twenty-four times.

The FBI noted that in November 1946 Lionel Berman, 'of the CP Cultural Section, New York City', stayed at Losey's Arch Drive home. On the 18th, Berman and an unidentified woman left the Losey home at 2.30 p.m., toured the city for an hour, then drove to CP headquarters at 124 West 6th Street. Berman's calls from Losey's house were tapped and transcribed. However, throughout this time Losey himself was in New York, where the Bureau was busily monitoring his movements.

HUAC descended on Hollywood in 1947. On 15 October Losey staged a rally at the Los Angeles Shrine Auditorium, sponsored by Progressive Citizens of America, to raise money for the defence of the Hollywood Nineteen subpoenaed by HUAC. The Nineteen sent Losey a letter of appreciation carrying their signatures: 'We cannot leave our home town without expressing to you our gratefulness for the unselfish and effective work you have contributed to the common defense of our Constitution and our industry.

You have launched a counterattack from which our people will profit. This is small thanks, but you don't need any anyway.'[4]

Soon after HUAC adjourned its hearings on Hollywood, the big studios and producers in the Motion Picture Association met at the Waldorf-Astoria, New York, on 24 November 1947. The upshot was an announcement that they intended to discharge or suspend without compensation those five of the Hollywood Ten still in their employ. ('The Ten' were those members of the subpoenaed Nineteen who, called to testify before HUAC, had sought protection under the First Amendment and generally hurled defiance at the Committee. As a result they were cited for 'contempt', and were later convicted in the courts and jailed.) Losey's head of production at RKO, Dore Schary, had boldly told the Committee that until it was proved that a Communist was a man dedicated to violent overthrow of the Government, he would judge him by his personal capabilities alone. But now, after the MPA's Waldorf-Astoria meeting, he too adopted the position that in order to protect the 'freedom' of the film industry, it would be necessary to blacklist anyone thought to be a Communist. In mid-November RKO fired Adrian Scott, one of the Ten, as producer of a film largely of his own devising, *The Boy with Green Hair*, which Losey was scheduled to direct. Although Scott was a friend, Losey did not resign in protest; he directed the film. Only after he and Schary had parted company did he condemn him for his capitulation in 1947: 'I think that the attack in the beginning . . . could have been stopped by just a little bit of courage from people, like Dore Schary and later Kazan, who were in positions of success and financial stability.' In Losey's lexicon, to bracket anyone with Kazan (who eventually named names), was a form of execution; as late as May 1988, he could not resist writing to inform Schary that he would always be grateful that they had not met for over thirty years. Schary replied with dignity.[5]

In 1948 the FBI noted that a three-year investigation under the character 'Internal Security – R' had been closed and had not substantiated the allegation that Losey was [two lines blanked out, but almost certainly 'a Soviet intelligence agent']. However, the investigation had confirmed that Losey was a member of the Communist Political Association (as the Party still styled itself in the immediate post-war period) and of the CPUSA. In March 1949 Losey was listed among the sponsors of the Cultural and Scientific Conference for World Peace, held at the Waldorf-Astoria, New York (but he did not attend). Also recorded were favourable reviews of *The Lawless* in the *Daily Worker* (23 June 1950) and in the West Coast Communist paper, *Daily People's World* (22 June).

On 13 May 1949 he wrote to Hella Wuolijoki: 'Fascism begins again. Unmistakably and faster. Also tougher and stronger. Little by little the liberals and honest democrats abdicate their prerogatives . . . Just now the prospects are horrifying.' In September the FBI noted that Losey signed an *amicus curiae* brief to the Supreme Court on behalf of John Howard Lawson and Dalton Trumbo who, representing the Hollywood Ten, had lost their first appeal against a one-year prison sentence for contempt of Congress. In 1950 he signed a telegram to Justice Robert Jackson, of the US Supreme Court, protesting its refusal to adjudicate on the case. Many of Losey's colleagues and friends also signed one or both of these appeals, among them: Daniel Mainwaring (who wrote *The Lawless*); John Hubley (the artist who worked on *Galileo* and three of Losey's films); Alfred Lewis Levitt (one of the writers of *The Boy with Green Hair*); Norman Lloyd (involved in the production of *Galileo*); Van Heflin and Evelyn Keyes, leading actors in *The Prowler*; Waldo Salt, who worked on the script of *M*; Losey's friend Edith R. Sommer; John Huston, associate producer of *The Prowler*; John Houseman; and several writers with whom Losey worked, notably Hugo Butler, Ian McLellan Hunter and Sidney Buchman.

Within the Screen Directors' Guild a battle was heating up. In 1950 Joseph L. Mankiewicz was elected President of the SDG, having stated his opposition to a mandatory non-Communist oath for members. Cecil B. De Mille, a leading anti-Red vigilante, took advantage of Mankiewicz's absence in Europe to persuade the SDG's board of directors to pass a bylaw imposing a mandatory oath. A ballot of the entire membership was required to endorse the innovation but every ballot paper was signed and few dared to resist openly. On his return Mankiewicz called for an open meeting. In October De Mille dispatched motorcycle couriers to all SDG members urging them to sign ballots to 'recall' (kick out) the recently elected President.

Mobilizing in Mankiewicz's defence, Fred Zinnemann, John Huston, Losey and a few others decided to call a special meeting to gather the required twenty-five signatures for a special meeting of the full membership. Setting up their headquarters in Chasen's restaurant at seven on a Friday evening (13 October), the rebels harvested the required twenty-five signatures by cornering directors arriving for dinner. Paradoxically Joe Mankiewicz himself now told his supporters: 'If we are going to oppose the loyalty oaths in principle then we have to make clear to everybody that none of us is Communist.' As Losey comments, this was entirely self-defeating: 'And I opposed it very strongly, as did Nick Ray, John Huston and Charles Walters.' But Mankiewicz had brought along the laywer Martin Gang, an expert in 'clearance', who reminded the rebels that as a result of the August ballot no member was by the rule 'in good standing' unless he had signed the oath and that each of the twenty-five signatories must therefore sign it in order to validate their petition for a full meeting of the membership. A hot debate raged. Gang, whose services Losey privately employed, 'leaned over me and said, "Go ahead and sign it. It's not perjury. It's not under oath."' The twenty-five were persuaded to sign what Losey called a 'disgraceful statement': 'I am not a member of the Communist Party, or affiliated with such a party, and I do not believe in, and I am not a member nor do I support any organization that believes in or teaches the overthrow of the United States Government by force or by illegal or unconstitutional methods.' John Huston was number one on the list of signatories, Losey number seventeen, Billy Wilder number twenty-one. To resist a mandatory oath they had signed a 'voluntary' one.[6]

In 1966, when six directors challenged the oath successfully in the New York State Court, Losey wrote from London to the president of the Directors' Guild of America: 'Don't you and your fellow members of the board think it's high time that the ridiculous and shameful loyalty oath is discontinued as a requirement of membership?'

In 1976 Losey was questioned by Michel Ciment, who suggested that what had happened in Russia had been 'much worse' than the trials and tribulations of the Hollywood Ten:

LOSEY: I doubt if it could have been – maybe worse but I doubt if it could have been.
CIMENT: People were killed.
LOSEY: Well, they weren't killed in court . . .
CIMENT: No – well they were killed in camps.
LOSEY: Yes, I know. I know now. They were also killed in a different way ['they' referring to Hollywood's victims].[7]

Ciment also asked whether American Communists had shut their eyes to what was happening in Russia. Losey replied: 'I don't think most of them knew it . . . I think the revelations of Khrushchev came as a profound shock.' Asked about the persecution of artists and intellectuals under Stalin, Losey replied: 'It was a horrible, disgusting period that all of us should be deeply and profoundly ashamed of . . . My attitude was closed . . . I at one point functioned on a Communist Cultural Committee that was grotesque.'[8]

On 18 March 1951, Harry Cohn, President of Columbia, announced a major collaboration with Stanley Kramer, who would make thirty top films over five years: 'This is the most important deal we have ever made,' said Cohn. Kramer, while using the Columbia lot, would remain independent and self-operating; his team of directors would include Carl Foreman, Fred Zinnemann and Joseph Losey.[9]

Almost simultaneously, HUAC opened a new act in the Hollywood inquisition. Hugo Butler was subpoenaed as he began work on the screenplay of Losey's *The Big Night*; Jean Butler recalls that he and Losey were 'driving around from one obscure little California motel to another, up and down the coast and inland . . . I don't remember where Louise was at that time. We were terribly paranoid; afraid to use phones . . .'[10]

The new HUAC hearings on Hollywood, the first since October 1947, began on 21 March 1951. On 23 April Losey was described as on loan-out from Kramer, and due to report for his first directorial assignment for Kramer, *The Four-Poster*, in late summer. In 1967 and again in 1973 Losey recalled that his three-picture deal with Kramer was to direct *High Noon, The Four-Poster* and *The Wild One*. But it is not clear from Losey's correspondence with his lawyers that he was definitely assigned to any Kramer picture. 'Sam Katz [chairman of the Stanley Kramer Company] . . . stalled . . . on the contract . . . Katz had heard that I was going to be involved in the Un-American hearings and he wanted me to sign an anti-Communist oath . . .' The lawyer Martin Gang suggested a formula: if Losey were subpoenaed by HUAC, Kramer could break the three-picture contract (worth about $200,000) for a payment of $10,000.[11]

Losey did sign a contract with Kramer in July 1951, but the document has not been traced. On file is a letter drawn up by Martin Gang, to be sworn by Losey before a Notary Public:

'I am not now, nor have I ever been, a member of the Communist Party. I feel that Communism is evil . . . that every decent American must oppose Communism and Communists in every possible manner . . . I feel that Soviet Russia is our country's greatest enemy . . . [And more in the same vein.] . . . I realize that you are employing me in reliance upon the statements I am making herein. I also realize that should any testimony be given before [HUAC], or any facts developed which contradict the statements herein . . . I give you the right to immediately cancel [the contract].' In that event Losey would keep the $10,000 advance.

There is no evidence that Losey actually signed this affidavit: the 'nor have I ever been' would in his case have been inhibiting. But it seems most unlikely that Kramer–Katz would have offered him the contract without the signed oath. According to Losey, Gang warned him that he was going to be named by two witnesses and advised him to testify in closed session to HUAC; if he did so, Gang believed he could persuade the witnesses to withdraw their testimony. 'I simply said, "I'll have to think it over," and three days later I was on my way to Europe.'[12]

The circumstances surrounding Losey's sudden departure from America were never reported accurately (or consistently) by Joe himself. Having interviewed him twenty-five years later, Victor Navasky reported that 'his last memory was of hiding in a darkened home to avoid service of a subpoena'. Yet in 1955 Losey told his lawyer, Sidney Cohn: 'However no attempt was made to serve [a subpoena on] me as far as I know . . . I did not evade a subpoena . . . No attempt was ever made to notify me of the existence of one . . . in any official or usual way . . .'[13]

In reality he was the beneficiary of a cock-up. On 13 June, HUAC dispatched his subpoena to the US Marshal, Los Angeles, demanding Losey's appearance on 19 July. The authorities, however, went to the Loseys' previous address. Joe's wife of that time, Louise Losey Hyun, takes up the story: 'The FBI had gone to the house we'd lived in before Maxwellton. Fortunately the parents of some friends of mine were living there and got a message to us. Two weeks before that Joe had gotten an offer to come to Rome to direct . . . *A Bottle of Milk* [*Stranger on the Prowl*] . . . He didn't tell me then – only after we got the news from our friends in the other house . . .'[14] This alarming news clearly arrived after 9 July, since a letter of that date to Hella Wuolijoki contains no hint of it, but he did express a more general disillusionment about Hollywood: '. . . the horrors and restrictions are so great that it would be worth much sacrifice to be free of the bondage and the ugliness.'

After the tip-off the Loseys clearly expected a knock at the door at any moment. Joe summoned the producer of *The Big Night*, Philip Waxman, from his bed. Having driven from Beverly Hills to the valley, Waxman was informed that Losey was leaving the country within hours and instructed how to cut the picture. Gavrik, who was just thirteen, remembers Louise taking Joe to the airport after briefing Gavrik that two 'insurance men' might call: he was to tell them that Joe was at the dentist. Losey reached Paris on 17 July. According to the FBI file, by the time the Deputy US Marshal belatedly arrived at 12045 Maxwellton Road – more than five weeks after HUAC issued the subpoena – he was informed by the gardener that both Losey and Louise had left town.[15]

On 29 August, Losey was staying at the Hotel Nettuno, Pisa, and was 'about to begin shooting'. His arrival in Italy was greeted by the Communist Party newspaper, *L'Unità*, with a two-page centre-spread. Representative Donald Jackson, of HUAC, announced the names of those still physically unserved with subpoenas to appear in Los Angeles on 17 September: Losey was among them. Martin Gang reported to him on 18 September: 'The inevitable happened today. You were named as a *former* member of the Communist Party. Fortunately, the witness who named you was a client of mine.' Leo Townsend, who described himself as a member from 1943 to 1948, and who named thirty-seven people as Communists, identified Losey as 'a member of the Screen Writers Branch of the CPA'. However, as a favour to Martin Gang, he added that, as far as he knew, Joseph and Louise Losey were no longer members of the CPUSA. On 20 September, Gang urged Losey to testify in private to HUAC – what moral problem could there be in naming those already named? Why martyr yourself for a cause you no longer believe in?[16]

Losey would not do it. Replying to Gang, he assumed Kramer and Katz would now abrogate the contract (an indication, though not conclusive, that he had indeed signed it). If they were later to change their mind, he promised to regard the retainable money already paid to him as an advance against services. Whether or not Losey had signed the 'I-hate-Communism' affidavit drafted by Gang – evidently it was required of him – it is clear that his contract gave Kramer–Katz an unconditional 'out' if Losey were named before HUAC.

On 24 September, Gang reported that the Stanley Kramer Company had indeed cancelled the agreement dated 1 July. Katz and Kramer would, however, renew the contract on the same terms 'if you cooperate with the Committee ...' Carl Foreman, a stockholding partner in the Kramer Company, was forced out with a large financial settlement (probably $250,000, according to his lawyer, Sidney Cohn) for his stock after he took the Fifth Amendment in qualified form in September. Foreman and Kramer were old friends from army days, yet according to Cohn, Kramer had Foreman locked out of his office at Columbia and told them both, 'I'll wreck this company before I have anything to do with Mr Foreman.' Clearly Losey, a minor figure on the Kramer horizon, stood no chance. Foreman was to become his most important friend after they both exiled themselves to Britain the following year.[17]

On 26 October, the *Los Angeles Times* reported that the right-wing 'Wage Earners' Committee' was picketing five local picture theatres, including Paramount's, currently showing *M*. The FBI took a clipping. On 8 September 1952, Losey wrote to Vappu Tuomioja from London: an attempt was being made to block distribution of his 'Italian film' in the USA. 'Fascists' were now running the show in America. 'It means that I cannot make films anywhere that depend to any degree on the American market ...' Losey later contended that the juvenile lead in *The Big Night*, John Barrymore, Jr, turned up in London during the summer of 1952, with time and money on his hands – and nine years later confessed that his family had pressed him to make the trip on behalf of the FBI. There is no independent corroboration of this story.[18]

On 12 October 1952 Losey returned to New York after fifteen months in Italy, England and Franco's Spain (where he'd made a Shell demonstration film, mainly about the training of police dogs to attack demonstrators, which sickened him). In New York he lodged with his wealthy patron T. Edward Hambleton, at 270 Park Avenue. An important call was on his lawyer, Bill Fitelson (Fitelson & Mayer, 580 Fifth Avenue), 'a well-known Trotskyist whom I always sort of liked ... I saw him at nine-thirty, and at three-thirty he called me and said, "I have your money from Spiegel ... and I'm taking 25 per cent of it."' (Fitelson took $2,500 out of $10,000 due on *The Prowler*.)[19]

He remained in New York 'for about one month', but found no work available in film, theatre, radio, education or advertising. 'At one moment there was a really very strong possibility that I would do *The Crucible* of Arthur Miller, but Kermit Blumgarten . . . got cold feet . . . because of the political situation.' (Losey's discovered correspondence of the time makes no reference to *The Crucible*.) He left for permanent exile in England in the first week of November, the day after the election of Eisenhower as President (as he stressed). In reality, there was another factor: tax. Having been abroad for less than eighteen consecutive months, he was advised by his business manager, Henry Bamberger, not to stay in the USA more than thirty days if he was to qualify for a new tax benefit.[20]

The process of informing continued far into the 1950s. On 12 March 1953, for example, Paul Benedict Radin, described as a radio and television agent, previously in advertising, testified to HUAC that he had been taken to three closed Party meetings in 1946 or 1947 by Joe Losey, who had invited him to join the CP – which Radin never did. The meetings took place at the homes of John Wexley, Leo Townsend and Jay Gonney, and (Radin continued) were attended by Louise Losey, Lester Cole, Richard Collins and Waldo Salt among others. 'I was impressed by the lack of independent thinking,' said Radin, 'by the almost automaton procedure', but (he explained) 'Joe was a friend of mine – I just was not too sure of myself. I didn't want to encourage Joe's displeasure.'[21]

Radin had figured among a list of Losey's chosen, and predominantly left-wing, guests to a trial-run preview of *The Boy with Green Hair* in the summer of 1948. Losey told Victor Navasky: 'I sometimes lent him my house when I was away . . . In those days he was a bachelor.' (As of September 1952, Losey owed Paul Radin $500 according to the list of creditors kept by Losey's business manager; the debt remained unpaid in September 1953, when Radin was running the Beverly Hills office of Ted Ashley, a New York agent.) Interviewed by Navasky in the 1970s, Losey added a mean aside: only later had he discovered that 'this man' had had an affair with Louise. When Louise read this in Navasky's book *Naming Names* (1980), she wrote an angry letter to Losey and Navasky – the offending sentence was altered in subsequent editions. However, Louise has dirt of her own to dish out about Joe and this informer: 'I found out *much* later that Joe had approached *him* "sexually" and been furious when not accepted. Reason enough?'[22]

Losey told Navasky that, 'roughly twelve years ago [*sic*: he is talking of the late 1950s]' this same man had called him from Switzerland to propose a project. They had breakfast at Brown's Hotel. When confronted, the informer protested his complete innocence. Afterwards Losey wrote inviting him to put it in writing that he had never named names. No reply was received. What, however, is to be found on file, among Losey's possible film projects dated 20 December 1960, is: 'Paul Radin – The Briar Patch'. Certainly Carl Foreman was prepared to work with Radin, as was the screenwriter Lester Cole (also named by Radin in 1953): 'no compunction' are the words used by Foreman.[23]

In later years a widespread myth invaded the press in France, Britain and even America – that Losey had been hounded out of career and country by Senator Joseph McCarthy. McCarthy became a synonym for the entire blacklisting era and for the complex purge launched under the Truman administration. The critic Raymond Durgnat regularly described Losey as a victim of 'the McCarthy Committee'. In 1967 a British television programme announced that Losey had been forced into exile 'by another man from Wisconsin, Senator McCarthy'.[24] In the French press '*la chasse aux sorcières* [witch hunt]' and McCarthy were one and the same. On occasion interviewers even added 'McCarthy' to Losey's own words, as with this sentence from an *Observer* interview in 1973: 'When I was blacklisted [by Senator McCarthy] in America over twenty years ago . . .' But McCarthy, as a Senator, could not be a member of HUAC, which had jealously cornered the Hollywood market.

The period of persecution and exile already described also involved the break-up of Losey's marriage to Louise. They had remained childless. He regarded her as without talent and told her so; his behaviour was such that, four decades later, she harbours a relentless animosity which stirs unease in the interviewer about her objectivity. But clearly he put her down: 'I was always the servant'. She still complains that he refused her a credit for some additional dialogue for *The Boy*, but denies, not entirely convincingly, that she resented his failure to cast her in his films. In 1949 or 1950 she suffered a miscarriage – 'I don't remember dates, only husbands' she explains, a very Hollywood lady. The loss of the child threw her into deep despair partly because, so she told friends at the time, and still confirms, a doctor examining the cause of their childlessness had diagnosed a 'low sperm count' in Joe which he attributed to alcohol. Jean Butler recalls: 'I guess at that time she gave up the idea of having a baby. Sometime not long after she began to spend vacation time in Mexico, and her liaison with the young Mexican (son of a police chief, we heard) may have begun about then.' Joe found solace from his brooding wife in the company of other women. He was the money-earner, Louise entirely dependent – and contemptuous of his obsession about screwing money out of her former husband, Jack Moss (whom he was trying to hunt down, via lawyers, well into the 1950s, though evidently for obligations confined to his mind alone). Joe was often away from home: in a small beach house at Malibu he coached the young John Barrymore, Jr, for the leading role in *The Big Night*, assisted by Frances Lardner, wife of Ring Lardner, Jr, at that time imprisoned as one of the Hollywood Ten. Gavrik Losey concluded that his father was having an affair with Frances Lardner. Louise came to the same conclusion, but less philosophically. When Ring Lardner emerged from Danbury prison, Losey offered him clandestine work on the script of *The Big Night* – all in the family.[25]

When Losey left for Europe in July 1951, evading his own subpoena, Gavrik, now thirteen, set off with Louise for Mexico, later joining his mother Lizzie in the Virgin Islands. Later yet, he drove back to California in an old Ford station wagon with Louise and her Mexican friend Arnold de Perez, calling on Joe's mother Ina in El Paso, where she was working as a shop assistant. Ina gave Louise a set of willow-pattern turkey plates and a set of bone china fish plates. What she made of Arnold de Perez we don't know.

In January 1952 Louise set out for Italy in pursuit of Joe, intending to take Gavrik along, but Lizzie intervened to forbid it. Gavrik was sent to school in New York. On arrival in Rome Louise was met by George Tabori with the news that Joe was in the south of France with the actress Joan Lorring. After Joe's return to the apartment at Via Gregoriana 20, Rome, she found a letter to him from Lardner congratulating him on having found true love at last (in the form of Lorring). Louise's visit was clearly a disaster; she left Italy in March 1952, intent on obtaining a divorce.[26] Evidently Lorring returned to America still in love with Joe, whose friend Ned Calmer carried her gifts from New York to Rome. On 20 June 1952, Losey mentioned her to Hugo Butler as keen to play in a film of a Butler script which he wanted to direct.

July 1952 he spent in Paris, staying with the Calefs in the rue des Martyrs, then headed for London. As already described, in October he was in New York, a cat on a hot tin roof, avoiding California where nothing remained except an unserved subpoena, the blacklist, a broken marriage, a vacated home. He telephoned Louise to insist that it was all over but did not tell her about the Sam Spiegel money. 'I am really quite worried about her,' reported his loyal business manager, Henry J. Bamberger, who arranged the purchase of a 1938 Cadillac for her for $250 (Losey's Jaguar having been sold to pay off debts). According to Louise's present recall, not only the Jaguar but furniture as well had callously been removed from the house in her absence, on Joe's orders, to be sold off. Bamberger denies it; indeed, on 12 December 1952, he informed Losey that Louise herself had sold off major items such as chairs and couches.*

*To save warehouse rental, Losey's papers were stored in Bamberger's garage. His more valuable books were rescued by Robert Presnell, the screenwriter. Fifteen years later Presnell reminded a puzzled Losey that, 'I didn't remove your whole library, bear in mind, but only a boxful. I wish now that I had – for those were troubled days and I knew those lovely books would somehow become lost. I think you have these only because I happened in that day.' They included: 11 volumes of Beaumont and Fletcher; 19 of the Harvard edition of Shakespeare; 3 of Marlowe; 3 of Dante; 5 of Samuel Pepys; 5 of R.L. Stevenson; 4 of Washington Irving; 5 of Poe; 4 of Fielding; 6 of Schiller; 3 of Molière; 9 of George Eliot; 5 of Goethe; plus Sheridan, Thackeray and a family Bible.

Losey spent Christmas 1952 in Rome, arriving in London on 4 January. He had given Louise $2,975 during an entire year of absence. In January she had a kidney operation. Two days after reaching London Losey set out his financial proposals for a divorce and complained about Louise's pusillanimity in obtaining one. 'Louise and I have managed quite well on what virtually was and is an illegal marriage' (a mail order Mexican divorce before the California divorce from Lizzie was final). A divorce decree was granted to Louise by the First Civil Court, Ciudad Juárez, State of Chihuahua, Mexico, on 14 March. This does not contrast boldly with the venue of Joe's divorce from Lizzie at the Court of First Instance, Provost District, Chihuahua, Mexico. In January 1954 the FBI noted that Louise Farnum Moss Losey was now Louise Farnum Moss Losey Perez – the lady who remembers her life by husbands. She is now Luisa Stuart Hyun.

13

An Exile Without a Name

In January 1953 Losey arrived back in London at the lowest ebb in his life – in England he was nobody. 'I was petrified. And I had physical attacks. I thought that I was going to die. I thought that I had a heart problem. I used to have to leave the theatre because I was suffocating. I had to sit down in the middle of the London traffic on the kerb because I couldn't stand up. I couldn't breathe . . .' A fine shard of our sympathy may be forfeit when he adds: 'I had no family: my wife had left me . . .' He was free to wheel and deal in a fairly liberal, tolerant society where the blacklist was abhorred. His fellow-exile, Carl Foreman, who had brought to London a handsome settlement for his Kramer Company stock, and who was now enjoying gambling (gin-rummy) and girlfriends galore, was 'extraordinarily warm and generous . . .' Walking in Hyde Park on a Sunday, they discussed projects. Adept with 'under the table' money, Foreman lost no time in introducing him to the American-born Danziger brothers, Edward and Harry, who set him to work, uncredited, on a television series for '$100 a week under the table so I didn't have to pay tax on it . . .' Pounds is more likely.[1]

In addition, Foreman offered him work on the *Aggie* television series; shooting was limited to half a day with no opportunity for re-takes. While at work he was struck down with flu. A letter to Hella Wuolijoki, dated 5 February, confirms that he found remunerative work within two or three weeks of arrival – but: 'England has been about as bad as England can be, climatically and in other ways monstrously depressing . . . I am proscribed in so many places and have so many dependants.'

He tied together the threads of his professional life with characteristic energy; by April he was represented by the Christopher Mann agency, 140 Park Lane, and by MCA. Edgar Anstey, a documentary producer with the Crown Film Unit, the Post Office, and British Transport Films, commissioned him to write £75 scripts for BTF. He was also befriended by the film director Anthony Asquith, 'a most enchanting and kind man, who went to extraordinary lengths to assist me when I first came to England'.[2]

Confronted by the work permit problem, Losey turned to Sir Leslie Plummer, a Conservative MP strongly attached to individual freedom – it was Plummer who arranged his official permit. In 1976 Losey told the journalist Sam White that there had been great pressure to expel him from the UK as an alleged Communist – he had been saved only by the direct intervention of Sir Leslie Plummer and another Conservative MP, Sir Walter Monckton. Losey's copious correspondence with two American lawyers, Sidney Cohn and Edward Mosk, offers not a glimmer of evidence that he was facing deportation. By his own recall he was obliged to report to Immigration every week, 'and I never got a permit for more than thirty days'. This is doubtful. Losey's Certificate of Registration B 22181 (Aliens Order 1953) shows that on first arrival he stated his profession as 'Story Consultant and Script Writer' and was granted a four-month stay. The registration book was thereafter stamped by the Metropolitan Police, Piccadilly Place, at irregular intervals.[3]

The 'un-Americans' formed their own circles of friendship. Joe was welcome in the home of the blacklisted writers Donald Ogden Stewart and Ella Winter, who (Leon Clore remembers) also lent him money. Another favourite hearth was that of the socialist playwright Benn Levy and his American-born wife, the actress Constance Cummings. A brilliant figure on the exiles' landscape was Olwen Vaughan, who ran Le Petit Club Français in St James's Street ('the French Club'), a scene of hospitality for Free French forces during the war and now a watering place for the show-biz left. A lady in senior middle-age, Olwen Vaughan was generous with credit, allowing Joe to chalk up his meals. When she died in 1973 Losey sent a tribute to *The Times*, recalling how single-handed she managed everything – cooking, serving, accounts – in between attending film festivals and helping her friends. He didn't mention their brief affair.

His natural sexual charm was reinforced by the romantic aura of an exile stripped of his family. It was at this time that Joe began an affair with the *jolie-laide* actress Jill Bennett. He took her dancing, impressed her, amused her, cooked for her and accepted her generous gifts of clothes. According to Bennett he proposed marriage but she was not inclined, as yet, to commit herself. Visiting his father in England during the Coronation summer of 1953, Gavrik was entrusted by Joe to Alexander Mackendrick's wife, Hilary. After two weeks of rain Joe decided they should visit the Continent. Arriving at Ventimiglia, Joe was pulled off the train as *persona non grata* in Italy and both father and son were deposited over the border. (Losey later misplaced this episode when reminiscing to Michel Ciment, cross-fertilizing it with his first journey from Italy to England in July 1952, imagining that he'd travelled from Italy 'to meet my son in Menton'.)[4]

But a foreigner resided in Britain on sufferance – every employment had to be negotiated. On 12 January 1954 his residence permit was extended until only 31 March: 'Allowed by Secretary of State to finish Sleeping Tiger.' When he re-entered Britain on 11 April, entry was granted only until 31 May. Living at 60 Queen Anne Street, he made a 'horrible' (he said) television series for the BBC, in collaboration with Richard Lester. Money was a constant preoccupation. In 1954 he had an operation on his hand, at a cost of about £60. 'National Health is slow and not particularly useful for my – probably – psycho-somatic ills.'[5]

Sidney Cole and Ralph Bond, both left-wing members of the film industry, talked to Losey in 1953 about the difficulties of entering the closed-shop Association of Cinematographic Technicians. The ACT records show that Losey was not admitted to the union (membership No. 15670) until 16 October 1954 (entrance fee £2 2s. od), which means that he must have directed *The Sleeping Tiger* under a waiver from the general secretary, George Elvin, whose good will he cultivated. In January 1955 Losey reported to his New York lawyer that he was one of only two 'aliens' granted membership.

Wealthy benefactresses of the Left were a feature of the blacklist era. According to Losey's assistant, Pieter Rogers, the American producer Ethel Linda Reiner put money into *The Night of the Ball* while the actress Phyllis Kirk, wife of a Hollywood mogul, arrived with wads of dollars. Losey's closest relationship was with Hannah Weinstein, a colleague from the New York theatre of the late 1930s, who bought Nettlefold Studios, set up Fountain Films, and produced the successful TV series *The Adventures of Robin Hood* largely on the basis of scripts by blacklisted writers. A shrewd businesswoman of the Left, she called Joe 'sweetie' and sent him letters of comfort before herself settling in London. 'If you suffer from any fleeting moments of nostalgia, don't take them seriously,' she wrote. Hollywood was hard and brutal. Losey consulted Weinstein before signing his non-Communist passport affidavit in 1956.[6]

Weinstein's British associate producer, Sidney Cole, is sure that Losey made no contribution to the *Robin Hood* series, but there are clear indications of an affair between Losey and Weinstein. 'I never regarded Hannah really as a friend, in spite of a brief period of intimacy, which, conceivably, gave her some cause for grievance,' he wrote to James Proctor in 1975. Ruth Lipton explains the 'grievance': the affair ended disastrously 'when Hannah became pregnant'. (By 1958 she was Hannah Fisher. She died in 1984, only weeks before Joe. The previous year they'd met for lunch in Hollywood, their first encounter in years. 'We had our froideurs and disagreements,' he told a friend, 'particularly in the period of her third marriage . . .')[7]

In March 1955 Losey was granted a new labour permit, valid until 1 April 1956. Stamping of his Alien Registration Certificate was intensive early in

1955 (1 February, 3 March, 22 March), followed by a gap until 3 October. Pieter Rogers wrote many of Losey's applications for renewal. On 24 May Losey complained to Hugo Butler: 'I'm so bloody sick of fog and gloom and cold and rain and muck.' He sent Bamberger the bill for a leg operation, hoping for cover from his Blue Cross insurance – and did Blue Cross cover 'psychiatric treatment, i.e. psycho-analysis or psycho-therapy'? The answer was no. On Dr Barrie Cooper's recommendation, Losey had begun regular psychotherapy sessions with Dr Joe Shorvon.

Despite two theatrical setbacks, *The Wooden Dish* and *The Night of the Ball*, he increasingly lobbied London theatre impresarios, particularly Hugh Beaumont, of H.M. Tennant. No luck. Paul Rotha recommended Losey to Peter Cotes, head of the drama department of Associated Rediffusion TV, but ARTV was too dependent on the American market to take the risk with a blacklistee. In November Losey reported that he expected to begin 'a 3-reel documentary for the London Stock Exchange. Salary would be £500.' A notarized statement from Bamberger of his American financial obligations included outstanding medical bills for some $2,000; insurance and taxes brought the total to $4,218. In May 1956 he moved to 8 Ennismore Street, Carl Foreman's residence, for three months.

Paranoid or not, Losey continued to suffer ongoing persecution by the American Inquisition: passport problems, blacklisting, oaths of loyalty. His last passport (331431) issued in America was dated 27 July 1950. In applying, he claimed that his previous passport, issued in December 1934, had got lost but, if memory served, he'd spent eight months in Switzerland, Finland and Germany. After his new passport was issued in 1950 the FBI checked the records and discovered that he had forgotten to mention Russia. Strangely, he was able to renew his passport in Venice in July 1952, without difficulty. We can guess that he chose Venice, rather than Rome, just as he chose Glasgow rather than London for further renewal.

On 23 August 1954 he obtained a new passport valid for two years. But the FBI took note and the State Department requested the Glasgow Consulate to endeavour to withdraw his passport. The American Embassy in London moved swiftly; on 31 December 1954, Walter M. Walsh, the American Consul, requested him to hand over his passport. Losey did visit the Consul, but without the document. Reading from State Department instructions, Walsh informed him that it was alleged by confidential informants that he was a Communist and a sponsor of the Cultural and Scientific Congress for Peace, held at the Waldorf in New York in 1949. Losey was required to make an affidavit under Section 51135, Part 51, and to surrender his passport forthwith. He was warned that it was illegal to use a passport after a formal request for its surrender.

Writing to Sidney Cohn, Losey questioned the legality of these threats and listed his influential friends: American Ambassador Winthrop Aldrich had attended the opening of *The Wooden Dish*, and only the previous week had come to *The Night of the Ball*. 'I have given all the facts to several of my MP friends who are aching to make an unwholly [*sic*] row . . .' In February 1955 Cohn – who with his colleague Leonard Boudin handled some of the crucial Supreme Court passport cases from 1955 to 1958 – advised Losey that he did not believe the State Department had acted legally in revoking his passport in view of the *Baur* case and lack of due process. But Cohn's advice was 'for God's sake don't travel outside of England until the situation is cleared up'. (Losey did travel to France on more than one occasion after receiving Cohn's warning. Dorothy Bromiley remembers his constant fear of not being re-admitted to the UK.) Cohn advised that going to court would cost $7,500, payable whenever Losey could find the means, but $1,000 would be required in advance to cover immediate costs. Losey decided not to go to court.

Not only his freedom to travel, but his UK residence permit – the precondition of a work permit – depended on a valid American passport. His main contact was H.W. Clark, of the Foreign Labour Division of the Ministry of Labour, Ebury Bridge House. Clark was helpful and so was Joe; in May 1956, he sent him two tickets for the trade show of *The Intimate Stranger*, explaining that he would need a permit to shoot *Time Without Pity*, beginning 25 June; editing would keep him at work until December. Max Bitel & Co., solicitors for Harlequin Films, heard from the Home Office that Losey's residence permit would be extended until December only if his passport, due to expire on 23 August, was revalidated. 'I am, gentlemen, your obedient servant . . .' the Home Office man signed off.[8]

This was the crunch: grovel or perish. According to the FBI File, on 25 April Cohn accompanied Losey to the London embassy. In May Losey signed a legal deposition or affidavit, presumably drafted by Cohn, stating that he had joined the CPUSA in 1946 and to the best of his recollection had left in 1947 or 1948 – 'I am not quite sure as to the exact date.' He added: 'I was baptized and confirmed in the High Anglican Church.' This evidently did the trick. On 28 May the State Department authorized the London Embassy to renew the passport. Losey could travel without limitation, apart from the routinely designated eight Communist countries.

He could stay (and work) in Britain. As if to punish him, Registration stamps become threateningly intensive: 15 June, 16 June (the day he was married), 9 September, 24 October. There followed a gap until 28 July 1957, then no entry until 14 January 1959. The final dated stamp is 4 April 1959 and the last entry is 'Aliens Order 1960. Exempt from registration with the police but should retain the certificate.' But the long struggle had deeply depressed him: 'I must confess,' he told Hugo Butler in December 1956, 'that I am often desperately homesick and that I am tired to death fighting forever against losing odds.' Yet when his friend, the writer Howard Koch, also depressed by exile, decided to return to America, Losey called it 'a grave mistake, personally, tactically and politically'.[9]

In 1958 the State Department again flashed its teeth by demanding the dates of his Party membership. By June 1962 the climate had changed and he was granted a passport in London 'within a matter of minutes' – but not until June 1968 was Cohn able to inform him that the State Department would no longer ask 'Are you now or were you ever .. ?'[10]

Joe's sister Mary, meanwhile, was suffering a political ordeal of her own closely, although not exclusively, linked to her brother's politics. In May 1954, while working as a film and television officer for the public information department of the World Health Organization in Geneva, she was confronted by a long, written 'Interrogatory' from the US International Organizations Employees Loyalty Review Board.

The first substantive question focused on Joseph Losey, indicating that he and 'his wife Louise Losey have been members of the Communist Party . . .' Mary was required to state the full extent of her knowledge about her brother's 'Communist sympathies, activities, associates, affiliations or member-ships . . .' Also: 'State whether or not your brother or Louise Losey ever attempted to induce you to attend meetings . . . or sought in any way to influence your thinking with regard to Communism or with regard to the Communist position on issues of the day.' She responded: 'I do not know whether my brother or his former wife were members of the communist party . . . Neither my brother nor his wife ever attempted to induce me to attend meetings . . . nor did they furnish me with communist literature or seek to influence my thinking with regard to communism.'

The Interrogatory then pursued her about many other questions affecting her own life, work and associations since the late 1930s, but not connected with Joe. A United States Civil Service Commission visited Europe to investigate Americans employed in United Nations agencies. She was not legally required to attend in person but decided to travel to Paris for a 'hearing' held at the American Embassy on 19 July. It lasted three hours and was tape-recorded. The panel included Grady Gore, a real estate operator from Potomac, New York; Edmund L. Tink, Superintendent of Schools, Kearny, NJ; and Henry S. Waldman, a lawyer from Elizabeth, NY. On the advice of her lawyer, Mary Losey Mapes repeatedly said she did not know whether her brother Joseph was or ever had been a Communist. The Board raised its eyebrows: could she honestly say that? 'Yes.' But was her answer not based on her lawyer's advice? 'No.'

'I went home and took a sleeping pill and slept for two days. I never got over it.' Not until June 1955 did she receive a 'clearance' from the Loyalty

Board and feel free to visit America without fear of losing her passport. And Joe's attitude to all this? 'Joe didn't even ask me about it. When I saw him he said, "Yeah, tough."' Although she admired Joe's talent, Mary had been since their earliest years a sceptical observer of what she regarded as his pretensions and affectations, his egoism and his sometimes casual cruelty. 'Joe put people down all his life.'

To his first wife he remained joined by more than their common parentage of Gavrik. 'Dearest Lizzie dear,' he wrote on 15 January 1955, 'I have been terribly upset by your letters and the waste and even sometimes apparent destruction of you. What a bloody mess we have made of our lives chiefly because of lack of realism and a combination of romanticism and perfectionism! About one thing I feel we have not done badly, either jointly or separately, and that's Gavrik. I want him here, I think it would be good for him here. I will not desert him, he will not go hungry. I will not be "running off". But security and certainty I *cannot* assure him of or guarantee.' At some stage Lizzie had entered a sanatorium. Joe later told Ciment: '. . . she wound up a hopeless alcoholic. At least once had herself committed to an institution . . .'

He was now about to marry for the third time. Was his divorce from Louise valid in Britain? On 15 June 1956, Losey heard from Breed, Abbott & Morgan, American counsellors at law in London, whom his solicitor, Emmanuel Wax, had engaged to validate his Mexican divorce. 'I understand that your former wife had resided for at least one year and possibly more in Mexico City' – and had therefore been qualified to bring a divorce action there which would be recognized by the State of California and throughout the USA. Losey's response is not found but must have been affirmative; in cold fact Louise had not resided for a year in Mexico; earlier correspondence with Edward Mosk makes it clear that Joe had pressed for the divorce-on-demand obtainable at Ciudad Juárez, which did not require the six months' residence demanded by Mexico City but which had proven its value when Joe and Lizzie were divorced. On 16 June he married Dorothy Bromiley in Kensington Registry Office.

Joe now summoned his son to London – though Gavrik had been accepted by Harvard and Dartmouth – explaining that his sterling funds were blocked and he could not finance a university education outside Britain. Joe asked his sister to persuade Gavrik to join him in England; Gavrik, she says, was extremely bitter: 'I don't owe him anything, et cetera.' The local draft board at Santa Monica granted him six months' deferment and permission to live abroad.

Joe wrote to Louise de Perez (who had restyled herself Luisa), at 2723 Laurel Canyon, Hollywood: Gavrik would be sailing on Monday and Joe wanted to know what had happened to: '(1) the family silver (i.e. from Mother), and (2) the family china, such as the fish set and the china from Aunt Fanny and the various bits of Fanny silver, such as the fruit bowl?' He also wondered 'What other things of mine – if any – are surviving the disaster . . . my dress braces . . . my father's revolver? . . . If you're not using all of the copper pots, I'd appreciate having some of them. We are starting a household from scratch and it is very difficult. Love . . .' The Loseys settled into 2 Montpelier Square in Knightsbridge. In August 1958 he negotiated a new three-year lease with his landlady, Luisa Parry, wife of Gordon Parry and mother of Natasha (who was married to Peter Brook).

14

British Studios and Stages

The Sleeping Tiger

Losey's first British film was made under deep cover. No American distributor would touch the credited work of blacklistees but Carl Foreman displayed genius as a fixer; adopting the pseudonym Derek Frye, he set to work with Losey on a screenplay by Harold Buchman (also blacklisted) from a novel by Maurice Moiseiwitsch, which Losey later described as 'a lousy cheap story, as bad as James Hadley Chase's *Eve*, worse maybe. A sort of bedtime reading for senile stags.' According to Losey the screenplay fee was £750, which he presumably split with Foreman. He later put the budget at 'around $300,000' and his own director's fee at £1,000 'plus a small interest'. The production company was Nat Cohen's Insignia, with Anglo-Amalgamated the distributors. Backing came from Dorast Pictures, otherwise known as Sidney Cohn, New York lawyer to both Foreman and Losey; the investment was probably $27,000. Losey received at least $1,500 from Dorast, 'under the table'. 'The English market wanted to employ me because . . . I knew my job . . . they got me very cheaply . . . they thought I would make pictures for the American market . . . they thought I would attract American stars . . .'[1]

What got the film off the ground was the participation of Dirk Bogarde. 'Certainly my career, and even the existence of a career at all, was made possible by Dirk's acceptance . . .' Bogarde recalls: 'Olive Dodds, head of artists' contracts at Rank, wanted to keep me busy. Every July Rank decided whether to take up their option on me.' Losey was shown his most recent film, *The Hunted*, which Bogarde regarded as his best and least Odeon-idol performance – and loved it. Dodds brought Bogarde to Pinewood to see *The Prowler*. 'It was freezing in the small theatre where the film was projected. Joe preferred to walk about outside in the slush like an expectant father . . .' After twenty minutes Olive Dodds called him in. '*The Prowler* was extraordinary, brilliant. I hadn't seen such genius.'[2]

When Alexis Smith arrived in England she was unaware that the director was a blacklistee. She and Losey were dining in the Ship hotel at Shepperton on 'the very first night' when in walked Ginger Rogers (Losey recalled – in fact it was her mother Lela, or both), 'one of the worst, red-baiting, terrifying reactionaries in Hollywood. And she knew me, and she knew Alexis Smith.' By Losey's account, he and Smith fled through the kitchen, primarily for the sake of her career. Bogarde recalls smuggling Losey from the studio to another hotel in Windsor: 'We got him two rooms . . . And when we proudly took him there, hidden in the back of my car under a rug (can you believe!) he complained bitterly about the noise from a gentle weir in the Thames some distance across the gardens. It was then, I suppose, that I knew I'd got a problem for a friend.'[3]

Harry Waxman was in charge of photography, while Reginald Mills came in to edit the first of many films for Losey. The music was by Malcolm Arnold. Losey accepted the pseudonym generously offered by Victor Hanbury, who held distant credits as a director from 1934 and 1936. The official art director was John Stoll but Losey's real, semi-clandestine collaboration was with Richard Macdonald, who lacked the requisite union card for official accreditation. After the day's shooting he and Losey would talk, drink and produce sketches through the night until Macdonald drove off in his battered station wagon at '85 miles per hour across the bridge . . .' and back to London. The film was shot at Nettlefold Studios, near Windsor, 'the oldest in England', during December 1953 and January 1954. Bogarde recalls weeks of discomfort making this 'hackneyed little thriller' in a run-down little studio where they had to fire a gun to frighten the sparrows from the sound stage before every take. But he shared the admiration for the director expressed by Alexis Smith: 'What's astonishing with Losey is that one has the chance of playing one's part at a very high emotional level, without ever having to limit the intensity . . . like Raoul Walsh, he generates enormous energy on the set, and it's this energy which really makes the film . . .'[4]

The film was first shown at the Odeon, Marble Arch, on 24 June 1954. The few reviews which appeared were unimpressed: in the *Spectator* (25 June), Virginia Graham commented that 'neither Victor Hanbury's brooding though intelligent directing, nor Malcolm Arnold's nerve-scraping music helps to dispel the illusion that under a cloak of psychoanalytical claptrap everybody is being very silly.' Appraising *The Sleeping Tiger* twelve years later, Raymond Durgnat commented: 'Losey's treatment was substantially at odds with British critical fashions of the mid-50s, which demanded he tone down emotional excesses . . . Instead he concentrates on . . . a structure of passionate actions and reactions . . . The film is a series of *"temps-forts"* . . .' But Losey himself admitted that the explicit psychoanalytical explanations, on which the story hung its own improbabilities, were a failure.[5]

The film was distributed by Astor Pictures in the United States in October 1954, opening at the Palace, New York. The identity of the director and screenwriters went unnoticed. So, unfortunately, did the film.

The Wooden Dish and *The Night of the Ball*

From Brecht's *Galileo* in 1947 (see Part V), until his production of *Waiting for Lefty* at Dartmouth twenty-eight years later, Losey directed only two stage plays: *The Wooden Dish*, by Edmund Morris, and *The Night of the Ball*, by Michael Burn.

A study of bitter and demented old age in rural Texas, *The Wooden Dish* starred Wilfred Lawson and Joan Miller, with Losey's future wife Dorothy Bromiley in a supporting role. This was essentially Lawson's show; for the first time in his career his name appeared above the title. E.P. Clift was the impresario. The stage designs were by Reece Pemberton. According to Peter Cotes, Losey had 'haunted' his production of Strindberg's *The Father*, starring Lawson, during its four-week run in 1952 at the Little Arts Theatre. The critic T.C. Worsley described Lawson's performance as one 'of such controlled power and scarifying truth that no one who saw it is likely to forget it'. Losey and Donald Ogden Stewart wrote several drafts of a screenplay of *The Father* in 1953, but apparently no one could be found to insure Lawson. Losey called him 'my idea of a really great actor', but in private correspond- ence he also described Lawson as 'a psychotic alcoholic for at least the last twenty years of his life . . .'

In *The Wooden Dish*, Clara Dennison (Joan Miller) is kicking against a flagging, passionless marriage – a handsome, sexy woman in her late thirties, anxious to harvest the pleasures of youth while there's still time. The scapegoat for her discontent is Pop, her semi-senile father-in-law (Lawson), whom she longs to get rid of. Pop firmly believes that you are quietly disposed of in old people's homes.

The play opened its provincial run in Glasgow and Newcastle, then moved to the Phoenix, in the Charing Cross Road, on 27 July 1954. Cotes remembers Lawson's impish, sadistic streak, his jokes about 'lousy Losey' during rehears- als. When a party was held on the first night at the Turk's Head, Newcastle, Lawson started tossing rolls of bread at Losey. At the Phoenix opening he went one better, sending Losey a piece of rotten fish wrapped in blue tissue paper, with the message, 'I forgive you, Joseph.' Dorothy Bromiley recalls that 'Joe was shattered, speechless. He'd been very patient with Lawson.'[6]

The anonymous critic of *The Times* (28 July) praised Lawson's performance as the grandpa forever scheming to get hold of the tobacco and beer denied to him by the doctor, and so prone to break crockery that he had to be fed out of a wooden dish – 'but the play unfortunately keeps him pottering about the fringes of a laboriously worked out and slow-moving domestic drama'. No mention of Losey's direction. The reviews were indifferent and the production was soon in trouble. On 7 August, barely ten days into the run, *The Times* reported that the cast had been forced to choose between closure or working for Equity's minimum wage. The production had already lost £6,000 since it opened in the provinces. However, on 16 August *The Times* reported that the management had put the cast back on full salary because of improved audiences. The fatal blow came when Wilfred Lawson accepted a proposal from H.M. Tennant to play in *Bell, Book and Candle*, by John van Druten, in the same theatre. *The Wooden Dish* closed on 11 September, after seven weeks. 'Lawson shafted us,' comments Dorothy Bromiley.

Losey's production of *The Night of the Ball*, by Michael Burn, began its provincial tour in November 1954. Burn – playwright, novelist and poet – had been a prisoner of war at Colditz, and later *The Times*'s correspondent in Central Europe. His father, Sir Clive Burn, had been helpful in obtaining Losey's work permit when he first arrived in London. Pieter Rogers recalls that the play got worse as it was endlessly re-written during rehearsal. The impresario Stephen Mitchell launched *The Night of the Ball* at the New Theatre, London, on 12 January 1955, with a glittering cast: Wendy Hiller, Gladys Cooper, Jill Bennett, Tony Britten and Robert Harris. Losey's salary may be inferred from a letter to Hugo Butler: 'The theatre pays a top fee of £300 for what proves to be three months' work.'[7] The reviews were poor; according to Eric Keown, of *Punch*, 'The only performance that stirs one faintly comes from Jill Bennett, as the unfledged niece, still groping among new emotions.' (Losey had recently parted from her in favour of Dorothy Bromiley.) The play was withdrawn on 2 April on the pretext that Miss Cooper had engagements to fulfil in America, a disaster for Losey – 'traumatic' is Bromiley's word – and for Stephen Mitchell, who turned a deaf ear to all of Losey's subsequent approaches.

The Intimate Stranger (Finger of Guilt)

Losey's second British feature film was again under a pseudonym; on this occasion he borrowed his own Christian names, Joseph Walton – a defiant gesture which the American distributors prudently excised. Losey had signed a contract with Alec Snowden of Anglo-Guild, a Nat Cohen subsidiary, to make two low-budget feature films a year at a salary of £2,000 plus 2.5 per cent of profits. (This was one of the rare occasions when Losey eventually received his percentage of the gross; by 1976 he reckoned his fee for *The Intimate Stranger* had been quadrupled – 'extraordinary'.) The film was shot at Merton Park and Elstree Studios in November–December 1955 in the space of only twelve days on a budget of approximately $125,000. Anglo-Amalgamated were again the distributors.

The script (originally *Pay the Piper*) was the work of the blacklisted American writer Howard Koch, writing under the name of Peter Howard. Losey and Koch were sharing a London house. Despite the low budget and a blacklisted director, the producers were able to import two American actors, Richard Basehart, a major box-office star, and Mary Murphy. Indeed Constance Cummings, although settled in England, was also American by birth. The British cast was headed by Roger Livesey, Mervyn Johns and Faith Brook (who had recently played in a revival of Sartre's *Huis-clos*).

Wilfred Arnold was credited as art director, but Richard Macdonald was again Losey's real collaborator on set design. The main settings were the home and office of a film producer, graced by loaned paintings, Henry Moore sculptures and Lurçat tapestries. The photography was by Gerry Gibbs, the editor was Geoff Muller and the music was by Trevor Duncan – none of whom worked with Losey before or after.[8]

Howard Koch's script is based on a nice idea: false identity and masquerade at work within a profession (acting, films) which regularly turns the unreal into the real, illusion into reality. The story also touches on Losey's own predicament; the geography of political ostracism is translated into sexual terms, and the passion of a film man, Reggie, for the world of films is powerfully conveyed. The compromising letters from the woman – the Informer – also echo the political persecution that was to haunt Losey until 1960: 'The past won't leave me alone,' says Reggie. 'It's been coming at me from all sides.' His predicament parallels Losey's: 'Out of pictures I was like a duck out of water, which is why I decided to come to London ... Here I was given ordinary pictures but I had the story tight as a drum ... I've learned to live according to the law of the jungle that we call the picture business.' Despite contrivances such as overheard conversations and banal dialogue – 'I'm going to find that woman if it's the last thing I do' – *The Intimate Stranger* is taut, professionally crafted, 'tight as a drum'.

The film was first shown in London on 25 May, 1956, but it was mainly ignored as a B movie, even by film journals. The BFI's *Monthly Film Bulletin* (July) offered lukewarm praise: 'The plot of this melodrama is ingenious ... a fair degree of tension. The dénouement, however, ends the story on a tame anti-climax. The direction is quite slick ...'

In October 1956, RKO released the film in the USA under the title *Finger of Guilt. Variety* (12 November) praised it as engrossing entertainment, despite a contrived climax: 'a better-than-average English import ...' As regards Losey's identity, *Variety* referred to the producer as if he were also the director: 'Alec C. Snowden, who gives film very good production, mounting, does a smooth job in his direction ...' Clearly the name 'Joseph Walton' had been removed from the American credits. *Film Daily* called it an 'absorbing story of intrigue filmed and directed by Alec C. Snowden'.

Time Without Pity

Time Without Pity offered Losey his first release from soul-corroding anonymity – but not from the fatal formulas of exiled Hollywood. For the third time in nine years his screenwriter was Ben Barzman, the result being a mixed-genre hybrid snared by webs of intrigue. The Hollywood exiles regarded themselves as the true professionals – tough, taut, tense – who commanded the patronage of British producers hungry for the American market. Barzman was paid accordingly: the production budget allocates £9,150 for 'story and script'. Losey's fee, negotiated by Cecil Tennant, was a leap up: £4,500, with a £2,500 deferment which (he noted) 'should pay off in full – or in part – sometime during 1958'. (The pound was then worth ten times its present value.)

The provenance of the script lay in Emlyn Williams's play, *Someone Waiting*, described by Losey as 'a straight thriller which Barzman and I and the producers, most notably Leon Clore, turned on its head' by revealing the identity of the murderer before the titles. On its backside, maybe – the film's hero must still thread his amateurish forensic intelligence through a fog of bafflingly tedious revelation.

Produced by Harlequin (Leon Clore, executive producer), the film was shot during June–August 1956 at British National Studios, Elstree – although Losey's references to the project date back to the spring of 1955. The budget was £100,412, most of it put up by the American brothers controlling Eros, Phil and Sid Hyams, who were prepared to credit Losey and Ben Barzman under their own names. If Losey felt gratitude he took care not to display it; Clore recalls the day when the Hyams brothers dared to visit the set and Losey blew a fuse – 'Who are these people!' – ordering them off and out. Phil and Sid Hyams promptly informed Clore that henceforward they would make no particular effort to promote the picture, and indeed the film opened at the Astoria cinema without a press show or normal promotion.[9]

Losey was now working with a distinguished all-British cast: Michael Redgrave, Leo McKern, Ann Todd, Alec McCowen, Paul Daneman and Renée Houston – indeed this was the only occasion during the 1950s when he did not inject American or European actors into a British scenario. Losey regarded Redgrave as a great actor: 'And his gifts are more than acting: he's an intellectual, a poet, a literary man, an innovator in the theatre. But he's completely destroyed by alcohol.' (Leon Clore comments that telephone calls to Redgrave in America before the production had to be conducted through an alcoholic haze. However, Redgrave was far from 'completely destroyed'.)[10]

Redgrave was paid £15,000, Ann Todd £5,000, and others on a daily rate, ranging from £150 for Peter Cushing to £40 for Lois Maxwell (a friend of Losey's), £35 for Alec McCowen, £30 for Paul Daneman, £25 for Joan Plowright in her first film. Alec McCowen knew little or nothing about Losey, but was delighted to be acting opposite Redgrave, one of Britain's 'big five'. When Losey took him to meet Ann Todd, the young actor was so over-awed by the glamorous star of *The Seventh Veil* that he promptly sat on and broke his glasses.[11]

Paul Daneman retains a clear memory of Redgrave as virtually directing the acting in his own scenes, creating an atmosphere of concentration verging on tension. The young Daneman, who'd recently made his name in *Waiting for Godot*, found Losey sympathetic when the cast gathered to eat 'strange mixtures of food' at Montpelier Square, but on the set Daneman (who was performing on-stage every night) was reproached for eating a bacon sandwich before playing an emotional scene. Daneman assesses Losey's prime skill as recruiting a first-class cast and crew: Frederick Francis, the lighting camera-man, and his operator, Arthur Ibbotson, were 'brilliant'.[12]

Freddie Francis, however, regards Losey as having been far from brilliant. He was surprised – given Losey's Hollywood background – by his nervous indecisiveness on the set: 'Arthur Ibbotson and I had to force Joe to move the camera.' Amused in later years to read about Losey's 'fluid camera move-ments', Francis describes him as 'a bit of an old misery, an unhappy person' who later 'fell in love with the publicity, particularly from France', adopted the grand manner, and 'was turning his back on people from that time on'. (Francis is not listed among Losey's admired cameramen in his conversations with Ciment.) Reece Pemberton, the production designer, picked up one or two David Hockneys at the Royal College of Art to grace the sets. As usual, Losey brought in Richard Macdonald under-cover, to work on drawings laying out the set-ups. The interior designs are bold, striking, heavily symbolic, seducing the eye as the ear labours with the clumsy dialogue and creaking story line.[13]

The film opened in London at the Astoria on 27 March 1957 and was massively ignored by the critics despite a notice in *The Times*. The lack of a press conference, Leon Clore points out, didn't help. One critic commented that Losey had failed to learn from 'the best of neo-realism', tending instead to 'neo-baroque', while another complained of 'exaggerated hysteria'. Both *Punch* and the *New Statesman* gave the film a miss, although the latter's William Whitebait found space in this and succeeding weeks for *Sea-Wife*, *Interpol*, *The Lost Continent*, and *The Smallest Show on Earth*.

The film was released in America in November 1957 by Famous Pictures, opening in New York at the Fifty-Fifth Street Playhouse. Losey's name surfaced for the first time in five years, without political consequence. In the *Herald Tribune* Paul Beckley was friendly, but no comfort was forthcoming from the *New York Times* (23 November), where 'R.W.N.' advised that on leaving the theatre the viewer was still asking 'who knew what about whom and why ... This is not the kind of crime detection the British film makers are known for, or that a tidy mind can abide.'

Eros showed negligible receipts in the USA, less even than in Israel, and Leon Clore remains to this day disbelieving about the account statements. 'No one knows what it earned.' In January 1959, Losey wrote to Clore's production company, Harlequin: 'I've been worrying lately about the fact that you haven't got your money back ...' What about a re-release next year?

In Raymond Durgnat's view, *Time Without Pity* represented 'a soul fight' of the kind that American regional puritanism bequeathed to Ince Westerns. 'The race against guilt, at speed, in the stony arena, is a superb example of tensions *physicalized*.' By contrast, Foster Hirsch concludes that the film 'tries to be too many things at once – a chase thriller, a psychological study of dominance, a statement against capital punishment – without having the force or clarity to treat any of its themes in depth'.[14]

Three years later this film aroused a storm of excitement among cinephiles in France.

The Gypsy and the Gentleman

Losey's status in the industry remained precarious or (as he felt) desperate. In May 1957, after gruelling delays, he finally signed a deal with Rank to cover (he thought) three pictures. He now plunged into the abyss – although for a handful of cinephiles *The Gypsy and the Gentleman* provided the ultimate proof of Losey's cinematic gifts: a riot of dynamic imagery and exquisite framing fashioned out of rock-bottom, literary hocus-pocus (a screenplay by Janet Green from the novel *Darkness I Leave You*, by Nina Warner Hooke.) Losey, who frequently despised the scripts from which he worked, called this one 'immoral, vicious, *déjà vu*, old fashioned and badly constructed'.

He had money to spend – 'close to $1 million' – which meant greater than normal access to location shooting, and a Regency house with Adam ceilings and fireplaces. Losey and Macdonald adopted a style influenced by Thomas Rowlandson's prints, but once again Macdonald could not be credited. 'So I was working every night, with Richard Macdonald, on ideas, on style . . . and then I was taking sketches to this highly resentful art director [Ralph Brinton] in the morning – who, by the way, couldn't even sketch.' Losey loathed Rank's designated producer, Maurice Cowan: 'He was on the set at eight-thirty every morning . . . with his watch in his hand, saying, "Where is everybody? Why haven't we started?" And that was his single contribution, really, to the picture.' Cowan was indeed a producer of 'the old school' – the director was merely the most important hired hand. Losey's most distinctive contribution was to break with sentimentality by making the gypsy beauty, Belle, an undeviating bitch – though marginally redeemed by love for one (bad) man. Remembering Melina Mercouri in *Stella*, he rang Jules Dassin in Paris to trace the tempestuous Greek actress. 'Well, curiously enough, she is right here beside me in bed,' responded Losey's fellow-blacklistee. Rank's boss, John Davis, regarded Mercouri's facial complexion as impossible, but Losey sent her to his own doctor, Barrie Cooper, who cured the blemishes.[15]

The film was shot in Technicolor at Pinewood Studios and on location at Oxhey from June to September 1957. Jack Hildyard was the lighting cameraman. 'He's not particularly my kind of cameraman,' Losey recalled, 'because he uses an immense amount of light and works slowly . . .' Losey's discontents accumulated. As in Hollywood, the studio insisted on controlling the editing and the music. The composer Hans May was signed up without Losey's consent: 'For the first and only time in my life I left the picture before it was finished.' Losey's relationship with John Davis and his producer, Maurice Cowan, deteriorated to breaking point.[16]

The Gypsy opened in London on 2 February 1958. The press ignored it. In *Films and Filming* (March), 'P.G.B.' described the first two reels as 'outstandingly good . . . Colour, character, movement, it all blends into a few moments of wonderful film making.' But the plot became increasingly bogged down in absurdities. It was sad to see Mercouri forced into such antics after 'those noble Greek and French pictures'. In the BFI's *Monthly Film Bulletin* (March), 'J.A.D.C.' expressed surprise that the director of *The Lawless* and of Brecht's *Galileo* should have chosen, as his first film for Rank, 'this florid barnstormer'. (But the choice had not been Losey's.) His loyal champion, Raymond Durgnat, later claimed: 'In retrospect [*Gypsy*] looks less like an *exercise de style* (though it has the visual impact of a good Western) than like a fancy dress draft of *The Servant*.' (But to compare the grotesque Belle to the subtle Barrett, as interlopers, is to give comparative analysis a bad name.) Foster Hirsch more aptly calls the film 'a Regency horse opera conceived on the level of Georgette Heyer on a bad day . . .' As for his leading actress, 'Losey clearly didn't know what to do with Mercouri, who works in an entirely different register from the director's stock company of subtle, modulated British actors . . . frankly gross . . . like a female impersonator . . . a vaudevillian high camp style that is ludicrously inappropriate . . .'[17]

The Gypsy was released in America in August 1958 with a shorter running time; however, the film was not shown in a major New York theatre and no American review is listed. Hirsch notes: 'Apparently withdrawn from circulation.' John Davis, an autocrat of intensely conservative disposition, having forced out Losey's patron, James Archibald, announced that Rank wished to 'settle' Losey's three-picture contract – 'And everybody in England knew that I had, in effect, been fired . . .' Losey despairingly drew up a list of projects, grouped against the names of their potential producers: one with James Lawrie, eight with Leon Clore, four with Nat Cohen and Stuart Levy, four with Julian Wintle, five with Sam Spiegel, nine with Daniel Angel.

Projects collapsed in rapid succession, including a Sydney Box film, *SOS Pacific*, about the evils of the atomic bomb, which was to star Hardy Kruger. 'And I didn't see any way out of the whole morass.' Indefatigably he continued to cultivate producers, including Sir Michael Balcon and (Lord) John Brabourne, of Mersham Productions. In January 1959 he described the British film industry as 'going to Hell on wheels . . . Subjects are generally conventional and unimaginative with many taboos.'[18] Now signed up to direct *Blind Date* by Sydney Box, he begged to direct *Conspiracy of Hearts* – a script which he'd recommended to Box the previous August – as well. But all Box could promise Losey was a percentage of the profits on *Conspiracy of Hearts*. In July Box declined to wine, dine and talk further projects. Swallowing pride and distaste, Losey even approached Rank's boss, John Davis, requesting a meeting 'to correct wrong attitudes and preconceptions. I would like to know whether we can work together again and on what basis ... I have no desire to recriminate or rehash the past.' There is no evidence of a reply.

15

Blind Date (Chance Meeting):
Return of the Blacklist

Losey's evolving signature is etched deep into *Blind Date*, a film distinguished by its design-values, luminous camera-work, ravishing love-affair with light, and by its portrait of the naïve but unshakeable innocence of the socially committed artist. Yet Losey remained fettered by the studio system's attachment to melodramatic formulas and dialogue heard only in cinemas. Enter, again, Barzman!

Two companies joined forces to produce the film: Independent Artists (Julian Wintle and Leslie Parkyn) and Sydney Box Associates. Wintle owned the rights on an Eric Ambler script based on a novel by Leigh Howard. Losey hated the script and turned to Ben Barzman* and another blacklisted writer, Millard Lampell. According to Barzman, Losey was never satisfied. A letter came through the door: 'As you know, I'm an asthmatic and ought to contain my feelings. My mother and my analyst died last week. I have to tell you you're an arrogant bastard. You've always given me the feeling of bestowing a gift when you condescend to work with me. You've treated me with scorn.' While Barzman was reading this the telephone rang – too late, Losey wanted him to tear up the letter without reading it.

The Losey–Barzman–Lampell team transformed the American oil engineer hero of the Ambler screenplay (dated July 1958) into a poor Dutch artist living in London. Bob, with his 'very masculine self-assured' manner and his red sports car, became the naïve, questing Jan who rode on the bus. Whereas Ambler's police inspector is neither here nor there, Barzman's Inspector Morgan is fiercely resistant to the blandishments of the Establishment: a plus. But what carried over into the Barzman/Lampell screenplay (March 1959) were lines like 'Wait a minute' and 'You have an unpleasant mind, Inspector' and 'Let's quit the guessing games. I asked you a simple question. Just give me a simple answer.' Political adjustments brought a limited literary dividend:

BOB: But this is England, isn't it, not Moscow. [Ambler]
JAN: . . . in this country the police can't just walk into a person's house without a good reason. [Barzman]

*Barzman's numerous film credits include *He Who Must Die*, *El Cid*, *The Fall of the Roman Empire*, *The Blue Max*, *Heroes of Telemark*, and *Rififi*. For *Blind Date* Barzman won the British Screenwriters' Award.

The primary calculation for the producers was the casting of Hardy Kruger as Jan. The part of Inspector Morgan had originally been written for Peter O'Toole, but (recalls the associate producer, David Deutsch), 'The matter went to the board of Sydney Box Associates . . . they wanted a star name . . . how about Stanley Baker? Much against Joe's inclinations the part was offered to Stanley who accepted it immediately.' This was the first of Losey's four films with Baker, whom he had known for some years and whom he described as a young man with 'dark, wavy hair and a great deal of arrogance and machismo'. After Virginia McKenna turned down the part of Lady Fenton, Losey and Deutsch visited Paris to sign up Micheline Presle, the irresistible 'older woman' of *Le Diable au corps*. They got on well, though Presle herself became distressed when he demanded more bare flesh in the love scenes than she'd been led to expect. The luminosity of *Blind Date* is a tribute to Losey's director of photography, Christopher Challis, who rates his visual genius alongside that of Carol Reed, Michael Powell and Stanley Donen. 'He was a craftsman to his fingertips.' Every shot was meticulously planned.[1]

The budget was about £140,000. Prior to production Deutsch visited Munich and concluded a deal with a German producer, Luggi Waldtleitner, who put up £40,000. Sydney Box Associates contributed £98,000. When Paramount bought the US distribution rights for £140,000 (or $480,000, according to the *New York Times*), the entire costs of the film were covered at a stroke. Losey's salary was £8,000 plus a £2,000 deferment.

The film was shot at Beaconsfield Studios and on location in London from March to May 1959. Although Losey got on well with Deutsch, he soon loathed the executive producer, Julian Wintle, an ailing haemophiliac often in great pain, with a short temper for Losey's exacting attention to detail – for example, the sitting-room was panelled, then wallpapered, then painted. Losey also had one run-in with Deutsch for presuming to view the unfinished cut when he was out on location. Thirty years later, Deutsch remembers Losey's storm as if it had occurred yesterday.[2]

Blind Date hangs on a murder so physically improbable as to be ridiculous – the elegant, slender Lady Fenton disposes of her able-bodied rival by stifling her! In only one Losey film (*M*) does murder hold its head up as more than a cinema cliché. Losey explained the murder as a metaphor for revenge and jealousy; encouraged by the acclaim of *Cahiers du Cinéma*, he also invoked Brechtian alienation; the *mise-en-scène* of *Blind Date* was 'a method of knowledge, and this method doesn't differ from that of the scientist or philosopher'.

The film was distributed by Rank in the UK. According to Losey, John Davis, who had cancelled his three-picture contract after *The Gypsy*, brought pressure to cut the 'class aspects' of the Police Commissioner's behaviour –

but to no avail. David Deutsch, however, insists that Davis raised no objection. 'There was no pressure at all.'[3] Either way, *Blind Date* opened on 20 August 1959 at the Odeon, Leicester Square. For the first time since his arrival in Britain, Losey's talent was recognized. In the *Spectator* (28 August), Isabel Quigly responded to the film's 'dash and polish', although she found the plot preposterous. She also admired the love scenes which 'clearly reflect the new British effort to face up to sex without blinking, and show an appreciation of touch . . . the quality of skin, for instance, the actual texture, as it were, of physical contact.' On the other hand, Losey had given Hardy Kruger 'far too much self-conscious hippity-hopping to do; a rather oversized Christopher Robin' who could not match Gérard Philipe's performance opposite Micheline Presle in *Le Diable au corps*.

Financial success at last! Earnings in the US were about £120,000, resulting in a profit of some £160,000 after the sale of American rights to Paramount. Then disaster: the film was due to open in New York under the title *Chance Meeting* in February 1960 when the blacklist again descended.

In 1951 *Legion Magazine*, organ of the American Legion, had commissioned from the veteran Hearst Newspapers expert on Commie subversion, J.B. 'Doc' Matthews, an article entitled 'Did the Movies Really Clean House?' They didn't. Matthews raised blood pressure among studio executives by listing the films recently made or released with the collaboration of Communists or their 'collaborators'. Among them was *The Prowler* – Losey and his leading actor, Van Heflin, having signed the *amicus curiae* brief to the Supreme Court on behalf of the Hollywood Ten.[4]

The Big Night was released in New York after publication of Matthews's article, and some picketing of the theatre was reported. In August 1952 the powerful boss of the International Alliance of Theatrical Stage Employees, Roy Brewer, a leading vigilante in the Motion Picture Alliance for the Preservation of American Ideals, called on HUAC to initiate legislation to ban the importation and screening of movies made by American Communists. He singled out Losey's *Stranger on the Prowl* (then called *Encounter*), naming Losey, 'a fugitive from a [HUAC] subpoena', its producer John Weber, a 'long active party functionary', Bernard Vorhaus, and Ben Barzman, 'identified as a party member'. United Artists was duly intimidated, but in November 1953 slipped the film into the World Theatre, New York under the new title and crediting Losey as 'Andrea Forzano'.[5]

Exiled to Britain, Losey was, as already described, recognized by Ginger Rogers's witch-hunting mum, Lela, while shooting *The Sleeping Tiger* – the result being a notice in *The Hollywood Reporter* (25 February 1954) that Losey was directing the film. The FBI duly clipped this report 'at the Seat of Government'. Nevertheless, the film was released in the USA without overt penetration of 'Victor Hanbury'. Convinced that his mail and that of his blacklisted correspondents was intercepted, Losey carefully avoided any allusion to the two films he made pseudonymously.[6]

On 18 January 1956 the blacklist reached out for him again when *The Hollywood Reporter*'s columnist, Mike Connolly, asked: 'Isn't the Joseph Walton who directed Tony Owen's *Pay the Piper* in England for RKO release really the Joe Losey who ducked the [HUAC] subpoena and fled to England?' (The original working title of *The Intimate Stranger* was indeed *Pay the Piper*; clearly Connolly had an informant in England.) But when the film came out as *Finger of Guilt* in America, in October 1956, the name 'Joseph Walton' had been erased from the American prints and credits.

In November 1957 *Time Without Pity* opened in New York under Losey's own name, yet not a single review remarked on it. Was the blacklist dead? By no means: on 20 August 1958 Losey informed the screenwriter John Collier that 'the old bogey has reared its ugly head again. John Woolf says he cannot get American distribution with Columbia if I do the picture and he refuses at the moment to honour my contract . . .' Losey was probably referring to a project called *SOS Pacific*, to be produced by Sydney Box, with backing from Columbia. 'There's no point in discussing the script because Columbia will not have you,' Box told Losey.[7]

But Columbia already had Carl Foreman. In 1956 Sidney Cohn had negotiated a new exculpatory formula on Foreman's behalf – testimony in 'executive session', with only HUAC's chairman and counsel present. Foreman described his Communist period but by pre-arrangement no naming of names was required. In a letter to his lawyer, Edward Mosk, Losey commented that Foreman had acted honourably but 'he went considerably further than I (or most people) would have been willing to go . . . Personally, I continue to . . . trust his basic decency, but not his ambitions.' Twenty years later Losey told Victor Navasky: 'He did not name any names . . . but he made statements that would have stuck in my throat . . .' Yet Losey continued to lean on Foreman for professional support, soliciting, for example, an introduction to Columbia's Mike Frankovich, 'casually at a cocktail level, as you once suggested'. This letter, dated 17 June 1959, coincided with Losey's hope that Hammer could secure a deal with Columbia for Losey to direct *The Pony Cart*. In July Columbia said no.

In 1960 the American Legion was again spurred into action by Otto Preminger's decision to hire Dalton Trumbo to write *Exodus* under his own name, and by Stanley Kramer's employment of a blacklisted writer, Nedrick Young, to write *Inherit the Wind*. In February 1960 the Legion picked up news of Paramount's impending distribution of Losey's *Blind Date* (*Chance Meeting*), written by two blacklistees, Ben Barzman and Millard Lampell. The Legion accused Paramount of violating the film industry's Waldorf Declaration of 25 November 1947. *Variety* obligingly ran a piece, 'Alleged

REDS, in partnership with ex-NAZI, sell BLIND DATE to Paramount.' The red-baiting columnist Leonard Lyons repeated the story in his syndicated column. Paramount's president, Barney Balaban, assured a press conference that it was not the company's policy to employ people blacklisted by HUAC. He cancelled both a projected tour of the USA by Hardy Kruger – and the film itself. Paramount then struck a deal with the Legion and in October the film was released quietly in New York as the second feature in a double bill on the Loew's circuit. In the *New York Times* (29 October 1960), Eugene Archer recommended 'the absorbing new murder mystery that crept quietly into neighborhood theatres yesterday'. Joseph Losey – by name! – was 'a strikingly adept technician with an alert and caustic personal style . . .'[8]

In despair, Losey inquired of Sidney Cohn whether a contract with a 'major' could now be cleared by a simple letter, involving no appearance before HUAC and no naming of others. A new Columbia Pictures deal was on the horizon; by 28 May Losey had a non-Communist statement on file, presumably drafted by his lawyer. But he hesitated to sign it. On 9 September, Carl Foreman wrote to Paul Lazarus, of Columbia, on Losey's behalf, stressing his tremendous cult following in France and his popularity with writers and actors – for example, Bogarde and Kruger would work with him at almost any time or price. Losey was now prepared to write Columbia the same kind of letter that United Artists had required from Jules Dassin.

Lazarus replied on 14 September: what Columbia required was a simple letter which was 'forthright, frank and dignified'. Losey then signed the following statement, dated 26 September 1960:

'TO WHOM IT MAY CONCERN.

'I make this statement voluntarily and under oath hoping thereby to put an end to any possible question about my loyalty to the United States and to remove once and for all any doubt as to my position and my opinion on communism.

'I am unreservedly opposed to anything that communism represents and espouses. I join with those who fight totalitarianism and I will support and defend the United States against its enemies.

'I am not a communist and I disavow any previous association which may have involved me in any way with the communist party or any front organization.

'JOSEPH LOSEY'

Losey sent his sister a copy: 'No announcement of [Columbia] deal yet, so say nothing.' Anxious to forestall rumour and misunderstanding, he sent copies to Dalton Trumbo, Adrian Scott, Daniel Mainwaring, John Hubley, Millard Lampell, Hugo Butler, Ben Barzman and John Berry. 'I only hope,' he told them, 'this step on my part will make it possible for me to live out the rest of my life as a functioning member of my profession, and . . . will bring no additional hardship to anyone else who has been engaged in a similar struggle to maintain his self-respect and principles.' The response was friendly, heartening. Adrian Scott wrote: 'Neither Trumbo nor I are [*sic*] at all critical of what circumstances compelled you to do.' Warm letters arrived from Mainwaring and the Butlers, Hugo and Jean. Losey had in effect disavowed his own past – but he had almost certainly done the same in July 1951, when signing up with Stanley Kramer. He never mentioned either episode in public.[9]

An American Legion meeting was held in Florida on 17 October. Films listed as aiding Communist propaganda included *Chance Meeting*. The FBI took note: 'Clipped at the Seat of Government.' In November 1960 a group of blacklisted film people, headed by one of the Ten, Lester Cole, appealed for money and names to support, as co-plaintiffs, 'a legal action aimed at the destruction of the blacklist'. The action was to be brought against all the 'majors' and demanded 7.5 million dollars in damages. On 5 December, Foreman's office in London sent Losey a copy of Cole's appeal. A month later, responding to a personal message from Cole, Losey regretted that he had no money to spare, and indicated that he was afraid of public gestures: 'I only hope that the anti-trust suit does not backfire in relation to the blacklist as [*Chance Meeting*] did with me.' No hero, he continued to worry about not having sufficiently cleared himself; Dalton Trumbo's widow, Cleo, recalls a meeting with Losey in London, at Joe's bidding, 'because he wanted Dalton's opinion about writing a letter to the Committee [HUAC]'. Trumbo, who politely advised him that he must make his own decision, regretted Losey's Columbia oath as a step in the wrong direction when the blacklist was beginning to crack.[10]

Fifteen years later, on 24 January 1975, Losey wrote to James Proctor, who had evidently accused him of 'opportunism' during a recent evening together in New York. A friend and political ally since the 1930s, Proctor had done the publicity for Arthur Miller's *All My Sons* and *Death of a Salesman*, and later for Losey's *The Servant*.[11] Proctor's charge of 'opportunism' – he cited Hannah Weinstein's similar opinion – must have related to the Columbia oath of 1960. 'I don't find it necessary to justify myself either to Hannah Weinstein or to you,' Losey declared, but then offered an obfuscatory account of what had actually transpired after he had consulted Weinstein about the passport oath he had signed in 1956: 'I did later receive a three-picture contract from Columbia . . . I was then invited to go to Washington and testify secretly as some others had done. I never entertained the idea and the contract was abrogated.' This is not correct: Columbia (Carl Foreman) financed George Tabori's screenplay and Losey's £12,000 salary for *The Horizontal Man*, which collapsed when Foreman became disenchanted with the project fully two years after Losey signed the Columbia oath. There is no evidence that Losey was urged to testify secretly in Washington.

Losey was anxious to gain admission to Italy, from which he was barred for political reasons which were never stated. In 1953 he had entered Italy (with Gavrik) as a tourist and been promptly deported. After attempting to pull strings in Rome, in 1957 he appealed for help to the American Embassy in London; the Embassy promptly informed the FBI. Curtis Gordon (Bill) Pepper, a *Newsweek* journalist and a friend, could extract no comment from the American Embassy in Rome, which duly reported his mission to the FBI, whose Washington file grew fat on Losey's efforts to enter Italy.

In the spring of 1960 Hardy Kruger drove Losey across the Italian frontier in his Cadillac convertible but when they returned at the end of the day the frontier police threatened Losey with arrest and detained him.

He turned to the underworld. While setting up *The Criminal* late in 1959, he had been introduced by Stanley Baker to a 'big, enormous, American Jewish man' who said he could fix the Italian Job for expenses and a percentage on future pictures. Charles (Dick) Duke duly took up residence late in 1960 at the Hotel Quirinale in Rome and reported that he was fixing things with the help of 'the Professor' (an Italian lawyer whose name appears regularly in correspondence). On 29 November Losey informed Duke that he was about to fly to Rome via Paris. Duke responded: 'Don't come yet! The Professor is a decent fellow . . . and a very important guy hereabouts. Wanted a few more bobs . . .' (A 'bob' was the pre-1971 British shilling.) A three-month permit (*permesso di soggiorno*) came through for a fee of $500. In January 1961 Duke presented 'the Professor' with a half case of champagne, some Scotch and assorted liquors 'just to nail things down tight . . .' In February Losey flew to Rome with his wife; on his return he profusely thanked Duke: 'Dorothy is writing to you . . . about her absolutely delightful watch and we were a bit overwhelmed by your extravagance and generosity.' By mid-March Losey was again anxious: having twice visited Rome on his three-month *permesso*, he faced its imminent expiry. He turned again to the Italian Embassy in London – which took umbrage and referred him to the Ministry of the Interior, where 'the Professor' had been working his expensive miracles. However, on 5 April the Italian Embassy in London issued Losey with a further *soggiorno*, this one for an aggregation of ninety days until the end of the year.

The American embassies in London and Rome were clearly in close contact with the Italian authorities. A memo dated 13 April 1961 from 'Legat' (Legal Attaché) in London to the FBI is mainly blacked out, except for: 'This case will be kept in a pending status awaiting Rome reply.' By June 1961, with *Eve* now in prospect, Losey was again pressing 'the Professor' (and also Stanley Baker, already in Rome) for a permanent visa. This was apparently obtained by the producers of *Eve*, the Hakims, after Losey's arrival in Rome at the end of August. Meanwhile the inventive Dick Duke was enlarging his field of operations. By May 1961 he was urging Losey to obtain full and final clearance from the film blacklist through an ex-Commander of the American Legion, whom Duke happened to know, but who would require a 'gift' of about $1,500. 'NOT A BRIBE. It's like paying a lawyer.' Losey was shocked: 'It is just that it goes against all of my principles . . .' Duke cheerfully urged him to be more of this world. Losey concurred: he would pay up if 'necessary'. In 1969 he was still in touch with Charles 'Dick' Duke, then living in New York.[12]

In 1965, when in Rome for *Modesty Blaise*, Losey received a summons from the police. 'And I finally discovered this was a plot between the CIA and certain fascist elements involved in the Italian production . . .' In reality, uncertain whether the police summons related to Losey's historic visa problems or to problems of marital status under Italy's archaic laws, Joe and Patricia took the next plane out.[13] The FBI did not abandon its increasingly spasmodic monitoring of Losey until 1966, its last clipping being a piece by Charles Champlin in the *Los Angeles Times* (20 April), who reported that Losey was 'totally unforgiving towards those he feels compromised their principles . . .'

Captive States

The Criminal (The Concrete Jungle)

Not until his sixth British feature film did Losey work closely and admiringly with a 'native' screenwriter (Alun Owen). The outcome was his best film since *M* but, despite a quantum-leap in authenticity, his Promethean talent remained chained to the rock of imposed studio formulas – the result being another hybrid. Merton Park Studios were again the producers, and Nat Cohen's Anglo Amalgamated the distributors. Losey kept within his miniscule budget of £60,000 by exemplary professional efficiency, speed and economy. His fee was £7,500.

Michael Carreras, of Hammer Films, had sent the original screenplay (by Jimmy Sangster from his own short story) to Stanley Baker, who showed it to Losey. 'I was given a script which I considered absolute nonsense . . . a direct plagiarism from practically every American prison film that had ever been made.' 'Baker suggested bringing in Alun Owen, a playwright of Liverpool–Irish origin with a gift for dialects and popular speech patterns, but the producers insisted on retaining the thriller storyline and the girl-factor beyond the prison walls. Owen's screenplay consisted of seventy-five scenes, which Losey broke down into 388 shots. As he explained to *Cahiers du Cinéma*, he did not like too detailed (*trop découpé*) a scenario, preferring to achieve his own break-down while shooting rather than by '*pré-découpage*'.[1]

Baker introduced Losey to a larger-than-life underworld character, Albert Dimes – 'a huge, staggeringly handsome man [who] drove around in a smashing, big white convertible with black upholstery'. The film's hero, Bannion, played by Baker, was modelled on the elegant, flamboyant and macho Dimes. Also helpful were the Prison Governor of Wormwood Scrubs and C.H. Rolph, the *New Statesman's* expert on police affairs. The film's title emerged only after shooting; the script contains a list of alternatives written in Losey's hand: The Feel of Money, You're on Your Own, The Syndicate, The Ring, The Clearance, The Cemment [*sic*] City. Losey didn't care for the eventual American title, *The Concrete Jungle*, with its echo of *Asphalt Jungle* and *Blackboard Jungle*. 'And it [the prison] isn't concrete, it's brick.'

Richard Macdonald designed a set, a prison interior, which the cameraman Robert Krasker was able to double in size for long shots by use of mirrors – a brilliant achievement. Another master stroke was to enlist Johnny Dankworth (who had worked with Karel Reisz on *These Are the Lambeth Boys* and

Saturday Night and Sunday Morning). Losey visited the Dankworths: 'He was living on the third or fourth floor of a slum where you crawled up dark stairways over rubbish cans and rats . . . he threw the keys out of the window . . . They sat me down on a settee. They were very nervous . . . and they put a glass of gin in my hands, and they began the song – and at that moment the couch collapsed and I went over backwards with the gin . . .' But Losey had no complaint: 'I had wanted a kind of jazz score in the style of Count Basie, and that's what I got.' Dankworth set to music Owen's lyric, 'Thieving Boy', for the haunting voice of Cleo Laine: 'All my sadness, all my joy/Come from loving a thieving boy . . .'[2]

Shooting began in December 1959 at Merton Park Studios. Tension is reflected in Losey's marginal comments on a list of potential unit members: Good, Bad, Abysmal, No, First Rate, Never ever, Useless, Impossible, Don't Know. 'Impossible' was the art director, Scott Macgregor. Losey was soon in conflict with the producer, Jack Greenwood, who was accustomed to making a dozen second features and Edgar Wallace thrillers per year and whom Losey regarded as a philistine. David Deutsch recalls: 'One day Jack rang up and said, "I've fired him".' There'd been an argument on-set – probably over shooting schedules. Arriving at the studio, Deutsch found everyone seated at various points on the stage, not talking. Greenwood was seething and determined to bring in a new director. Deutsch spoke to Stanley Baker, who was ready to walk out if Losey was fired. Losey, meanwhile, was obsessionally rearranging pencils on a table, in a blind fury. In the end it was Greenwood who had to climb down. Deutsch believes that Losey was 'very Machiavellian' and deliberately provoked such conflicts when he wanted to rid himself of someone, whether a superior or a subordinate.

Losey overran the alloted running time of ninety minutes to an impossible 130 minutes; the distributors insisted on thirty-five minutes of cuts, resulting in poor continuity, but *The Criminal* bears the stamp of masterly direction. The film won a Diploma of Merit at the Edinburgh Festival, where it was first screened on 28 August 1960, with Losey in attendance. The *Observer* (28 August) called it 'the most trenchant and observant British entry at [the Festival] for a long time'. Released in London by Anglo Amalgamated on 27 October, *The Criminal* had the ill fortune to open the day after Karel Reisz's *Saturday Night and Sunday Morning*, which dominated the attention of the press. The *Guardian*'s anonymous critic called Reisz's film 'the genuine thing . . . *The Criminal* is not.' Yet the film was made with 'a mastery of tension and tempo which any British director ought to envy'.

But among film-buffs the wind was now blowing from Paris. In the Losey archives is to be found a typed-out copy of a long, unidentified article entitled: 'Notes Towards a Definition of Losey', subtitled, 'The Criminal'. This article echoes exactly the analytic categories of *Cahiers du Cinéma* and

Positif. 'How to define the nature of the art so triumphantly exemplified by *The Criminal?*' Like Klee and Stravinsky, Losey 'fines reality down to a clean hard core', infusing it 'with the maximum intensity' and discovering 'its true complexity'. Where Fellini studied society, Losey studied a gesture, and discovered society expressed in it. Above all, Losey respected 'cinema as an autonomous art, the art of mise-en-scène, in contrast to the majority of British directors who still approached a picture as the illustration of a script. 'Framing, lighting, the position of the camera, the placing of objects, the lines of walls, ceiling, furniture . . . are conceived to lead the eye straight to the core of significance.' The review ended with a quotation from Nathalie Sarraute's *The Planetarium*: 'The language has not yet been discovered which could express at one stroke what is perceived in a flash, a whole being and its myriad little movements rising from a few words, a laugh, a gesture.' But that language had now been discovered – 'the language of mise-en-scène, as practised by Joseph Losey'. Losey evidently agreed. Talking to Ian Cameron, V.F. Perkins and Mark Shivas, of the magazine *Oxford Opinion*, he explained: 'I hate naturalism – as much as Brecht did . . . the only way to approach reality is to break down the thing, clean it out . . . then select the reality symbols that you want and place them back in the scene . . . in a heightened way.'

Retitled *The Concrete Jungle*, the film did not reach the USA until May 1962. Distributed by Fanfare, and apparently further cut from 97 to 86 minutes, it was sold (Losey complained) badly and cheaply. According to Hirsch, it was not shown at any major New York theatre. The film crept around the USA in fits and starts. Not until 1965 did it reach the West Coast. The few reviews it received were enthusiastic. Kevin Thomas, of the *Los Angeles Times*, concluded that 'a Losey film half heard is worth twice as much as a movie made by most other directors'. Nineteen years later, Losey complained: 'I hold a percentage in the film and continue . . . to get occasional statements saying the film is still in the red.' He didn't believe it.

The Damned (*These Are the Damned*)

Despite its cinematic clichés and contrived love story, *The Damned* is a film of original imagery and refreshing integrity. In common with *Time Without Pity*, *Blind Date*, and *The Criminal*, *The Damned* conveys a humanistic warmth missing from many of the 'big-name' pictures of Losey's later years. Losey made *The Damned* for Hammer and Columbia in May–June 1961. The budget was £160,000 but no contract is discovered: indirect evidence (Fontped's given list of receipts for Losey's services) suggests a fee of only £5,000 but £10,000 is far more probable.

The chosen location was Portland Bill, in Dorset – 'absolutely bleak and wild and very ancient . . . a place where the British were developing germ warfare', Losey added, perhaps confusing Portland with Porton Down in Wiltshire, a Ministry of Defence biological research laboratory. The story makes its point by edging science into science fiction, the seminal myth being that human organisms – the radioactive children imprisoned underground by the State – can transmit radiation without being destroyed by it.[3]

Barzman again! Early in April 1961 Losey visited the Barzmans in the south of France to work on the screenplay, adapted from H.L. Lawrence's novel, *The Children of Light*. Then Losey made the break, long overdue, however wounding at the personal level – he rejected Barzman's script and called in the playwright Evan Jones at short notice for the first of their four collaborations. Jones had made his mark on television; Losey had admired *In a Backward Country*, a play about land reform in Jamaica (where Jones's own family were landowners). 'We had a certain political kinship and we got along very well in other respects, too,' Losey recalled. Meanwhile, he attempted to placate Barzman but without success: on 4 June Losey wrote to him, 'Your violence and unreasonableness I have known for as many years as I have known you and I can predict them to the day and the degree – and yet they never cease to astound me.' Barzman would still get equal money and percentages with himself – so what had he done that was so unforgivable?

Losey and Jones noticed that the Teddy Boys seemed more conspicuous in the otherwise placid setting of Weymouth, a Victorian-Edwardian seaside resort, than in a big city. 'They came out of poverty, unemployment' and roamed the streets lined by old hotels frequented by 'dying *rentiers* . . . I wanted to exploit Rock, which was very primitive then . . . very much associated with direct violence.' Losey juxtaposed the flamboyant aggression of the 'Teds' with the 'indirect, secret, hidden, hypocritical violence' of the military-scientific establishment practising its Orwellian experimentation deep under the cliff.[4]

The pivotal figure of the liberal sculptress Freya was invented in response to Losey's discovery of Elisabeth Frink's bird-men sculptures; a statement of affirmation in a sterile civilization. Casting Viveca Lindfors as the sculptress was a stroke of genius on Losey's part – though her 'method acting' taxed his patience to its limits and beyond. Otherwise he was less fortunate with his cast: despite a riveting performance by Oliver Reed as the gang leader, the impact of the film is blunted by performances either routine (Alexander Knox, Macdonald Carey) or mediocre (Shirley Ann Field).

Shot in black and white, this was Losey's first film in Cinemascope. 'You have to learn also about equivalent lenses because while it gives you the width it distorts the depth . . . much less depth of focus . . . I could do long sustained . . . movements across landscapes and across rocks with boats and with helicopters, partly because of the shape of the screen.' The photographer was Arthur Grant.[5]

Columbia shelved the film because of its political message. Shown at the Spoleto Festival, it was noticed by *Films and Filming* (September 1962) in a disparaging review by John Francis Lane – 'clumsy space [*sic*] fiction with noble pretensions that remain submerged in the heavy symbolism'. Finally it was released in London on 20 May 1963, without a press showing, as the second half of an X-certificate double bill with Hammer's *Maniac*. 'They decided it should be slipped out quietly as a second feature,' reported the *Evening Standard* (23 May). Display ads showed Shirley Ann Field wearing tight vest and trousers, with the motorcycle gang leering.

The reviews were varied, some laudatory. In the *Observer* (19 May), Philip French hailed it as 'one of the most significant recent British movies, a disturbing work of real importance'. According to French's obituary for Losey, written twenty-one years later, Losey regarded this review as the turning point in his reputation. The film won the Associazione Stampa Guiliana Trieste Premio della Critica at the second International Science Fictional Festival in 1964, but was not released by Columbia in New York until July 1965 under the title *These Are the Damned*, its running time cut from eighty-seven to seventy-seven minutes. (Losey himself had offered and strongly recommended cuts to Columbia's Mike Frankovitch.) The reviews were good, though few. Writing in the *Herald Tribune* (25 July), Judith Crist concluded that Losey was at his best not with mirrors or symbolism, but when he told an engrossing story in powerful cinematic terms. Here was 'a remarkable movie unsold and, when finally sold, unsung, to be discovered in a 42nd Street grind house or on the bottom of a light-topped double bill . . .'

As usual, Losey didn't believe the financial returns. In April 1967 he wrote to his accountant: '. . . it is very fishy that a picture I was told cost £160,000 is now listed at £170,000 cost, and is in the red to the extent of £139,166.'[6]

King and Country

Here we jump three years and two films: *The Servant* (already discussed) and *Eve* (to come). Losey's sequence of low-budget 'issue' films, charged with social commitment, carries us from *The Criminal* and *The Damned* to the minor masterpiece he shot in 1964, before pitching into his 'baroque' phase. This was *King and Country*, his last film in black and white. Total confinement to the studio reined Losey in to an awesome display of discipline. The social passion inherited from his theatre work of the 1930s, which had surfaced like a thrashing swimmer in his mixed-genre melodramas of the 1950s, finally found its immaculate expression. With the female sex banished, Losey's own psychic war yields to a tender, penetrating portrait of another war. The humanist is astride the magician.

Dirk Bogarde was shown the TV script of *Hamp* (as it was then called) when in New York with Losey to promote *The Servant*. Bogarde's father had served at Passchendaele and the Somme, bringing home neuroses – he couldn't watch an egg boil because of memories of men slipping off duck-boards into drowning mud – which affected the young Bogarde, an obsessive painter of war scenes. 'Joe called me "piss elegant" and liked to torture me about the rain and mud. I sent him to the Imperial War Museum, of which he'd never heard.' Bogarde gave him a copy of *Covenant of Death*, with grim photographs and the paintings of William Orpen.[7]

John Wilson's stage and television play, *Hamp*, had been adapted from James Lansdale Hodson's book and radio script. Losey called in Evan Jones, who decided to work from the radio script. The first screenplay (27 April 1964) is called 'Glory Hole' and carries no credit. Every point is spelled out, hammered home. Bogarde didn't think much of it: the confrontation between Hargreaves and his CO was 'appallingly written'. 'Hargreaves was meant to take a dagger and attempt to stab his superior – sheer nonsense.' Bogarde (an officer in the Second World War) says he rewrote the scene. Losey's own improving hand is strongly evident in script revisions. He altered sections of Hargreaves's summing-up speech to the court martial, changing, for example, 'He has a blazing honesty' to 'an embarrassing honesty'. Hargreaves's plea that 'if justice is not done to one man, all the brave men are dying for nothing', is changed to 'other men are dying for nothing'. In addition, he corrected a number of amateurish inconsistencies in the script: for example, in the final execution scene, Hamp is said to be blindfolded; two lines later he is said to show 'no recognition' when Lieutenant Webb approaches him.[8]

Daniel Angel served as executive producer for BHE Productions. The budget was only £86,000. No contract is on file; Fontped Securities evidently received only £5,000 for Losey's services as director. Bogarde, too, was performing for love, as was the young star, Tom Courtenay. Losey had seen him in *The Seagull* at the Old Vic, introduced himself, and taken the actor to Boulestin's for dinner (paintings of bulls and Clydesdales on the walls, engraved glasses, an air of opulent formality, and over-priced food and wine). Courtenay didn't know much about Losey but was impressed by this large, 'sad-faced' director from Hollywood. Losey's first film project for Courtenay, Alun Owen's *No Tram to Lime Street*, was still-born, but Courtenay went on to make *Billy Liar*. 'Joe seemed so touched by what I was doing. He liked me a lot. I was always comfortable with him. Always relaxed. Once he suggested that I should keep my head up – then retracted it, as if to say, "Do it your way." '[9]

On the design side, Richard Macdonald came up with the idea of flooding a studio in mud and dragging in a dead mule. His preliminary sketches – Private Sparrow's ear bitten by a rat, Sparrow's scream – focused Losey's eye. Battle scenes were out of the question, but Losey was tempted to begin with the final shot of Lewis Milestone's *All Quiet on the Western Front*: a young soldier's hand reaches out from the trench to trap a butterfly; a shot is fired, the hand falls back into the trench. Losey's idea was to begin his film by freezing the hand reaching out; and to end with it falling back. Replying promptly to Losey's 'Dear Milly' letter of 30 March 1964, Milestone pointed out that only Universal could grant permission – but he himself advised against it. 'Why take chances with the critics?'

Following ten days' reading rehearsal, during which the subsidiary actors (the 'Greek chorus' of Tommies) were mainly in the care of the American actor/director Vivian Matalon, the film was then shot in eighteen days, during May–June 1964, at Shepperton Studios. Denys Coop was director of photography, his first and only film for Losey, but Chic Waterson, the camera operator, and Reginald Mills, the editor, had both worked on *The Servant*. From first script and casting to final editing and mixing, the film was made in only three months. 'It's the most rapid operation I've ever taken part in.' This despite labour disputes – the crew insisted on 'wet money' half-way through shooting. A meeting of the ETU electricians was held at 12.45 with one scene to shoot before lunch. 'Joe argued with them,' Bogarde recalls. Surely they should hold the meeting in their own time. 'It's the only time I've seen him in true despair, tears filled his eyes. "Don't you really want to make movies?" he said.' Shooting conditions were not pleasant. Losey recalled: 'Each time one called for rain there was an error in the tap, we were inundated, and the projection lamps went out . . . mud got into everything . . . rats were everywhere, the [stinking] body of the horse . . . we began to hate the studio, hate the film, hate each other: and I believe the film benefited.' He had fifty caged rats at his disposal, some too fierce to use. Chic Waterson remembers the dead mule beginning to stink but reckons it added to the atmosphere.[10]

King and Country was first shown at the Venice Film Festival on 5 September, with Losey in attendance. He had cultivated Professor Luigi Chiarini, who selected the Festival films: 'I do hope that [the film] contributed a little to winning your battle . . . it is important for us all that you win your battle.'[11] The competition for the Golden Lion of San Marco was stiff: Antonioni's *Il deserto rosso* was the Italian entry and the eventual winner. Courtenay received the Volpe award for 'Best Actor'.* Then came selection as the 'Outstanding Film of the Year' for presentation at the London Film Festival. Losey's list of celebrities to be invited included the Prime Minister,

---

*The following year the film won the La Critica Cinematografica Milanese award. Success, also, at the Film Festival for the Working People in Czechoslovakia: a delegation presented the British Embassy with Losey's prize – a plaster-cast angel's head taken from the baroque Trinity Column in Olomouc's main square.

Harold Wilson; Charlie Chaplin; the French ambassador; Field-Marshal Montgomery; the millionaire N. Gulbenkian. 'The following people we may ask you to invite later if we have room for them: Vivien Leigh, Graham Greene, Lady Pamela Berry.'

Released by Warner–Pathé at the Carlton cinema on 3 December, the film was greeted with critical acclaim. In *Sight and Sound*, Penelope Gilliatt noted Losey's precise ear for 'the awful intricacies and humiliations of the English class system'. This was a 'black-hearted and implacable' film based on Losey's sense of form, 'stiff as a grid . . . a shape that you could lift off the story like a motif imprinting wet cement'. *King and Country* was nominated for the 'Best Film' award of the Society of Film and Television Arts, but the prize went to Stanley Kubrick's *Dr Strangelove*.

Having been screened at the New York Film Festival in September 1964, the film was distributed in America by Ely Landau's American International Pictures through Allied Artists, but not released until September 1965. The New York opening was further delayed until 27 January 1966. Losey fretted – no one was telling him anything. The press was good: according to *Life* (1 October 1965), 'Losey has an ability to make each frame a matter of conviction and feeling, to make the screen seem darker, like an etching.' But the film was showing only at the Rendezvous, a grade A art house owned by Landau. Women were reluctant. Despite good reviews, their references to tragedy and rats had had (Landau's Director of Public Relations sadly reported to Losey) 'a somewhat negating effect'. The film failed at the box office. Gross receipts to 31 July 1966 were £48,322 for the UK, £43,269 for foreign sales (including the USA), plus the £22,213 grant from the BFFA, making total receipts of £117,358. But those involved in the production – and on a percentage – smelled a rat as large as any seen in the film. Having cost only £86,000, 'How could it fail to make a profit?' asks Losey's co-producer, Norman Priggen. Losey raged: they had been (perhaps) 'gigantically diddled'. Bogarde isn't convinced, either.[12]

17

Joe and the Women

Joe's first important sexual relationship after he settled in London was with the actress Jill Bennett – although no comment, private or public, is discovered in his papers. Their friendship outlasted the affair. Losey later confided both to Dr Barrie Cooper and to Patricia Losey that he'd also had an affair with Hanna Axmann, a German painter whom, in Cooper's words, 'Joe rather handed on to Nick Ray'. An Axmann painting hangs in an upper hallway of 29 Royal Avenue, Losey's last home.

Born in Manchester, the young actress Dorothy Bromiley returned to England in 1954 after a spell with Paramount in Hollywood. Her agent rang with the news that an American director (whose name meant nothing to her) was searching for someone who could play an American teenager in Edmund Morris's *The Wooden Dish*. She met Losey in the office of the impresario E.P. Clift. 'He was wearing dark navy blue clothes and looked like a wrestler.' He was also her senior by twenty-one years. Their relationship began on tour, in Newcastle and Nottingham; in Newcastle they watched ballroom dancing – 'Women in sequinned dresses, he'd never seen anything like it.' Twenty years earlier, according to Joe's sister Mary, 'he couldn't dance for nuts'; Dorothy describes him as 'a wonderful dancer, very elegant, tiny feet'.

In 1955 Dorothy moved in with Joe to Queen Anne Street, followed by a succession of rented accommodation. In June 1956 they were married in Kensington Registry Office. Joe's mother Ina came. Carl Foreman's wedding present was fifty pounds. 'Dorothy was a child,' Dr Barrie Cooper recalls; she remained in awe of Joe, paid homage, and would refer to him as 'Mr Losey'. Leon Clore and his wife invited the Loseys to dinner with Nick Ray and his young wife. Losey asked Ray, 'Is your wife ever referred to as a child bride?' David Deutsch remembers Joe seated at the head of the table, serving the food. 'I wasn't a home maker,' Dorothy says. Losey's friend Ned Calmer wrote of Dorothy as 'your little Renoir bride'.

Joshua was born in the Middlesex Hospital, on 16 July 1957, when Losey was shooting *The Gypsy and the Gentleman*. Fraülein Sybille Kappe was employed for a year from August 1958 to look after the infant. It was Joe who chose the (mainly Conran) furniture for their new home in Montpelier Square (Leon Clore calls it a typical outburst of extravagant spending). Dorothy was impressed by the furniture's 'hard masculine lines then in fashion. He bought the furniture as a surprise. He did it for me.'

Arriving in England, Gavrik was informed by Joe that his high school diploma was no use and that he must sit British O and A levels. 'Joe told me as I got off the plane. He bundled me straight into a crammer.' Dorothy remembers that the night of Gavrik's arrival Joe had two tickets for the Bolshoi Ballet, with Ulanova, but felt he must be at home with Gavrik. The Loseys occupied the second, third and top floors of the house in Montpelier Square. Gavrik, whose strong visual memory extends back not only to the Jane Street apartment but to West 48th Street, noticed that the ceiling of Dorothy's dressing-room had been covered in black stars. 'Joe,' he adds, 'was one of the first people to buy Conran furniture.' There were strains. Gavrik was now a long-haired Beat-generation rebel youth in shabby jeans. 'In Joe's house things were run for Joe – there was no give and take.' Joe was obsessively tidy, orderly, neat. You didn't invite your friends in or play music. Dr Barrie Cooper recalls: 'Gavrik would face and outface him.' On one occasion Cooper was driving Joe, Gavrik and his own daughter down to the country. Gavrik was playing pop music and Joe's fury was rising: 'I stopped the car on a country road and they got out and fought, almost physically, about Gavrik's "savagery, primitiveness, vulgarity".' Gavrik had a girlfriend who was studying psychology, but Joe would not allow them to sleep together in Montpelier Square (he later confided to Patricia that he felt guilty about this). 'I got on better with Joe's wives than Joe did,' comments Gavrik, who is only eight years the junior of Losey's third and fourth wives, Dorothy and Patricia.

In 1955, Barrie Cooper had sent Joe to the clinical psychiatrist Joe Shorvon. Losey told Cooper that he'd received psychotherapy when a student, and later when living in Hollywood, at a time of political and marital stress. The analytical content of these sessions with Shorvon, says Cooper, was subordinate to helping to release him from his inner storms and to enable him to function, work, write, direct. The problems were booze, mood swing, anger, obsessionality, perversity, depression, anxiety, sexual ambiguity, relations with women, his perceived problems with his mother, all augmented by rage at his own situation. Losey visited Shorvon twice a week for four years; he died late in 1959 at the same time as Joe's mother.

Joe's ailing mother Ina had become more than a physical burden to him – according to Mary, by the end of her life Joe couldn't stand her. Ina, it will be recalled, had left La Crosse for El Paso, Texas, where a friend owned a department store and got her a job. Although sixty-five in 1945, she convinced the management that she was younger. She remained in El Paso for ten years, paying occasional visits to Joe and Louise in Los Angeles. In 1950 she spent a summer with Mary's family in Geneva, taking herself off to Monaco to have herself a good time gambling, a lifelong pleasure (mainly bridge). Dependent on social security cheques of about sixty dollars a month, Ina remained a source of humiliation, even of shame, for Joe, the source of his Puritanism, the ghost in the machine of his art. He did admire her courage. From 1955 Ina came to live with Joe and Mary in alternation, six months of the year in England, six months in Geneva. 'Joe couldn't abide it. He mostly put her in a Torquay old people's home,' Mary comments. Dorothy remembers Ina as a very handsome woman who was always short of money; Dr Barrie Cooper describes her as a 'difficult, cantankerous, domineering, ungracious, ungrateful, not very likeable woman'. She was 'determined, tough, self-willed and demanding'. In June 1958 she developed a cancerous growth of the mouth. Eighteen months later, when Joe was shooting *The Criminal*, he asked Gavrik to make daily visits to the hospital. 'I'd hold her hand and she'd say, "I don't want to die."' During her last night Ina woke up and said, 'Mary, Joseph, God help me.' The nurse thought she was praying – but she merely wanted her children. Mary recalls: 'The night she died Joe went to the nursing home but couldn't stand to go upstairs to see her.' Ina's ashes were buried beside their father's in La Crosse.

Joe could not tolerate disorder, his sister emphasizes: 'He'd come in when Dorothy and I were talking, to clear out the ashtrays and make everyone feel uncomfortable. I kept looking at him. Dorothy said, "And one day he'll tidy me away too."' The Loseys had a housekeeper, Mrs Macdonald, who 'did everything', and Mrs Francis Smith who did the cooking, making meat pies, quiches and tarts, doting on Joe's expensive tastes. 'She thought she was a better hostess than I, and that she should have married Joe,' says Dorothy. Joe wrote home to La Crosse asking Mary's godmother, their wealthy 'Aunt' Grace (Mrs Gysbert van Steenwyck) to send some local recipes for Dorothy. 'I keep telling her about Hannah's chocolate cake, for instance.' He also mentioned pepper pickles and pickled peaches – and may have been flattering his honorary aunt before explaining that Montpelier Square was costing £800 a year in rent, which the landlords were threatening to double. (In reality he had signed a new three-year lease the previous year, 1959.) As an American he was ineligible to apply for loans to building societies or banks. Would Aunt Grace grant him a mortgage or loan of about £7–8,000 ($20–23,000) if he could find a sixty to ninety-nine-year lease? There is no discovered response. Meanwhile, in March 1959, he had bought a ninety-nine-year lease on a run-down shepherd's cottage with an acre of land in North Wales, the landlord being his friend Sir Osmund (Michael) Williams.

During their marriage Dorothy worked intermittently: a television dramatization of E.M. Forster's *A Room with a View*; Peter Brook's *Heaven and Earth*; a small role in *The Criminal*, a vignette in a Soho café. 'He wanted me to use the Lancashire accent that Central had got rid of. I didn't feel good playing with Jill Bennett. I knew he'd had an affair with her before *The Wooden Dish*.' When on tour she used to receive love letters from Joe with suggestions that she contribute more to the household expenses. In March

1960 she opened in an Agatha Christie play which transferred from Edinburgh to the Duchess Theatre, London. Frequently Joe wrote to producers and friends on her behalf. She retains a pained, hurt love for him. 'He educated me. I was very much in love with him. He had an appetite for life, he laughed a lot, there were good, lovely things.' He gave her books to read, including Freud's *Interpretation of Dreams* – apparently Joe had a recurrent dream about Brecht similar to one recorded by Freud. 'He called me his child bride. I felt we had a Pygmalion and Galatea relationship.'

During his first year at University College, London, Gavrik helped organize a refectory sit-in to challenge the feudal attitudes, the lack of basic status and rights considered normal by American students. Joe reminded him that he was in university to study. No doubt one Communist in the family was enough. And one womanizer, too. A former girlfriend of Gavrik's wrote to Joe, begging information. Joe advised her 'not to worry too much about deep emotional involvements at your age'. When she wrote again, asking for a photograph of Gavrik, she received a message of no comfort: 'I should perhaps tell you that he [Gavrik] is accompanied on his walking tour to Greece by two young ladies. Perhaps there is safety in numbers – although I doubt it.'

The following spring Gavrik got himself in deeper trouble: as editor of the student newspaper, *Pi*, he delivered an attack on Sir Ifor Evans, Chancellor of London University – and was summoned. Acutely concerned, Joe spoke on his son's behalf to the film producer (Lord) John Brabourne, a member of the Board of Trustees and a friend of the Chancellor. Gavrik then failed four of twelve second-year exams and was sent down. Lizzie wrote to console Joe: 'I am sure that his time in England hasn't been wasted – degree or not . . . I have a hunch if you could be horribly nice and patient it would pay off in a big way . . .' Some months later Joe conveyed his own view to Lizzie: 'He is as yet, and maybe always will be, pretty unaware of other people unless he needs them.' Worse, Gavrik had never shown any real sense of direction. On top of that he looked 'like a Jewish Balkan peasant, hunch, goatee and all'. The real problem was Joe's inability to convey overt affection to his sons. 'I had to come to terms with this,' Gavrik says. 'Joshua never has.'

Summers, Dorothy, Joshua and his German nurse, Rosemarie, flew to Nice, Joe driving down to join them. Dorothy felt confident of his fidelity. 'It didn't seem to me he was a womanizer. I always felt his work came first'. Reading his conversations with Michel Ciment twenty years later, she was astonished to learn of his 'fifty to sixty relationships'. His affair with the film producer Hannah Weinstein seems to have extended into the early period of their marriage. When Louise came to London in 1959 both Dorothy and Ruth Lipton had lunch with her – separately. Joe's women converged morbidly, the dead passing notes to the dying. Joe himself refused to meet Louise and didn't want anyone in his life to see her, either.

Dorothy remained in the dark about his affair with Ruth Lipton until she had 'a flash of intuition' and he admitted it. 'I was very angry. I came from the lower middle class and I took marriage vows seriously. You should work through problems.' By the time he went down to Weymouth to make *The Damned* in May 1961, they had already begun a 'trial separation'. Leon Clore remembers Dorothy putting a brave face on it: 'My ulcers have cleared up since I left him.' Quite a few of their friends had the impression that the decision had indeed been hers.

Gavrik reports three occasions when he and his father had mistresses in common. While working in the cutting-room on *Eve*, he had a fling with an assistant editor who was also Joe's lover. By Gavrik's account, he and Joe both had an affair with a young actress who visited Gavrik's 'commune' home in Spring Street. One night, when Gavrik was visiting Montpelier Square, Joe told him, 'I don't like this sharing. I'm twice your age and any woman you have I can take away from you.' Ruth Lipton writes: 'The sexual competition between Gavrik and Joe was very ugly. Joe often accused me of having an affair with Gavrik, who is only a couple of years older than my son Peter. Although Joe was pathologically jealous, it was inconceivable to me that he could really think I was sleeping with his son . . ."Gavrik speaks of his father's 'life of honesty and integrity', but he also points out that Joe's Midwestern 'Puritan' morality, a rigid sense of responsibility, went hand in hand with the hedonistic existence of a bon viveur of international show business – an unresolvable contradiction.

By 1962 Joe was living in Bywater Street, 'a small, chintzy house', then moved to Markham Street while Dorothy and Joshua settled in Prince of Wales Drive, Battersea. Dorothy insisted on going to the opening of *Eve* in Paris in 1962, 'because I wasn't willing to let go'. Patricia Tolusso, destined to be the next wife, flew in from Rome to act as interpreter and – as it turned out – to be told that he couldn't work on his next film without her. A few months later Dorothy accepted a small, wordless, part in *The Servant* while Patricia, on renewed leave of absence from her young family, made coffee for the cast. The single day Dorothy spent on the set was 'humiliating – only Dirk knew who I was. The crew didn't and I was treated like an extra. I felt Joe might have protected me from that.' But at this juncture he was not protecting any woman from the unhappy queue of love; no numbered tickets were issued.

Dorothy's divorce was No. 74, 1963, granted by the Probate, Divorce and Admiralty Division of the High Court. Custody, care and control of Joshua were awarded to Dorothy. Joe's maintenance payments to her on behalf of Joshua under a court order issued in April 1967 were £1 a week to Dorothy and £18 a week to Joshua. 'He'd made clear to me,' she recalls, 'that he'd never paid alimony to Lizzie or Louise and I was intimidated. I was also a feminist.' Ten years later the amount had not altered (£936) despite 250 per cent inflation during the decade. However, the court order represented only a fraction of what Joe spent on Joshua's schooling, travel, holidays and much else.[2]

In Losey's conversations with Ciment he refers to Dorothy as 'my third marriage' – no name – adding that his only important relationships were to Lizzie and Patricia. This stung Dorothy to the core. She never got over it. Anger flashes: 'And I think Joe and I have unfinished business.' She fights with tears. 'People have been tremendously disloyal to him over the years and you won't find disloyalty in me. I feel Joe needs protecting. He rampaged through life making enemies. I am not his enemy – even though he hurt . . .' She recovers herself: 'There is pride, isn't there? It calls into question an important period of my life. It's hard to come to terms with.'

When Joe died, the *New York Times* awarded Dorothy's son Joshua to Elizabeth Hawes, while the *La Crosse Tribune* reported that 'Losey was married twice'.[3]

Ruth Lipton first met Losey at a Thanksgiving dinner given by the film producer Hannah Weinstein. Eight years younger than Joe, Ruth was a woman with a dedicated radical record. Politicized while a student at NYU, she had worked for the American Labor Party and the charismatic Marxist politician Vito Marcantonio. She had also inherited a considerable insurance payment after her first husband was killed in an air crash which took the life of the Soviet ambassador to the United States, Ousmansky. Now married to Peter Weingreen, an assistant director and a vice-president of the ACTT (formerly ACT), she worked in film publicity for Rank at Pinewood Studios then served as unit publicist for Losey's film *Blind Date* – although promoting the German actor Hardy Kruger stuck in her throat until she got to know him and 'had a very discreet affair which ended when I became involved with Joe in 1960'. She was divorced from Weingreen the same year.[4]

Joe, she comments, was 'charming and exciting and charismatic to most women' but terrible to those he put down. 'He talked about Louisa [Louise] in the same kind of tone that he talked to Dorothy . . . It was a total put-down.' Viveca Lindfors, who stayed in Ruth's Queen's Gate flat while filming *The Damned*, recalls: 'She was mad about him, adored him . . . He had a mystery about him, a lot of women fell for it.' Lindfors is sure that Ruth wanted to marry him but Ruth's version is otherwise: 'He wanted us to have a child, a girl to be named Clea after the Durrell character . . . but I retained enough sanity to know that marriage to him would be fatal for me . . . He wouldn't let me live!' Yet she went to Rome some six times while Joe was making *Eve*, unable to resist the pull of 'this enormously, attractive and very unboring man'. He was generous and thoughtful with gifts and was always interested in her clothes. 'In Rome I tried on a suit and he said, "You've got a middle-aged figure." I lost twelve pounds in a week. When I was in need he was wonderful. I was hospitalized, and Joe put back a commercial [in the Canary Islands] until after I had come through the operation.'

Joe was a long-standing friend of the journalist Curtis Gordon (Bill) Pepper and his wife Beverly Pepper, who loaned sculptures and pictures of her own for *Eve*. Visiting her studio outside Rome, Lewis and Irma Fraad observed the solicitous way Joe treated her, his concern about the dangers of welding. Barrie Cooper, who possesses several pictures by Beverly Pepper, believes that Losey may have been in love with her. Patricia Losey confirms an affair which had ended by 1962. Joe acquired an oil painting, 'Buddhist Priest and Boy', signed B. Pepper 1960 (professionally valued at £730 in 1969) and 'B. Pepper, 56, water colour sketch, Figures', valued at £150. Many years later he conveyed his admiration for Beverly Pepper to Charlotte Aillaud.[5]

Ruth Lipton was still on the scene when *The Servant* was being set up. She looked after Joe during the bad bout of pneumonia which temporarily put him out of action, and regularly telephoned his notes to Bogarde in the studio. After his recovery they had a row, and she walked out. His secretary/assistant, Patricia Tolusso, moved in but only after, she recalls, an entr'acte involving another woman whom she doesn't name, but probably the actress he shared with Gavrik. Patricia would overhear difficult phone calls from Ruth. After Patricia's fateful return to Rome, Joe attended the wedding of Ruth's son – their final encounter. 'Joe came with broken ribs . . . Much later that evening when all the guests had left Joe telephoned me and pleaded with me to go to him.' She thought about it, but didn't.

Though older and tougher than Dorothy, Ruth then made the same mistake as she: trying to revive a dead connection through work. On 23 September she wrote to Joe, reminding him that Richard Macdonald was in debt to her for the time and effort she had expended on his problem with the ACTT – yet Macdonald had never brought her company, Omega Films, any work. Perhaps he was under the impression that to do so would be disloyal to Joe? Would Joe kindly disabuse him of any such misguided sense of loyalty? She ended by asking if Joe himself would be interested in working on commercials for her . . . No answer is on file.[6]

The most elusive of Joe's long-term friendships was with Vappu (Wuolijoki) Tuomioja, for whom he'd fallen heavily during two brief meetings in Finland in 1935. She later told her son Erkki (who became Deputy Mayor of Helsinki) that she was 'the great lost love' of Joe Losey's life. 'If my mother is to be believed, he proposed to her on several occasions over the years, each time his relations with other women faltered.' But Joe and Vappu did not meet again for twenty years, until her husband was appointed ambassador to the United Kingdom in 1955. Hers was a family of almost ferocious achievement. Her mother Hella (with whom Losey resumed correspondence after the war) was everything that he admired: a talented intellectual with wide international literary contacts, a wealthy entrepreneur (he once wrote to her as 'Lady Wuolijoki') who was also a Red and a Member of Parliament. Vappu's husband, Sakari Tuomioja, was finance minister, Governor of the Bank of Finland, foreign minister, prime minister, ambassador and high-level UN functionary. What Joe made of Mr Tuomioja we don't know; evidently he suffered from melancholia. In September 1959 Vappu confided to Joe that her marriage was an unhappy one and that she had sought solace in love affairs. 'I've never since my youth been so depressed and unhappy. It's horrible to "keep face" when you are screaming inside. I wish I could be an alcoholic but I just can't drink.' She then came to London for two weeks but they failed to meet. On 1 December, Joe wrote: 'I was horrified to get your note this morning and to find that I had missed you. Gavrik, out of a Freudian block or negligence, never mentioned the fact that you were in town.'[7]

In November 1972, two months after lunching with Joe and Patricia in London, Vappu wrote to him: 'The time for a serious, deep discussion with you, that I would like, will probably never take place.' He had been shooting *A Doll's House*. Had she understood him, only deliberate provocation could have accounted for this: 'Also in Stockholm there is just now a wonderful performance of the [*sic*] *Doll's House* in Dramaten with Bibi Andersson as Nora, which I think you ought to see.' Joe's response was predictable: 'I am no longer interested in Noras of any nationality . . . Next time give me more time.' (A curt reference to not seeing her while shooting in Norway.)[8]

That their mutual 'moment' of 1935 created an almost romantic bond became apparent when Joe agreed to take part in a 'This Is Your Life' Finnish TV programme about Vappu. She described the experience: 'I had a feeling you tried to embarrass me. You positively chuckled over some things you said and left the audience wondering what was unsaid. You said lots of nice things but I could have kicked you . . .' To this Joe replied: '. . . anything that was unsaid was unsaid only for the large audience and perfectly explicit to you . . .' Promoting *Mr. Klein*, he told the Finnish newspaper *Uusi Suomi* (21 October 1976) that he had fallen in love with Vappu on first sight, 'but nothing more came of it' because 'her future husband was already on the scene'.

18

Eve

Losey – artist and man – poured himself into *Eve*, his moment of release from the confines of the British studio system, his entrance into the world of Antonioni and Resnais, the high art of Europe. Venice! Rome! *Eve* was to be the most traumatic disaster of Losey's career.

The Hakim brothers, otherwise Interopa (Rome), had approached Jean-Luc Godard to direct a film of James Hadley Chase's novel, *Eve*. Godard was keen on Stanley Baker but not on Jeanne Moreau; in 1973 François Truffaut accused Godard of having deliberately humiliated Moreau by pulling out. It was Baker who proposed Losey as replacement director, with Moreau's approval. By agreement the setting of the story was transferred from Hollywood to Venice and Rome. Losey's contract, dated 1 August 1961, provided for not more than forty-eight days principal photography, to begin by 15 October, his fee to be $50,000 plus first-class expenses. Classified as an Italian National Film, *Eve* was to be made by an Italian crew and technicians, apart from Richard Macdonald as production designer and Reginald Beck (originally Reginald Mills) as editor.

Chase's pulp story had been published some fifteen years earlier by Phoenix Publishing Company, Paris. Set in California, it is a dreary, lifeless tale, full of sentences like 'it had been an interesting, if expensive, quarter of an hour'. The hero, Clive Thurston, is the author of three successful novels and a stage play when his powers dry up. He passes off a dying man's play as his own. By chance the prostitute Eve and her lover of the night take refuge in Thurston's Big Bear Lake cabin during a storm. He tosses the man out into the night then offers Eve 100 dollars for her services. Thus begins his pursuit of a gambling woman who treats men like dirt: ' "Did anyone tell you that you've got a swell pair of eyes?" I said, grinning down at her.'

Pursue this rubbish further we need not. But Losey did. The first screenplay was written by his old colleague and fellow blacklistee Hugo Butler, then living in Rome, with whom he'd conducted a warm and voluminous correspondence throughout the 1950s. Ruth Lipton warned him that Butler was wrong for this one but Losey insisted. Director and writer then worked together in London but 'he never could quite get at what I wanted . . . Hugo was being very Hollywood in his writing, very cold in his approach to it . . . I was ridden with guilt toward an old colleague, an old comrade, but I did not feel I could do this film with him.' The two men fell out, Butler acutely wounded. Losey brought in Evan Jones.

In Butler's script, the writer was American. Losey and Jones cast him as a Welsh charlatan, Tyvian – a mean, macho, suddenly-rich pretender blending guilt with arrogance – for the benefit of Stanley Baker. (Yet one may wonder whether a single filmed book could endow a Welsh miner's son with apparently limitless wealth in Venice and Rome.) Losey had flown to Brittany to talk to Moreau by firelight – her chauffeured Rolls met him off the plane – to discover that her enthusiasm for Billie Holiday matched his own. After working on the script at Hardy Kruger's house in Lugano, Losey and Jones presented it to Moreau in Rome. She was delighted with it but the Hakims weren't. After further rewriting everyone concerned signed every page of the script.[1]

For Moreau, the film was about exile: Eve's, Tyvian's, Billie Holiday's loneliness and longing for harmony – and Joe's several exiles. According to Jones, Losey cared more about the settings than the story; enjoying his newfound fame in France, he was about to parody his gift for mise-en-scène to the point where it became the true 'subject' of the film. Revised script pages demonstrate how frequently his attention was arrested by a wedding or a funeral on the Grand Canal, the angel being lowered to pass under the bridge; a sunken yard in Rome inhabited by cats – the following morning Jones had to write a scene involving Moreau and stray cats. The violent bedroom encounter between Tyvian and Eve occupied only a half-page of script but became a four-minute continuous shot. Losey was personally infatuated by Moreau and determined to direct her as no other director had done. Paul Mayersberg comments that Moreau in *Eve* 'is conceived as anti-Malle and anti-Antonioni, as well as anti-Truffaut'. Losey had gone 'auteur' on a James Hadley Chase potboiler: 'The first time I saw it being filmed I went into shock,' recalls Jones. 'It had this tremendous lush quality.'

Losey took a small, pre-production unit to film establishing shots of the 1961 Venice Festival while his director of photography, Gianni di Venanzo, was still finishing Antonioni's *Eclipse* (likewise for the Hakims). Di Venanzo was Fellini's cameraman, a maestro, who taught Losey to line up through the camera, not the 'finder', and then trust the operator, on this occasion the gifted Pasqualino de Santis. (At that time the sightlines through the viewfinder and through the lens were not identical. Nowadays refraction allows a lens-eye view during shooting.) The autocratic di Venanzo also spurned equipment that Losey had flown out from London including 'a new centrifugal device for a tripod to be used in boats and places where there is a lot of vibration', and a gyroscope attached to a hand-held camera to offset movement. 'And we went down to the canal-level door to look at them . . . Di Venanzo flew into a violent rage, jumped like Nijinsky and stamped on all the equipment. So that was the end of that . . .' Losey learned much from di Venanzo, whose crew of nine or ten were 'absolute slaves to him, whom he treated miserably, abominably, screaming and yelling and abusing them . . . but five minutes later it was all forgotten'.[2]

Losey's conflicts with the producers were so abrasive that shooting came to a halt several times. Gavrik Losey, who had joined Beck in the cutting-room, marking up tracks and assembling the dolly reels, remembers how on the first day of the dubbed projection the film ran out of image-sound sync after three minutes. No one could explain it. Then Robert Hakim walked in: 'What do you think of my new cut?' During the weekend Hakim had brought in an editor of his own to cut the master picture. Losey then 'hit' (probably seized) Hakim. In Reginald Beck's phrase, 'Joe dusted the wall with him'. Work stopped for a week; lawyers arrived from Paris. 'It was a real crisis,' recalls Beck.

Likewise with the music. Negotiations with the great jazz musician Miles Davis broke down over money. Of the 'thirty' Billie Holiday songs that Losey had hoped to use, the rights to only three were acquired. John Dankworth was invited to Rome but (as he recalls), 'My agent and the brothers Hakim didn't seem to find any common ground, and eventually Michel Legrand, who is a very good musician, did the score . . .' This was Legrand's first film for Losey; he remembers meeting the Hakims in a huge bar. 'We don't talk to Losey,' they told him. 'We decide on everything.' Losey himself was equally stubborn: 'Tell those shitheads . . .' Legrand found him secretive and reserved: 'Everything was a problem.'[3]

Losey stayed on in Rome until July 1962. This brings us back to the Hakims and on to Losey's loudly publicized and often repeated version of the depradations they inflicted on the film. Clearly his own preferred cut was self-indulgently (and self-defeatingly) long. Even after recutting and remixing the entire picture with Robert Hakim breathing down his neck, it ran for 155 minutes. The Hakims then withdrew it as the official Italian entry at the Venice Festival. 'Robert Hakim came to me one day on the Via Veneto and said, "I have a Siritsky distribution in France . . . So will you waive . . . your rights to control the negative in Italy, so that I can get the print to Paris."' Losey signed the release. 'I was never paid in full – neither was Legrand, nor Moreau, nor Macdonald.' (In reality legal files show that Losey got most of his money, his main complaint concerning expenses and Italian taxes. Legrand comments that he was never paid at all.)[4]

And so to Paris. According to Pierre Rissient, a private screening of *Eva* was held at the Club Publicis for a small audience. Sensing non-responsiveness, Jeanne Moreau visited Losey in the Lord Byron hotel and advised him to cut the film drastically. By Rissient's version Losey (wrongly) lost his nerve and himself took twenty minutes out of the positive print while asking the Hakims to cut the negative. As shown at the Paris gala opening (a disaster) the film was down to 116 minutes. When talking to the press Losey tended to mask his own role in cutting the film, but a letter to Robert Hakim, dated 17 October 1962 and found in the files of Losey's London solicitor, reveals the truth. 'I do not blame you or hold you responsible for those cuts because I went along with them, and even beyond them in a couple of instances, when we discussed them and worked together in Paris . . .' But he now listed cut scenes that he wanted restored; Losey had convinced himself that excessive cutting was responsible for the hostile reception in Paris. The Hakims did meet one request, to restore the casino scene, but in preparing an English version for distribution in the UK and the US (where, unlike Italy and France, the director enjoyed no authorial copyright), the producers slashed the film by a further ten or fifteen minutes, redubbed it, substituted an American voice for Virna Lisi instead of Anna Proclemer's, inserted lines and exclamations in silences, butchered the musical cues, inserted opticals and fades, and left much of the picture, particularly the post-sync, out of sync.[5]

Losey saw this version, distributed by Gala in Britain, on 23 April 1963 and called it 'a common, tawdry, little melodrama – unclear, pretentious, without rhythm and taste . . .' Incensed, he drafted a press release announcing that Moreau, Baker and other principal collaborators withdrew their names from the film. Solicitors representing the Hakims and the distributors issued a stream of threatening letters under a contractual clause forbidding harmful publicity and warning that Losey had no right to withdraw his name. Alarmed, Losey adopted guerrilla tactics, briefing sympathetic critics off the record while his solicitor, Laurence Harbottle, and his agent, Robin Fox, defended in private correspondence his right to answer questions from the press. From Paris, Hakim was quoted: '*Eve* was originally supposed to be an action film, running for about one-and-a-half hours. But Mr Losey let it run much longer and made it into something else. I had to cut it back to get to the basic theme.' As late as February 1965, when the film reached America, Crawley & de Reya, representing the Hakims' Interopa Film company, were issuing warnings about the damage inflicted by Losey's press statements.

To understand the point of turmoil at which Losey had arrived, in both his life and art, we must look at *Eve*. Losey later admitted that this film 'was almost an orgasm, and it was probably . . . self-indulgent . . . it was like a coitus interruptus . . . a moment when I really broke out, exposed myself and was completely vulnerable'.[6]

The print viewed, the longest version available, begins with a prologue which is the first part of the epilogue to the main story. It is winter in Venice, a misty view of the island cemetery; Eve (Jeanne Moreau) is seen passing in a water taxi. A lamenting trumpet is heard, evoking Miles Davis, then yielding to Billie Holiday. The credits appear.

Eve's water taxi arrives at Harry's Bar. The camera pans across the clientèle: Losey himself is glimpsed on a bar stool. The loud voice of Tyvian (Stanley Baker) is heard, off: 'My brother was first of all a drinking man . . .' Enter Branco (Giorgio Albertazzi), producer of the film of Tyvian's novel, visiting Venice for the second anniversary of the death of Tyvian's young Italian wife, whom Branco loved. Branco angrily confronts Tyvian, who rushes from the bar, clutching his face in self-disgust. Adam and Eve are glimpsed on the corner of the Doge's palace.

The main action begins two years earlier, in contrasting Venice Festival summertime, with Tyvian water-skiing across the Lido in a straw hat, skimming between two streaming banks of foam, while his voice is heard overlaid: 'I entered Babylon, on a chariot of fire.' Eve observes this stranger from a yacht then turns away. Tyvian is enjoying the success of the film of 'his' novel; in black tie and dinner jacket he moves easily but contemptuously among celebrities and fauning critics, relating a whimsical Welsh story about his brother, religion and sex, to an admiring group.

Branco's beautiful and adored assistant, Francesca, has fallen in love with the Welsh brute. Tyvian and Francesca are leaning over the balustrade when Eve passes in a gondola with her small, middle-aged escort. Tyvian callously abandons his invitation to Francesca to stay in Venice rather than travel to Rome on business.

A downpour brings Eve into Tyvian's life. We cut to the lagoon and muddy bank near the Torcello farmhouse he has rented. Eve's escort has broken into the house for shelter from the storm. The interior is décor-intensive: a giant reproduction of Masaccio's fresco, *Expulsion from Paradise*, hangs at the top of the staircase. Eve – observed only by us – makes herself at home in Tyvian's bedroom with Billie Holiday's 'Willow Weep for Me'. After a luxuriating bath the hollow-cheeked vamp dances around, fingering objects, hugging pillows, wriggling her toes, discarding garments, wrapping herself in towels, studying herself in the mirror. Enter Tyvian. She shrugs insolently. He is captivated. After enjoying the glow of his big fire (and some clever photography) she declines his offer of food and takes herself back up to his bedroom. The little escort makes to follow his prize upstairs but Tyvian strong-arms him out of the house into the storm.

'EVE is in her nightgown, on the bed, lying in the Goya position' (says the script). Tyvian joins her. 'You must be pretty good.' He then demands his portion: 'Let's see what you can do.' As in Chase's novel, she knocks out the predator with a heavy glass ashtray.

Now we find Eve in her Rome apartment, 'lying on a chaise longue in the position she takes when relaxed . . .' This is – again – 'reminiscent of Goya's "Maja" (the Duchess of Alba)'. Full-length mirrors abound. Tyvian, it transpires, also has an apartment in Rome. So here he comes, pursuing Eve and her next escort to a 'chi-chi' night club where a black male dancer is writhing and extravagant masques indicate decadent *dolce vita*. Eve is treated to another of Tyvian's folkloric Welsh tales in his open sports car as he proceeds up the Via Marche towards the Villa Borghese. Establishing shots of the city skyline, with St Peter's dominant, confirm that this is Rome. The script reflects Losey's priorities:

> CAMERA *ends on the huge bust set in the old Roman wall which extends along the street from the Porta Pinciana. Then a line of marble men behind the hedge which surrounds the Foro Italico. Then* CAMERA *moves to show the many lines of male busts on pedestals which line the Cianicolo.*

Eve and Tyvian keep walking (Eve tottering on high heels in a long Cardin coat) for the benefit of the busts. She speaks to one of the statues and imitates its expression; she slaps another, kisses a third and leaves a stain, then begins to paint the statue's eyebrows and lips with her lipstick and eyebrow pencil. Tyvian, in quick close up, is grinning with a kind of vulgar shamefaced joy. With spit Eve slaps a 10,000 lira bill on the statue's forehead. On second thoughts, she retrieves the note, allows Tyvian to drive her home – then slams the door in his face.

A visit to the office of the film producer Branco offers an opportunity for more *objets d'art* and top-deck tourist views. The jealous Branco informs Tyvian's fiancée Francesca that he has launched an investigation of the writer's credentials back in Wales. A heated exchange in Italian ensues.

Now Tyvian is once again with Eve in Rome – the Square of the Knights of Malta. She offers a short biography of herself: two orphaned sisters, poverty in Lyon, illness, sexual abuse by a neighbour ... Then she laughs. 'You'd believe anything.' A long way off, sunlight glints on the dome of St Peter's. There follows a scene written in at Losey's behest: an ancient excavation of temples, overgrown with trees and creepers, a garbage bin for centuries, the home of hundreds of wild cats. Enter Eve, playing hide and seek with Tyvian. She trips over an old beggar woman and screams.

EVE: I think I'm ugly.
TYVIAN: I might persuade you that you're not. I like to make women happy.

Back in her apartment, Eve rips open the buttons on his shirt: 'Let's see what you can do.' They drop out of frame straight into the reflection from the bevelled edge of a Venetian mirror, thence into the distorting lens of a fish tank.

Tipped off by the jealous Branco, Francesca makes Tyvian admit he saw a woman the previous night.

TYVIAN: I love all women, six to sixty.

As a character Eve is scarcely credible but the real problem is Tyvian, forever posing like a blockheaded male model (despite bursts of Dylanese Welsh poesy), constantly competing with Eve/Moreau for the swaggering camera quality of sexual insolence. For scenic value, Eve goes walkabout, passing through a small squatter camp then coming into view of some architecture – don't high-class prostitutes worry about their ankles? A distraught Francesca goes in search of Eve, to lay eyes on the woman who commands Tyvian's passion. Eve is found seated in the Parco Balestra, reading Tyvian's novel, *Strangers in Hell.* The camera moves in to VCU of Eve's fantastic sunglasses. Reflected in them, Francesca approaches, her image growing larger and larger.

Tyvian begs Eve to go away with him for the weekend. Her condition is 'the best hotel' in Venice – and gambling. This can only mean the Danieli, with 'the full grandeur of the baroque stairwell'. Entering the bridal suite, Tyvian advises Eve that this weekend has cost him 'a few friends, thirty thousand dollars and a wife'. Later he becomes maudlin, confessing that his book was written by his dying brother. 'He loved me, Eve . . . I stole my dead brother's soul. Help me, Eve.' She responds warmly, covering him with kisses, but soon the contracted weekend has ended, so far as she is concerned: 'I want to be paid. Cash.' Tyvian empties his wallet. She counts the big notes slowly, cigarette in mouth (brilliantly done), decides her going rate is higher, and throws the money back, scattering it on the bed and floor. Here Stanley Baker works hard on Tyvian's wounded indecision: he kneels to gather the scattered notes then thrusts them back under the bed in self-disgust. Eve, lying in the bed, orders him out.

Tyvian and Francesca are married in Venice, transported by gondolas decorated in white flowers. Eve looks sourly down on them from a window of the Danieli. Briefly they share a kind of married happiness, working in a vineyard near his farmhouse until Branco turns up with a document from Wales proving Tyvian to be a fraud. Francesca hesitates then hands it back unopened. Work takes Francesca away to Rome.

We find Eve in winning form at the baccarat tables in the company of her 'husband'. Enter Tyvian, pursued by a prostitute and fated to lose (what's left of) his money to Eve's 'husband', who turns out to be more a gambler than an engineer – or a husband. A humiliated Eve commits herself to Tyvian's outboard motor launch at dawn. When he slips in the mud on landing, and cuts his hand, she laughs: he knocks her down with a slap across the face.

EVE: Only my husband can do that.

Revenge ensues. She takes herself into his bed, naked, but warns him not to follow (an absurdity). Cowed and awed (and no doubt afraid of more glass ashtrays on his head), Tyvian backs off, taking refuge in alcohol and finally falling asleep on the floor. Arriving home next morning, Francesca finds Eve in front of a mirror and wearing her own negligée. Francesca's horror and betrayal are powerfully conveyed by Virna Lisi against a painting depicting similar emotions. She runs across the farmhouse garden to the pier, with Tyvian in pursuit, leaps into the motor boat and roars away across the water towards a dredge – we hear the splintering impact.

Cut to the carved angel on the prow of Francesca's funeral boat as the cortège makes its way along the canal to the island cemetery. In black suit and crisp white collar, Tyvian displays his normal expressionless wooden face to the churches of Venice. Branco follows in a second boat. Everyone is in white and black in black-and-white.

That night a demented Tyvian climbs into Eve's empty apartment overlooking the canal, then hides on the balcony outside her window. Eve is observed (a long take) alone in her suite, undressing, talking to her cat, smoking a cigarette, crying a bit, listening to Billie Holiday, smashing the record across her knee, then sleeping. Enter Tyvian. After some struggling and pleading, Eve reaches up to the wall for her riding crop: again and again she lashes the cowering hero.

EVE: Loser, loser, loser.

He staggers down the stairs and falls into a pile of refuse in the courtyard, his face and eyes as swollen as if he'd been roughed up by a streetgang. This brings us (none too soon) to the epilogue, two years after these events. Eve is about to leave Venice with a wealthy, relaxed Greek, a kind of *doppelgänger* of the suitor whom Tyvian had thrown out of his house into the rain. But Tyvian is now a broken Welshman, a pathetic supplicant bereft of pride.

TYVIAN: You'll be back in ten days? I'll meet the boat . . .
WIDE ANGLE *as in* MASTER SHOT *to include part of Piazza San Marco and part of the esplanade . . . Reminiscent of the Giacometti group, each goes his lonely way.*

One final speech for Tyvian (off) is added in Jones's hand: 'And he placed at the East of the Garden of Eden, cherubim, and a flaming sword, to keep the way of the tree of life.' The siren of Eve's departing ship is heard – a last glimpse of Adam and Eve.[7]

Visually – design, camerawork, composition (Macdonald, di Venanzo, Losey) – *Eve* is a ravishing *tour de force*. The black-and-white settings of Venice, shaded to wintry grey in subtle dawn shots, are matched by Moreau's alternation between Pierre Cardin costumes in glittering white and the darkest hues. Arresting images abound: a symbiosis of shapes between musical drums and brandy glasses; a gaggle of priests sweeping down steps; interior views of the world beyond the walls; Francesca mirrored in each lens of Moreau's sunglasses; the horrified Francesca approaching her own marital bedroom, as if she were the intruder, watched by the insolently smoking Eve through a mirror; a high shot plunging down the rectangular marble staircase of a luxury hotel in Venice; a circular camera movement in the Danieli suite when Tyvian collapses and confesses; a high vertical shot of Tyvian and Eve emerging from a lift in her Rome apartment block – they appear to walk out of the wall; Eve stretched on a sofa, beside a vast bowl of roses and a painting, the black telephone her weapon, cigarette in mouth, listening to Billie Holiday – Ingres after Goya; the rhythms of water and sky, marshland lagoons at dawn, all enhanced by Michel Legrand's hypnotic score.

Kenneth Tynan imagined, regretfully, a genuinely complex hero worthy of a Montgomery Clift.[8] But Tyvian/Baker was Losey's fall guy – not merely a purgation of his own, contemporaneous, sexual riot, but also an affirmation of his superior culture and sensitivity. Paradoxically, it was the actor, rather than the director, who was granted Moreau's passing favours, though many involved in the production regarded the film as a deliberate humiliation of Stanley Baker. But the wound inflicted by Stefan Brecht in a letter to Losey (29 October 1962) could not be deflected on to the actors: 'Most people consider it a piece of chi-chi crap, but I don't quite see it that way. My impression is you intended to give a flashy-hollow-phoney-tinsel treatment of ditto persons and relationships, but were so caught up in the formalisms and excitement of the thing that you overdid it and thus gave people the impression that you intended something profound and significant . . .' Almost with one voice, the French and Italian critics shredded the film. (The French critical response is described elsewhere.) *Giornale d'Italia* commented that Losey wanted 'to remake Antonioni, and perhaps also Fellini, without possessing their talent . . .'

Distributed by Gala, the hated, 100-minute version of *Eve* opened at the Cameo-Royal on 18 July 1963. The cinema's portals carried a notice: 'You are warned that *Eve* is a nasty, sick, exciting and sensual film.' Writing in the *Guardian* (19 July), Richard Roud lamented that the Hakims had left only 'a rather adolescent sex drama of a man destroyed by a prostitute'. In the *Sunday Telegraph* (21 July), Philip Oakes described the film as grotesque, banal, repetitious, and Baker as 'Wooden Indian of the Year'.* Released by

---

*In the course of a Losey season at the NFT (April 1966), *Eve* was shown in its fullest available version, with Swedish and Finnish subtitles. Losey had viewed this print in Helsinki in January 1965 at the invitation of Jussi Kohonen, of Suomi-Filmi Oy, and was delighted to find the original soundtrack. Kenneth Tynan rated the longer version alongside *The Servant*.

Times Film Corporation, *Eve* did not appear in America until 5 June 1965, opening at the Little Carnegie in New York. Bosley Crowther reported in the *New York Times* (5 June) that the film had made him want to climb up the wall. 'Mr Losey said the producer ruined it by cutting. The rejoinder is: he didn't cut it enough.'

Losey never kicked the habit of building sociological pyramids over the embalmed corpse of *Eve*. Fifteen years later he came up with this: 'The Bible, the shame, the heterosexual/homosexual aspects . . . the beauty destroyed, the impurity, the blasphemy, the destruction of icons, the bells, all these things . . . I was talking about a marriage . . . which wasn't a marriage . . . which a bourgeois audience would never accept if you said it directly . . . [but] would then go home and say, "But Christ, that's precisely what happened between me and my wife last week" . . .'[9]

Eve was followed by the film which brought Losey his international reputation with critics and audiences alike. In *The Servant* he severely disciplined his inclination to extravagance and self-indulgence, harnessing his camera to a script of geometrical severity. The outburst of mid-life sexual adventurism now yielded to an almost self-effacing fusion of professional talents (Pinter, Bogarde, Macdonald, Dankworth). Only in the final 'dope scene' did this discipline crack open. Yet the lineage between *Eve* and *The Servant* is evident from the first mirror-shot – the jaded, jazz-cool urbanity, the passionate quarrel with the libido. But what followed *The Servant* was unpredictable; the sparing naturalism of *King and Country* seems like a penance, an expiation, for the indulgences of *Eve*. Savanarola Losey resurfaced; Venice, Rome and Chelsea yielded to the stench and slaughter of the trenches, urbane modernism to nail-down realism.

But the vows were precarious: *Modesty Blaise* followed. When Raymond Eger, of Les Films EGE, Paris, proposed a film version of Poe's *Histoires Extraordinaires*, Losey was excited. The setting was again Venice in winter. Maria Callas was to play a speechless *marchesa*. Calling it 'a high romance in the baroque manner', Losey had in mind a cinematic style 'encrusted with fabled lavishness . . .' Happily he made *Accident* instead.

PART V

The Sublime

19

Galileo with Brecht

In 1947 Losey became the first director to stage a Brecht play in English while working in collaboration with the playwright himself. It was a distinction that he neither forgot nor allowed to be forgotten – although he was not destined to direct any other work by Brecht on the stage or screen. By a poignant digital reversal, his film adaptation of *Galileo* was not achieved until 1974.

Written by Bertolt Brecht in 1937–8, *Life of Galileo* [*Leben des Galilei*] was first performed in German at Zurich's Schauspielhaus in September 1943. The central theme of Brecht's play concerns Galileo Galilei's discovery that Copernicus had been correct that the earth revolves in orbit around the sun – a challenge to the Church's insistence that the earth (and its Church) occupies a unique position in God's universe.[1]

Losey had not been in contact with Brecht since 1936–7, despite his fruitless request, through Brecht's collaborator H.R. Hays, to broadcast *Trial of Lucullus* on NBC's *Worlds at War* radio series. In November 1944, after leaving the army, Losey moved to Hollywood: 'There I found [Brecht] in his little box of a frame house in Santa Monica, all the familiar stamps of his personality and [Helene] Weigel's, the scrubbed wood floors, the Chinese prints, the battered upright piano, the chest of manuscripts, the beautiful iron-stone china which Hellie had collected from the flea markets.'[2]

In April 1946, visiting Boston to see Orson Welles's production of *Around the World in 80 Days*, Brecht opted for Welles as director of *Galileo*. Excited by the challenge, Welles hoped to begin rehearsals by 1 August, but adamantly would not work with Mike Todd as producer. Welles pulled out; his last recorded meeting with Brecht occurred on 5 May. After discussions with other potential directors, including Elia Kazan and Harold Clurman, Brecht had by the end of August 1946 settled for Losey. '*Es wird wohl Joe Losey*,' he told a colleague. Losey inherited an English-language version of *Galileo* based partly on Brainerd Duffield's translation, massively revised by Brecht working in collaboration with the actor Charles Laughton and with Ferdinand Reyher, his most intimate friend during the years of American exile. An FBI telephone tap reveals that, 'After Losey's first conference with Charles Laughton, he told [Hanns] Eisler, on August 11, 1946, that Laughton was "afraid that I am a little left wing and that I will be influenced by Brecht . . ."' 'A midnight meeting took place at Universal Studios with Mike Todd, Laughton, Brecht, Howard Bay . . . and Jack Moss [Louise Losey's former husband] . . . then assistant to Todd.' A Losey memorandum written soon after the event noted that 'Todd talked of how he would dress the production in "Renaissance" furniture from the Hollywood warehouses . . . Brecht only listened and giggled nervously.' He would have nothing more to do with Todd.[3]

Losey spent three months in New York after flying east from Los Angeles on 17 September. As he recalled: 'The Hegelian legend '*Die Wahrheit ist konkret*' was 'tacked in bold letters to the bare wall above the draftsman's bench which constituted Brecht's desk in his New York apartment, where I lived for six months in 1946'. The décor was characteristic: 'white plaster, the book case hand-fashioned of piled bricks and unfinished planks. The couch and curtains were of burlap [hessian]. The light was high, white and exposed. The decoration, a single, perfect Chinese scroll . . . the eye never tired in that flat.' Losey remembered Brecht as always accompanied by two or three devoted women friends, and that 'he ate very little, drank very little, and fornicated a great deal'.[4]

Did Losey live 'for six months' in Brecht's apartment? In April 1971, James K. Lyon, then Assistant Professor of German at Harvard, wrote to Losey: 'Your statement that you worked on the Laughton translation with Brecht and Laughton for "many months". . . comes as news to me and the rest of the world of Brecht scholarship. I have already tried it out on two or three who should have known, and they did not.' Indeed Brecht's home was in Hollywood and the New York apartment was almost certainly that of his mistress, Ruth Berlau. FBI surveillance shows Brecht living on the West Coast from May until 16 September 1946, when he visited New York. During the period from September to January 1947, the FBI's continuous reports on Losey's movements and residences in New York provide no hint of residence with Brecht/Berlau: he stayed with Lizzie at 52 Jane Street until 24 September, then moved into the Barbizon Plaza; on 27 October he transferred to the Chelsea, where he remained until leaving for LA on 2 January. The FBI regularly observed him emerging from the Chelsea in the morning. Although he frequently visited the Brecht-Berlau apartment for work sessions involving Laughton – his son Gavrik vividly remembers the apartment – it seems that Losey's role remained peripheral, mainly confined to his theatrical connections and his ability to get a show on the road, including casting (he was present at Laughton's hotel when they interviewed the actress Eda Reiss Merin for the part of Signora Sarti).[5]

On 7 January 1947, Losey reported to Hella Wuolijoki: 'It is such a great play, so clean and clear and architectural . . . So frighteningly above all of the little squirming talents in our commercial theatre. So much what I have worked for and prepared for all my life.' On 17 January, Brecht and his wife Helene Weigel lunched with Losey at the Gotham Restaurant, Los Angeles (the FBI tells us). The Loseys left the West Coast on 22 January, travelling east on the Santa Fe 'Chief'.

Losey was under contract to T. Edward Hambleton to direct a play by Robert Presnell, but after this production was cancelled he stayed in New York until 16 April, directing a new play by Arnold Sundgaard, *The Great Campaign* – a farmer runs for president and is cheated out of victory – which opened at the Princess Theatre on 30 March, sponsored by the American National Theatre Academy and by Hambleton, whom Losey later described as 'a rich man's son from Baltimore of considerable taste and wry humour'. The play ran for only five nights – 'successful but not commercial', recalled Losey. One of those who saw the production was Brecht's collaborator, Ferdinand Reyher, who commended Losey's direction to Brecht, at that time in California: 'He was back in the spirit of the Federal Theatre . . . He knows casting, he has the feel for it; he knows what to do with actors; he can get a crowd sense without numbers, and movement that isn't just confusion, and keep the whole of a play in mind . . . I'd trust him with *Galileo*. He'll help you greatly. Don't let Laughton get him down.' Losey's active collaboration resumed after his return to California, where his benign production boss at RKO, Dore Schary, kept him on paid leave of absence.[6]

It was Losey who involved the wealthy and idealistic Hambleton in *Galileo*. Laughton and Hambleton each invested $25,000. Hambleton approached Pelican Productions, managed by John Houseman and Norman Lloyd, with both of whom Losey had worked in the 1930s. Pelican had been set up to present plays of quality at the recently renovated, 260-seat Coronet Theatre on La Cienega Boulevard, in Hollywood. (Norman Lloyd rather casually recalls: 'Joe Losey came to me with a story for a film he wanted to show to [Lewis] Milestone . . . I read it and turned it down, but I suggested that Losey contact Laughton and Brecht with a view to directing *Galileo*.' If correct, this must refer to the previous summer of 1946.)[7]

On Losey's recommendation Brecht engaged John Hubley, a former Disney animator, to draw sketches designating the decisive moments of *Galileo*. Hubley (who later collaborated with Losey on three Hollywood films) came up with a cartoon strip of pre-production drawings which Brecht admired and which corresponded to his own emphasis on a fast, ironic, theatrical tempo. In the event, Hubley's sketches were used during rehearsals, but not in the show; they were also exhibited in front of the theatre. Again on Losey's recommendation, Brecht engaged Robin Davison, who had designed the sets and costumes of *The Great Campaign*, and Anna Sokolow, the choreographer in the same show. (She quit under Brecht's iron rule.)[8]

When rehearsals officially began in June, writes Lyon, Losey was director in name 'but it soon became apparent to the cast and producers that Brecht, with assistance from Laughton, was really directing, and that Losey functioned as his mouthpiece'. Brecht, meanwhile, prowled in his habitual denims, reeking of cheap cigars. One member of the cast, Stephen Brown, told Lyon: 'Brecht and Charles were the real directors and everybody in the cast knew it.' John Houseman agreed. According to Lyon's interview with Lotte Goslar, Losey had said, 'I'm learning' when asked if he felt redundant. Alexander Knox recalls: 'Charlie Laughton called [Losey] "morose". Galileo, in my recollection, saw Joe at his gloomiest . . .' John Houseman described Brecht's attitude as 'consistently objectionable and outrageous . . . harsh, intolerant and, often, brutal and abusive. The words *scheiss* and *shit* were foremost in his vocabulary . . .'[9]

Helene Weigel offered her services as wardrobe mistress; his non-wardrobe mistress, Ruth Berlau, took many of the production photographs, including twenty minutes of moving-picture footage which came to rest in East Germany's Staatliches Filmarchiv. During rehearsals Laughton had developed a habit of thrusting his hand deep into his pocket and playing with his genitals. Brecht telephoned Norman Lloyd, who replied, 'Joe Losey is directing the play. Get him to speak to Charles.' Brecht replied: 'He will not do it.' Nor would Brecht do it. Helene Weigel then had the pockets of Galileo's trousers sewn up – an appropriately scientific solution – but Laughton had them unsewn.[10]

Losey's description of the greatest of twentieth-century dramatists at work is striking: 'The twinkle and often malice never left his eyes. He giggled like a girl and his laughter was a great, high-pitched bawdy hacking laugh that emerged around his chewed, foul cigar . . . Stillness was another aspect of Brecht and his work. He could listen. He could sit still, though sometimes his impatience to clarify or correct or supplement led to an agitated, frenetic walk, words starting then deferring . . . Objective in all his visions, but . . . the most passionate of men, full of anger and sudden rages, utterly enthusiastic, but opinionated and intolerant of the unseeing or unfeeling . . . In his life he was a man-eater, every person, every thing served his art . . . small, electric, balanced like a dancer, eyes jabbing like his hands, quick in all things.'[11]

Apparently Brecht and Losey had two rows during rehearsals. In the first Brecht provoked Losey to the point where he threw his script at him and threatened to quit. Losey told Lyon that he went home and started gardening.

Laughton rang urging him to return. Losey demanded an apology from Brecht. Later Laughton rang again: 'Brecht says please come back, and he also says you should know Brecht never apologizes.' The second row occurred shortly before the opening when Brecht complained about the colour of the scenery. Losey also had to take the flak from actors frustrated by Brecht's constant rewriting – in Scene 9 he persistently changed the lyrics. However, Losey too was capable of giving offence. Eda Reiss Merin walked out after he continued to find fault with the way she played Signora Sarti – Helene Weigel persuaded her to return.[12]

Brecht and Laughton, meanwhile, were engaged in a struggle over the final exchanges between Galileo and Andrea. Fired up by the Cold War and the 'treason' of the atomic scientists, Brecht now wanted Galileo to condemn himself savagely: 'Then welcome to the gutter, dear colleague in science and brother in treason.' Laughton preferred the more subtle contradiction between Galileo's cunning and his apostasy, found in Brecht's original (1938) version (some years later, in East Berlin, the Marxist actor Ernst Busch put up a similar resistance to Brecht). Losey's film notes on *Galileo*, of the 1960s and 1970s, indicate that he adopted Brecht's harsher position.[13]

The performance took place on what Losey later described as 'a very bare wooden set with Shakespearean areas and a few carefully selected objects, as well as projections of Galileo's drawings'. This conformed to Brecht's 'clean eye' and dislike of clutter. After a week's postponement, the play opened on 30 July – the talk of Hollywood, which was full of European emigrés who idolized Brecht. The glamour set attended the opening: Chaplin, Charles Boyer, Ingrid Bergman, Anthony Quinn, Van Heflin, John Garfield, Gene Kelly, Billy Wilder, Richard Conte, Howard da Silva, Sam Wanamaker, Lewis Milestone, Frank Lloyd Wright. Marlon Brando and Burt Lancaster also saw the play. Every performance was standing room only: 'Turn out for the Theatah,' mocked *Variety*. The main problem was the blistering heat. '*Ich muss ein 7-Up haben*,' declared Brecht, leaving the theatre.

When the play closed on 17 August, 4,500 spectators had attended the seventeen performances; the friendly reviews confirmed that the West Coast press was still in a progressive frame of mind. By the time the play opened in New York on 7 December, Brecht had fled from America. The witch-hunts were on. In May 1947, HUAC had held closed hearings in Hollywood. Hanns Eisler, who was composing incidental music for *Galileo*, was summoned to testify. Brecht himself appeared before the Committee in Washington on 30 October. Six days earlier the *Hollywood Reporter* had announced: 'Joe Losey and Charles Laughton train out for New York today', but Losey accompanied Brecht to face the Congressional music in Washington – a striking demonstration of moral courage and personal loyalty. Fearing that hotels were bugged and full of informers (Losey recalled), they walked 'the empty streets of that mausoleum city' at night, discussing tactics.

Brecht likened the experience to a zoologist being cross-examined by apes. His performance before HUAC was an exercise in cunning – many would say duplicity. Although his English was by now good, he used an interpreter on Losey's advice. David Baumgardt, of the Library of Congress, volunteered, but his English was so thick that the Committee Chairman, J. Parnell Thomas, complained, 'I cannot understand the interpreter any more than I can the witness.' Losey also advised Brecht to request permission to smoke cigars on the witness stand. Thomas smoked cigars, they all smoked. Brecht charmed and baffled the apes by turns. Losey recalled: 'Anyway, I flatter myself that my advice was good and . . . he conducted himself so extra-ordinarily that at the end, about three o'clock in the afternoon, the Committee congratulated him (number eleven of the "Hollywood Ten", so to speak) and adjourned the hearings. And there were no more hearings for two years.' (*Sic*: three-and-a-half years.) Brecht later reflected: 'It was in my favour that I had almost nothing to do with Hollywood, had not been involved in American politics, and that those who preceded me on the witness stand had refused to answer the Congressman.'[14]

But Brecht was not sanguine. He was frightened. 'It's like a *Bauern* court in Austria under Dolfuss,' he told Losey. He took a train to New York with Losey and T. Edward Hambleton. (Losey did not mention Hambleton's presence in his numerous public recalls of the event – but in September 1975 he wrote to Raymond Wolff: 'No one other than Mr Hambleton and I attended the hearing . . . not [Brecht's] wife, who was in New York, nor any of his children.') On the following day Brecht flew to England, thence via Paris to Switzerland. Hellie Weigel, shortly to depart herself, gave Losey a present wrapped in tissue paper: 'You know Brecht, he can never do these things himself . . .' With the present, 'an absolutely exquisite carved ivory opium pipe which he had picked up in China', came a message from Brecht: 'You should relax.' Losey never saw him again.[15]

Brecht sent Losey an undated letter from Gartenstrasse 38, Zurich, soon after his arrival: 'Dear Joe, Now, from a distance, I feel even more satisfied with the Coronet performance. L.'s creation of a dialectical (contradictory, changing) character on the stage, the first of this kind, I believe, is a case of sheer genius. I am sure, the basic content of the play, the *treason* of Galileo, will come out even clearer in N.Y., after our experience of the audience in Hollywood, which led us to the alterations of scene XIII. Of course, it will be necessary to lose nothing of the gay character of the first scenes when the end will be more sinister. On the other side, the more repugnant, criminal, destroy [?destroyed?] Galileo (in XIII) will appear, the more positive the highest achievement of this great brain, his self-condemnation, will appear.

'There is one thing which is of great importance: I must have fotos (*sic*) and a film of the New York performance; no consideration of the audience, nor even of Charles should be allowed to stand in the way – you have no idea how deep the impression of Ruth's [Berlau] prints on the actors and director here has been. *Please*, prepare Charles and get his agreement *now*. I am sure he will find a way to give Ruth the opportunity. Herzlich Your b.

'Please send copies of the scripts!

'greetings from helli!'

This may be the only substantial letter that Brecht ever wrote to Losey. In the 1950s, the correspondence he received from East Berlin was written by Helene Weigel or Elisabeth Hauptmann. James K. Lyon wrote to Losey in 1971: 'I have read a large number of letters you wrote to Brecht which are on file in the Bertolt Brecht Archives. But there is no indication that Brecht ever replied to any of them (typical for him – he's the world's worst letter-writer). Did he ever write you?'[16]

The New York production went ahead in Brecht's absence. In line with Brecht's latest notes, Laughton, Losey and the writer George Tabori slightly modified the last scene. Sponsored by the American National Theatre Academy, the play now had, except for Laughton, a new cast. Again minimal salaries were paid. In the *Herald Tribune* (8 December), Howard Barnes greeted 'a fascinating and brilliantly articulated production'. Laughton performed with 'relish and artifice'. As for the director, 'Losey has directed the work with tremendous resourcefulness, introducing boy singers in intervals between scenes and putting on a strolling show on All Fools Day, 1632, magnificently.'

According to the programme, the play was scheduled for only six performances, from 7 to 14 December, at the Maxine Elliott Theatre. However, in 1957 Losey told Robin Fox that the New York run was for two weeks extended to three; by 1976, a Falstaffian expansion had occurred: 'We had a sold-out run and standing room only for four weeks. There were always fifty to a hundred standing. We tried then to get the money to transfer it commercially, and we raised it by a series of readings that Laughton did. But we couldn't get a theatre guarantee for more than two weeks . . .'[17]

On 13 March 1949, Brecht signed an agreement to give Rod Geiger exclusive film rights in *Galileo* for all languages except German. Brecht hoped the film would be made in Italy, and Losey was urging him to let George Tabori work on the screenplay, but the project died largely because of Laughton's unwillingness to become involved. Brecht would contemplate no other English-language actor. Laughton had previously been innocent about what his association with Brecht and Losey entailed in terms of constant FBI surveillance. Now, with his application for American citizenship pending, the actor had become alarmed by the adverse political publicity surrounding Brecht's subpoena. His attorney, Loyd Wight, warned him that he might suffer investigation himself.[18]

20

Galileo: The Barren Years

Charting a director's life solely in terms of his realized achievements may be justified in critical terms but is entirely at odds with his desk diary and his dreams. Between 1947, the play, and 1974, the film, Losey's relationship with *Galileo* and with the Brecht establishment in Berlin was one of acute frustration – a bitter chronicle from which Losey himself emerges unhappily.

In July 1951, a few days before he himself went into exile, he received 'a magnificent picture of herself from Helene Weigel Brecht as she appears in the role of Mother Courage'. Not until 1954 did he begin seriously pursuing his dream of directing *Galileo* in London, or, failing that, an alternative Brecht play. But Brecht and Weigel sensed his lack of status in England, regarding Laughton, not Losey, as the crucial custodian of the 1947 production. On 16 October 1954, Losey accordingly wrote to Laughton, proposing an approach to the Old Vic. A cursory reply from the actor's Hollywood agent indicated that Laughton had 'no desire to do anything further with *Galileo*'.[1] Reporting to Hellie Weigel, Losey enclosed a copy of what he called this 'formal – not to say offensive – response . . . This I assume relieves me of all moral responsibility and Brecht of all legal responsibility if I go ahead and approach Michael Redgrave . . .' (But how interested was Redgrave? On 14 February Losey complained to the actor that he hadn't answered his two last letters.)

Presenting himself as the true keeper of the Brecht seals, Losey now began 'telling teacher'. On 3 March he asked Weigel whether she and Brecht were aware that the Theatre Workshop was planning an English version of *Mother Courage* at Stratford East, with Joan Littlewood in the title role. 'I do not wish to be in the position of trying to set up productions while other people control the rights, or are peddling them.' Losey's letter seems to have had its desired effect; on the 8th Brecht sent Losey a copy of an admonitory telegram to Theatre Workshop: 'Final decision on any Brecht production rests with me.'

Losey found himself isolated and outbid in a suddenly booming Brecht market. On 4 March, Brecht's collaborator, Elisabeth Hauptmann-Dessau, informed him that Christopher Sykes, of the BBC Third Programme, had been in East Berlin to talk to Brecht about broadcasting his plays. Would Losey provide Sykes with a copy of the Brecht–Laughton English version of *Galileo*? The affront was not lessened when Sykes wrote to Losey, asking for the English text but not hinting at any involvement by Losey. Sidestepping the provision of a text, Losey indicated that he would like to produce a radio version himself, 'if it were possible, considering my American nationality'. No reply from Sykes is visible.

Losey had approached Hugh Beaumont, impresario of the highly commercial theatre management, H.M. Tennant, without informing East Berlin. On 21 March Beaumont turned down *Galileo* and *The Good Woman of Setzuan*, 'two very fascinating plays but I fear would be regarded by English audiences as "experimental"'. Losey now made the mistake of transmitting Beaumont's letter to East Berlin; a stinging reply (31 March) came from Elisabeth Hauptmann: 'With the situation being difficult in England – please, don't ram Brecht down anybody's throat.' Losey's experiences, she went on, were merely deepening Brecht's caution about English theatres and audiences. (Losey's own theatrical status can scarcely have been improved by his recent production of *The Night of the Ball*, a bauble of high-society voyeurism which soon sank with all tiaras on board.)

Was Joan Littlewood aware of Losey's damaging intervention against Theatre Workshop's planned production of *Mother Courage*? On 27 April Losey sent Brecht a surprising report: Littlewood had now approached him to direct the play. But Theatre Workshop went ahead without him. On 8 August he wrote irritably to Weigel: had the production been 'authorized' by Brecht 'or merely suffered by him'? Meanwhile Losey was again trying to interest Beaumont in a *Galileo* performed by Olivier, Redgrave or Eric Portman 'in that order'. But Beaumont was not prepared to reconsider until after the Berliner Ensemble's scheduled visit to London in August 1956. Lunch at Scott's was all he could offer.

On 12 November, Losey approached the impresario Oscar Lewenstein: was it true that he was planning a production of *Galileo*? Would George Devine direct it? 'If not, obviously, I should be very disturbed not to be considered.' Lewenstein replied that he had no plan to do *Galileo* either on his own account or with the English Stage Company at the Royal Court. Three months later Losey learned from Hauptmann that George Devine was to direct *The Good Woman of Setzuan*, with Peggy Ashcroft, at the Royal Court. Devine and Ashcroft had visited East Berlin the previous year, 'and we have been in contact with them ever since'. On 5 May 1956, Losey wrote to Weigel in an obvious attempt to scotch the Royal Court production: he doubted whether George Devine would obtain a satisfactory adaptation, 'and so his tactic is to wait the matter out . . .' (Devine had a small part in Losey's film, *Time Without Pity*, which was concurrently going into production.) Losey further advised that the Royal Court was not good enough for this 'or any other Brecht play'. Still a semi-stranger to the left-wing theatre companies, he preferred the commercial constellation. On 2 July, he wrote asking Brecht whether he was planning to visit London – Losey wanted to arrange 'a small evening' for him to meet such people as Beaumont, Tynan and Redgrave. Weigel replied on the 13th: 'Please do nothing because B. is not coming.' Brecht was fatally ill.

Losey's brooding possessiveness about Brecht was mirrored in an odd letter he sent to the *New Statesman*, challenging a published letter from Brecht's translator, Eric Bentley, about Brecht's appearance before HUAC in 1947, and insisting that, 'It was to *me* that Brecht likened the proceedings to *Bauern* courts in Bavaria under the Nazis.' Losey concluded: 'For reasons that, I regret to say, are still as operative as they were in 1948, I must ask you to withhold my name ... TEMPORARY EXILE.' (But an anonymous 'me' letter is hopeless.)

When Brecht died on 14 August, Losey immediately wrote to Beaumont: 'His death will undoubtedly result in lots of activity, climbing on and off the bandwagon – mostly "on" I should think ... Can we not, while Helli Weigel is in London, set-up a definite production?' On 16 August, Losey received a telegram from the Ministry of Culture in East Berlin, requesting his presence at the Schiffbauerdamm Theatre, two days later, for an international gathering in honour of Brecht. 'I couldn't go because my passport at that time was not valid,' he recalled in 1976. The invalid passport is a puzzle – a new passport had been issued by the American Embassy at the end of May. With his three-picture contract for Rank under negotiation, and the film blacklist still operating in America, he perhaps judged it wiser not to be seen in East Berlin. Yet he later expressed indignation about Laughton's behaviour at that juncture. 'I sent [Laughton] a long cable asking him to at least send a message. He never did. Now in that case I know that he asked the FBI what to do.' Laughton's biographer writes: 'The FBI indicated that there would be no repercussions if Laughton sent a telegram of condolence. So he did.'[2]

The Berliner Ensemble arrived in London to perform three productions at the Palace Theatre. Losey called on Weigel at the Shaftesbury Hotel, accompanied by his new wife, Dorothy, who recalls chatting, one actress to another, about ways of pinning one's hair. More to the point, Weigel commented on a weakness among British actors – they didn't want to be disliked or hated in their stage characters. The inference is strong that she still wanted Laughton for Galileo; on 27 August Losey sent Laughton a long letter emphasizing that he was the only English-speaking actor capable of doing the role justice. On 3 September, Losey invited Weigel to supper and suggested a Sunday drive to the country – nothing came of it. Two weeks later he approached John Gielgud about playing Galileo. A secretary replied: 'Sir John thanks you for your letter but ...'

On 14 December, having once again got no joy from Laughton, Losey decided to send Weigel an extract from Kurt Singer's *The Charles Laughton Story*, published in 1954. Singer claimed that Laughton had been innocent of the Communist affiliations of Brecht and his entourage; and that Laughton 'had more or less been kidnapped by the Communists . . .' Losey insisted that Laughton himself had 'inspired' Singer's book. The following month, writing to Ken Tynan, who unlike Losey had recently seen the Berliner Ensemble's first production of *Galileo*, he made an interesting claim: 'Brecht gave me the rights to the English production some years ago . . .' In 1961 Losey told Bertha Case, a New York agent representing Stefan Brecht: 'I have a personal letter from Bertolt Brecht giving me the rights for *Galileo* in England. I have never used the letter and never intend to.' Talking to Tom Milne in 1966, he remarked: 'Anyway, Brecht gave me the exclusive rights to do *Galileo* in English for many years . . .' This is hard to reconcile with the painful petitioning already described.[3]

Back to Laughton. In 1958 the actor was in London, enjoying a great success in Billy Wilder's film, *Witness for the Prosecution*. Losey reported to Weigel: 'I overcame my everlasting repugnance at his betrayal of Brecht and the rest of us in his [*sic*] biography and rang him to welcome him, and ask him to dinner. He was cordial enough on the phone – having been caught – but I have not heard from him since . . .' Losey advised Weigel that Laughton was now physically past playing Galileo: 'He is an utter wreck.' After a long interval Hauptmann replied that Weigel 'would rather know Charles Laughton's final stand on the matter before having any other actor approached'. What was needed for Galileo was 'this huge hunk of contradiction' – Laughton.

In May Losey made a point of meeting the 'utter wreck' when he and Elsa Lanchester were playing in Jane Arden's *The Party*, and again when Laughton took part in the Stratford season of 1959. During the ensuing two decades Losey rarely missed an opportunity to disparage the actor. In the course of an interview with the *Los Angeles Times* (March 1975), he went further, claiming that, 'Charles Laughton turned my name in to the FBI. He went to them and denounced Brecht and me.' The actor's widow, Elsa Lanchester, promptly published a denial: Losey was guilty of 'self-aggrandizement' and of the smearing tactics he deplored. Losey privately informed the reporter who had interviewed him, Guy Flatley, that his allegation was hearsay and impossible to prove – yet seven years later, in 1983, he warned Lanchester that he'd never said anything about Laughton that he could not 'document'.[4]

Losey regretted to Weigel that he had not managed to get to Berlin or Milan for the productions of *Galileo* and *The Good Woman*. But why hadn't he? Given his prodigious epistolary labours to achieve a new Brecht production, he seems to have been insensitive to the need to keep abreast of the Berliner Ensemble's own, 'definitive', performances. Losey's claim on Brecht was curiously petrified and airtight; it was a once-and-for-ever kind of knowledge; and the conclusion is difficult to resist that Losey's ego made him resistant to the work of other directors.

In the autumn of 1960 he drafted at least nine versions of an article about Brecht, 'L'œil du Maître', for the December 1960 edition of *Cahiers du Cinéma*. He attacked the idolators and all those who would attempt to '"fix" Brecht in some permanent niche, define him, enshrine him, establish him . . . how absurd the canonization of him and his theories . . . In all of my times with Brecht he never once mentioned theory.' Losey added: 'His sardonic death mask looks down at me from above my desk. He is the most undead man I know.' Losey sent the typescript of this article to Weigel. Gently she rebuked him for denigrating Alberto Cavalcanti's film of *Puntila* as 'a catastrophe'. Losey changed his text to 'Cavalcanti's *Puntila* was muddy'. Meanwhile, he also succeeded in offending Lotte Lenya. Reminding her that they first met twenty years earlier in the company of Kurt Weill, he claimed that the Royal Court wanted him to direct a production of Brecht's *Mahagonny* – but 'the Auden adaptation is of course impossible'. She rebuked him for his impropriety. 'By the way,' he wrote to Weigel, 'the Auden adaptation . . . is disgraceful. I don't understand your addiction to him.'[5]

When *Galileo* finally arrived on a London stage, at the Mermaid Theatre, it was directed by Bernard Miles and Josephine Wilson, with Joss Ackland as Galileo. Talking to Tom Milne in 1966, Losey swept the British productions of Brecht into the bin: '. . . totally inferior, over-intellectualized, misunderstood productions . . . All the productions here at the Mermaid, Royal Court and the rest were ludicrous . . .'[6]

As Losey recalled it, while in Italy shooting *Eve*, he received a telephone call from the actor Anthony Quinn about playing Galileo: '. . . he had a million dollars for it, could I get another million'. Brecht's son Stefan 'said yes' but Helene Weigel intervened. Losey travelled to East Berlin for the first time and was shown the Brecht archives, the room in which the playwright died, his grave. Losey impressed on Weigel that none of them was getting any younger, but she was firmly against Quinn: 'I'm in no hurry.' She walked Losey to the border checkpoint. 'It was a bitter personal disappointment to me . . .'[7]

However, soon after he finished cutting *Eve* he managed to interest the Italian producer Alfredo Bini, of Arco Films, in a *Galileo* film. Discussing casting with Tynan, he called Olivier (his first choice when lobbying Beaumont in 1956, and recommended by Tynan) 'the most empty, desiccated actor in the English theatre'. He preferred Paul 'Schofield' (*sic*). Tynan replied: 'The trouble with Paul is that he would look as if he had been through the Inquisition before the film began.' Circumstances offered Losey an inspiration: Burt Lancaster was currently in Italy for Visconti's film, *The Leopard*. To Losey's approach Lancaster replied cordially, remembering the Los Angeles production in 1947. On 17 September 1962, Weigel cabled:

'Burt Lancaster is fine with me and Barbara [Brecht] already said yes.' Elated, Losey gave Lancaster a script, sketches and notes, and completed a long memorandum on a *Galileo* movie. The film must be shot in colour and Cinemascope on location in Florence, Padua, Venice, Rapallo and Rome – but it must not be a 'period or costume piece'. Stefan Brecht set out the family's terms and conditions: $100,000 against 5 per cent of the net, approval of screen writer, Eisler's music to be used, Brecht's daughter Barbara to be given 'a production test ... for the part of Virginia'. However, on 4 November Patricia Tolusso wired the bad news from Rome: Lancaster had said no. 'Maybe his dislike of *Eva* had also some weight in his decision though it was not the only reason.' As with Laughton, Losey now decided that Lancaster had been, during the blacklisting era, an 'informer' at second or third remove (a reference to the Hecht–Lancaster production company in 1951).[8]

Losey told Lee Langley of the *Guardian* (9 January 1965): 'I've been quoted as saying I learned a lot from Brecht. This is true. But I also said that he learned a lot from me.' In February 1968 he declined an invitation from the Berliner Ensemble inviting him to take part in a 'Brecht Dialogue' – Brecht was a property not a seminar. In August he asked Weigel if it was really true that the film rights in *Galileo* had passed to Paramount, with no cast or script approval required by the Brecht estate. It was true. Weigel suggested that Losey make contact with Stefan Brecht and Paramount ('*Warum setzt Du Dich nicht in Verbindung . . .?*') In October 1973 Losey's old flame, Vappu Tuomioja, wrote from Helsinki with typically bruising candour: 'I've been reading Brecht's *Arbeitsjournal* and find it fascinating ... But how is it possible that he doesn't mention you at all, he writes a lot about Laughton and once he mentions that Orson Welles will direct *Galileo*, and leaves it at that. Could it be that he deliberately erased you out of his diary?' Losey replied: '[Brecht] was, of course, very much stage-struck and extremely fascinated by Laughton and also to some extent by Orson Welles, though he, naturally, let him down. I suspect Helli of a lot of editing – maybe even Barbara [Brecht] but, even so, it's a little inexplicable.'

Finally – after twenty-seven years – the apple fell into his hand. Paramount's rights having lapsed, the Ely Landau Organization was able to buy them cheaply from the Brecht estate for its American Film Theatre series (which included Pinter's *The Homecoming*, Genet's *The Maids*, and Osborne's *Luther*). Landau approached Losey in April 1974, with shooting due to begin at Elstree only two months later. Losey alone knew how long, hard and painful the journey had been.

Galileo – The Film

Losey called in Barbara Bray, whom he justly described as 'a distinguished literary lady, winner of the Scott Moncrieff Prize for the best translation of last year; collaborator of Pinter, Beckett and others . . .' She suggested a screenplay based on a fusion of sources: the 1947 Brecht–Laughton text; changes made in Brecht's final German version; and Brecht's intention as described by Ralph Manheim and John Willett in their edition of Brecht's plays. Bray sent a copy of her first draft to Willett, who responded with a wide range of helpful scholarly points, archival indicators and rewritten passages; but Losey was in haste.[9]

Each production in the AFT series was budgeted at about $1 million, with uniform fees for directors, a modest $30,000, plus 2 per cent of the gross to $1 million, 3.5 per cent thereafter. Landau advanced Losey (but no other director in the series) a further $15,000 against these percentages.

In April Losey resumed his obsessive pursuit of the elusive Brando, with whom he could make absolutely no contact, even when visiting LA. Imagining – as he had to – Brando's objections, Losey sent a long cable (6 May): 'Yes . . . yes . . . I know about the . . . Indian picture and I respect and believe in your conviction and I also know that you make it a principle not to work in the summer.' But – 'I implore you . . .' Not even an Actors' Studio murmur came back. Otto Plaschkes, European head of production for Landau's AFT, advised Losey to take Topol, star of *Fiddler on the Roof*. Losey acquiesced, turning down a better actor, Nicol Williamson. Plaschkes insists that he recommended Topol only because he knew that Williamson would not long tolerate the kind of personal behaviour that Losey was displaying. On the set Losey displayed frequent irritation with Topol – Plaschkes and Barbara Bray both comment that he showed no inclination to respond to the easy-going, extrovert actor's desire to talk through his part. Losey distributed some notes on Brechtian acting (precision, economy) which may have raised some eyebrows among seasoned British actors who did not normally flap their hands vicariously or skip about like amateurs.*[10]

Losey had ideally wanted to shoot the film on location and to abstract certain aspects of the décor by means of lighting – 'one piece of ornamentation, one fragment of décor, instead of a whole palace or a vast room which would distract the eye . . . in a studio it is virtually impossible to get the kind of texture needed'. Location filming being impossible on such a budget, he brought in Richard Macdonald and opted for 'a vast composite set on a revolving stage. But instead of the stage revolving it was the camera that would move around the set.' Brueghel the Elder was taken as a source for the colours and compositions, and for the choreography of the Carnival scene.

*Patrick Magee had appeared in *The Criminal* and *The Servant*; Michael Gough, Edward Fox and Margaret Leighton in *The Go-Between*; Edward Fox in *A Doll's House*; Georgia Brown in *Secret Ceremony*; Michel Lonsdale would soon play in Losey's *The Romantic Englishwoman* and *Mr. Klein*.

Macdonald produced simple, uncluttered sets conveying depth of texture, using wood, metal and stone: the impression is of deep focus enhanced by Losey's normal attention to exactly achieved naturalistic sound effects – bird song for a scene in a stylized garden. Losey's stated aim was, 'Not the photographic record of a theatre performance, but a new kind of cinema that is, in a very real sense, theatrical. Perhaps some comparison with Cocteau and *The Cabinet of Dr Caligari* is in order . . .'[11]

Losey had maintained intermittent contact with Hanns Eisler – 'Dear Hansel' – since leaving America. But Eisler was by now dead and in his archives only a score for five instruments could be found, with no trace of the score for seventeen he had written for the 1947 stage production. Losey called in Richard Hartley to compose a new arrangement, using a harpsichord, percussion, a range of woodwind, with cellos, horns and bass. The effect of the missing range of middle strings, violins and violas (Hartley comments) was to reinforce a 'brittle' effect as the score 'progressed vertically'.

The film was shot on the biggest EMI stage at Elstree Studios, one composite set representing Padua, Florence, Venice and Rome. A second stage was used for the ballet and a third for the recantation scene. Starting on 8 July 1974, working six days a week, with eighty-six speaking parts, Losey cut as he shot – shooting was completed in four-and-a-half weeks. Gerry Fisher being unavailable, Michael Reed was the lighting cameraman. Losey gave him a rough ride: when the tests were screened 'Joe wiped the floor with Mike', Losey's assistant Richard Dalton recalls. Of Losey's films, *Galileo* was the one that Dalton least enjoyed working on. 'Joe was at his most embarrassing . . . a sort of awkward outlandishness. Going around the studio in a kaftan. Nodding off on the set. He was drinking. I felt protective and resented any snide comments.' Otto Plaschkes, finding Losey fatigued, hungover, sometimes late on set, aloof from the cast, 'offering no help to the younger actors', felt chilled by his lack of visible affection for his colleagues.[12]

Of particular interest is Losey's approach to Scene 14, which covers the last decade of Galileo's life (1633-42) as a comfortable prisoner of the Inquisition while secretly writing his *Discorsi*. When he reveals this to Andrea, the young disciple's attitude reverts to adulation. Galileo demurs: 'I taught you science and I denied the truth.'

ANDREA: You concealed the truth. From the enemy. Even in the field of ethics you were a thousand years ahead of us.

Galileo is humorously sceptical about this. 'I was shown the instruments' – pure fear had prevailed, no plan. Brecht had called this the 'pelican scene' – the pelican pierces its own breast to feed its young, just as Galileo humiliates himself in order to offer Andrea the full truth. During the period of post-war rewriting, Brecht had imposed on Laughton an unequivocal condemnation of

Galileo out of his own mouth: 'Welcome to the gutter . . . I surrendered my knowledge to those in power, to use, or not to use, or to misuse, just as suited their purposes.' Paradoxically, Brecht had turned the moral point away from Marxism; history now depends on the courage, the heroism, of an individual (Galileo). Losey concurred: 'One man of genius' could make a 'critical difference to the history of a specific time'. (Order of Lenin, or Stalin.)

The Brecht–Losey parallel between Galileo and the 'betrayal' by the Manhattan Project atomic scientists is unconvincing. Both Brecht and Losey discovered their own disgust about Hiroshima only some time after the bomb was dropped. Losey's 1974 statement exposes a degree of Party line opportunism: 'By the time the play was revised, the emptiness of the fascist defeat, the disillusionment about the oppressive society in the East and the full significance of the scientists' abdication . . . had become historical fact and now must weigh heavily . . .' There is a certain sleight-of-hand in Losey's slipped-in phrase, 'the oppressive society in the East'. On the contrary, the Soviet atomic bomb remained 'progressive': it meant 'Peace'. Brecht chose to settle in East Germany; as late as 1958 Losey and Elisabeth Hauptmann were in strong agreement that the playwright had chosen the 'progressive' face of Germany. Addressing his cast in 1974, Losey ventured beyond Brecht's own (1946–7) position to embrace also the case of J. Robert Oppenheimer, a *cause célèbre* in 1954: but Oppenheimer was punished for past political associations he now repudiated, not for his physics.

Losey began by promoting his film to the promoter himself, Ely Landau. This, he wrote in November 1974, was the first film since Pabst's *Dreigroschenoper* in the 1920s 'which in any way mastered the problem of transmuting [Brecht's] work from theatre to film'. He urged Landau to show the film at Cannes, 'where I have, at one time or another, won all of the prizes as well as having been President of the Jury'. Landau invited 100 film critics and entertainment editors to the Magno screening room on 55th Street, New York, to view 'product reels' from American Film Theatre's coming season. A dinner party at Mamma Leone's followed, with a wounded Chaim Topol on hand to reverse – deliberately, one cannot doubt – the political perspective of Brecht and Losey. Speaking as an Israeli with military service under his belt, he chided Europe for crawling on its belly while the Russians rubbed their hands. As for Brecht, he had fled from the Nazis and McCarthyism, then sold himself to East Berlin. 'But he wrote good plays,' Topol added. Losey, who invariably rebuked his actors for publicly stepping out of line, evidently remained unaware of Topol's heresy. Arriving in New York to promote the film, on 15 January 1975, Losey stayed at the Algonquin, appeared on CBS's *Mike Wallace Show*, and gave the usual newspaper interviews, roaming across the years and settling scores ancient and modern, including Laughton and Nelson Rockefeller. To Topol he reported: 'I have just come back from New York, where I did the kind of arse-breaking, mind-boggling, soul-destroying press campaign, which you undoubtedly know so well.'[13]

As AFT's vice-president, Mrs Edythe Landau, explained, the target audience was mainly the college-educated people and the professional classes. AFT made available teachers' guide books for its filmed plays. The subscription price for the season's five productions was $20 (evenings), $12.50 (matinees), $10 for students and senior citizens. Each film was screened locally twice on Monday, twice on Tuesday. A Telepictures Corporation statement dated 30 September 1982 put the US receipts at $1,185,607 gross, $1,084,626 net. Losey's (Launcelot Productions) 2 per cent share was $21,693.

Galileo won critical acclaim across the hinterlands of the United States as never before in Losey's career, and never again. The superlatives rained down: '. . . the vitality, radiance and grandeur of a Renaissance painting . . . superb . . . a triumphant marriage of theatrical and cinematic art'. In the *New York Times* (27 January), Vincent Canby described Losey as 'a man who knows more about film and more about Brecht than possibly any other film director at work today'. But less luck across the water. In May 1975 *Galileo* was shown at Cannes, without subtitles, in the newly created 'Les Yeux fertiles' section devoted to filmed stage plays, but distribution of the film in the UK by Seven Keys ran a full year behind the United States. A limited season finally opened at the Curzon Cinema in May 1976 after bitter internal rows about its promotion and scope, under the title 'British Film Theatre'. From France Losey raged against the British distributors' incompetent packaging of six films in a manner likely to destroy viable publicity for each. Critical responses were subdued and there was no general release.

21

Pinter Again:
Accident

Accident is Losey's best film. Pinter's screenplay, Losey's direction and Bogarde's performance are close to perfection.

Jud Kimberg, who was working for Sam Spiegel's Horizon Pictures, first drew Losey's attention to Nicholas Mosley's novel. Spiegel bought the rights and Losey enlisted Harold Pinter to write the screenplay. Intensive discussions took place in Amsterdam and Sicily, where Losey was filming *Modesty Blaise*. Pinter produced a draft script in July 1965, a second draft less than two months later. Losey recalled: 'It was quite clear immediately that Spiegel wanted to impose himself on the script . . .' The great producer invited Pinter to work on his yacht – Pinter pleaded a family holiday. 'And Spiegel, who was obviously very anxious to isolate him from me, said, "I will send a helicopter for you!"' Spiegel also challenged the prospective casting: '"What do you want Bogarde for? Who's ever heard of him?"' Losey claimed that on leaving Spiegel's office Pinter vomited in the street – but Pinter flatly denies it. In Losey's view, Spiegel had become 'a megalomaniac and I think impossible for most directors to work with now because he wants his films to be "Spiegel" pictures'. Eventually Losey and Pinter persuaded Spiegel to relinquish the rights, but the price was high: 'We paid him above the cost, $30,000 and a percentage.' (Later, working on another project, Pinter came to admire Spiegel's intellect, broad culture, and the film-intelligence with which Losey, he feels, was not at ease.)[1]

In Nicholas Mosley's novel, *Accident*, two married Oxford dons (Stephen and Charley) fall in love with the same student, an Austrian aristocrat (Anna), who is also pursued by, and finally engaged to, another student, a young English aristocrat (William). But he is killed in a car crash, late at night, driving with his fiancée to the home of one of the dons.

Mosley explained the central proposition of his novel: 'The [car] accident is a catalyst in the sense that people can carry on with the difficult, tormented relationships, keep them spinning, until something disastrous . . . makes them think about their responsibilities. They are intelligent, civilized people who are aware of their violence and their neuroses.' Losey agreed: the characters knew the questions but not the answers.[2]

Pinter set himself two transformational tasks. Mosley's novel is written in a jerky, staccato style, very short sentences, often without verbs, in an attempt to capture the jumps of consciousness. But Pinter concluded that free association is a technique which does not work in the cinema. The novel is narrated by a philosophy don, Stephen Jarvis. As with both *The Servant* and *The Pumpkin Eater* (and later *The Go-Between*) Pinter smartly disposed of the narrator, turning the 'I' into 'he', the knowing observer into the objectified protagonist. But how to turn streams and leaps of consciousness into film language? Pinter's solution is indicated by a letter to Losey accompanying his first draft (19 July 1965): '. . . this is *not* the script we have been talking about in our provisional discussion . . . it does not explore in that way the various levels, of memory, association, etc. . . . Almost every possible mental "flash" seemed to me, while working, unnecessary, too explicit, even crude. I couldn't see *room* for them, I don't know. Eventually I decided to go the whole hog and tell the thing without them . . . In principle I've tried to concentrate on a hard appraisal of the happenings and keep it objective . . . The camera is everything . . .'

The second draft script is dated 2 September. The general impression is of few changes. For example 377 shots becomes 378. The structure of Pinter's original screenplay remained far more intact than with *The Servant* – and rather more so than with *The Go-Between*. The reason is not obscure: the screenplay of *Accident* achieves structural perfection. Losey's notations and proposed amendments were few, and mainly confined to minor inflections of dialogue. For example, Stephen's line, 'Aristocrats were made to be . . . murdered' becomes, in the second draft, 'slaughtered' (though in the film itself the verb is, inexplicably, 'killed').

Pinter clearly experienced some nervousness when, in April 1966, the moment of sending Mosley the screenplay could no longer be delayed. Having taken some drastic liberties, he enclosed a '*mea culpa*' letter which he urged Mosley not to read until he'd read the script. In Mosley's novel Stephen describes his colleague and friend, Charley Hall, as almost an *alter ego* or *doppelgänger*: but Charley's is the less inhibited libido. This *doppelgänger* dimension disappears in Pinter's script; Charley is amalgamated with another don, the much-televised Tommy Parker, described by Mosley as a 'historian wearing Italian-style trousers and a black polo-neck sweater'. Charley thus becomes the extrovert TV-don whose brash, macho personality is the foil for Stephen's introversion. In the novel a real crisis – William's death in the car accident – rapidly brings Stephen and Charley together, their sexual rivalry over Anna subsumed by tragedy. Their common moral concern is whether Anna was driving at the time of the accident – and whether they can justifiably conceal that she was with William in his car. But in Pinter's script the almost childlike Charley, a helpless prey to his own lust, is the last person Stephen could turn to or confide in.

From this basic alteration more followed. In the novel Stephen's attraction to Anna leads to no physical contact between them. Charley has done what Stephen would like to do – but can't. In Pinter's version, Stephen harbours and conceals Anna, confides in no one and, by the end of that terrible night, virtually imposes sex on her in his empty house. In short, Stephen was no longer the man that Mosley had created.

At least two factors were at work here. The first was Pinter's own inclination towards rough play. In Pinter's universe (at that time) indirection, sublimation, euphemism were the preludes to the big bang. The other factor has more to do with dramatic form. A first-person narrator who is also a philosophy don is almost 'set up' to dither and hesitate. Those who can, do; he writes because he can't. As soon as Stephen becomes an objectified character in the screenplay, he gains a potential capacity for action, for having the girl he wants.

So the scrupulous Stephen ends up by more or less raping his own student while she lies in a state of shock – by any criterion a degrading, demeaning, squalid thing to do. The critic Raymond Durgnat offered his own explanation, while avoiding moral strictures: 'Stephen ... thinks of himself as very civilized, but he's a "can't quite" man. He can't quite declare himself to Anna, he can't quite get Charley's TV job, he can't quite be honest, and in a fury of despair he concludes with a near-rape ...' Losey's production notes dated 29 July 1966 attempted to clarify and justify this episode for the actors' benefit: 'Even as [Stephen] does it he is aware of guilt, shame and hypocrisy ... It is a battle of the sexes in the fullest sense of the phrase, and humiliation, subduing, sadism and all other such animal aspects of physical sexuality when it is without love ...' (Reading these production notes for the first time, twenty-four years later, Mosley is appalled by directives foreign to his novel – but typical of film people, he adds.)[3]

Pinter's covering letter to Mosley sought to explain: 'As you see, there's one major deviation, change – it might be said distortion – the fact that Stephen sleeps with Anna and that Charley knows nothing about anything at the end.

'I must tell you that I worked very hard to follow your ending at all points to begin with ... But there was something wrong ... the long debate between Stephen and Charley [immediately after the car crash] simply did not work, convince, sustain itself in dramatic terms ... Stephen, ultimately, must be alone in final complicity with Anna. Or so it seemed to me. And, in many long discussions, to Losey ... Possibly the main point and the point you might consider an inexcusable distortion of your intentions is that Anna and Stephen now sleep together. All I can say is that this, from what I've been saying, sprang into being as inevitable (their complicity, their closeness, etc.) ...'

Mosley's written response was evidently sympathetic, even grateful. He was impressed by the grace, symmetry and sensitivity of Pinter's screenplay. However, he did (he recalls) object to Stephen having sex with Anna after the car crash. As he now puts it: '. . . even in terms of Harold's "Stephen" it was a false note to insist that he was such a shit as to sleep with Anna in the circumstances'. But how much of this Mosley actually conveyed to Pinter or Losey is unclear. Mosley was in awe of Pinter–Losey. 'They were the best' (a view he still holds). Besides, having seen *The Birthday Party*, *The Caretaker* and *The Homecoming*, 'One knew that if one was going to have a Pinter script one was going to have one's characters brutalized.'[4]

Dirk Bogarde comments on Stephen's conduct. 'I resisted it as far as I could but Joe wanted it desperately badly . . . It was bestial and I think Joe wanted that crudity.' He also recalls that Losey dressed Sassard in a hooped skirt for the night of the car accident, 'so that it would blow up all the time and show her legs'.[5]

The search for an alternative producer to Spiegel began. Losey later remarked, 'I cannot tell how many people turned down *Accident* and the people who finally made it said, "we don't like it, we don't understand it, but go ahead and make it".' The production got off the ground mainly because of the sympathetic intervention of John Terry, of the National Film Finance Corporation (NFFC), who persuaded Sidney Box's London Independent Productions to enter into a matching investment, £150,000 from each party, each to receive 33⅓ per cent of net profits (the gross minus 'negative costs' and the distributor's commission). From LIP's 33⅓ per cent of the net profits, 7½ per cent would be paid to Horizon (Sam Spiegel) and 2½ per cent to Harold Pinter. The total production budget was, finally, £272,811 – roughly double the corresponding figure for *The Servant* and roughly half the budget for *The Go-Between*. Royal Avenue Chelsea Productions agreed to lend Losey's services as director and co-producer for £17,500 plus 9 per cent of the net profit. Bogarde's fee was the same. The most striking inequality is found in Horizon's agreements with Mosley (£2,700 for the rights), with Pinter (£12,000 for the script) and with Mrs Pinter (the actress Vivien Merchant) – £8,000 for 'collaboration' on the script. Even if Mosley's fee was doubled on first photography (as possibly suggested by LIP's allocation of £25,627 for 'story and script'), novelists might be excused a growl.[6]

Mosley's was an Oxford story but, painful to reveal, in February 1966 Losey approached a number of Cambridge colleges, including Clare, to whose Senior Tutor he explained that he would prefer to shoot the film in Cambridge 'for many reasons'.[7] This heresy is buried in an unmarked grave. The main location for *Accident* was to be St John's College, Oxford. Losey's correspondence with heads of Oxford colleges and other dons followed a pattern, each letter validating his project by mentioning the 'many international awards' won by *The Servant*, and the distinguished cast already assembled for *Accident*. The screenplay had been 'enthusiastically approved by Lord Ravensdale' (this was Nicholas Mosley). Losey's clear preference was for the more spectacular quadrangles, towers and façades of Christ Church and Magdalen.

Approaching these ancient corporations can be daunting; drafting a letter to the President of Magdalen, T.S.R. Boase, Pinter worried about the correct form of address – 'Dear President' or 'Dear Mr President?' – and feared that his letter was 'a little cold'. What turned out to be freezing was the President's room when Losey called on Boase bearing a copy of the screenplay; a single electric fire fought a losing battle with the damp chill. But Losey was treated to lunch and returned to Lowndes Square in a mood of confidence. The phone rang as he closed the door. President Boase had been warned that the female undergraduate in the film slept with a don – but two dons was altogether one too many. Magdalen said no. Losey wrote to the Warden of Wadham, visited the President of Trinity and the Warden of New College and thanked the Dean of St Edmund Hall for his hospitality.[8]

One don in particular was prudently shown a script before shooting began. This was the Warden of St Antony's College, Raymond Carr, a close friend of Mosley's since the post-war years. Mosley had loosely hung his novel on an actual incident: Carr and his wife had given a party for New College undergraduates at their country home, after which a student had suffered a quite serious car accident. No sexual dimension was involved and the doctors reported that the victim had not drunk excessively, but the Warden of New College had taken a severe view of it and Carr was asked to resign his fellowship. Mosley had asked his friend whether he minded him building on the incident, and Carr – rather liberally – raised no objection. In the novel reality is reversed: the head of the College urges the don not to blame himself for the undergraduate's death. But now New College turned Losey down after the 'Carr factor' in the story became a talking point. 'The annoying factor,' Carr wrote to Losey, 'is that my connection with the film is now public property and I presume I shall be identified as, at one stage of my life, rather drunken and slightly lecherous.' But Carr also offered his 'weedy tennis court with a battered net which might be useful to you'. Sir Raymond Carr writes: 'This was a mercenary act. I thought that I might make some money by hiring out the tennis court.' He adds: 'I didn't take to Losey . . . I had been told that he was "left wing" and opposed to the establishment. I came to the conclusion that he was a cinematic Proust, fascinated by the upper crust, even a snob.' Indeed *Accident* focuses exclusively on Oxford's country-house set, although by the mid-sixties half the undergraduates came from state schools. Carr was also critical of Pinter's dialogue: 'Both the wives are made to use language which the models would never have used and which, I don't think, would be used by dons' wives in general . . .'⁹

Indeed the script is full of minor anachronisms: *He watches* ANNA *walk across lawn towards college.* WILLIAM *crosses lawn, at a distance from her.* In general Oxford quadrangle lawns are not to be walked upon. Aristocrat though he is, young William's cocky insolence towards his tutor Stephen sounds odd.

> WILLIAM: Well, come on! What do you think of her? [ANNA]
> STEPHEN: I don't think.
> WILLIAM: I thought thinking was your job!
> STEPHEN: Not about *that*.
> WILLIAM: You're not past it, are you? Already?¹⁰

Before the film's release, Losey invited Mosley, Carr and their wives to lunch, followed by a Wardour Street screening of the film, in order to cover himself against any future objection from Carr. Mosley remembers that Carr, in a teasing and 'rather tiresome' way, kept calling out objections. Finally Losey turned round: 'Shut up, you're disturbing other people.' Carr comments: 'It was a farcical caricature of a small social set – and even that not accurate in detail. I made my comments to Dirk Bogarde who was sitting in front of me. He mistakenly thought he was playing me in the film . . .'¹¹

Dirk Bogarde insists that he was not Losey and Pinter's first choice for the part of Stephen, but prefers not to mention the actor they had in mind. 'I wanted the part badly, so they were landed with me.' According to Losey, bringing Bogarde and Stanley Baker together was not easy – Bogarde derided Losey's 'passion' for 'bullying actors', particularly Welshmen like Baker and Burton. On location Bogarde loyally described his collaboration with Baker as 'a marvellous matching of contrasted flavours, a real strong country Cheddar and a more delicate, insidious Demi-sel . . .' Today Bogarde is less diplomatic: 'I was against Stanley – too thuggish, too much the working-class lad. He arrived in a toupée and a lot of make-up and was always doing his eyelashes for the first two days. After that he was terrific, wonderful – I was very fond of him.'¹²

Stanley Baker was an elemental actor relying on a brooding menace, physically assertive, a cocksure style with women, always himself. Losey invariably used him as the outsider, the loner who never fully belongs and does not wish to. Baker regarded Tyvian, in *Eve*, as a 'pseudo-intellectual', and likewise Charley in *Accident*, but in *Accident*, Baker remarked, everything lay beneath the surface. 'To play it you really have to dig deep into yourself, and bank all the fires down.' The critic Margot S. Kernan found his nervous, furtive gestures 'perfectly suited [to] a pop intellectual ill at ease among his tweedy betters in the Common Room'. Baker held that he learned the seriousness of acting from Losey – Losey agreed. After Baker's death his widow, Ellen, wrote: 'You were the man who made Stanley . . . the complete caring professional that he became, and he never forgot this.' Losey sent her some poems 'written on clay tablets, fifteen hundred years older than the Homeric epics (the story of Gilgamesh)'.[13]

Losey cast the French actress Jacqueline Sassard as Anna, the beautiful Austrian undergraduate for whom two dons and a young aristocrat are in competition. Although – and because – Anna is virtually a non-part, Losey's casting notes visualized an unusual combination of incompatible attributes: her 'animal-like' quality should be 'backed up by a good but very female mind. She seldom speaks but it is clear that her mind is never at rest. There is an extraordinary energy in her body in spite of its apparent lethargy and an enormous reserve of passion in spite of an apparent surface calm.' This was a tall order and Jacqueline Sassard wasn't it. Losey called in several coaches to improve her English diction, testily describing her as 'almost inaudible' and 'a very "touch and go" proposition in all departments'. Both Bogarde and Ann Firbank describe Sassard as nice, warm, merry, outgoing, and keen to please Losey, but, according to Bogarde, 'Joe terrified Jacqueline, petrified her – literally'. By the end of filming Losey was considering bringing in another actress to post-sync the part. When the film opened in New York, the *Times* quoted him: '*Accident* will do for Jacqueline [Sassard] what *Darling* did for Julie Christie, *Morgan* did for Vanessa Redgrave, *Georgy Girl* did for Lynn Redgrave.'[14]

Losey persuaded Delphine Seyrig to play the cameo part of Francesca, the Provost's daughter and an old flame of Stephen's. The star of two Resnais films, *Marienbad* and *Muriel*, Seyrig was 'enormously busy and expensive and I got her to play the role on the basis of spending two days shooting in England'. Given Mosley's description of Francesca, via Stephen, as 'a middle-aged matron with her teeth going', the choice of Seyrig was blatantly box-office.

Accident launched young Richard Dalton's long collaboration with Losey. It was May bank holiday weekend and Losey needed a first assistant. Dalton, who had been working on the TV series *The Avengers*, was summoned to a flat in Knightsbridge. 'There wasn't a technician or artist who wouldn't have worked with him.' He encountered a huge man in a huge settee in a huge lounge. Losey handed him a script. 'I was glad to get away.' On the set Dalton noticed that Losey lacked only one quality, humour. Dalton and other football fans in the crew were using portable radios to follow the fortunes of the World Cup. Joe didn't like it: he referred to soccer as 'this homosexual game where they all hug each other'. Later he gave Dalton six goblets and placed him in his pantheon of first assistants, along with Bob Aldrich and Carlo Lastricati.

The unit travelled to Oxford on Monday 4 July 1966. Stephen's fictional Oxford college uses exteriors from both St John's (including the bell tower and clock) and the river frontage of Magdalen. The window side of Stephen's study is a room in St John's; the rest of the room was built in the studio. Stephen's country home was likewise a combination of exterior locations (Norwood Farm House, Cobham, in Surrey) and interior studio sets, designed by Carmen Dillon. William's country home, 'Codrington Hall', was filmed at Syon House in West London, owned by the Duke of Northumberland, who invited Losey to dinner and whom Losey addressed, when writing, as 'Dear Duke'.[15]

As usual with Losey, the design dimension of *Accident* was of the highest standard. Eschewing extravagance, he and Carmen Dillon settled for a naturalism which drew its peace and beauty from English settings wrapped in unostentatious privilege. Dillon explained that 'the look' of this film was intended to be 'not having a look, avoiding any obtrusive stylization . . . making everything look used, lived in, believable'. Bogarde expressed his gratitude, as an actor, to her meticulous attention to detail.[16]

The film was shot in Eastman Color. During the early 1960s, most European 'art-house' films were made in black-and-white, and colour was still associated with more popular entertainment. By the mid-sixties, however, the industry had reached a turning point. Losey still preferred black-and-white but the money-men insisted on colour. 'I suppose the colour in *Accident* is more like that in [Resnais's] *Muriel* than anything else . . . and with none of the frantic signalling Antonioni indulges in *Deserto Rosso* . . .' As for Oxford itself, it was 'a grey place, the sandstone is largely grey, so it's grey and green, and green of course is the most dangerous colour in films because it is the most obtrusive'. As Virginia Starr, TV librarian of KABC-TV, Hollywood, astutely noted, 'In *Accident* the emphasis is on softness, the almost too-warm color, and clutter, the cloying too-closeness in Stephen's life – the deadliness of suddenly unwanted serenity producing feelings akin to the panicky depression of over-sleep.'[17]

This was Gerry Fisher's first film for Losey, the dawn of a remarkable collaboration. Having served in the Royal Navy during the war, Fisher had been promoted to principal cameraman for a portion of David Lean's *The Bridge over the River Kwai*, later impressing Losey when working as operator on *Modesty Blaise*. His major break arrived when Douglas Slocombe became unavailable for *Accident*. While shooting in Ireland, Fisher received a call from Losey, followed by Pinter's script: he was given twenty-four hours to decide. (Promotion from operator to lighting cameraman was the one excuse accepted within the industry for leaving a production.) 'It was a plunge into deep territory to do a Losey film. Some people thought I was foolish.' There was also a scarcity of good operators and Fisher lacked the status to engage one of his own choosing. When they viewed the first rushes in an Oxford cinema, Losey gave him a rough ride. In certain scenes, notably the punting sequence on the Cherwell, Fisher had to perform as his own operator. Fisher lives in a house overlooking the Thames; on the day this writer visited him a swan flew low over the water, evoking memories of the beautiful swan-shot in the film – hearing the beat of wings behind him he had swung his camera. When filming was completed Losey congratulated him, offering him the very best light meter he could want. Fisher praises Losey as a complicated, caring, graceful director (and man), who always created an atmosphere in which his camera team could concentrate on and enhance the central images he wished to achieve. He encouraged a discreet use of skills, not as an end in themselves.[18]

The weather was terrible: summer scenes were constantly interrupted by cloud and rain. One shot of Vivien Merchant in a deckchair on the lawn, supposedly in mid-afternoon, had to be done at night. Fisher comments: 'You learn to analyse everything you're looking at – then to restore it photographically.' The outdoor scenes in *Accident* were 'a kind of chaos which ended as totally natural, relaxed . . . made up of tiny stitches and pieces like a tapestry'.

Of the three films that Losey made with Pinter, *Accident* was the most restrained stylistically. Taking the chessboard analogy, in *The Servant* Losey corralled his few pieces into a tight corner, the camera trapping them in shifting silhouettes and highly stylized permutations of claustrophobia. In *The Go-Between* the camera soared across the board, zooming in on every corner, targeting the scattered pieces in a succession of triumphant strikes. In *Accident*, by contrast, Losey moved quietly, self-effacingly, like a player whose orthodox defence would in time yield its winning positions. In fact he used a zoom for the first time on *Accident*, but with discretion. Pinter commented: 'I think you'll be surprised at the directness, the simplicity with which Losey is directing this film: no elaborations, no odd angles, no darting about. Just a level, intense look at people, at things. As though if you look at them hard enough they will give up their secrets.'[19]

Michel Ciment noted the striking use of sound in *Accident* – the noise of oars (a punt pole, surely?) in water, the hissing of a kettle. Losey was using both sound booms and telescopic 'rifle' microphones. The Oxford of *Accident*, and the neighbouring countryside, is sonically expressive: bells, planes, cows. In Losey's view the primary sound was of bells – 'and of course bells are an obsessive thing with me. And that began, as far as I know, with *Eve*. And then later in *A Doll's House*, bells over snow.' Peter Handford, the sound engineer, recorded the abrupt stutter of the unseen train which has been puffing away steadily in the distance during the run-up to the opening car crash. Losey's post-shooting notes on the overturned car read: 'Sounds of steam, dripping oil as well as ticking ignition and rolling and pouring of whisky bottle are important to clarify image . . .'

22

Accident:
The Film

Night. As the credits come up, the camera stands at the gate of Stephen's elegant Oxfordshire house, gazing up the gravel drive to the façade like a burglar weighing a break-in. Only one window is lighted (our protagonist's); a typewriter chatters faintly as we hear the distant growl, then roar, of a car approaching at reckless speed. A dog barks, a distant train answers, a splintering, screeching crash signals the fate of the car. A light comes on in the hall; the camera jolts down the country lane, the baffled eyes of a man running to help. Cut to the grotesquely overturned car, lying on its side, and to Anna (Jacqueline Sassard) and William (Michael York) slumped inside. Our view of the disaster is Stephen's – images of the two young people framed by the open door, shattered mirror and windows. William's face is covered in blood.

Stephen (Dirk Bogarde) climbs on top of the car and wrenches the door open. Blood, glass, a whisky bottle. Anna is alive, William dead. Stephen hauls her out; in her stunned state she stands on William's face. Stephen carries her to the side of the road; she lies inert in her short white party dress, then manages to walk to the house. For reasons we shall later discover, Stephen is tonight alone at home.

Her dress covered in oil and blood, Anna slumps in a chair, dumb and stunned. He asks her if she was driving, she doesn't answer. Stephen summons the police. By the time they enter the drawing-room Anna has vanished; Stephen stumbles into the personal peril of not revealing her presence in the car. But does silence in the face of an unasked question amount to perjury? A good question for a philosophy don. When the police have gone Stephen discovers Anna stretched across a bed upstairs (her dress riding up her thighs – his erotic response is not in doubt).

Here – towards the end of the story – we enter the long flash-back which constitutes the central narrative, culminating in the car crash. One further close-up of the dead William explodes into a close-up of an animated, smiling William, the camera slowly pulling back to show him, very much at ease, ending a tutorial in Stephen's college. The sherry comes out and their conversation is no longer about philosophy, settling on Stephen's pupil, Anna von Graz und Loeben, whom William has noticed 'walking about' and intends to meet.

Bogarde establishes his donnish incarnation with a pipe, a sherry bottle, and a well-observed manner of lightly holding books and tossing them away. At the end of this scene the camera looks down from Stephen's window to observe Anna fondling a goat tethered on the grass in the centre of the quadrangle. She is Stephen's next tutorial and Stephen's nemesis: John Dankworth's jazz score, with its throaty saxophones and heavenly harps, will reinforce alternating moods of elegiac delight and guilty anguish.

An establishing scene at Stephen's extremely comfortable country home follows: his pregnant wife Rosalind (Vivien Merchant), his two small children, the dog, the lovely garden – happy family. (Has Stephen inherited wealth? His salary, or 'stipend', would not be sufficient to sustain this lifestyle and Rosalind clearly has no job. Some prosperous colleges offer a house at a nominal rent to married dons ('fellows') – but such a house?)

A pair of feet brashly resting on the back of a chair introduces Charley (Stanley Baker) who, as the camera pulls back, is discovered in the Senior Common Room alongside other reading dons, including Stephen and the taciturn Provost (Alexander Knox). Charley reads aloud some statistics concerning the sexual activity of students in an American college. Evidently 0.1 per cent had sexual intercourse 'during a lecture on Aristotle'.

PROVOST: I'm surprised to hear Aristotle is on the syllabus in the state of Wisconsin.

Next we discover William with Anna as he punts her along the Cherwell river, passing Magdalen, weaving between summer foliage – an idyllic afternoon gorgeously photographed with a rapid alternation of perspectives and views which enhance the punt's romantic progress. They pick up Stephen, who is carrying his gown. The tutor self-consciously seats himself in the punt beside Anna, clearly preoccupied by her lazing body in its summer frock (it can't be her intellect or wit which fascinates him). William's punt pole carves through the water, the younger man in control, his tutor faintly humiliated to be a mere passenger alongside the girl. A swan flies low over the river.

When Stephen tries to be useful by standing up and fending off a tree hanging low across the water, he manages only to fall overboard. In the film (but not the script), Stephen yells and both William and Anna laugh loudly. Losey told his editor, Reginald Beck: 'For the obtuse . . . I think it is important to cut in Stephen's cry and loud splashing in the water . . . ' The point is hammered home in the next scene. William accompanies the soaking Stephen back to college and advises him to 'do an hour's squash every morning'. Stephen invites William home to lunch the coming Sunday, with Anna. It's Anna he wants but William's presence may validate the invitation in the all-seeing eye of Rosalind. Stephen is a philosopher of self-deception.

Losey found much in the sandstone quadrangles of St John's College to absorb the camera-eye. Occasionally he unfettered his relish for ornamental detail – a tutorial in college is intercut with a zoom-lens journey along the gargoyles and carved signs of the zodiac adorning the quadrangle, synchronized with the tolling of the hour and supported by Dankworth's clarinet and two harps. Stephen is giving Anna a tutorial. He offers her a book by his colleague Charley Hall, 'the archaeologist' (in the film) and 'the zoologist' (in the script), which apparently 'expresses another point of view' – but about what? Neither with William nor Anna does Stephen venture into the subject of the tutorial; what counts is the sexual politics, as when Stephen asks Anna whether she knows Charley Hall.

ANNA: Yes, I think . . .

Charley will turn up uninvited to the Sunday lunch; Stephen will not guess that he is already having an affair with Anna. But how did Charley pick up Anna in the first place? The archaeologist (who also writes novels and appears on television) clearly does not give her tutorials. These questions are not quibbles: they reach into the basic implausibility of Stephen and Charley as dons, although Bogarde has a magnificent shot at it.

Losey's camera-eye catches the wider context of Stephen's tranquil, but now sexually subverted, existence. It roams across the roof-lines of the college, dips into roaring traffic along the ring road round the city, pursues Stephen down a country lane as he posts a letter and rescues his young son from a tree – a happy man fatally hooked. In the build-up to the traumatic Sunday lunch, we observe Stephen and his wife fencing:

STEPHEN: About Sunday. I can put them off.
ROSALIND: Why should you put them off?

The inner emotional core of *Accident* is found in the interaction of this married couple. Seven years into marriage, breeding steadily, a don's wife would typically develop chips on her shoulder, a feeling of exclusion from the sparkling conversations, the new ideas, the fast displays of wit. When Stephen first mentions to Rosalind that he has a new pupil, an Austrian princess actually, she asks sardonically: 'Has she got golden hair?' (In Mosley's novel she says, 'Yum Yum.') Better, perhaps, would have been, 'Is she very clever?' Although in the last month of pregnancy, with very young children, Merchant's Rosalind seems almost fatalistic about Stephen's impending infidelity, as if she knows her husband better than he does.

On the Sunday Charley turns up in his sporty car, cigarette in mouth, romps with the children, throws a ball at William through the open kitchen window, and generally imposes his restless, muscular presence. Only the glasses he wears indicate an affinity with the world of books. After lunch, in the garden, with pylons rising in long shot, the sexual voltage rises. Stephen alone is too tense to lie on the grass; crossly he tends his garden, his back turned to the others, his ears cocked. This is the 'master scene' of the film. Closely following Pinter's lay-out, Losey wrote his camera instructions: 'The camera starts close on zoom revealing clutter of coffee cups, brandy glasses etc. Zoom back slightly to reveal central group of Charley and William and more clutter. Zoom slowly back to high long position to include entire groupings. William and Charley lying on grass . . . not far from each other. Rosalind in deck chair under apple tree. Dog lying down beside her, cool. Camera left. Her back slightly to camera. Clarissa and Anna making a daisy chain under a fruit tree right. Stephen and Ted weeding flower bed in foreground. In the far background the cornfield, a copse and the path leading away from the garden.'

Charley is describing the plot of a possible novel to William:

CHARLEY: Rosalind is pregnant. Stephen's having an affair with a girl at Oxford. He's reached the age when he can't keep his hands off girls at Oxford.

Losey's directives moved on to some swift camera movements and cuts:

'Anna Clarissa 2-shot. Zoom to cu Clarissa.

Time lapse

Cut to:

Close Rosalind in other deck chair. Zoom back to show dog beside her panting with heat. Pull back to high shot of tennis as in tennis notes. Zoom into Rosalind's empty chair. Dog gone.'

The sexual battle continues on the tennis court (with a cat crouching under the net). Charley keeps hoofing or heading tennis balls in a display of iconoclastic physical exuberance, then deliberately serves into Anna's backside. The three males on the court slash at the ball with increasing pugnacity. Losey sketched a series of foolscap diagrams showing the positions of the players at various phases of the game, which was filmed by Gerry Fisher hand-held. 'So we did the game in cuts and master shots, four or five times, with Gerry entirely free with camera, operating it himself, of course, in this case using the zoom – and it was done in the cutting room after that.'[1]

At tea-time Anna accepts Stephen's invitation to go for a walk, having refused William's. Her willingness is implicit, but Sassard's personality does not suggest a young woman happy to poach every man in sight. Losey begins a shot with a close-up of Stephen's hands and Anna's resting on a gate, very close but not touching – then pulls back to show the countryside across which they are gazing, together but apart, their backs turned to us. A plane is heard; distant bells. Stephen lets the moment slip:

ANNA: Shall we go back now?

They return to the house, but the camera lingers momentarily on the empty landscape – a fine metaphor for a lost opportunity. Pinter's extraordinary instinct for the light tap which hits like a fist is displayed by sending Anna off for a sedate walk with Rosalind – in Anna's mind Stephen has blown his claim on her.

Anna's dumbness lies partly in Sassard's personality, but more fundamentally in Anna's reduction to a sexual object – albeit one with breeding – by Mosley, Pinter and Losey. During her first snapshot philosophy tutorial with Stephen she says nothing at all; during her second one she speaks only two words: 'I'll try.' Mosley had asked his friend Raymond Carr to introduce him to the 'most sought-after girl in Oxford'. Carr promptly obliged. She wasn't an Austrian aristocrat – Mosley added that 'for glamour' – but 'very blank, passive and available: people fell in love with her'. Carr denies that the girl, 'the much sought after daughter of an earl', was 'blank', likening her to a beautiful 'cross between Hedy Lamarr and the young Princess Margaret'. Whatever version we take, Anna moves through the film like a sleepwalking clothes-horse.[2]

As evening approaches all three men are getting drunk. William now feels spare, sidelined, outmanoeuvred. Charley is sulking, loutishly surrounded by beer bottles. Stephen, whose undischarged lust impelled him to invite Anna and William to stay to supper, has been preparing the meal. Tensions again over-spill. Charley talks about his wife (whom Stephen and Rosalind know well), then the sparring turns to Charley's performances on television, Stephen's envy surfacing to the moment when he reveals: 'I have an appointment with *your* producer next week.' After shooting, Losey instructed Reginald Beck: 'This is where the boil breaks and all the subtleties disappear ... all of Stephen's niceness should have disappeared and we should use the ugliest shots of him, including the two where he has a piece of potato stuck in the side of his mouth.' Stephen's occasional stutter, signalling moments of guilt, derives from Mosley's own, was Charley's in the novel, then eliminated from the screenplay. Its revival by Bogarde when Stephen is lying could have offered Charley an obvious strike when Stephen announces his intention to compete with him on television:

[CHARLEY: You intend to t-tell the t-truth on television?]

William is too drunk to drive home and Anna does not possess a driving licence – a fact revealed by Charley, the first indication that he knows her better than Stephen, and foreshadowing the final car crash – who was driving? The genius of *Accident* largely resides in its carefully planted yet unobtrusive repetitions and 'moments', each complementary to the other. For example: Stephen's head beside Anna's in the punt, Stephen's head beside Anna's on the bedroom floor in the 'rape' scene.

So everyone is staying the night. Charley corners Stephen:

CHARLEY: Which room . . . is everybody in?
Pause.
STEPHEN: How the hell should I know?

The point of this exchange is reinforced moments later when Stephen, entering his bedroom, imagines for a delirious moment that the woman lying in the fourposter is Anna. He whispers her name, then realizes it's Rosalind asleep.

STEPHEN (*whispering*): Rosalind. (*Pause.*) Darling?
ROSALIND (*blurred*): What?
STEPHEN: I love you.

For the first time we break away from Oxford and Oxfordshire. A bemused but ambitious Stephen finds himself an unregarded visitor in the brash, daylight clatter of the television building with its huge plate glass window overlooking the city and its vast stairwell. Carefully dressed and nervous, Stephen is thoroughly humiliated by a former pupil, Bell (played by Harold Pinter). The producer whom Stephen had an appointment to see, it transpires, is very ill in hospital. In this brilliant sequence, conceived by Mosley, Pinter's gift for placing contingency and absurdity at the heart of a familiar reality is stunningly realized; Losey achieves a rapidity of motion conveying the heartless impatience of the media world, and perfectly contrasted to the slow, secure comforts of Stephen's college and home.

For consolation, the humiliated Stephen telephones an old flame, the Provost's daughter Francesca (Delphine Seyrig). As he arrives outside her house, wrapped in hesitation, we hear her voice over: 'I was in my bath when you phoned.' Throughout their sequences together – her flat, a restaurant, in bed – Pinter adheres to voice-over. In order to marginalize this brief encounter as an 'unreal' moment in both their lives, as 'time out', Pinter detached the characters from their voices. 'The words,' his script advises, 'are fragments of realistic conversation. They are not thoughts. Nor are they combined with any lip movement on the part of the actors.' It's a painful scene; why is the elegant Francesca alone and so immediately available; why does she tell him, 'I'm very happy'? The camera tracks along the window of the restaurant where they are eating, studying them through rain-splattered glass, she gold and green and bathed in a soft, misty light – then cuts abruptly to them lying on their backs in bed, like recumbents on a tomb, she smoking and gazing into some private pain of loneliness or regret.

Returning home from Francesca, Stephen finds Charley and Anna spending the night in his empty house (Rosalind has taken the children to her mother's prior to her confinement). The camera is on the first-floor landing as Stephen enters through the front door – an abrupt objectification of the hero (and an amusing reversal of the parallel scene in *The Servant*, where it was Bogarde who played the sexual intruder in the master's bedroom and the camera gazed up the stairs). Charley appears, insolently wrapping himself in Stephen's dressing-gown.

STEPHEN: Hullo. I've just come from London.
CHARLEY: I know.
STEPHEN: To see the television people.
CHARLEY: Did you see them?
STEPHEN: I'm hungry.

Stephen makes an omelette – the camera close to the frying-pan and to Stephen's busy hands, with Charley and Anna at the kitchen table. Charley has purloined and opened a letter from his wife Laura to Stephen. 'You might even hint that sooner or later he'll be bored to death by her,' Laura has written. Charley reads this aloud then bites into Stephen's omelette. The 'omelette scene' was shot in a single take, 'to hold the enormous tension between the two men and the girl . . . It's always to me stronger than if you intercut close-ups and medium-shots and insets and so on . . .'[3]

Stephen goes upstairs and surveys the rumpled sheets which Anna is trying to straighten.

STEPHEN: What are you doing?
ANNA: Making the bed.

(An echo of the 'Sunday lunch' scene:

STEPHEN: What are you doing?
ROSALIND: Making tea.)

Noticing Charley's flung-off clothes, Stephen leaves the room and sets in motion the rocking horse in the corridor. Later Charley dresses himself in front of Stephen and informs him that he's been having Anna for weeks. 'We used to go to her room in the afternoon . . . Where can I damn well take her?' Stephen gives Charley the key to the house so that he and Anna can spend the weekend together. Why does he do this? In a vain attempt to prove that he doesn't give a damn? Out of perverse masochism? As a kind of tribute to the do-er? To prove his good nature to Anna?

Stephen stutters guiltily when he visits Charley's wife Laura (Ann Firbank), with the purpose of consoling her. She'd asked for his help but now she has nothing to say. Her sense of dignity briefly ousts her pain. This scene was not intended to be shot in the rain, but it kept raining. To highlight Laura's own distracted state of mind, Losey introduced an absurd touch: she is watering the garden in the rain. 'Everybody thought I was crazy, but it worked.'⁴

This scene is intercut with Stephen visiting Rosalind on a sunny day at her mother's house, the two sitting in garden chairs by the Thames (the scene that Bogarde most enjoyed). Stephen, as usual, is confessional, though he clips the edges off total truth. Rosalind's fury on hearing about Charley's affair with Anna is a splendid example of Pinter's capacity to express deflected anxiety:

ROSALIND: How pathetic . . . Poor stupid old man . . . Stupid bastard . . . I've never heard of anything so bloody puerile, so banal . . . That poor stupid bitch of a girl.

William has been out of the picture but now he invites Stephen to his grandiose family home. After dinner the men take off their black jackets to play a traditional family version of the Eton wall game in the great hall. Participation is obligatory, although Stephen doesn't fancy it.

WILLIAM: You must play. Only the old men watch. And the ladies.

Losey decided that this sequence 'should be shot in a highly formalized way matching the symmetry of the Adam house'. There were to be four master shots, two essential and two 'luxury'.

'1. A high long shot from a rostrum on the left hand side, which will include most of the hall, using a wide angle lens. If possible cutting left on the main doors and cutting right on the doors leading to the inner court, and including the Apollo statue, the rotunda at the far end, and part of the gallery and stairs leading to the main apartments.

'2. A second essential master shot from behind the green marble figure of a dying gladiator . . .'

Losey's camera indications here are typically detailed: 'Continuation scene 263 and 264: . . . Cameras pan to include gallery. Hold them in close two profile framing gallery entrance as dowager types like full-sailed schooners float down steps to platform beyond Apollo figure. Blood lust the same as women at boxing matches although it has different veneer.'

During the long flashback which forms the central part of the film, William dies and revives three times in Stephen's imagination. In the second of these, Stephen 'kills' William in the wall game scrum. As William breaks free of the scrum, clutching the cushion which serves as a ball, only Stephen (positioned 'in goal') stands between his pupil and a score. He leaps on William's back. We see William's bruised face sinking out of the frame. This shot cuts quickly to a close-up of William in a cricket cap, cheerfully striking the ball. Charley, batting alongside him, gets himself clumsily 'out' to a slow delivery – a metaphor for the sexual defeat that follows when Anna tells Stephen (they are watching the game) that she's getting married to William. Stephen gnaws his pipe. The enraptured William says he wants to come and talk to Stephen later that night, after taking Anna to a party. Stephen insists that he bring Anna. As William and Anna run off hand-in-hand, Charley watches from a deckchair, then gives Stephen a defiant wink – Stanley Baker has his moments.

And so to the fatal accident where we began. The flashback is completed, the crash has taken place, the police have come and gone, and Stephen is standing over the recumbent Anna in a darkened bedroom, his lust overwhelming. This scene divides itself into two acts and an interval. In the first Stephen forces his (clothed) body on to Anna's despite her shock and cowering disgust, unable to control his desire yet masking it in an angry attempt to shake her out of her coma into revealing whether anyone knew that she was in the car with William. 'You *were* driving, weren't you?'

From downstairs the telephone is heard. In no hurry to respond, Stephen has reached the upstairs landing when it stops.

When he returns to the bedroom Anna is standing by the window. He pulls her to him and kisses her. She stands rigid in her bloodstained crash-soiled dress. He withdraws to turn off a bedside lamp. Slowly she moves across the room to stand beside him. Why does she submit in this way? Because she's in his power concerning the car crash? Or simply 'droit de seigneur' when no other seigneur is to hand?

At dawn Stephen learns by telephone that Rosalind has had a difficult birth; his reaction is minimal, his main concern being to explain that he'd been 'asleep' when the phone rang during the night. Anna is leaning, exhausted, beside the open front door. The sense of degradation is overwhelming.

Stephen drives her into Oxford while the world is still asleep. 'No one must see you.' In the script Anna, unable to climb over the wall, asks Stephen to help her, and he becomes stupidly unpleasant about it before getting out of his car. In the film he is merely stupid (given the height of the wall) and only belatedly realizes that she needs help. As her legs and body come over the wall, the whole sexual theme is in reprise.

Stephen gazes at his new baby in a hospital incubator.

ROSALIND: They phoned you. You weren't there.

Later that day Stephen and Charley visit Anna at the same moment. Charley has heard that William is dead, but is otherwise ignorant of what happened during the night. Finding Anna packing to leave, Charley grows desperate – whereas Stephen, whom Pinter depicts as increasingly calculating, is probably relieved that her departure may shield him from scandal. But he also takes pleasure in Charley's despair.

STEPHEN: There is nothing to keep her here.

A taxi carries her away. Then a moment of normality – Stephen walking across his college quadrangle. But now the soundtrack carries the approach of the car before the crash. The last shot of the film shows the same frontal façade of Stephen's house as in the opening shot, this time in broad daylight. Stephen's small children, Ted and Clarissa, are seen running over gravel towards the front door. Clarissa falls, cries. Stephen comes out and carries her into the house, the dog following. The front door closes. Sound of a car drawing nearer, skidding, crashing.

That was the intention, but when it came to it, there was a cock-up. The dog bolted down the drive and past the camera towards the unseen road. The sound of the car crash therefore suggested a new accident, caused by the dog – or at least left the audience in some confusion. On top of this, the camera pulled back exposing visible tracks on the foreground. Worse, the sequence could not be retaken. In order to extract genuine tears from the small girl playing Clarissa, Losey had secretly assigned a member of the crew to hide behind a hedge and pull a string trip as she came running along the gravel path. It was done – the girl's tears were genuine. According to Gerry Fisher's account, Pamela Davies, Losey's devoted continuity woman, was so shocked that she slammed her script shut and climbed into her car. Bogarde's account of this incident has Pamela Davies striding across to Losey 'like a colonel', yelling at him 'like a sergeant', and even striking him.

In March 1967, *Accident* was selected as the official British entry for the Cannes Film Festival. Losey was in Cannes from 4 May, for nine days, staying with Patricia at the Colombe d'Or at the expense of Alliance International (the Festival refused to contribute; the regulations of the Ministères de Tutelle allowed subsidized accommodation only at designated Festival Hotels). *Nice-Matin* (25 May) carried a picture of a bronzed Losey coming out of the sea – apparently the Loseys bathed only '*chez Madeleine*' on '*La Plage Sportive*'. Antonioni's *Blow-Up* won the Palme d'Or after a tumultuous jury meeting during which some jurors pressed the claims of *Accident* or a Yugoslav film by Alexsandar Petrovic, both of which were awarded the Grand Prix Spécial du Jury. *Accident* also took the UNICRIT prize and, later, the Premio de Selezione di Sorrento Award, 1967; plus the Grand Prix de l'Union de la Critique de Cinéma in Belgium (January 1968).

When *Accident* opened on 9 February 1967 at the Leicester Square Theatre, BBC 2 was about to screen a programme, 'The Movies', on the new film and Losey's work in general. The press was almost unanimous in its praise. 'The Joseph Losey film everyone has been waiting for,' announced *The Times* (9 February). Despite Mosley's 'rather untidy, overwrought novel', full of Dylan-Thomasy free association, Pinter had achieved a perfect adaptation. 'Mr Losey has directed the film with a restraint bordering on the heroic: no cadenzas of technique, no big effects . . .' *Accident* was 'the clinching masterpiece in which all his gifts work together to produce an entirely satisfactory work'. Other critics lauded Losey's genius for extracting 'performances of incredible clarity', his 'classical precision', his 'infinitely subtle direction'. However, Robert Robinson's unique view, in the *Sunday Telegraph*, that 'The camera doesn't suit [Pinter], it lets the wrong sort of light in', was described by Pinter, writing to Losey, as 'a cuntish exhibition'.[5]

Despite the reviews, Losey was incensed by the performance of both Rank and London Independent Producers, which (he reported to Pinter) 'effectively consists of [Bill] Gell and [John] Hogarth here in that order . . . [Sydney] Box of course unfortunately is no longer active . . . Gell is an ignorant, vicious, stupid peasant (without commenting on the darker corners of his character) . . . Hogarth, for instance, a day before the opening said he "preferred the trailer to the picture" and Gell said he was "not going to sit through that dull bit of shit again" . . . These people as you know represent 50% of the finance and from now on control the distribution . . . which includes advertising, prints, everything.' Losey listed his complaints: '. . . ever since the notices we have had a constant battle to get them to run display ads . . . The poster is bad. The press book is bad . . . Distribution has been rushed . . .' Within a week of opening in the West End, *Accident* was double-billed on general release with *Just Like a Woman*. The film came off after three weeks at the Leicester Square cinema. Losey heard that it was being moved to the Berkeley, 'a crumby theatre [which] indulges in cheap sensational advertising'. Why not the Prince Charles, the Odeon–Haymarket, or the Ritz? 'Spot advertising on television is not only usual but de rigeur [*sic*] for any important general release. This is apparently not being done because it entails an expenditure of £1,000 – £2,000. Who makes these decisions?' *Accident* ran for only one week at the Odeon, Kensington. Visiting Glasgow for the local première, he was interviewed by the BBC and used the word 'virgin' several times, flouting the Corporation's ban on the word. Cut off, he walked out and proceeded to Scottish TV, a commercial station, where he again spoke of virgins.[6]

The competition in the quality-film market was fierce during 1967. Claude Lelouche's *Un homme et une femme* filled the Curzon for weeks. Godard bombarded the screens with *Alphaville* and *Pierrot le fou*. Also on show were Truffaut's *Fahrenheit 451* and Pasolini's *The Gospel According to St Matthew*. Buñuel's *Belle de jour* won the Golden Lion at Venice and enjoyed a huge commercial success throughout Europe. *Accident* was nominated for the 'Best Film' award of the Society of Film and Television Arts, together with Antonioni's *Blow-Up* and the eventual winner, Fred Zinnemann's *A Man For*

All Seasons. Bogarde was nominated for 'Best Actor' but the award went to Paul Scofield; Pinter was nominated for 'Best Screenplay', but the winner was Robert Bolt. Once again Losey was left empty-handed.

At this time Pinter was living in an apartment on Fifth Avenue, and enjoying the Broadway success – a Tony Award – of his play *The Homecoming*. He was optimistic on behalf of *Accident*. 'The screening I had over here brought a truly remarkable response. All sorts of people literally bowled bloody over . . .' But Pinter's optimism was belied by events. On 8 March, Michael B. Bromhead, of LIP, informed Losey that the majors had all seen the film – Paramount, United Artists, Warner, MGM, Columbia, Embassy, Walter Reade, Universal and Fox – and none would take it. The film had been finally taken on by Cinema V Distributing. Losey's response was to blame LIP. On 10 March he complained that the American advertising and publicity were appalling. The poster was 'unspeakable'. Pinter took an entirely different view, praising the LIP-Cinema V team for their enthusiasm and commitment. The première party was held at the New York Hilton, the film opening on 18 April at Cinema II, on Third Avenue at 60th Street. Alliance International (LIP) sponsored Losey's stay at the Plaza Hotel from 12 April, and his visit to Toronto on the 18th. His attitude towards Alliance International and Cinema V brightened considerably: 'I have never before in my professional career been absolutely happy about the hands in which my film's future rested.'[7]

Losey and Pinter faced the college editors in a session chaired by Andrew Sarris. Ross Westzsteon reported the resulting disaster in the *Village Voice* (20 April). 'I'd like to hear your explanation of the accident at the end of the movie,' one of the student editors asked Pinter. 'I'd like to hear yours,' Pinter replied. When someone asked if perhaps the relevance of the movie could be fully understood only by an English audience, Pinter reportedly blew his nose, pocketed his handkerchief and said, 'What do you mean by "relevance"?' The questioner offered 'message' instead. 'Message?' The same fate befell 'significance'. Andrew Sarris commented that everyone had been asking what *The Homecoming* meant. 'I'd understand questions about meaning,' Pinter replied, 'if I knew what the word "meaning" meant.' And later: 'I have no explanation for anything I do at all.' By nature explicatory, Losey nevertheless joined in the Trappist ritual. Was there any social comment implicit in the large quantities of alcohol consumed in the movie? 'I don't know what you mean by "social comment".' To this Pinter added: 'People just had bottles around and drunk out of them.' Asked if he edited his own films, Losey replied: 'It depends what you mean by "edit".' Well, did he make the final cuts himself? 'It still depends on what you mean.' The critic John Simon accused the pair of insulting their audience and asked Pinter if he was aiming to do away with words altogether. This got a reply: 'I love words. I just try to use them selectively.' Simon then pointed to some factual errors about Oxford in the film and spoke of 'the sublime disregard for reality that pervades the entire film'. After the meeting he announced: 'I've seldom seen two such consummate phonies on stage at one time.'

Indeed, of the three Losey–Pinter films, *Accident* was the most ambivalently received in America despite the enthusiasm of critics like Judith Crist. Writing in the *World Herald Tribune* (18 April), she described 'two master craftsmen at work', Pinter showing his genius for capturing the essence through small talk, the needle pricks which barely blemish the skin. 'And the vivid camera eye of Joseph Losey's direction observes, implies and challenges us to see beyond its own visualization.' Pauline Kael announced that she had enjoyed the film but 'I don't really *like* it.'

In July Losey learned that the box-office had been a 'disaster' in Los Angeles, despite breaking records at Cinema II in New York. Twenty-three weeks after opening, the film had still not penetrated any major city beyond New York and Los Angeles. NBC Network, whose buyer wanted the film, recoiled from the 'pervasive atmosphere of adultery'. *Accident* won the Movie Critics of the Foreign Language Association Award for the Best British Film, 1967 – Losey made bitter play of the association's inappropriate name. The film came away empty-handed from the New York Film Critics' Awards.

Despite international critical success, *Accident* was a financial disaster. The royalty statement from LIP to 31 May 1968 showed gross UK receipts of £43,010 and gross overseas receipts of £95,153. Clearly the film was not going to earn back a budget of £272,811. As usual Losey distrusted the account statements. In May 1970, hearing that LIP planned to sell the television rights to the BBC for £10,000, his immediate response was 'outrageous' – it should be at least £50,000 – but Robin Fox advised him that Independent Television had made no offer at all: '. . . the liquidator has a duty to raise whatever monies he can'. Norman Priggen feels that Losey was unrealistic in his expectations. 'London Independent didn't really believe that *Accident* was box-office, and they were right. It was an art-house picture.'[8]

Later, in the year of *Accident*'s release, Pinter wrote from New York: 'The [*Homecoming*] is in good nick here. I think it can be said that it's made something of an impact on this little old town. But I don't like New York. . . .' His reactions to Losey's probes concerning future projects were guarded: 'Of course we'll work together again. That's something I *want* to do. But we've got to accept the fact our timings cannot always meet. I've started to write a play – *just*. The year is a mist, a solid mist. I don't know. I would do *The Go-Between* like a shot, of course, because I'm in residence there. I'd do that under the table, if necessary. But the thought of anything new at the moment . . .' (Pinter's dots – eleven of them.) The impression is of slight unease: 'You I think a great deal about. And me and you. And you and me. And us. What I said to you that night at Twickenham always holds. But I don't know that I could, or should, or would ever make this year. What with One thing and Another.' On 13 September Pinter reported that he was now writing film scripts of *The Birthday Party* and *The Homecoming* (but Losey was not invited to direct). He would also be directing Robert Shaw's *The Man in the Glass Booth* in New York – 'So next year is going to be a little busy for me.' Losey was also busy; this was his Burton-and-Taylor time, and his recurrent irritability was softened by an underlying confidence. Pinter offered a short-stemmed olive branch by indicating that he would be happy for Losey to direct a film version of his play, *The Lover*, but did not see any way of legitimately expanding the sixty-minute text to film length. Losey never directed an original Pinter text, and it may well be that he feared excessive confinement – textual, directorial – if he did. Though sporting metaphors were more Pinter's than Losey's, it would be an away match.

These were years of growing wealth and hedonism. On 10 November, Losey – 'just back from our long tour of duty [*Boom!*] and a bit dazed' – proposed an evening of Noël Coward, the Dankworths, the Pinters and the Loseys. The drink level was rising sharply. On 5 January 1968, Losey sent 'abject apologies' for having mislaid a Pinter script. 'I am appalled by the wastage and effect of excessive alcohol consumption. The night we were with you had evidently been preceded by such a day for me and I remember very little of what was said after we got to the brandy . . . [Patricia says] I refused to get out of the car and sat there for some time after she had gone into the house. The script was of course in the car but I was totally unaware of it. The car was put into the garage the next day for the duration of our trip abroad . . .' The mislaid script was of Pinter's one-act, two-character play, *Landscape*. By 15 January Losey had read it: 'I think it is beautiful and complete in a way that exceeds all of your previous work. Although it is relatively short its simplicity . . . hides the most fantastic complexity.' In short, Losey was baffled.

At this time Pinter had a clandestine relationship with a woman prominent in the media – neither wished to end their respective marriages. Pinter tells the story: 'One day Joe said, "I hear you're having an affair." Of course I denied it. "Who told you?" Joe admitted it was David Mercer.' Pinter telephoned Mercer (he reproduces their rather different accents) and summoned him to a pub. Mercer pleaded forgetfulness, drink. Later, when Pinter was performing in *The Homecoming* at Watford, he invited both the Loseys and his lady on the same evening – but separately. After the performance he offered them all a ride home in his chauffeured car, dropping off the Loseys first. Next day Joe called: '"So that's who your girlfriend is! I could tell by the way you ushered her into the car."'

In May 1969 Losey was hoping to bring Richard Burton and Pinter together on a film adaptation of James Kennaway's *The Cost of Living Like This*. Pinter found it powerful and compelling, 'but finally I simply don't share your overall enthusiasm . . . It's strange that our reactions should be so different . . .' On 5 August, Losey sent a telegram: 'From one ex Aryan to an honorary Aryan congratulations stop love anyway Joe.' The reference is obscure and Pinter has no recollection of it, but Losey always felt that his own mid-Western origins combined well with Pinter's Jewish provenance to create a uniquely alert eye and ear for England. In February 1970, writing from Dartmouth College, Losey reported to Pinter that he was about to run a film called *The Pinter People* – a television documentary – for the Film Society after a screening of *Accident*. 'I think it is quite a brilliant piece of work, extraordinarily animated and you are extraordinarily caught – if you will excuse the expression – if only in your evasions.' Pinter was delighted by this news. 'I did my interview for *your* programme here a few weeks ago.' This was Thames TV's film, *The Moviemen*.

PART VI

Boom and Baroque:
The Burton Years

23

Op and Pop

Modesty Blaise

Modesty Blaise is Losey's junk-film, the prelude to the baroque glitter of the 'Burton years'. Joseph Janni – producer of *A Kind of Loving*, *Billy Liar* and *Darling* – had long admired Monica Vitti in the films of Michelangelo Antonioni, and, himself Italian by birth, had hopes of inducing her to play in an English-language film. While visiting Vitti in Rome to discuss a project, Janni inadvertently left behind a screenplay of the strip-cartoon 'Modesty Blaise'. Not long afterwards he received a call from a Sardinian harbour-master – Vitti was on the line from a yacht, to stake her claim to Modesty. Arriving in London, she did not achieve a meeting of minds with Sidney Gilliatt, Janni's intended director. 'She spoke the new film language,' Janni recalls. Losey had been turning out commercials for Janni's Augusta Productions: 'He was obsessed with Vitti,' Janni recalls – a view confirmed by Patricia Losey – and expressed his ardent desire to be considered. Like Jeanne Moreau before her, Vitti admired Losey's work and was delighted at the prospect.[1]

The $3,000,000 venture was pre-financed by 20th Century-Fox. Losey's contract (July 1965) with Janni's Modesty Blaise, Ltd was for an ungenerous fee of £22,000 (given the budget and a higher fee for *Eve*). Once hired, he rejected the original settings, crew and script, brought in Norman Priggen and Richard Macdonald, set off to reconnoitre locations and started shooting within a month. John Dankworth was engaged to compose and record a jazzy score blending dark drama, irony and romance.

Peter O'Donnell's strip cartoon, launched in 1962 and carried daily in the *Evening Standard*, had achieved syndication in, reportedly, sixteen countries. O'Donnell announced that he had taken the name Blaise from the Arthurian legends – 'the master mentor of the magician Merlin' – then applied it to an orphan of the Second World War prison camps, who made her way 'across the Continent and the Middle East growing up meanwhile into the extraordinary girl you all know'. Losey again brought in the screenwriter Evan Jones, who insists that he provided a clear, hard narrative in the James Bond tradition – 'You knew where you were.' Jones may be right that Losey had begun to despise narrative: 'If the point of a scene was the theft of a ruby, he'd say that doesn't matter, what matters is the wig or tattoo.' 'Grossly self-indulgent', is Jones's verdict. 'Many of the Amsterdam joke-visuals fell flat.' The industry, meanwhile, wanted a James Bond with legs, 'more deadly than the male' (as the ads said). According to Jones, he was paid £800 for writing the screenplay of *King and Country* and £2,200 for *Modesty Blaise*. His agent was not allowed to negotiate his fee because Losey had already tied it up as part of the deal. 'It doesn't matter what you get this time, working for me will build up your reputation.' Jones speaks with bitterness: 'After he started working with the Burtons I don't think he made a movie that's worth watching.' At the time, however, Jones was keen to continue working with Losey, who was offered his screenplay *All the Angels* in 1968 and initially interested Elizabeth Taylor in *The White Witch of Rose Hall* the following year.[2]

Evidently Modesty Blaise herself was not pleased with the screenplay. On 23 July 1965, Beaverbrook Newspapers' legal department dispatched a threatening letter to the solicitors representing 20th Century-Fox and Modesty Blaise, Ltd, arguing that the script dated 25 June broke the contractual undertaking to preserve the essential strip-cartoon characteristics of Modesty and Willie. Some twenty instances followed, with the 'trade mark' factor upfront: in the screenplay Modesty was fair, not dark; she did not carry her hair in her 'trade mark' bun; she did not wear her 'trade mark' combat suit; she didn't carry or use her 'trade mark' kongo stick; she did very little 'superb acrobatic fighting'; she did not carry a gun belt or engage in gun fighting; her elegance and refined taste were replaced by 'stuck-on tattoes, frills and suspenders, etc.'; her strategy and tactics, normally brilliant, were 'of the lowest level'; instead of drinking 'only rough red wine' she drank 'only cold white wine'; and so on. The finished film includes a few gestures, mainly concerning Modesty's costumes and wigs, to meet these objections. Jones says he did not take the O'Donnell strip seriously. 'It was boring.' In the event, Beaverbrook Newspapers serialized the film in the *Evening Standard*.

Monica Vitti – she who had drifted captivatingly through Antonioni's slow, sensuous, sunlit spaces of the heart, a darling of the European art cinemas – came to believe that Losey and Jones had fatally altered the myth of Modesty, a view echoed by Antonioni: 'She suffered very much making that film. You cannot make a film about a myth while destroying it.' But it's doubtful whether Vitti's own, profoundly non-violent, personality was compatible with the myth of a fighting adventuress – you can't beat the world's worst villains while avoiding being photographed in profile. Dirk Bogarde also had his difficulties with the script. The dialogue was 'dire, it had no style about it'. Losey 'had a tin ear for that kind of dialogue. I had to write the funny lines – "Champagne! Champagne!" instead of "Water! Water"', when staked out on the desert sands.[3]

The Motion Picture Association of America, represented by Geoffrey M. Shurlock, registered strenuous objections to the sequence in which Willie talks to Modesty on the phone while in bed with another woman – in particular his line, '. . . Just in the garage getting serviced . . .' Shurlock went on and on about dressing, undressing, beds, nudity, suggestive lines. When Modesty poses in a brothel window as a prostitute, 'she must be decently covered' – i.e. her dress not unzipped to below her navel. Losey agreed to cut Modesty peeling a banana. The British censor, John Trevelyan, was no more keen than Shurlock on 'Just in the garage getting serviced', but gratuitous violence bothered him more than sex. The shooting down of the plane was acceptable but Trevelyan disliked the emphasis on the pilot having young children.[4] In February 1966 Losey reported that giving Trevelyan lunch at the White Tower – 'very useful, pleasant and expensive' – had cost him almost £9. (Just in the White Tower getting serviced.) Writing to Elmo Williams, of 20th Century-Fox, Trevelyan called *Modesty Blaise* 'a 1975 picture made in 1965 – a delightful and enchanting fantasy . . .'

Shooting began on 12 July in Amsterdam, moved to Sicily during the second half of August, proceeded for a week to Naples for the harbour and boat scenes, moved briefly to Rome, then wound up during September–October at Shepperton Studios. Jack Hildyard (winner of the 'Best Cinematography' Oscar in 1957 for *The Bridge on the River Kwai*) was making his second film with Losey, who became vexed by his slow, painstaking caution. Hildyard refused to shoot a scene in Amsterdam because of poor light; he was within his rights; Losey called up Gerry Fisher, Hildyard's operator, as lighting cameraman for this sequence. Fisher was soon to become Losey's primary cameraman.[5]

The production was fraught with conflicts, most notably Losey's running quarrel with Vitti and a blow-up with an assistant director in Amsterdam, continuing into the post-production phase, when Losey complained bitterly to Janni about the quality of the processing and printing at Technicolor, Rome, which frustrated his aim of creating 'quite separate colour palettes for . . . London, Amsterdam and the Southern Mediterranean'. Re-prints were generally inferior to the original rushes; defects in density caused 'softness and blurring to the point where the whole film seemed to be out of focus'.[6]

Losey's editor, Reginald Beck, describes Losey's animosity towards Janni as obsessive, and Norman Priggen remembers his abrasive attitude towards Michael Birkett, assistant to Janni – and a peer: 'This isn't the fucking house of Lords,' Losey would growl, before stalking out of the room. Janni took a boat trip with Losey off Sardinia, the temperature over 100 degrees, not a breath of breeze; when Janni instructed the boatman to speed up, Losey exploded: 'You guys think you own the fucking world.' Janni is forgiving – 'He hated anyone in authority' – and describes Losey as 'the most honourable and honest film maker', though over-intense, nervy and liable to slave-drive his crew for hours without a meal-break. Janni also comments that Losey displayed respect verging on humility when faced with the legendary Hollywood mogul, Darryl F. Zanuck, President of 20th Century-Fox. This is confirmed by correspondence: on 16 May 1965 Losey wrote to Zanuck, explaining that they would both be in Paris on 19 May and begging for 'ten minutes'. Janni chuckles in recall: 'Zanuck saw the film in London and loved it. Joe was like a small boy, "Yes, Mr Zanuck, thank you, Mr Zanuck." '[7]

Modesty Blaise is not without its pleasing sequences. These begin with Gabriel (Dirk Bogarde) on his yacht in an amazing pair of slit glasses and the ash-blond wig he'll wear until things get desperate. With Bogarde's entrance comes the first genuine moment of humour (as distinct from lacklustre camp). Gabriel's castle was once a monastery bestriding Sardinian clifftops and quite a climb under azure skies. Walls and courtyard floors are decorated in dazzling op-art designs. Gabriel's castle boasts classical statues, a revolving radar scanner, a baroque organ played by a resident monk, an obsessively cost-conscious Scottish accountant (Clive Revill), and Mrs Fothergill (Rosella Falk), a tall, handsome sadist – nicely dovetailed into sixties porno-masochism – putting her sweating prisoners through press-ups till they flop. Gabriel finds her appetites faintly amusing if excessive, himself preferring to monitor her feeding time from the shade of his op-art, Bridget Riley-style interiors. (Falk was a good enough comedienne to put herself both inside and outside the part; she grasped the idiom. Losey, who had fired another 'Mrs Fothergill', Annette Cavell, after shooting began, adored Falk.) Beyond the psychedelic patterns and violent colours, the verandahs are sculpted in luscious curves, with glorious seascapes stretching below. Happily we find Gabriel with his silver wig and pastel parasol, sipping lilac liqueur with a goldfish floating in it. Two lobsters are held up for his selection. 'Oh decisions, decisions,' he moans. Later comes a comic shot of Gabriel not noticing Mrs Fothergill's body swinging by the neck from the top of a well. When he does notice he rips off his wig and races down the castle steps with terrific agility and no stand-in – though Losey told Ciment that Bogarde was so unathletic he could scarcely walk. The Sicilian landscapes and seascapes become ever more outrageously beautiful. The finale is amusing: the desert camp of Arabia, with the captive Gabriel impaled on the hot sand under a merciless sun, a radio playing full blast in his ear. 'Champagne! Champagne!' he cries, then: 'McWhirter, my God, I thought you were Mother.'

Modesty Blaise is ravishing to the eye – the composition and depth of the frames, perfectly focused long-shots behind close-ups. Losey unfurls a city (Amsterdam) through an op-art interior, expanding his horizon by swift, subtle movements – or begins a shot with an astonishingly complex mirror reflection, making the eye work and marvel at reflections in canals. Macdonald's designer-intensive décor is a feast. The film exploits the 'Avengers' culture – cool, amoral, gorged on gadgetry and playfully disguised lethal weapons, with English gentlemen who may not be quite gentlemen in bowler hats, extravagant villains, grotesque settings involving transvestites, masks, clowns, gargoyles. If Losey intended a critique of consumerism, of the new, sleek capitalism peddling marketable decadence, he succeeded only in proving that 'post-industrial' capitalism is happy to turn criticism into one more commodity. At this level outright victories are elusive.

Modesty Blaise was the official British entry at Cannes and the first film to be shown. The guests of honour, Princess Margaret and Lord Snowdon, arrived an hour late for the screening while the audience, including a galaxy of stars, waited in stifling heat. *The Times* (5 May 1966) somewhat desperately advised that the answer to Britain's problems at international festivals was to abandon films wearing their cultural pretensions on their sleeve, 'and hit them instead with James Bond, or the Beatles, or both, relying on the surprise of sheer entertainment value to sweep the board'. But no board was swept.

20th Century-Fox laid on a grand dinner. Presiding was a member of the Hakim family, André, who was related by marriage to Fox's President, Darryl F. Zanuck. Placed next to Hakim, Losey took his plate and sat elsewhere; Janni remembers his fury, Bogarde remembers taking care of the 'royals' while Losey 'sat there nursing his suppurating wound of hate for the Hakims, those Jewish Egyptian bastards or whatever they were'. Patricia Losey remembers 'no such scene' and is sure the royal couple were not at the dinner.[8]

The British première, held at the Odeon, Leicester Square, on 5 May 1966, was preceded by a flood of feature articles and prominent photo displays (a bare-chested Losey, still in excellent shape, showing Vitti how to hold a gun). The reviews were generally scornful or derisive. Kenneth Tynan (*Observer*, 8 May) called it 'a triumph of inventive obsolescence with instant rust built into its modish glitter . . . The whole show is like a travel brochure animated by a surrealist with a sharp eye for witty bric-à-brac.' The US première took place at the Woods Theatre, Chicago, in July 1966. Losey complained to Zanuck that the advertising was 'disastrously miscalculated' and that the film had been 'largely dumped' in the USA, England and France. In response,

Zanuck claimed that it was the press, not Fox, which had created the notion of a female James Bond. With few exceptions, the American critics hated it. Bosley Crowther shot it to pieces in the *New York Times* (11 August) as 'slam-bang and splashy' 'and sophomoric clowning'. Losey had extracted poor performances from talented actors and was responsible for taking on a lousy script. Jonas Rosenfield, of 20th Century-Fox, confessed to Losey in September that he hadn't sent him all the American reviews because he would have found them too depressing.

Of *Modesty Blaise* Losey himself later said, as usual, many things. Some were touchingly self-critical, but most weren't. 'I believe, with a little bit of perspective, it will be seen to make its own comment on a particularly empty and hideous era of our century.' Arriving in Algeria for a Losey retrospective, he was taken to task for the film's portrait of Arabs; 'slightly angered' (*légèrement agacé*), he came up with the explanation that 'the derision is aimed at the sheikhs of the Arabian Peninsula and their petrol'. In 1976 he admitted to Ciment: 'I sort of fell between the high camp spectacle comment that I wanted to do and plotty-plot that has to have every knot tied.' But again he blamed the script, the producers, the advertising.[9]

Although he told Tom Milne that the film was 'on its way to grossing six million dollars', two years after release 20th Century-Fox's accounts showed a gross take of only slightly over $3 million (which was what the film cost). Losey would therefore never see his 'deferment', since the film would be contractually 'in profit' only when Fox had recouped 2.5 of its negative cost.[10]

For Hire – Commercials

In pursuit of cash, Losey had launched a second, parallel career as a director of commercials. In 1957 he was hired by the J. Walter Thompson agency on the recommendation of James Archibald, who had been responsible for his feature-film contract with Rank. In March and May 1958 he made four fifteen-second films for Roses Lime Juice: 'Girl in Deckchair', 'Girl in Bathing Suit', 'Girl in Slacks', 'Girl in Trapeze Dress'. The photographer was Freddie Francis. Losey's retainer from J. Walter Thompson was £181 per month. Six forty-five-second films for Horlicks followed in June and August, their titles cryptic: 'Shop', 'Conveyor Belt', 'Tea Party', 'Front Door', 'Shoe Boxes', 'Clothes'. In June there was some bad blood about an extra payment to him of £50 for a Lux Liquid commercial. On 1 September he signed a new contract with the J. Walter Thompson agency, requiring him to make nineteen 'units' of film during a six-month period, for a payment of £2,375 – a unit being measured as anything between seven seconds and 2.5 minutes.

Pieter Rogers recalls that many of Losey's commercials were shot both in black-and-white for television and in colour for the cinema. Rogers, an expert in casting, remembers Losey searching for an actress with perfect hands – probably for Lux Liquid, a fifteen-second advert made in 1958. Rogers found the hands. In May 1959 it was six sixty-second commercials for Addis Domestic Brushes, involving twenty-three different products. The script was set out in two columns, Audio and Visual. For the Addis gift set a unison of voices announced: 'When the invitations start to come each day/We like to think of presents we can give away. For weddings and ... for birthdays too/We know what to get for you.' There were also shorts for Rowntrees Aero chocolate, Brooke Bond tea, and about a dozen more for Horlicks ('Curtains', 'Buttons', 'Policeman', 'Café'). At this stage he entered, with James Archibald's consent, into a new arrangement with Joseph Janni's Augusta Films, with J. Walter Thompson retaining first call on his services. For the Horlicks commercial he shot at Merton Park Studios in September 1960 he called for a 50 or 40mm lens to achieve the largest possible close-ups of the eyes. The subject should then turn towards the camera with a gesture of fatigue, perhaps leaning against the door as her children went out. 'Highlight in the two eyes as the camera comes into close-up. The light at the end of the tunnel is made by a merging of the two highlights in the eyes.'

In 1960 he made a fifteen-minute publicity film, 'First on the Road', to advertise Ford's new two-door family saloon car, the Anglia. Produced by Leon Clore for Graphic Films, the film was shot in Technicolor and bathed itself in jaunty jazz music. The Anglia turns slowly on its exhibition stand as gleaming stars cover the screen. Graphics illustrate the car's streamlining and rear wings balancing traditional American and British designs. A model climbs into the car, crosses her legs, then bounces joyfully on the well-sprung seats. Impressive piles of luggage wait to be loaded. There is no commentary – information is conveyed by texts or flashing arrows. Comfortably holding two couples, the car is driven along a pleasant country road empty of traffic and bathed in sunshine, to the accompaniment of a fast jazz piano. The Anglia is put through its turning test in a field, is driven through water, and emerges from these ordeals, like Donald Duck, OK. There follows a strange interlude during which the car, back on its exhibition stand, is toasted in champagne by various models, a picnic table loaded with fruit to hand. This motif is then repeated, but now it's bathing costumes, lilos and beach-balls. Finally the gleaming car runs out of ideas, turning and turning on its stand. Leon Clore recalls that when the art director, Reece Pemberton, produced some champagne glasses, Losey wanted to know where he'd obtained them. They had been hired. 'I'm not shooting these glasses,' Losey said. 'Go get some in Fortnum and Mason's.' The same happened with a basket of fruit – back to Fortnum's.

In 1971 Losey recalled the general scene. 'Karel Reisz inherited the Kellogg's account from me, and Lindsay Anderson inherited it from him. I inherited the Horlicks from Sandy McKendrick, who was the grand-daddy of us all, having begun in advertising before television. I forget whether Clive Donner or I came first on Nimble Bread. The most prolific of all of us was – and is – Richard Lester . . . Probably the best TV commercial I ever made, before my scruples overcame me, was one for a cigarette (which one, I now forget) for which John Dankworth did the score. Everybody admitted its virtues but it was considered too far out for distribution.' (The brand, in fact, was Players Bachelor and the date was January 1961. Losey called it 'highly sexual', and shot mostly in close-up.) The James Garrett agency sent three lovely models to him at Markham Street on a Saturday morning to be auditioned. 'Marie Louise. Tall, good, tough. Hannerle Dehn. Good figure somewhat less lush – available only Friday.' Following three fifteen-second films for Nimble and six thirty-second commercials for Fray Bentos (stewed steak, corned beef, steak and kidney pie), he did a thirty-second Aero chocolate commercial in December 1961: 'We took one of the largest stages in England – the whole floor was glass – and the whole commercial was one shot and one dancer on the glass floor, with a score by Richard Lester, "Seventy-Six Trombones".'[11]

A December 1963 agreement between Fontped Securities Ltd (in which Losey had vested his services) and Augusta stipulated payment of £210 for every day or part-day during the first three days in a week, £100 per day or part-day for the rest of the week. The previous month Augusta paid £540, representing half of his fee for Gibbs Fluoride and Colgate Espri. His fee for Stork Margarine interviews, filmed in October, had been £450 (about £4,000 in current values). He mildly rebuked Peter Madron, of J. Walter Thompson: 'I hear you've been doing some Horlicks without me. I hope they were very bad. It would serve you right . . . Haven't you some new product for which you would like me to establish the format for later . . . directors?' In February 1964, he reported making TV commercials in London but without mentioning the product – in fact Senior Service cigarettes – and he also made a Gold Leaf commercial.

By this time he was working for a variety of agencies, including London Press Exchange, TV Advertising Ltd, Rank Advertising Film Division, Augusta, and James Garrett. This pluralism led to acrimony – for example, James Garrett expressed disquiet about the conflicting claims on his services and his sporadic availability. In December, Losey's solicitor, Laurence Harbottle, pointed out to him that he still owed J. Walter Thompson twelve more £200 units, while Augusta was claiming his exclusive services through its solicitors, Crawley & de Reya. Garrett – also reaching for his solicitor – was complaining that he'd advanced Losey £1,000 against a payment of £225 per

unit. In 1967 Losey told the *New York Times* that he was the highest paid
TV commercial director in Britain. 'It's not a great distinction, but it's a
great comfort.' While in Cannes to promote *Modesty Blaise* in May 1966, he
filmed a liquor commercial involving a pretty girl and a white horse for
Cossack Vodka, through the S.H. Benson agency. John Dankworth wrote the
music. On 6 September a Benson executive sent him a cheque for £275,
'being the rest of your Cossack fee. As you can see, it is drawn on a personal
account to avoid detailing your name against the payment in our company
records . . .' A week later Losey regretted that his 'advisers' would not pass
this arrangement.

Losey evidently made no commercial between 1966 and 1973. In October
1973 he received £500 from Young & Rubicam for a fifty-minute
'Architectural Heritage Film' sponsored by Rank Xerox 'to awaken the
interest and pride of European peoples in their *common* architectural heritage
. . .' The idea was to focus the film on a group of fourteen-year-olds at work
on a research project, but the agency finally decided not to proceed with the
documentary approach proposed by Losey. In 1980, when Volvic mineral
water made inquiries, Losey's agent, Georges Beaume, replied that this
client's asking fee was 50,000 dollars. (The file contains nothing further on
this.) In December 1982 Losey returned to Vicenza, the setting of his *Don
Giovanni*, to make a thirty-second commercial about hair care on behalf of
L'Oréal-Elnett. The fee negotiated between Camera 6, Publicis Conseil–
Services Comptables and Losey's Launcelot Productions was £15,000.[12]

Projects

Losey's career is littered with unachieved projects; by his own account sixty
to seventy. This chapter of frustration might take its title from one of them,
Save Me the Waltz. The fickle lady with the marked card and swivelling eyes
was the industry.

Losey had known George Tabori since the 1940s – a wry, comic, endearing
personality who, according to legend, had almost married Greta Garbo; in the
event he married a younger Swedish actress, Viveca Lindfors. Tabori's
screen-writing was brilliant, urbane, witty, but with a tendency towards
meretricious flippancy and gratuitous vulgarity. *The Horizontal Man*,
originally called *The Holiday*, was a project for which Losey signed up in
May 1961 with Carl Foreman's Highroad Productions, for release by
Columbia. Set in a Greek village, it was to be made on location with Hardy
Kruger and (Losey hoped) Melina Mercouri. Kruger was critical of Tabori's
script but in a constructive spirit; Tabori called the actor's comments 'half-
assed and picayune'. The project collapsed when Foreman – a powerful figure
after the recent worldwide success of *The Guns of Navarone* – rejected it in

what Losey called 'a very high-handed way, and very stupidly. Since then I've had very little to do with him.' In January 1962 Losey managed to extract the second half of his £12,000 fee from Columbia.

The next Tabori collaboration, *The Intruders* (from *Les Inconnus dans la maison*, by Georges Simenon) 'was a total disaster. I went to Cabourg, rented a house and worked throughout the summer [of 1964] with Tabori. Finally he showed me a disastrous screenplay.[13] Tabori finished *Black Comedy* in November 1964, written in his smartly elliptical manner. 'Comedy,' he warned in a preface, 'is not a matter of laughs but of criticism. Tragedy, pure and grecian, is no longer probable, and the loss is not ours.' Losey sent the script to a host of producers, without dividend. John Osborne sent it back on behalf of Woodfall Films: '. . . silly beyond words, forced, facetious and fake profound . . . I really *am* sorry.'

Concurrently there was *The Man from Nowhere*, an original screenplay by Daniel Mainwaring, written for Hardy Kruger and Losey. In 1963 Losey brought in Tabori to write a new version: Corona-Film Produktion of Munich were to pay Losey $30,000 in cash plus a further $20,000 on a deferment basis. Tabori wanted to 'abstract' the Spanish village into 'Swiftian Fantastica' with 'an unrecognizable lingo'. Having discussed the project with Gina Lollobrigida, Simone Signoret and, later, Burton, Losey turned to the young producer John Heyman, who found himself involved in acrimonious exchanges over rights and credits with the German agent representing Kruger and an embittered Mainwaring who had been suffering from appendicitis when Losey fired him – 'I turned around to spit in my bed of pain in Zurich and found myself replaced by George Tabori . . . Under the rules of the Writers Guild I am entitled to writing credit. Sadly, Dan.' When Hardy Kruger came to Mainwaring's support, Losey's response was impatient: it was nineteen years since Mainwaring had given him a script that he 'really wanted to do' (*The Lawless*). Mainwaring's recent letters had been 'childish and hysterical'. (They were gentle and dignified.)[14]

As early as July 1961 Losey had been interested by Marguerite Duras's script, *Ten Thirty on a Summer Night* (*Dix heures et demi du soir en été*). At that time Simone Signoret held an option which then passed to Anatole Litvak (one of Losey's colleagues during service with the Army film unit on Long Island in 1944) and Jules Dassin. In 1964 Litvak and Dassin invited Losey to direct *Ten Thirty on a Summer Night*, with Melina Mercouri, Peter Finch and Romy Schneider. Accompanied by Priggen and Macdonald, Losey made a recce of Spanish locations with Dassin in November. On 22 December Dassin assured Losey (or himself) that 'our little frictions are stuff for amateur psychiatrists . . . Melina is happy'.

In response Losey was uncharacteristically contrite, but four weeks later following a meeting with Losey and Litvak, Dassin exploded: 'Let's forget it

– drop it – it has become a nightmare to me. I cannot support your condescending attitude, your *tolerating* me, your insults. I told you some months ago that I did not believe we could collaborate ... In Spain you offended me time and again.' This time Losey's reply was not contrite – Dassin was 'hysterical'. Mercouri wrote from Athens, 'My position is impossible. You know how much I wanted to work with you.' Dassin went ahead. Priggen recalls: 'The film finally came out at the same time as *Accident* and got minor notices, which pleased Joe.'

24

Public and Private Ceremonies

Boom!

In *Modesty Blaise* Losey satirized the universe of the fabulous with patchy humour; in *Boom!* he entered it unsmiling. After twenty years of low-budget films, he would now show the world what a director of unique vision could do with the super-stars Taylor and Burton. He would teach 'the money' how to make money out of art.

Boom! (the title came late, after shooting) was an adaptation of Tennessee Williams's play, *The Milk Train Doesn't Stop Here Any More* (which had failed on Broadway, twice, in 1963 and 1964). Williams himself wrote the screen adaptation, according to which Flora ('Sissy') Goforth, who has buried six husbands one way or another, owns a volcanic island in the Mediterranean, ruling it like a despot and refusing to admit that she is dying. Chris Flanders, a mellifluent adventurer, 'poet' and lady-killer, arrives – he normally calls on a lady a step ahead of the undertaker. Originally, the project had nothing to do with Taylor or Burton. Losey and Williams wanted Sean Connery as Chris and Simone Signoret as Mrs Goforth, interpreted as a former Paris cabaret girl who swore in 'gutter French'. By the end of 1966 Losey would have preferred a pairing of Ingrid Bergman and James Fox, but Bergman courteously turned it down. Mrs Goforth was too tough and vulgar for her: 'I can't say the word "Bugger" without blushing.'

Negotiations had dragged on while Losey was making *Accident*. The film rights in Williams's play had been acquired by Lester Persky, of Cineventure, 527 Madison Avenue. Both Williams and Losey wanted to get rid of Persky. By June 1967 John Heyman had entered the scene and Persky was bought out. Twenty-five years younger than Losey, an infant refugee from Nazi Germany, Heyman had set up World Film Services and made his first feature film, *Privilege*. Now, with financial backing from Universal, Heyman agreed to pay $50,000 for motion picture rights plus $100,000 for Williams's screenplay.[1]

What brought the money in was Elizabeth Taylor. Visiting Rex Harrison's Sardinian home, Losey was lamenting his failure to raise the $1.4 million required to make *Boom!* when Taylor declared that she would like to play Mrs Goforth. She admired Tennessee Williams: in *Suddenly Last Summer* and *Cat on a Hot Tin Roof* she had been inspired to extend herself. Seduced, Losey lost his head; Taylor's vindictive, dying old widow (played on Broadway by Hermione Baddeley and Tallulah Bankhead) looks all of thirty-six, a slightly plumper version of Cleopatra. Richard Burton came with the package and the budget leaped to $4.5 million, with each of the Burtons on $1.25 million. Losey's remuneration of £40,000 ($96,000) plus 13 per cent of net profits was the highest of his career to date.

Williams laboured on his constantly revised scripts – two in 1966, two more early in 1967, a 'final' script dated 19 June, but with new pages, versions, endings arriving well into the shooting period. Two years the playwright's senior, Losey got on well with 'Tom' Williams (and his companion Bill) during their conferences in Rome and London. 'Bill and I enjoyed so much the brown crystal coffee sugar this morning,' wrote Williams in March 1967. But, Losey recalled, 'he was good for me only three or four hours a day, what with drugs and drink, and when he was too far under, he was terribly abusive'. When Losey enlisted George Tabori to rewrite a scene in Williams's style, Williams exploded: 'What fucking god damn shit has been trying to imitate my style?' The style was distinctive: 'You're not so tough that some day you're not going to need something that will mean God to you.' Losey concluded that Williams was 'a dealer in words, he doesn't know what the image conveys . . .'[2]

The character called the 'Witch of Capri' was originally a female wearing a pearl-studded, cone-shaped hat and a costume suggesting Fata Morgana. Losey had invited Katharine Hepburn to play the part 'and she was greatly insulted'. The sex of the Witch was changed and the part offered to Bogarde. ('No, thank you!') Noël Coward said yes, thank you. The transubstantiation of the Witch excited Williams to suggest he should be called 'Cher': 'In my early New Orleans days I recall that all elegant queens . . . would call each other "Cher". There was one in particular who would never go out in the evening except with a small cut-glass container of ammonia. Whenever "she" sighted a girl approaching along the side-walk, "she" would cry out "*Poisson!*" [Fish] and lean against the wall of a building, holding the little cut-glass container to "her" nostrils till the girl had passed by . . . Personally I have always found girls to be fragrant, in all phases of the moon . . .' This 'Cher' reminiscence became a revised page in Scene 53, related by the Witch.[3]

What disturbed the American production code body, the MPAA, was Mrs Goforth's 'He mounts me again' and 'Goddam', and the Witch's use of the word 'bitch'. No such problems arose with John Trevelyan, the Secretary of the British Board of Film Censors, who assured Losey, 'I think it will be a great film and the best you have ever made.' For his part Losey assured Trevelyan, 'I believe your tenure . . . has been a major service to independent and responsible British film-makers . . . less that of censor . . . and more that of educator, protector of standards and taste . . .'

On 16 June, Losey returned from a ten-day search for locations, ac-

companied by Patricia and Richard Macdonald. The location chosen was at
Capo Caccia in Sardinia. The Goforth residence was built from scratch.
Macdonald and his assistant moulded bas-reliefs in plaster, with fifty-six
workmen toiling up with marble, sand and cement, striving to erect Easter
Island heads. At seven in the evening Macdonald would plunge into the sea,
covered in plaster, emerging cleansed and ready (Losey complained) for
anything. Goforth's imperial wealth was also conveyed by a large retinue of
guards, servants, a masseuse, a manicurist, a hairdresser, a nurse and doctor,
a 'court' musician. The set was blown down by a gale, leading to a four-day
suspension of filming.

The music had to match both the display of opulence and the booming sea.
'The system of push-button Hi-Fi throughout the villas and terraces makes
possible music wherever desired . . . Music should be Wagnerian and horrifying. It should merge with the sizzling sound associated with the image of the
sun and the diamond. Sound effect of the sun should be musical and
probably mechanically distorted . . . an intense, hissing, almost supersonic
sound, which recurs with variations throughout the film.' Losey had commissioned a score from John Dankworth, who engaged musicians (mainly strings)
and presented his tape to the director – who then turned it down on the
pretext that he didn't like strings. Wounded, Dankworth wondered whether
the decision had been entirely Losey's; that it indeed was is indicated by
Universal's insistence that Losey reduce his fee by £2,500 'on account of the
amounts paid . . . in connection with the composition of two scores . . .' Losey
called Michel Legrand to Sardinia. Heyman and Legrand missed their flight
at Rome airport; finding no small plane for hire, Heyman chartered one with
100 seats. Legrand stayed on the island for only one day, but long enough to
observe Taylor approaching Losey on the set: 'Do you want me to shout or
whisper this line?' Losey replied that shouting and whispering was all she was
capable of. Legrand departed and heard no more; the music was finally
written by John Barry.[4]

Noël Coward arrived, bearing unflattering tales of the sex life of Elia
Kazan and fanciful stories about how rapidly he'd written his own plays.
Burton described Coward's entrance, 'looking very old and slightly sloshed
and proceeded to get more sloshed . . . He is a most generous man but sadly
he is beginning to lose the fine edge of his wit . . .' Losey reported to his
agent: 'Coward is here and is an absolute delight and a great boost of
professionalism . . .' (In November the Loseys visited Coward at his home in
Montreux and enjoyed themselves.) The international press crowded in.
Arriving each morning with a special caravan and retinue, Burton and Taylor
began work with a large bloody Mary. On the day they were shooting the
funicular scene, the caravan fell over the cliff, narrowly missing the crew
below, broke up, and emitted a sinister red ooze. The crew scrambled down,

horrified, to discover that the red liquid was the tomato juice for the bloody Marys. The sexual temperature was high under the blue skies of Sardinia, and the legend (fostered by Losey) still persists that for personal reasons John Heyman imposed Joanna Shimkus, for the part of Blackie, on his sceptical director. But Heyman explains that Shimkus was somewhat foisted on them by MCA Universal. Losey described Heyman's behaviour as better 'than any producer I have ever experienced in terms of efficiency, fulfilment of promises and non-interference'. During the eighth week of shooting Heyman's marriage broke up, his wife left the location, and he started an affair with Shimkus. The best action on *Boom!* was normally off-camera.[5]

A Universal Studios press release in May 1968, doubtless reflecting residual nervousness about the blacklist, announced that 'Losey is as American as apple pie'. The New York première was held on 25 May, and the film was shown initially at the Sutton and Trans-Lux West theatres. Slaughter in the press – with a few voices dissenting – ensued both in America and Britain. Vincent Canby was damning in the *New York Times* (27 May): 'For all its overtones of Indian mysticism, Christian theology and Greek mythology, the movie is essentially the sort that Baby Doll would have hitched a ride into Blue Mountain to see – a tale of the very rich that tells the miserable critters in Dogpatch that money can't buy happiness.' Taylor was so 'plumply ripe and healthy looking' that she must be dying of 'some dread plot device . . .' On 8 June Losey reported to Tennessee Williams: 'I'm appalled for all of us, but particularly for you, by the scurrilous, ignorant and personalized reviews which I've seen . . . I am struck by the fact that the "critics" seem determined to destroy successes and idols that they have built up – namely the Burtons, you and me.'

The cost of production had been $4,592,762. According to the Universal statement of account rendered to 30 March 1974, total gross proceeds were $2,898,079 and the film was in the red (Producer's Deficiency) by $3,795,452. Theatrical box-office returns in the USA were $514,725, with a further $1,207,681 from US television. The most outstanding failure was in Britain, where the film grossed only $20,719. Losey blamed everyone but himself: critics (1969), audiences (1973) and distributors (1975): '. . . these people did not understand what I was getting at . . . they put it in the wrong theatre at the wrong time with the wrong promotion'. Losey's producer, John Heyman, recalls falling asleep while screening the film for Universal's Jay Kantor and Lew Wasserman.[6]

Secret Ceremony

Late in 1967, while in Rome dubbing *Boom!*, Losey was with the Burtons in the Grand Hotel when Elizabeth Taylor remarked (it is said), 'Why don't we do something again?' No more inclined than Winnie the Pooh to remove his head from the honey pot, Losey remembered a script by George Tabori originally written for, and turned down by, Ingrid Bergman. For Losey and his flamboyant designer Richard Macdonald, *Secret Ceremony* was to be the second décor-riot of the 'time of the Burtons'. John Heyman's role was to raise the money and assemble the cast: 'I was the shoehorn,' he says. Universal again put up the cash – several months before *Boom!* hit the rocks. The budget was $3,173,212. Losey's fee had more than doubled to £93,750 ($225,000 exactly, or £700,000 in current values), plus a percentage of profits after the film had grossed a sum equal to 2.6 times the negative cost (which it failed to do).

After seventeen actresses had been interviewed for the part of the disturbed orphan Cenci, Mia Farrow was signed up for $125,000. Twenty-two years old, Farrow had (like the Beatles) of late visited Maharishi Mahesh Yogi on the banks of the Ganges. Her most recent film, also for Universal, was *Rosemary's Baby*, as yet unreleased – Losey had seen her only in *Peyton Place*. Her English accent was equal to her total dedication to the role, but his directorial approach to her – by his own account – was initially aggressive, even sadistic: 'I kept the camera going all the time and shot her in bits and pieces. We did 15 to 20 takes – unheard of for me. She said she found it humiliating – and so it was. It was a kind of public execution – but instead of dismembering her, it shook her together into one piece' (Losey was happy to confide to the *Evening Standard*'s Alexander Walker, by way of pre-publicity). Farrow sent Losey a note in a large, girlish hand, dated 19 March, eight days after shooting began: 'Dearest Joe, My flowers so *so* beautiful and so are you!' Then a line-drawing of a smiling face. 'I will try my absolute hardest to be good Joe . . .' The impending end of her marriage to Frank Sinatra involved her in further tensions: on 17 April, Heyman wrote rebuking her for her refusal to give press interviews.[7]

Losey wanted a box-office actor for the repulsive stepfather, Albert. In the event Robert Mitchum was paid $150,000 for three weeks' work. More friction for Losey: 'But from the moment [Mitchum] arrived, he was on the defensive . . . hostile . . .' Heyman recalls that his worst exchange with Losey occurred on the set when Losey was determined to get rid of Mitchum.[8]

But the main 'character' in *Secret Ceremony* is the vast residence in Addison Avenue, Kensington, which Cenci has inhabited alone, without a single servant, since her mother's death. Losey chose an art nouveau mansion tiled in peacock blue and emerald green, and filled with the lustre of light and dark woods, leaded glass, Venetian galleries, post-Pre-Raphaelite stained-glass windows and De Morgan mosaics. Built between 1896 and 1904 by Halsey Ricardo for the furniture magnate Debenham, the house was now empty and dilapidated, offering Richard Macdonald a pretext for extravagant restoration and furnishing. Richard Rodney Bennett's music catches the tinkling mode of a Victorian music box – but a taut, taunting, ethereal tinkling. Shooting started at Elstree Studios (bedroom interiors) on 11 March 1968, then moved to London locations during April, May and June. The seaside scenes were filmed at Noordwyck, in Holland.

One must agree with Foster Hirsch: 'In its Pirandellian oppositions between appearance and reality, the story palpitates with fashionable modernist themes; but the parade of intellectual motifs has less substance than show . . . Losey at his most pretentious.' Here we must distinguish between the film script and Marco Denevi's original novella, *Ceremonia Secreta* (winner of the '*Life en Español*' prize) which introduces us to Leonides, a woman of intensely religious disposition engaged in a permanent ceremony of purgation, leaving flowers for a paralytic child, leaving periwinkle for three virgins, leaving a bouquet of nettles for a prostitute. 'The tall, gaunt figure of Miss Leonides . . . might have been taken for a Greek orthodox priest who, under cover of darkness, was fleeing a Red massacre . . .' The traumatized girl she encounters, Cecilia, is 'short, plump, with short, stocky legs . . . Her face, which was broad and somewhat coarse-featured, radiated innocence and kindliness, like that of a peasant.' Tabori and Losey turned this 'peasant' into Mia Farrow. In Denevi's story Cecilia remains in a zombie-like trance throughout – there is no element of sly, psychotic calculation, the kittenish teasing woven into her character by Tabori and Losey. Nor is there any trace of 'Albert'. An envious female relative who covets Cecilia's fortune has sent three leather-jacketed men to murder her – it is this assault which accounts for her permanent trauma.[9]

The first of Tabori's five draft screenplays was dated March 1967; the final revised script, 12 February 1968. Pages of rewrites reflected ongoing confusions of identity, relationship and motivation. As so often, Losey abandoned the tripartite schematizations found in his production notes: 'Miss Leonora should remain a bit of a mystery. Her background is projected in a series of images: Recall, Fantasy and Dream. Stylistically, these moments will be separated from the main body of the narrative. For instance, the Recalls may be sequences of "stills", isolated by reducing their frame; the Fantasy, slow motion; the Dream, a fragmentation of images.'

Leonora's oscillation between English and American idioms is inexplicable. 'I never got a red cent from her,' she tells the aunts in their junk shop – correcting herself to 'not a penny'. Moments later she threatens them: 'If ever you drag your ass down to that house . . . I'll set the cops on you.' Tabori sent Losey some revised lines to de-Americanize Leonora: 'Watch yer manners' for 'Mind your manners'; 'Pack it in,' for 'Cut it out'; and so on. The result was a vernacular dog's dinner. In Tabori's relentlessly facetious script (but not in the film) she meets God, 'a quickchange artist, now a cop, now a

priest'. God wants to know where she comes from: 'You don't sound like
Shoreditch to me. Or even Trenton, New Jersey.'

Geoffrey Shurlock, Director of the Production Code Administration of the
MPAA, asked for verbal cuts: 'frigging', 'holy shit', 'uncork her in the
morgue', 'screw standing on your head!', 'rain like piss', 'drag your ass', 'kiss
my ass', plus an undue quantity of the expletives 'Christ!' and 'goddam'. John
Trevelyan, the British censor, also objected to many of the same words and
phrases, but more sympathetically: 'I found the word "frigging" rather
unsatisfactory, since it is not an honest word . . .' On 25 March 1968, Losey
sent him a limerick 'endorsed' by Norman Douglas and beginning:

> There was a young man of Siberia,
> Who of frigging grew weary and wearier.

Satisfied, Trevelyan in turn sent Losey a limerick about 'a plumber of Leigh/
Who was plumbing his girl by the sea'. On 25 July Trevelyan was shown the
work print of *Secret Ceremony* and had dinner with Losey afterwards.

Losey told a French magazine that the film was about the terrible need that
human beings have for other human beings, and the impossibility for certain
people of fulfilling that need when the opportunity arises. Having rejected the
High Church ritual of his childhood, he had later discovered that the human
spirit requires ritual. The 'Money', however, was in no mood for limericks or
ritual. On 5th August, John Heyman informed Losey that Jay Kantor, of
Universal MCA, was in a panic 'because of his somewhat insecure position
and all his eggs are, so to speak, in our basket'. Heyman passed on Kantor's
notes, a catalogue of confusion: 'Why does Elizabeth go to the refrigerator . . .
Why is Elizabeth washing her face? . . . It seems to me we see Mia at a
balcony staggering down a hall and then ending up at what apparently
appears to be the same balcony she left . . .' Heyman recalls that 'Universal
hated it' and confesses that he found both Tabori's script and the finished
product 'obscure to the point of inaccessibility'. Losey later publicly accused
him of having tampered with the film. Heyman denies it. Losey complained
to the press that Universal's ads made it look like a lesbian picture – 'the
cheapest, most degrading kind of exploitation'.[10]

The film opened in New York on 23 October 1968, at the Sutton East and
New Embassy theatres. Mary Losey Field went to see it with her brother.
'We walked up the street. "You didn't like it, did you?" he said. "Not
really."' *Variety* (23 October) predicted the film could 'break through to hot
b.o. response in general adult situations, after proper launching in semi-arty
theatres'. As if in confirmation, a big ad in the edition of 30 October
announced record gross receipts of $28,658 and $22,964 at the two theatres.
But the reviews were generally derisive, none more so than Pauline Kael's:
Losey needed Pinter to fill his 'overfurnished vacancy'; this one resembled a

'Jungian decorator's dream'; the film was 'truly terrible . . . a Gothic version of folie à deux' or 'folie à how many?' Judith Crist bracketed *Secret Ceremony* and *Boom!* as 'disasters'. On 17 November the *New York Times* carried an Andrew Sarris interview with Losey, who explained: 'The tragedy nowadays is that it's easier to raise four million dollars for a movie with stars than to raise one million for a movie with ideas. The trouble is I'm a one million dollar movie director in a four million dollar movie market.' Stung by hostile reviews, he denounced American film critics on French television as wittily destructive 'ignoramuses'. And women were taking over.[11]

The film opened in London at the Odeon, St Martin's Lane, but was screened (with an X certificate) in only two other British cinemas: six weeks at the Curzon, Mayfair, two weeks at the Odeon, Kensington. The takings at the Curzon during the first week were good but slumped thereafter, although some critics accorded the film a surprisingly warm reception. Derek Malcolm, of the *Guardian* (20 June), found it 'quite beautifully made . . . full of that particular chemistry a top-class film-maker can summon up when really in the mood.' Taylor had never been better.

The UK gross takings were a minuscule $35,989, joining *Secret Ceremony* to *Boom!* in the disaster league. But the box-office weather was brighter in America. Records were broken in Dallas and Fort Worth. Universal's statement dated 30 March 1974 showed total gross proceeds of $5,232,905. Of this, theatres accounted for $2.7 million, television for $1.2. But 'profits' and percentage payments there were none – Universal enjoyed the contractual right to retain 2.6 times the cost of production, or $8,250,351, plus interest, giving a 'negative balance' of $3,562,365.[12]

The $1.2 million from American television involved the grimmest assault on his integrity that Losey had suffered since *Eve.* In December 1969 Morris Davis, of Universal, informed him that both *Boom!* and *Secret Ceremony* required special editing for network television, neither film having performed up to expectation in the theatrical market. Losey was invited to undertake the task himself and refused. Nine months later Universal agreed to remove Losey's name from the prints and advertising of the TV version. Losey circulated a letter of protest to film critics and enlisted the support of the Burtons to protest the cuts and the addition of a reel-and-a-half of what Universal called 'explanatory conversation between an American psychiatrist and an English barrister' which had been inserted at the beginning and end of the film – a stilted and bogus attempt to explain the characters' motives and actions. Herbert S. Stern, of Universal, sought to assuage Losey's anger: the situation had been so complex that 'adding a new voice [i.e. Losey's] to the many discussions would have resulted . . . in a waste of all that had gone before'. Losey turned to the Directors' Guild – which declared itself 'painfully aware of this situation for many years now', but remained supine: both

George Stevens and Otto Preminger had taken similar cases to court, and neither had prevailed.[13]

Liz, Richard and Dirk

The Burtons changed Losey's life. '1967 was the year when the Burtons began to go publicly mega-rich . . .' writes Richard Burton's biographer. Losey recalled: 'We all flew off to Portofino to talk to the Burtons . . . went up to their hotel, were not received, had a nice supper, went to bed . . . sat on the terrace of the hotel having an early breakfast, and as we looked, the yacht steamed away!' Visiting the Burton yacht, Losey and Norman Priggen were conducted to the bar, where Burton pulled a cheque out of a script: it was for $1.25 million − one quarter of a year's profits on *Who's Afraid of Virginia Woolf?* The yacht was called Kalizma after the children Kate, Liza and Maria.

Losey loved it all and took responsibility for procuring a salt substitute for the overweight stars from the French manufacturers. He took delight in reporting his early clashes with Elizabeth Taylor, feeding scenes from behind the scenes to the columnists. 'My working relationship with Elizabeth had begun with absolute hell.' Taylor didn't like her clothes, couldn't sleep, shooting was delayed by three days. Losey had to shoot Taylor's first scene 'thirteen' times. 'Elizabeth was belligerent towards me at the start . . . She didn't know what I was doing and it was a struggle . . .' The Burtons' wealth was seductive. On 10 October 1967, Losey wrote from Sardinia to his accountant, Gerry Burke: 'The Burtons seem to be extraordinarily lucky in their land speculations and have now doubled a quarter million investment in two years in Tenerife. They are buying some land around here and have asked me to come in for a small amount. I do not believe in speculations but perhaps . . . I can afford £1,000 or so.' To Brigitte Bardot, prospective star of *La Truite*, he sent telegrams and flowers; concurrently, with *Boom*! Losey entered his 'Dearest' phase: 'Dearest Elizabeth, Dearest Noël [Coward]'–gifts, flowers, yachts, Swiss châteaux. Taylor gave him an oval-shaped Cartier watch, inscribed 'To Joe with love − Elizabeth'. Dirk Bogarde remembers, 'with constant joy, the day he showed me his Cartier watch from Liz Taylor and his enraged reaction when I said that I liked him better in the days when he wore a "tin one" . . .'[14]

Burton and Taylor became an obsession: one or both figure on virtually every cast list for every Losey project of the late 1960s and early 1970s. Burton was gallant, courteous. In an undated letter written from a hotel near Elstree Studios, presumably after *Boom*! had been reviewed, he told Losey: 'I do not give a tuppenny bugger what the immediate critics say but we'll all be proud of it one day. It contains what I consider to be − tho' I may be alone −

a magical combination of words and vision.' Arriving at the Dorchester Hotel in February 1968, Taylor was invited by Losey to help audition a number of actresses for the part of Cenci in *Secret Ceremony*; he also looked forward 'to showing you our fabulous house'. Before shooting the film, she wrote: 'Joseph Darling, Not to worry – God is love! and all that. I bet we have even more fun than the last one.' But clearly Taylor harboured worries of her own, as indicated by a letter from Losey on 5 January: '. . . you will note that the massage sequence you were worried about being considered as lesbianism has been removed'. He assured her that the relationship between Leonora and Cenci was strictly of the mother–daughter variety. 'When she gets into the bath or snuggles up in bed it is like Liza or Joshua with you and me.' But he gravely reminded the actress that 'sex is an element of every close human relationship . . .'

Donald Wiedenman described a scenario on the set of *Secret Ceremony*. Taylor appeared wrapped in a white towel, ready to do the bath scene, but nervous because there were too many people on the floor. She stood in the shadows, on the edge of the hot studio lights, while nobody moved, everyone watching her. 'Suddenly a figure eases out from behind the camera and advances slowly, almost painfully . . . He touches her arm gently, whispers something briefly into her ear, then motions behind his back signalling for everyone to clear the set except a skeleton crew . . .' 'My wife and Joe Losey are having a professional love affair,' Richard Burton told Wiedenman.[15]

After shooting Taylor wrote: 'Dearest Joseph, Till the next one – please make it *soon*. All my love. P.S. Don't forget all about me.' He didn't. On 28 December 1968 he wrote to her in Switzerland: 'The roses from the Burtons were beautiful . . .' Two months later he cabled her in Las Vegas: 'Happy Birthday hurry back stop I hope the sound of rattling coins is improving your back.' 19 March 1969: 'Elizabeth – I am having a ring made for you by the man who did the crazy ring of Patricia's which you admired on the opening night of *Eagles*.' Taylor responded, 'I'd so love you to work with me this summer! I miss you terribly if you're not there to help me.' On 14 May Losey returned to the subject of the ring, which was the work of a young American sculptor living in Rome. 'Since he has seen your display of stones, he has decided in the case of your ring to put all the stones facing inward – rather like wearing a sable coat inside out. Very chic!' Later that month Taylor dined at Royal Avenue for the first time – after which 'tons of flowers and a Mexican dress' arrived for Patricia.

But the failure of two films was too much. The Burtons drifted over the horizon in a politely determined way, although throughout 1969 Losey constantly sent scripts to Burton. On 11 December, George Davis turned them down on the actor's behalf. At the end of the year Losey wrote to Taylor: 'The silver toad has a lovely pout and we have a new dining-room as a setting for it – all translucent and pink – and the ivory pepper mill is standing beside it. We loved your presents but mostly because they remind us of you.' The good will was there – but no further commitments. In February 1970, referring to the crisis over the television version of *Secret Ceremony*, Burton cabled: 'Dear Joe will support you to the death . . . Elizabeth says she loves you passionately tell Trapicia the same goes for me about her.' (The Burtons regularly called Patricia by this anagram, while Losey occasionally addressed Taylor as 'Mabel'.)

Losey continued to pursue Burton about Nicholas Mosley's *Impossible Object*, 'which you read and "didn't understand"'. And what about *Nostromo*? What about Paul Jarrico's screenplays on Leonardo, the Borgias, Machiavelli? In October 1970 Losey appeared with the Burtons at the Round House before a showing of *Boom*! and 'made an angry speech' about the additional scenes in the American television version of *Secret Ceremony*. John Heyman explained the legal position but Patricia felt it was the wrong occasion; a young and lively audience (she noted in her diary) wasn't interested. Afterwards she 'endured a long evening' with the Burtons at Keats' restaurant: Burton was 'belligerent and unattractive' while Taylor was 'okay except of course for her usual flirtation . . . with Joe . . .' When Patricia tried to explain to Burton why the Round House audience gave him a hard time, he quickly cut her off: he despised audiences. Clearly wounded, she reflected in her diary that the Burtons were 'stuck in their sealed world and determined not to receive new ideas or change themselves in any way . . .'

'I don't seem to have much luck connecting with you or Richard,' Losey wrote to Taylor, then resident at 11 Squire's Mount, NW3, 'and I don't even get much information as to where you are or what you are doing.' Although he was mixing *The Go-Between* at the time (December 1970), he urged her to 'have somebody ring and I will run quickly down the side-streets to see you'.[16]

From 1967, when Losey embraced the Burtons, he made no further film with Dirk Bogarde. An almost perfect collaboration soured into rancour; after three masterly films with Losey, in rapid succession, for love and little money, Bogarde felt discarded. Of *Boom!* he relates that Taylor 'dragged' Burton to the Audley Street cinema for a special screening of *Accident*, then told Losey: 'You've given Bogarde class, give it to my husband.' Losey and Bogarde represented two contrasting temperaments which are reflected in their correspondence: Bogarde cheerfully outrageous, Losey gloomily responsible. Both were vulnerable but Bogarde wears his scars on his sleeve in a bright patchwork quilt; Losey's wounds bleed through layers of armour.

It had been Bogarde's participation which got Losey's first British film, *The Sleeping Tiger*, off the ground. The 'Early Joe', Bogarde later recalled, was an 'exceptionally fine looking American in blue jeans and a jean-shirt, standing about in cowboy boots and wearing a red handkerchief knotted at his throat'. In 1956 Bogarde's influence was helpful in getting Losey his ill-fated, 'three-picture' contract with Rank, through Olive Dodds – but the film they were to make together aborted. *The Servant* was Bogarde's thirty-eighth film and a turning point in his career. His respect for Losey was boundless: 'We work together without any words ever being spoken. I never ask him anything, he never tells me anything. It's a completely mutual marriage of minds.' With the exception of *Modesty Blaise*, each of the films he made with Losey – this 'tense, harried perfectionist who looked like a Red Indian' – was a 'bitter, exhausting, desperate battle . . . the battle between Art and Profit'. But Losey was always determined, optimistic, courageous.[17]

Bogarde had lent Losey money and lovingly presented him with an old, rare, ivory netsuke of two frogs, which Joe carried in his hands until it was worn bare, and which he lost (Bogarde recalls) while making *Modesty Blaise*. Losey lamented, 'Everything's going to be bad from now on.' The memory moves Bogarde. He also bought Losey a foot off a Greek statue, which Losey used as a doorstop. Losey, however, connects these happenings with *Boom!* He took his good-luck talisman to Noël Coward's dressing-room and invited Coward to touch it. 'Oh, my dear boy! Thank you so much!' – Losey never got it back. After he'd explained this to Bogarde, 'which wasn't very easy', Bogarde arrived 'with the foot of a Roman statue which weighed about 40 lbs and he said, "This is something you can't lose . . ."' (Greek or Roman, *Modesty* or *Boom!* – how can history be told?)[18]

Accident, Bogarde reckons, was the most demanding film he ever played in. At the end of shooting, Losey wanted a dawn shot of Bogarde looking up at Anna's window and seeing her light go out. Bogarde killed time all night while 'Joe got pissed and slept'. Just before dawn it was bitterly cold. Bogarde saw a child's tricycle lying on the pathway. He moved it. Losey screamed, 'For Christ's sake, can't you just once in your life be professional!' A property boy who'd lent Bogarde his coat against the cold gasped, 'You know, the bugger meant that.' Bogarde got in his car, drove home, and refused to answer the telephone. 'He never thought I would do that, walk off. But the only thing I cared about was my professionalism. I got an apology – but we never got the dawn shot.' Immediately after shooting *Accident*, Bogarde telegrammed Losey: 'Stephen has completely gone away and it seems has taken part of myself with him. I have never felt quite so abandoned and destroyed . . . There is great happiness that you trusted me to play him . . . Love you more than you could ever guess.' Bogarde insists that Losey and Pinter had wanted 'another actor' for the part (but it was Spiegel, not Losey, who had wanted Burton).[19]

After *Accident*, Bogarde reflected some years later, 'there was nothing more for us to say together; weary, drained almost, and to some extent disillusioned, we realized that we must separate for a time and go our own ways ...' In reality it was less clear-cut. Bogarde remembers Losey, glowing with his new Burton-affluence (December 1967), turning up at his suite in the Bristol Hotel, Vienna, bearing extravagant gifts: a bowl of white cyclamen; a huge pocket watch, solid gold, with filigreed face, to be wound with a long key; a box of old Chagin leather containing gold-plated fruit knives. All this was in aid of requiring Bogarde to read two scripts, different projects, and asking him which one he should make. 'No one was allowed to contact me. The phone was cut off. He sat there drinking water and watching me while I read.' Joshua Losey and Bogarde's niece were playing with new toys in the adjoining bedroom until Joe lost his cool and yelled at them about the noise. Bogarde can remember nothing about the scripts – except that he advised Losey to make neither. It's odds-on that the scripts were *The Man from Nowhere*, by George Tabori, for which Losey was considering Bogarde in a supporting role – Burton was to play the lead – and Tabori's *Secret Ceremony*. Unknown to Bogarde, his name was among a dozen that Losey had pencilled in to play the ghastly stepfather, Albert.[20]

In the autumn of 1969, Joe and Patricia spent two days with Bogarde near Labaro (Rome). Bogarde had been working with Visconti on *The Damned*; for Bogarde, Visconti was the Old Master, the prince of the profession with a castle in Ischia: 'Joe was the king of directors, Visconti the emperor. Joe was of peasant stock, Visconti of princely stock, the *grand seigneur*. His intellect lay in the genes and the blood. Visconti told me, "Everything I do is 'bravo!' and like an opera."' Visconti knew from direct experience how families like the Krupps lived, whereas in *The Go-Between* (claims Bogarde) Mrs Maudsley appeared in an evening gown from breakfast on and the little boys sounded 'Earl's Court'. Pieter Rogers remembers Losey's irritation when Bogarde publicly declared that Visconti was the greatest – but Bogarde retains an ironic view of the political pretensions of both directors. 'Visconti claimed to be a Communist but I found it hard to accept among the palaces, Picassos, footmen and cooks ...' He was not impressed by Losey's political pretensions, either. 'I read an interview he gave in France about being a Marxist. I said to him, "Come on." He told me I didn't understand because I was a cunt. I don't think he was a Marxist at all.'

While shooting *Death in Venice*, Bogarde wrote from Rome on 6 March 1978 to his 'Dearest Josieposie'. Having explained that his daily routine involved driving from Rome to Venice in his new BMW 'in 4½ hours flat', he turned to Losey's displeasure about a recent Bogarde interview in the *Telegraph*: '... the fact that I did NOT say you were pissed out of your mind, and disgusting, the night I walked off the set ... and took ALL the blame; you choose to ignore ... If one thinks one is God one must behave as

God . . . I don't, honestly, see how we could work together again . . . and you decided, a while ago, to take another path my dear . . . the one with the lolly and lushness . . .' His animosity towards Burton surfaced: '. . . you say that you are going to California to "help Burton" get his Oscar, as if it were some noble deed . . . Jeasus! [*sic*] What are you doing for the Welsh bastard?' As for *Boom!* and *Secret Ceremony*, 'wheather [*sic*] you like it or not they were both simply AWFUL [scripts]! And dated.' Bogarde then took a poke at Losey's as yet unreleased *Figures in a Landscape*, which he had not seen: 'And perhaps [Robert] Shaw is JUST what you deserve. As far as I remember he only does two things really well . . . shout above rain and wind and stand with his legs apart.'

Today Bogarde denies having felt personal rancour towards Burton, but it sounds otherwise: 'I resented Burton for his betrayal of his own talent and craft. I don't give a shit about Burton personally. You can't be jealous of anyone who's better than you are and he was better than me – but not as a screen actor. He couldn't act on the screen and I hated the way Joe was selling himself so cheap.' Bogarde's letter went on to discuss *Death in Venice*: 'I know that you have long wanted to make it . . . you told me until I was blue in the face . . . but you never asked *me* to do it . . . or offered me the chance, or remotely thought that I even could! . . . You have never even said that I was more than passably good . . . I was lucky to have you. Instead of the other way around, sweetie!! . . . you know full well that you are deeply loved by me . . . But . . . there is no helping you . . . you eat love like candy and vomit it straight up again: like a dog.' Bogarde continued to hand it out: 'Incidentally, I did "This is Your Life" for Wendy Craig last week . . . I thought I *had* to after all the filthy things you have said about her in print . . . and I must say I was pissing myself at the thought of your poleaxed face, had you heard!'

Although Losey's reply cannot be traced, he was evidently not amused. On 18 April, Bogarde wrote from the Hotel Hassler, in Rome: 'Sad that my "paranoid" letter caused you such anger it WAS supposed to be funny . . .' In both collaboration and conflict Bogarde and Losey retained their distinctively feline and canine qualities: Bogarde's spine delicately arched, claws visible; Losey growling, pawing the ground.

On 22 August 1971, Losey sought to soothe Bogarde's feelings about *The Assassination of Trotsky*. Evidently Bogarde had accused him of stringing him along while knowing he was going to choose Burton. 'Dearest Dirkel, I received your sweet letter . . . For your information, the part was offered to you, then to Brando, then to Burton, then to Scott and then again to Burton when he rang asking if the role was still open.' When offered the part during the Cannes Festival, Bogarde had been 'quite emphatic about not being interested'. Forgetting that the screenplay of *Trotsky* had got worse, not better, Losey told Michel Ciment: 'I had originally cast Bogarde who stupidly refused on the basis of the script which was very bad. He demonstrated a fatal lack of confidence in me in refusing a script which I would never have shot anyway, whereas Burton demonstrated a confidence in me in accepting a script which I didn't shoot.'[21]

Then Burton again! Losey reported to Pamela Davies in January 1972: 'Relations with Dirk at the moment [are] strained because he thinks he should play [*Under the*] *Volcano*.' A week later Losey wrote to Romy Schneider from Rome: 'I was very much distressed by the evening with Dirk, as I know you were.' What had occurred? Desperate (as he himself says) to play the Consul in *Volcano*, Bogarde kept visiting Losey in Rome until he grasped that Burton, not he, was going to get the main part. 'I phoned and Patricia said he wasn't there, but I could hear him breathing and gasping in the background. He was always a coward – why couldn't he tell me that he wanted Burton?' Furious, Bogarde retaliated by hosting an expensive evening, beginning with drinks at the bar of the Haessler, then on to dinner at El Tula, his guests including Romy Schneider, Helmut Berger and the Loseys. Dropping off the Loseys at their apartment, Bogarde said goodbye, 'And I didn't mean *au revoir* or *auf wiedersehen*.' On the day of Losey's death he wrote to Patricia: 'I remember the business in Rome about *Under the Volcano* . . . and found it hard to rid myself of anger and bitter amazement . . . He had a quite extraordinary capacity for making enemies of those who liked him, and respected him.' Here one must add that Patricia Losey contests Bogarde's narrative of events, doesn't remember the Haessler dinner, and regards Bogarde as an 'unreliable' historian.[22]

Bogarde is uncomplimentary about the Loseys as a couple. They bickered – or Joe was foul to Patricia. 'When he was very drunk she got uppity – she'd done this and that and her script was better. She was into fashionable clothes and much transformed.' (Patricia comments: 'Not my behaviour. *Ever*. In any case, at the time we frequented Dirk I hadn't written any scripts.') As for their mutual friends, the Roux, owners of the Colombe d'Or (which Bogarde had been visiting since 1947, enchanted by Madame Titine Roux, who had 'the best legs in France' and who could remember the era when Scott and Zelda Fitzgerald fell out over Isadora Duncan), Bogarde insists that Francis and Yvonne Roux had become fed up with Losey's 'drunkenness, his rudeness, and unpaid bills'. Joe would occupy the tiny sauna of the hotel clutching a bottle of vodka. 'Nobody could go in there.' Bogarde depicts him as booking tables but cancelling 'if something better came up' – a phrase he often employs about Losey.[23]

On 27 March 1974, Losey told Daniel Angel that he was thinking of Bogarde for *The Romantic Englishwoman*. Bogarde was cutting: 'It sounds like a Georgette Hayer [*sic*: Heyer] Novel . . .' As for Glenda Jackson being romantic in the main part: 'Well. About as romantic as a Games Mistress with an off accent.' On 14 April he had another bite at *The RE* but clearly he was unhappy and insecure, claiming that he had *chosen* to make only three movies in six years, and then reversing an obvious truth: '*The Servant* was hardly a movie . . . it was bread.' A month later Losey phoned him in the south of France to cancel a supper engagement. 'Something better had come up.'[24]

Bogarde tells another story. In the late 1970s 'a tame millionaire' whom Bogarde liked apparently offered to invest five million pounds in virtually any Bogarde–Losey project. When the three came together at the Georges V, 'Joe spent half an hour on the telephone complaining about the service, the drink, and finally the blasted Poule au Pot!' Later the millionaire called Bogarde to explain that he found it impossible to 'get on' with Losey – 'but I must not worry about it, because he knew that Losey was a genius . . . Nevertheless, we lost five million smackers!' It's a tall story, but the best guess at its roots may be a millionaire called Jeffrey Pike who wanted to invest a lot of money in Losey directing Graham Greene's *Dr Fischer of Geneva*.[25]

Bogarde was on the telephone to Royal Avenue at the moment Losey died. The letter he wrote to Patricia that day was generous and honest – he had not expected this 'grumpy, cantankerous, loving genius' to die. He recalled good times. He would always remember Joe for 'his blinding courage, his passion for the Cinema and good players and good Crew, and his burning determination to make magic on the screen'. Bogarde felt that Losey had at least four great movies to his credit, and these four 'altered the way all Movies were to be made in the future'. It was Bogarde who organized the memorial for Joe, the gathering of the clan, in September 1984.

Figures in a Landscape

The money alone can have seduced Losey into directing a film as morally barren as *Figures in a Landscape*. Faced with the collapse of his hopes of directing *The Go-Between* during the summer of 1969, he disastrously responded to a call from his friend, the actor-writer Robert Shaw, and signed up with Cinecrest Films for a project financed by Cinema Center Films, a division of CBS. Another director had already been fired when Losey flew to Spain to collect a fee of $250,000 (though the figure was modified when he chose to reallocate a proportion of it to location expenses).[26]

Stanley Mann and James Mitchell had written a script adapted from the novel by Barrie England. Losey 'detested' both the political perspective and the gratuitous violence of the novel but convinced himself that he could create 'a statement that was mine' about 'the present condition of the world –

a fable, an allegory' about the 'brutalization of human society and the human animal'. He and Shaw worked on a succession of desperate revisions from May to September, well into shooting, their relationship warm and stormy – both were drinking heavily. Separated from his usual production team, Losey sacked the director of photography and called in Henri Alekan, who soon concluded that he was aiming at a *mise en scène trop grandiose*.[27]

Conditions in Málaga were arduous: Losey suffered acute asthma attacks when working in low temperatures at high altitude. Human relations were no better – Cinecrest's producer, John Kohn, issued a stream of memoranda on the script and the first rushes which infuriated Losey. He blamed Kohn for the work of the second unit, 'a waste of stock and completely unusable . . . all of our interests would be better served if you could . . . let me get on with the direction'. On 21 June he sent Kohn nineteen notes under the heading, 'Robert Shaw's complaints in no particular order, and after a great deal of patience'. Among them: '9. That there is a rotten undercurrent of discussion and secret behind-back grumbling. 10. That the Producer [John Kohn] seems lackadaisical in important production matters and stubbornly mistaken in artistic ones . . .'[28]

On 9 November, Losey flew to Los Angeles in a state of dread to show *Figures* to CBS executives. His fears were justified: Gordon Stulberg, of Cinema Center Films, already angered by Losey's habit of rubbishing Barrie England's novel during press interviews, complained about 'total audience disorientation as to what is happening and why'. On 11 March, Losey hurled at John Kohn a list of complaints about the print, the opticals, the titles, plus some personal insults: 'Considering that, in effect, we had no producer on the picture . . .' This was too much for Kohn: 'I've had a year of your witless sarcasm and I'm sick of it . . . your total lack of grace . . . just a grumpy old man.'[29] In April the Loseys attended a preview given by the University of Chicago Film Society and hated every minute of it. The audience came and went in quest of popcorn and the Loseys were shocked by the lines of helmeted police outside and inside the Civic Center – there was nothing allegorical about Chicago violence in the 'Days of Rage' era.

The landscape is spectacular but the 'figures' are abysmal. Shaw's twisted and often sub-Pinteresque dialogue is both monomaniacal and banal. 'I hate a messy killing,' rages the veteran mercenary MacConnachie (Robert Shaw), warning young Ansell (Malcolm McDowell) that this isn't going to be 'a Boy Scouts' picnic'. The only hero of this faintly pretentious adventure-travelogue is the pursuing helicopter, whose low-level acrobatics and menacing swoops from mountain to valley are photographed both from outside and inside the machine – a technical feat, no doubt, but to what end? Losey's evocation of flying dust, fire, downpour, rockscape and verdant valley, of physical endurance and callous murder, is admirably executed – but where the fable, where the allegory?

One scene only summons Losey's powers. By night the two fugitives stealthily enter a village, searching for a knife to cut their bonds. An interior shot shows Ansell staring in through a window; in the forefront of the frame we glimpse a supine face, eyes closed. Breaking in, MacConnachie and Ansell duly free their hands and load up with food. Cut: a seated peasant woman is staring at them, pale-grey, Goya-like. We mistake her silent trembling for fear: it is outrage. The camera pulls back to show her holding vigil beside an open coffin; the supine face we glimpsed belongs to her dead son, whose legs are loaded with her offering, a pile of recently baked loaves. The famished MacConnachie grabs one and bites into it – the woman screams. In the next shot the two men have taken refuge in a church alongside a line of carved saints. OK.

Losey's allegorization of the story merely results in a void. The personality conflict between the two fugitives is a mixture of 'me tough, you soft' and unconvincingly obsessive sexual hang-ups: the murderous MacConnachie keeps changing his tall tales about his wife, while Ansell maddens him with a swinging sixties vision of unlimited Norwegian crumpet in (apparently) Fortnum & Mason's. Declares MacConnachie: 'My wife had to have her face built up. Plastic jaw . . . alsatian dog sprang across the bar at her just as we were saying goodnight.' But later he says, 'She's got a pretty face but she's not got good legs . .'

This a pale shadow of the tension and social substance achieved by Stanley Kramer in *The Defiant Ones*, a story of two chained convicts, one white, one black, on the run. Losey's helicopter and the goggled soldiers of the pursuing State suggest an abstraction of totalitarian power; 'depth' is a matter of unanswered questions, but one notes that on page 1 of the script Losey himself asked why the helicopter pilot did not open fire on the escaped prisoners. The frontier guards of the neighbouring state where they seek refuge are discovered in mountain snow above the tree line, grouped like automatons in attitudinizing postures. Approaching the frontier, Ansell says, 'Come on Mac, we've made it . . . It's all over, we're free . . . Leave it Mac, you've got no chance.' MacDowell's acting gifts are reduced to a single, pained expression plastered in mud.

Figures was refused a showing at the Cannes Festival, and did not reach the public until it opened in London at the Carlton in November. Losey complained to Goddard Lieberson, President of CBS, that there 'was no première' and no discussion after the press show. Classified AA, this version shown lasted 1 hour 49 minutes. It ran for three months in London then disappeared. Most of the reviews found the script unworthy of the director; Derek Malcolm commented in the *Guardian* (19 November): 'Losey photographs the land from the sky and the sky from the land in such a way as to set the whole story tingling in the eye . . . But the mind, alas, needs something more than visuals.'

On 28 December, Stulberg blandly informed Losey that the film had failed commercially in London and was to be cut to 'a reduced running time version' for general distribution in a number of countries, 'more specifically, England, where the circuit is always double-billed'. Would Losey like to recommend cuts? Losey responded by blasting the distributors for shoddy promotion and advertising. If the film was cut, he would remove his name. By 24 March, he knew that fourteen minutes had been cut. Not until January –March 1973 did the abridged version reach Birmingham, Yorkshire and Wales – directed, apparently, by no one.

The New York première was delayed until July 1971, Cinema Center Films timing its release to precede by ten days that of *The Go-Between*. On arrival Losey promptly accused the distributors of taking advantage of his Cannes award. *Figures* was withdrawn soon after it opened. Joy Gould Boyum, of the *Wall Street Journal*, protested the burial of a film which she placed on the same artistic level as *The Go-Between*, but the majority of critics went along with Jay Cocks, of *Time* (9 August), who reckoned *Figures* was Losey 'at his worst . . . It's all very pompous, the kind of vague allegory that is open to any number of interpretations and able to sustain none . . . inflated metaphors bob about on the surface of the action like a collection of lopsided inner tubes.'

Dartmouth

In January 1970 Losey arrived at Dartmouth College, his Alma Mater, as Visiting Professor of Film and Drama. The Loseys were accommodated in the distinguished visitors' residence known as Choate House – but the plumbing and heating were out of order and the first night was spent shivering under piles of eiderdowns. To avoid tax, Losey had negotiated that his stipend of $5,000 should be taken in the form of expenses, with free accommodation. His host, Peter Smith, Director of the Hopkins Center, recalls that the business of eating out became an unexpected burden on the budget. Very few places in Hanover provided the kind of first-rate meal Losey required. 'We expected to pay the local grocery but not the best of restaurants and the best wine and brandy.'

The Loseys' arrival coincided with the Inauguration as President of a Jewish mathematician, John Kemeny, a liberal Democrat who later introduced co-education to Dartmouth but whose inaugural address did not impress Losey (though it sounded 'pretty reasonable' to Patricia). Losey showed a film a week for the ten weeks – including *Dr Strangelove*, Fellini's *Satyricon*, and *Eve* – to a class of ninety-seven pupils, then broke them up into discussion groups of twenty. He also experimented with soundless projection – 'All you can teach is an attitude, a certain sensitivity, a certain awareness of standards.' Scripts of *The Go-Between*, his next project, were distributed and became the focal point of discussion. Writing to Pinter, he described the Dartmouth Film Society as having 'an element that is satisfied by very little other than nudity and pop, and has a childish need to express itself vocally'.[30]

Losey's course, 'The Role and Responsibility of the Director', laid emphasis on the political purges of the 1950s. 'Losey worked as a teacher in the same way he works as a director,' recalls his pupil Jonathan Sa'adah, 'with interminable energy, patience, and focused intensity.' Radical by instinct, he issued a memo to the students of Drama 64. Since grades were required, 'I have adopted the following procedure: (1) Anyone who has done no work at all will necessarily fail the course. (2) All of those who have met minimum requirements will grade themselves. (3) All of those who in my opinion have demonstrated exceptional capacity and interest will be given specific citations in writing by me, and these will be matters of record with the Registrar . . . I will not in any way review, change or comment on each student's self-grading . . .' However, Patricia soon noted that Joe was having grading problems – the first batch had consisted of nothing but As from 'the most undeserving', and Bs from the 'really outstanding'. Losey wrote to the *Dartmouth Alumni* bulletin after eleven weeks in residence, describing the current undergraduates as 'gentle, aware and perceptive', and urging the College to go co-ed as soon as possible. 'Desperately needed' was a universal Pass/Fail/Citation system of grading.

There were trips into Boston, to New York (a big celebrity party given by Noël Coward), and to Montreal. On 16 March he spoke to John Grierson's class at McGill. Following the unhappy visit to Chicago for a preview of *Figures*, they were joined by Joe's sister Mary and her husband Osgood Field, who drove them through snow to La Crosse. The *Tribune* reported that Losey was paying his first visit to his native town since 1945. It was cold and bleak when they inspected the graves of the Loseys, Higbees and Eastons, then went up to Grandad's Bluffs to look out over the town and the river and view the bluffs. They had dinner with the wealthy Mrs Grace Van Steenwyck, from whom Joe had solicited recipes and loans when married to Dorothy, then visited the house in which Joe grew up – not a very big house, but 'pleasant', Patricia thought. On the porch was a swing bed that had survived since Joe's childhood. The Loseys flew on with Joshua to Nassau, pink sands, warm sea.[31]

25

The Baron's Court

Towards the end of his life Losey filled in the 'Marcel Proust Questionnaire'. His heroes were 'Lincoln, Lenin, Joris Ivens'. He hated most of all 'hypocrisy' as well as 'opportunists and betrayers'. His motto was 'Die varheit ist Koncrete (Hegal)' [*sic*! '*Die Warheit ist konkrete*' (Hegel)].' His own main flaw? 'Exaggeration.'

From our earliest evidence of him, Joe was a climber with a shrewd eye for ladders. He was ambitious, energetic, tenacious. His self-confidence was immense yet also fragile; when late in life he named his own vice as 'exaggeration' he was compressing a cluster of related defence mechanisms which expanded to meet a challenge, grievance or received complaint like a swelling ectoplasm; these complaints were normally dismissed as 'hysterical'. Choked by subordination and dependence, he sought ascendancy: 'He needed to dominate,' comments Michel Legrand, whose commissioned scores were greeted with a studied lack of enthusiasm: 'Yeah, maybe, yeah . . .'

A young esquire from the shires (Wisconsin), Joe fought with partial success for his knighthood in New York and Hollywood, and again in England, only to find himself in the mid sixties acclaimed a baron, perhaps a king. But 'always a king across the water,' recalls Lee Langley, 'always giving an impression of discomfort, shifting in his chair, kind or hurt in the same unemphatic, Olympian way – always complaining.' The bankers withheld the crown and in his last years he became an old cavalier of the cinema, still honoured but deprived of a horse.

'You always felt he was in some enormous conflict,' Viveca Lindfors comments. 'Always in a crisis, always frustrated. There was something self-destructive. He made it harder for himself. If only he'd been more human.' Joe reminded her of Strindberg, who never gave up. 'Stubbornness – he has this quality.' She weeps. 'Thank you for your courage,' she says to Joe's departed spirit. 'I liked him for that struggle quality. His courage, his consistency – he never gave up.' Richard Roud emphasizes 'a driving sense of urgency'. Patricia recalls: 'His greatest passion was his work – and he perceived it in extended terms. He was very puritanical in some ways, very driven; he couldn't relax. And he was never happy where he was – it was always the next place.' 'He was bigger than life,' says Florence Malraux, 'like a man from another epoch. When he entered a restaurant he was intense, *douleureux*, always wounded, *déchiré*, lacking *savoir vivre* – in despair when not working. A puritan.' Indeed Joe stated his own greatest misery as 'unproductiveness'. His charm was abundant when at ease, but he lacked, one might say, inner grace. Irritated, vexed or opposed, he became sullen, morose, and often brutally dismissive. He was more 'adored' than liked. Ann Firbank was his part-time secretary in the early 1960s, and an actress with a small part in *Accident*: 'I adored him. He was a monster, a father figure, demanding and giving.' According to Rosine Handelman, Gaumont's representative in New York, 'He had presence, charisma, he had "it" – there was complicity in his smile.'[1]

Many of his contemporaries or 'peers' speak of him with asperity. Sam Wanamaker: 'He was morose, bitter, very interior – self absorbed. Something of an intellectual snob. Embittered.' David Deutsch: 'I didn't find him easy or relaxing. Holding back his anger, restraining himself with deliberation, a calculating man struggling with his own innate turbulence, he would obsessionally rearrange the objects on a coffee table, over and over again.' The director Peter Cotes: 'A decent, difficult man, you couldn't relax in his presence, he didn't give of himself, you didn't feel he cared about others, there was a coldness, a lack of humanity. He was hungry and dry-mouthed and humourless. But when success came in the 1960s, he was "the American in Paris".' In the late 1950s Joe walked Betsy Blair along the Embankment. As a very young actress, married to Gene Kelly, she had been prominent on the Hollywood Left, and she shared Joe's friendship with Nick Ray. Now, on the Embankment, she was shocked by his proposition: that he should direct a film set up for her by Adrian Scott and John Berry, Losey's friends and fellow-blacklistees. 'He told me he *had* to direct it. I told him it was Jack Berry's. He said, "He's all right. He has work."' The film was never made, Joe moved on. His brutal dismissal of his second and third wives in his published interviews with Michel Ciment might suggest – as it did to the victims – an ego coarsened at the core.[2]

The writer Frederic Raphael was invited to visit Losey in Chelsea at the end of 1964: 'He lay on a *chaise longue* like some candidate for imperial orgies he was too weary to organize for himself and told me there could be no honest relationships between men and women . . . in a bourgeois society.' At some stage, for some reason, Losey turned against Raphael; writing to Roy Fuller, Basil Davidson and others Losey regularly denigrated him as 'a dangerous fake' et cetera. When Raphael asked him to second his application to join the ACTT as a potential director, Losey refused. 'I am bound to say,' Raphael complained, 'that to use blacklisting techniques to stifle the weak voices of those who have seen you so ill-treated and to want to silence those who would never be heard above the sound of your own major works strikes me as being less than magnanimous . . .' Describing Losey as a 'pacemaker', Raphael added: 'I regard the day you once asked me to come and see you and talk about a movie as one of the supreme highlights of my cinematic career, such as it is.'[3]

However, in Raphael's current view, Losey was 'a lot more typical of Hollywood filmmakers than he liked to think'. In October 1991 Raphael unleashed a vehement attack in the TLS: '. . . his long career of ruffled feathers and rumpled bedclothes hardly merits exhumation . . . the elevation of the director to quasi-divinity . . . Craving for veneration and the need to condescend . . . to give and procure favours and jobs . . . Stalinist myopia . . . the man of integrity endorsed guilt by association in a particularly craven way . . . Losey was long on accusation (and boozy paranoia) and short on loyalty himself . . . his tearful bullying, his frequent cowardice . . . his drunken conceit. His treatment of women was savage and patronizing' – and more. Raphael did not mention his own Losey-scar – and indeed denies that he carries one.[4]

Even John Loring, a devoted friend, reflects: 'Joe encouraged empty, charged atmospheres, oppressive and painful. He liked to close things off, let people destroy each other. There was a touch of cruelty.' But also of fragility; Peter Smith first met him at Dartmouth in 1970: 'I felt I could never question his work in his presence. He was hurtable and vulnerable for all his strength – a very winning person.' From Dr Barrie Cooper he invited a verdict on the rough cut of *Modesty Blaise*; the result was their only intense quarrel (Cooper recalls) and no communication for several months. 'Whenever Joe tried to solve a personal problem with a man-to-man talk, he emerged hurt and threatened and quickly revoked the pact by which the frank exchange had taken place.' Perversity led him to demolition, what Cooper calls 'the reduction of everything to nothing', with a concomitant blow-up of self-pity. 'He needed a feeling of omnipotence, could not tolerate inconvenience, and was incapable of sustaining a peer relationship. But this did not affect his ability to get on well with creative people and performers whom he respected.'[5]

Losey described himself as insecure and apprehensive but his list of the likely causes reads more like a text-book than genuine self-scrutiny: childhood, college, the blacklist and sex 'at a certain point'. It seems that Joe's ego, as it developed, resented the autonomy of others to an unusual degree, and took little joy in any human activity which could not be absorbed into 'Joe'. He tended to disparage the work of his colleagues – actors, writers, designers, cameramen – for other directors. He needed to expropriate the world. By the 1960s his wider view of the world was more a fixed inheritance than a process of inquiry; his production notes are remarkable for their repetition of worn formulas concerning characters and society. Intellectually, he was most confident when seated in the master's chair – interviews and press conferences: Losey on Losey. But he was wary of sharing the limelight or engaging in open-ended discussion with his peers. In August 1968, he turned down an invitation to appear in *Late Night Line Up* with people he claimed 'he didn't know': John Bowen, Nemone Lethbridge and Gore Vidal. He certainly knew Bowen and Vidal. 'He wasn't fast with people,' comments Paul Mayersberg. 'He couldn't catch a point and turn it around. There were pounds of wool in the interviews.' Barbara Bray likens him to 'a beached whale, out of his element'. In the company of more sophisticated, or articulate, people, he tended to withdraw into morose melancholy – or to leave abruptly. Alexander Knox remembers him at dinner with the writer Mario Praz: 'When Joe left, Mario asked me if he had ever been known to speak.' Both Patricia Losey and Monique Lange contest that he needed to be the focus of attention: 'But he couldn't bear chit-chat or small talk.' In the spring of 1969 Nicholas Mosley and his girlfriend dined with Peggy Guggenheim in Venice. 'Things began to go wrong when my friend engaged in conversation with Peggy Guggenheim about art,' Mosley recalls. 'Joe looked disgruntled. The next day we all went on a sight-seeing tour, possibly to the Accademia; Joe was in a foul mood and was unbelievably rude to my friend. She'd somehow upstaged him.'[6]

In December 1968 Losey and Dalton Trumbo engaged in a public argument through the pages of the *New York Times* following some disparaging remarks by Losey in the course of an interview. Private correspondence between them followed. Trumbo's letter to Losey, dated 20 January 1969, is strikingly perceptive:

'You ask me, apparently with astonishment, "Why didn't you write *me*?" Are you truly naïve enough to believe that you can derogate men's work in the most public fashion possible without some sort of public, rather than private, response? . . . I think you are a fine director. I have seen a good many of your films and found them deeply interesting. I feel that there is a strain of what I call "purity" which runs through them all, and I regard purity as something which cannot be learned, something which exists within the personality and proves itself by simply being. On the other hand, I think in your interviews – some of which I have read – you make a bit too much of the difficulties which have attended your career.

'One gets the impression that you tend to lay the blame for whatever personal weaknesses or mistakes have been revealed in the course of your professional life . . . on malign circumstances alone, or upon this person who failed you or that one who betrayed you . . . I feel that you lack kindliness to others when you are confronted with an interviewer . . . Why, for example, gratuitously kick Laughton, in the Sarris interview?'

Losey replied on 29 January. 'I take the personal points that you make and shall endeavour not to be guilty of them . . .' If he sounded arrogant or cruel in interviews and press conferences, 'I do not feel at all arrogant and I seldom set out to be cruel.' We do not know how he 'felt' – the psychic life is built on split vocabularies. Joe was easy to hurt but difficult to disable. Criticism never hindered his work. He kept going. Hostile reviews were rapidly dismissed as 'imbecile' or 'philistine' or sinisterly motivated.

Did Joe perceive his own anger as a problem? Not in those terms, Dr Cooper answers. 'He hung his anger on a target or situation. It was sometimes self-evidently contrived.' In Cooper's view, the dominant factor was 'perversity to the point of perversion'.

Joe had always been vain about his appearance – 'every hair in place', comments his son Joshua. The Fraads recall that in the late 1930s he put himself through an elaborate sun ritual to combat skin blemishes (he suffered from acne). A photo taken in Italy in 1951 shows Joe tanned, bare-chested, no flab, youthful for his forty-two years. In the 1950s and 1960s his clothes were made by the fashionable tailor Douglas Hayward, of Mount Street. Pieter Rogers describes Losey as constantly glancing at his reflection in mirrors.

That Joe's sexual energies were mainly directed towards women is beyond doubt, but he also adored handsome, cultivated men, and the legend of his bisexuality is a long one, extending back at least to the 1940s. But specific stories of actual homosexual activity evaporate on inspection. Only one interviewee claims direct evidence of it: this is Evan Jones, who wrote four of Losey's screenplays in the 1960s. '. . . Losey was a promiscuous bi-sexual with a potency problem. He made homosexual advances to me while we were working together in Lugano, which I declined. I forget the date.' Here we are into deep waters running far beyond the alleged incident. Dr Barrie Cooper more or less discounts the likelihood of actual homosexual practice. Dirk Bogarde adamantly dismisses it. Joe told Florence Malraux that he'd never had sex with a man: 'But he had a feeling and a look with a Stanley Baker, a Delon, a Laurent Malet, a sort of pleasure in looking at them,' she adds. And there were others, handsome, cultivated young friends like the artists John Loring and Richard Overstreet – but Loring cites 'that pathetic and trivial little scene in La Truite which is ample proof that Joe had only the most superficial interest in and no firsthand knowledge of male relationships'. Particularly in France, observant ladies felt a sexual ambiguity. 'Lightly oriented towards pederasty,' comments Charlotte Aillaud. In Nicole Stéphane's view he preferred the company of men – but, she adds, the truly homosexual directors like Cocteau and Visconti liked women and were good at directing them. 'Joe wasn't.' Losey's working relationship with actresses is discussed elsewhere; while working on La Truite he remarked that in any working relationship, whether with male or female colleagues, he had found that 'there needs to be an actively felt sexual attraction to get the best results on both sides'.[7]

He suffered wretched health. Joe remembered childhood insomnia to the point of panic: 'And I got over the insomnia with asthma, and I have the worst asthma attacks anybody could possibly have . . .' Although his bronchilators imposed a slight load on the heart, 'I use them unstintingly.' The main threat to his health and mental balance was alcohol. Dr Barrie Cooper describes a progressive deterioration: 'I have never known Joe when drinking was not a major force.' Alcohol had been diagnosed as the cause of his low sperm-count when he was approaching forty and living in Hollywood. According to Pieter Rogers, this condition persisted – and worried Joe – twenty years later. By then King Vodka ruled. He was drinking heavily while shooting *Figures in a Landscape* and again when he began filming *The Go-Between* in Norfolk. Dr Barrie Cooper warned Patricia that unless Joe stopped drinking he risked cirrhosis of the liver, but she felt powerless to intervene. 'So I'm drinking along with him for the moment,' her diary records. In October 1970 she reflected on the general drunkenness within their circle: 'Oh dear, it does seem to be so much part of our life.' Losey paid a visit to Buxted health farm, returning full of enthusiasm for yoghurt and a saltless diet, but was soon sitting around 'getting soused'.

He worked, he wheeled, he dealed. In October 1971 his quarterly phone bill was £226.33 (£1,300 in current values). Long-distance calls might be made from the sitting-room with everything going on around him. He relished an entourage, a court, secretaries, family, visitors, continuous business day and night. 'One wanted to help this driven man,' Patricia says. 'His enthusiasm and energy were infectious.' All his long-serving secretaries agreed, but others went faster than they came. Pieter Rogers, secretary/assistant for five years from 1953, received a half-a-crown an hour (12.5 new pence). Despite his low wage and devotion, Rogers was always 'the last to get paid'. At one time, when owed about £50 (representing 400 hours' work), he was presented with a splendid volume of Egyptian paintings, costing perhaps £5 (or forty hours' salary). Rogers remonstrated and Losey's eyes filled with tears. 'He behaved like a pasha but he always felt guilty.' In 1968 Rogers returned, briefly, to be Losey's production assistant, evidently a galling experience: long hours, always on call, lack of consideration. Losey was shooting *Secret Ceremony* (in which Rogers has a brief role as a hotel manager) at Elstree. Each morning Rogers would wait on cold street corners – for up to half an hour if Losey had woken late – to be picked up by the studio car. 'He wouldn't consider calling at my flat, although it was on the route to the studios.'

According to Rogers, Losey behaved 'appallingly' in getting rid of Philippa Drummond, his secretary from *Modesty Blaise* to *Secret Ceremony*. The relationship she had formed with Richard Macdonald on location for *Boom!* may have been a factor, but both she and Rogers point a finger at Patricia – the two women shared in common French, Italian and a sense of class (Drummond fixed the country house in *Accident* through a cousin). For her part Drummond felt that she worked long hours and was taken for granted; she was sacked while on holiday and in her parting memo to Joe wrote, 'Sometimes I thought you thought I was doing nothing.' Patricia's glamorous life as Mrs Losey inevitably inspired some resentment among other women but this was compounded by the fact that Losey's home was his office and that Patricia was increasingly involved not only in his scripts but in correcting his assistants' spelling. Drummond, she recalls, 'couldn't spell. I used to hear the thump of her dictionary on the floor.' Rogers describes Patricia leafing through the letters he had typed up from Joe's dictation, making spelling corrections. Patricia comments that this should have been 'perfectly routine', adding that Rogers's 'possessiveness about Joe dated back a long way'. 'I walked out,' recalls Rogers. 'A crate of brandy arrived from Joe the next day.' Patricia expresses surprise about Philippa Drummond's hard feelings: 'I feel I asked her why she wanted to leave. Joe didn't fire her.' Rogers says Joe instructed him to type out the letter of dismissal.[8]

Joe lived with Patricia for twenty-one years, her husband for fourteen, until his death. Their liaison was bonded in circumstances of emotional turbulence, jealousy, physical violence and family tragedy.

Born in 1930, the daughter of a clergyman, Patricia Mohan had worked for the magazine *Encounter* and as a translator in Paris. After moving to Rome, where she was much in demand as a bilingual secretary, she became by coincidence assistant and secretary to Losey's colleague, the blacklisted screenwriter Hugo Butler. In 1957 she married Bruno Tolusso, an Italian businessman working in the film industry. Through Butler, the original but discarded screenwriter for *Eve*, she met Joe Losey in 1961. Already a mother and pregnant at the time, she worked as his secretary and assistant in Venice and Rome. Jean Butler remembers her as 'an absolutely delightful, bright, literate young woman'. When Michel Legrand arrived in Rome to compose the score for *Eve*, Patricia was assigned as his translator; he found her vivacious, warm, friendly.

And Joe? Charismatic, in turmoil, he was much gossiped about. Patricia heard 'dreadful things' about him, but after she saw the rushes of *Eve* she was bowled over by 'the intensity of the vision and the emotions expressed'. Like Ruth Lipton, she soon discovered Joe's obsessive need to talk about past marriages and affairs. Patricia didn't want to know. After Joe left Rome and the birth of her second child, she joined him at Hardy Kruger's house in Lugano, returned to Rome, and in October 1962 travelled to Paris for the opening of *Eva*, working as Joe's interpreter for interviews and press conferences. She was deeply moved by Joe's wounds. 'We had a sad dinner together the last night, before I went back to Rome. He said he couldn't make *The Servant* without me. Joe said things like that.'[9]

Her husband, meanwhile, was suffering from deepening jealousy. 'I remember the early sixties as very difficult, frustrating years, reading technical books in an empty office or trying to uncover potential clients.' Hugo Butler warned him against long separations. 'Therefore, when Joe summoned Patricia to Paris for the opening of *Eva*, we both decided it should [be] the last time.' At 3 a.m. on the night the film opened, Tolusso telephoned the house outside Paris where his wife was staying, but to no avail; the following day she called to explain that Joe had rented her a room in his hotel for that single night. According to Tolusso, after Patricia's return to Rome it was agreed between them that she would not work for Joe again.[10]

But a few months later Joe called Tolusso, 'explaining how sorry he was to take her away from home, from me and the children, how he had counted on her and couldn't replace her . . .'[11] Again leaving her children in the care of an Italian-speaking German woman, Marianne – the Tolussos also had a live-in maid, Nella – Patricia arrived in London a week before Joe began shooting *The Servant*. She did typing and made coffee for the cast in Joe's Markham Street flat but lodged elsewhere. (Ruth Lipton had also been summoned from New York.)

In her divorce petition to an English court Patricia later admitted to adultery with Joe as of September 1963 – but the given date was dictated by legal tactics. Patricia is clear that when she returned home to Rome after *The Servant*, although Joe was now her lover and the affair with Ruth Lipton was over, she had no intention of leaving her husband – although she meant to go on seeing Joe. 'I never thought it would last.'[12]

Tolusso's recall is different. His letters urging her to return to Rome were met by 'apologies, explanations, recriminations or plain rebuttals. Even the [first] anniversary of our daughter Flavia, on April 15th, was forgotten.' Arriving in Rome at the end of May, she dined with Tolusso in a trattoria 'and she explained to me she realized "our big love story", as we used to call it, was over. That she wasn't in love with anybody, less [least] of all with Joe.' Tolusso was unconvinced. (Patricia does not mention this fateful conversation.)[13]

Disaster ensued on the following day, 1 June 1963. Tolusso observed that she spent the morning typing letters. By his account she asked him to post the letters, since he was heading for his office, while she took an afternoon rest. By her account, she didn't; Tolusso opened all of the dozen letters she had left lying on her desk, burst into their bedroom livid with rage, and began to beat her, although all the letters were 'innocuous' messages, 'simple thank-you notes'. 'This was the reason for Bruno hitting me for such a long time. He was trying to get me to tell him which of the people I was sleeping with.'[14]

However, two months later Patricia told her London solicitor, Mr Barr:

'He brandished a letter I had written to my doctor which, he said, proved that I had been unfaithful to him.' Tolusso himself insists that he opened only one letter, addressed to Joe's doctor, Barrie Cooper. According to what Tolusso told Jean Butler soon afterwards – which he now confirms – Patricia's letter asked Cooper 'to contact Joe and tell him that she thought it would have been unwise for her to write directly to him but she had already started talking to me and that it seemed that I would accept the separation'. She hoped she would be able to rejoin Joe in a few months' time.[15]

Tolusso entered their bedroom and locked the door. She recalls: 'He was in a blind rage and . . . gave me the first few blows in quick succession.' He suggested that she throw herself out of the window, then brought pen and paper and began to dictate a 'confession'. 'There were more blows, more discussion.' He picked up a heavy Chinese box from her dressing table and threatened to kill her with it. Then he sat on the bed and made her write a second 'confession'. 'At some point he gave me . . . the blow which broke my nose . . . blood poured out . . . and I fell off the bed . . . He said I was a whore and therefore it was better for me not to be dressed.' They heard the children, Ghigo and Flavia, return with their nurse. 'When I looked in the glass I was horrified and asked him to keep the children away . . .' Tolusso himself drove her to the hospital. The ordeal had lasted about two hours. She suffered a broken nose and cheekbone, post-concussion and acute nervous shock. 'She was hardly recognizable,' Jean Butler recalls. Tolusso later sent her 'confession' to Dr Barrie Cooper.[16]

After a period in hospital Patricia took refuge with the Butlers and other friends, sheltered in a convent, then travelled to London for more medical treatment. Jean and Hugo Butler were doubtless not alone in admiring Joe's support for Patricia in her hour of catastrophe; he behaved well. Yet the evidence suggests that for both Joe and Patricia there was only one villain on the stage: Bruno Tolusso. Dr Barrie Cooper recalls: 'She always presented herself as a victim, pure and simple. There has never been any hint of self-examination, let alone expressed guilt. It was as if her life with Tolusso – and Tolusso himself – began with the moment he beat her up. The play starts in act three.'[17]

In July the German nurse returned home. Tolusso took the children to Trieste, entrusting them to his sister and parents. In March 1964 he returned from a business trip to Brazil, discovered that he was being sued by Patricia for criminal assault, and rapidly removed himself and the children to Brazil. Unaware of this, Patricia arrived in Rome for a preliminary separation hearing. The judge began to dictate a separation order without asking any questions. 'He said he felt that as a general rule children should be left with their mother and immediately dictated an order to the effect that the father was to have custody.' Patricia was granted limited rights of access but only within Italy. In October 1966 Tolusso turned up at the Italian Consulate in São Paolo to renew his passport, but declined to disclose his private or work address. He was refused the passport on account of the criminal sentence (nine months in prison, suspended) passed against him in Italy. Not for seven-and-a-half years was Patricia to see her children again.[18]

How committed to Patricia had Joe been before she returned to Rome? 'I never thought it would last,' she says. But for the violent rage which seized Tolusso when he read her fateful letter to Cooper, one may wonder whether Patricia Tolusso would ever have become Patricia Losey. That Joe could have tolerated her children permanently under his roof seems out of the question. Tolusso's retrospective view of her dilemma is plausible: 'She couldn't go back to Joe because of the children and she couldn't go on living with me because of Joe. If she didn't go back in a short time, Joe, who was never consider[ed] a faithful lover, would probably get involved with someone else.'[19]

Joe put Patricia on the payroll of Launcelot Productions at a salary of £2,000 a year. She changed her name to Dickinson but increasingly signed herself 'Losey'. She was attractive, vivacious, sociable, dressed with flair, knew the clothes for every occasion, and (a friend recalls) had a capability for laying on instant carnivals, 'one tableau vivant after another'. She had the sixties gift for living – Joe himself flirted with the pop stars, the Beatles, the Jaggers, the model Jean Shrimpton. Not only did Patricia's language capability delight him but she loved to travel and was ardently keen to involve herself in his projects: Sardinia for *Modesty Blaise*, Sardinia for *Boom!*, Algeria, France, Morocco, Spain, Mexico, America. Patricia's diaries chart the pleasures of a star-studded existence with its gastronomic delights. On location for *Boom!* they were woken each morning at six with grapefruit juice and scrambled eggs. After Joe departed she would read Karl Marx in bed until it was time for a swim. Noël Coward arrived and called Losey 'Director baby'. Evan Tiziani, who designed Elizabeth Taylor's clothes, made Patricia a gift of towelling dresses. Frequently they escaped to the Colombe d'Or at Saint-Paul-de-Vence, where the local taxi driver, César, drove them to the Plage Sportive.

In November 1968 they flew to the Acapulco Film Festival in Mexico, via Bermuda, Nassau and Mexico City. Patricia's diary records her delight with Las Brisas: a bungalow on top of a mountain overlooking the sea, a swimming pool, a fridge stocked with fresh fruit and drink. She took her first swim in the Pacific, consumed 'a delicious avocado purée', then it was a rush to buy some 'delicious Mexican clothes' and to lunch with the man from the Rank Organization.

In London she frequented Anshell, a King's Road junk-jewel shop. She liked to be entertained, amused, and enjoyed the talented company they kept.

Whether consciously or not, she sometimes sent Joe into a tailspin of jealousy, not with cause but because he was intensely possessive. At her 1968 birthday party she wore 'a perfectly horrid little black shining mini dress and my new long wig . . .' but the party ended with David Mercer aggressively drunk and Evan Jones (also drunk) inflaming the situation at five a.m. Joe behaved 'marvellously', Harold Pinter 'sensibly and bravely in the face of abuse . . .'

Patricia's diairies also chart her persistent efforts to regain contact with her children. From time to time she received photographs, but unaccompanied by any letter. She sent presents which were sometimes returned. She dispatched a letter to Ghigo via the British Consul, São Paulo, in the hope that he might deliver it – 'and if you could get a glimpse of the children, that would be marvellous'. In January 1970 a woman member of the Embassy staff visited Ghigo and Flavia in São Paulo and told them how much their mother missed and loved them. Three months later, arriving in Málaga for *Figures in a Landscape*, it took her two hours to settle the house, engage a cook and a maid, and give the gardener his instructions. A few days later she was back in London, dining with quarrelling Burtons and viewing two new Pinter plays. Then Spain again and dinner with Gina Lollobrigida. But life with Joe was hard: her diary for December 1969 regrets his lack of sympathy, help and kindness. In Paris he lunched alone with Jeanne Moreau; Patricia persuaded him to stop off at the Galerie Iris Clert – Jack Loring had described Le Secrétaire Fantastique by Brö, which was just what she wanted for her new work room in 27 Royal Avenue. But it was currently in Italy and would cost thousands of dollars. Back in London it was more shopping and parties. 20 December: 'Joshua is now with us, so life is a little limited.'

Norman Priggen recalls Joe's brutally impatient tone with Patricia: 'Do it! Don't dither! Get it done!' She would walk away. Minutes later Joe would have forgotten about it. 'One wanted to look away', recalled Richard Roud. 'She'd say, "Joe, Joe," with dignity.' Peter Smith, who invited Losey to lecture at Dartmouth, noticed that he took it for granted that he would always have his way, the last word. Patricia stood up to him but he continued to treat her like a doormat. Mary Losey Field regards Patricia as mercurial, quixotic, difficult and highly-strung – but the only wife who said, 'Don't talk to me that way.' Jack Loring found her resilience remarkable. Patricia herself feels that Joe longed for an ideal, solid relationship: 'He wanted it to be equal, but he'd do everything he could to prevent it being equal.'

They were married on 29 September 1970 in King's Lynn, Norfolk, an epilogue to *The Go-Between*. After collecting Harold Pinter off his train on the dot of 10.45, and a drink at the Duke's Head, they were 'sanctified in a notary's office by a sweet clerk who bore the name of Mr Sanctuary, believe it or not'. (This is Joe.) When it came to the ring, 'Joe took my hand and took off the ring he gave me so many years ago. Then we all laughed . . .'

(Patricia.) The occasion was later celebrated at the hotel on the Tuesday Market Place with a champagne lunch. After putting Moura Budberg and Pinter on the 3.25 train, 'Patricia and I retired to our fabulous honeymoon cottage, Marsh Barn, at Brancaster.' (Joe.) That evening the Loseys played chess and read. Joe made hamburgers – a reversal of roles. The Pinters' present was a mirror in a velvet frame.[20]

As an infant living in Montpelier Square, Joshua Losey experienced a succession of nannies, or 'nurses' (as Joe called them). His parents separated two months before his fourth birthday and his memories begin with visits to Joe's flats in Markham Street and Lowndes Square. Visiting the set of *The Servant* at the age of five, he felt anxious in case he did something wrong: 'Being careful, being quiet, not getting in the way, in case I got shouted at.' This general apprehension on entering his father's world was to persist. Joshua never felt fully at ease when staying with Joe and Patricia, fearing that he would break some precious object, and meals were 'a bit of an agony' – some rebuke from Joe was never far away. In 1964, Joe was indignant to receive a bill from a Cabourg hotel alleging that Joshua had broken a glass door; Patricia, in whose care the boy had been at the time of the incident, blamed the wind, not Joshua. In Joe's universe everything must be in place, perfectly ordered. On the other hand, Patricia (as she recalls) extended her maternal feelings to Joshua, who had his own room full of his own things, and there were outings, games, with young friends like Daniel Chatto and Vicky Cooper.

'"Take your elbows off the table!"' When Joshua – himself an actor – imitates his father he calls up the accent. 'He had this American drawl which was, you know, so much part of him – a huge laugh, a great smile. He had a strong presence . . . incredible energy. I used to feel very inhibited about speaking.' Patricia emphasizes that Joe loved Joshua and remained a devoted and conscientious father, after his fashion. According to Ruth Lipton, 'He wasn't nice to either son. They failed to measure up to his expectations.' Philippa Drummond recalls Joshua at the age of ten: 'Patricia found him difficult. He was reserved. It was a non-family.'

Arrangements concerning Joshua's regular weekends and holidays with Joe were increasingly made between Dorothy and Patricia, since Dorothy and Joe could not talk without rowing. Clearly Joshua was given what money could buy, especially expensive clothes – the more expensive the more he liked them, in Patricia's recall, although Joshua says he never felt comfortable in his beautiful Italian outfits. Travelling to Sardinia for *Modesty Blaise*, at the age of eight, is his most vivid memory; he returned to Sardinia for *Boom!* Richard Burton gave him a set of encyclopedias but Elizabeth Taylor's kids were 'awful'; one of them tied him up and threatened to push him off the yacht. As Andrew Sarris was leaving Losey's New York hotel suite in November 1968, he noticed 'a ghastly Mexican death doll. It was a gift for Joshua. "I want him," Losey remarked, "to get an early start on morbidity."' The mask hung in Joshua's room in Royal Avenue.[21]

A brief appearance in the *Secret Ceremony* cemetery with Mia Farrow was followed by a flight to the Marchioness of Bute's castle in Scotland (where he remembers huge palatial rooms, driving a Mini-Moke, and breaking his tooth while cleaning the swimming pool – Dorothy had to be called). Taken to Spain, where Joe was shooting *Figures in a Landscape*, he flew in the helicopter, skimming the sea, a special treat, and saw huge shots of the sugar-cane fields. 'I was very privileged. The sixties were full of excitement.' In March 1970 he flew to Boston to join Joe and Patricia at Dartmouth College. On each trip Joshua tended to suffer initial lethargy, indifference, which in turn brought admonitions and a crisis. That summer, in Norfolk, he coveted one of the main boy's parts in *The Go-Between*, but Joe said, 'As your father, I can't take responsibility in case something goes wrong.' On 26 July they celebrated his thirteenth birthday at Marsh Barn. Joe gave him a watch, which Patricia regarded as a bit of a sick joke, Joe having given him one the previous year and then shouted at him daily because it wasn't going and Joshua wasn't treating it properly, etc. Joshua had perhaps developed a masochistic, attention-seeking, streak which invited Joe's rage – for example, getting himself soaked by walking into ponds and marsh. Jack Loring recalls that Joe would exasperate Joshua into tears. Joshua once piped up, 'He's going to make it [his next film] as excruciating as humanly possible.' According to Joshua, Joe seemed unaware that he suffered from a broken home: 'He always said, "Guilt is a useless emotion." His work was his life.' It was Joe's conviction that everyone should have a passion in life and that Joshua lacked one; he tended to blame Dorothy for the boy's lack of discipline and motivation. Aware of recriminations between his parents, Joshua sometimes plunged into piles of letters in search of the truth. He recalls Gavrik and Joe 'always at each other's throats' whereas he himself backed away from conflict. 'Joe respected the people who stood up to him – as Patricia did. He'd see reason later. Behind the big, bold, rather aggressive façade he was open to conviction.'[22]

In 1966 Joe had acquired the lease of 29 Royal Avenue from the previous resident with the help of a loan of £16,000 from Leslie Grade, on the security of the house. For furniture and fittings he borrowed from his sister (Mary Losey Field comments that it took 'close to six years to recover the loan and there was some fiangle about money being transferred from the Bahamas . . . I remember an amount of $15,000 but am not sure'). Joe subsequently acquired a lease on the bottom two floors of the adjoining house, No. 27, which were converted into an extension of 29. In April 1970 Joe heard from William H. Gill, a barrister and chartered architect, that both 27 and 29 Royal Avenue had recently been added to the Minister's statutory list of 'Houses of Special Architectural and Historic Interest'. The listing would enhance their value. The houses were now situated in the Royal Hospital Conservation Area, 'which has been so designated by the [Kensington and Chelsea] Council on my advice'.[23]

Joe's father had left his widow and children without an inheritance, and Joe was determined to avoid a repetition. At that time the estate of a deceased husband was subject to death duties. His will, dated 26 September 1970, speaks of his intended marriage, in consideration of which, 'I hold the properties described in the Schedule here upon trust for the said Patricia Dickinson for her own use and benefit.' The will establishes that at that date he had a 'leasehold interest' in both 29 and the bottom two floors of 27. He also gave his wife 'all the furniture effects and moveable property'. In effect, Joe handed Patricia the balance of his worldly assets in a remarkable act of love and trust; under the marital law then prevailing he would have been down the drain in the event of a divorce. As Patricia recalls, when the deed was executed before a Commissioner of Oaths and Joe was asked to hand her the symbolic keys to the house, he growled and slammed them down grumpily.[24]

Norden & Co. informed the Inspector of Taxes that on 14 June 1971 Patricia Losey purchased the freeholds of 27 (in part) and 29 Royal Avenue for £20,462 plus costs. This was achieved by a loan of £25,000 from the Trident Insurance Company (with Joe effectively servicing repayment and interest). In February 1972 Patricia bought a leasehold interest in the upper maisonette at No 27 for £12,750. Ownership of the adjoining houses was now complete. Joe also had his eye on Marsh Barn in Norfolk, originally rented from Anthony Russell-Roberts; in March 1973 he made inquiries about the freehold but nothing came of it.[25]

PART VII

The Palme d'Or

26

The Go-Between:
A Norfolk Summer

The Go-Between marks the summit of Losey's career and the climax of his collaboration with Pinter: the 1971 Palme d'Or at Cannes.

L.P. Hartley's novel wonderfully evokes the unwritten codes and certainties, the repressed longings, of a distant imperial heyday. Greeted by Evelyn Waugh as 'outstandingly the best work of fiction' of 1953, *The Go-Between* won the Heinemann Foundation Prize of the Royal Society of Literature. Writing to Losey, Hartley explained the genesis of the novel. He was sixteen years old when he personally experienced his own Norfolk summer of 1911. 'The house where I actually stayed as a boy was Bradenham Hall in Norfolk, somewhere between Wendling and East Dereham ... It belonged to the Rider Haggard family, who had let it to some well-to-do coal merchants called Moxey: their son was my school friend, who asked me to stay ... All I can remember of the house was the double staircase, the cedar tree in the garden, and the Deadly Nightshade in an outhouse.'[1]

Hartley set his story in the year 1900; Bradenham became Brandham Hall. Leo Colston is twelve (not sixteen) when invited to spend three weeks of a very hot summer as the guest of his school friend Marcus. Well-bred but poor, Leo is slightly overwhelmed by the huge house, the servants, the elaborate picnics, the grand dining-room, croquet on the lawn – and by Marcus's beautiful elder sister Marian. In making an adoring slave of young Leo, Marian has an ulterior motive. Unknown to her family and her fiancé, Viscount Trimingham, a scarred veteran of the Boer war and everything a gentleman should be, she is conducting a passionate affair with a handsome tenant farmer, Ted Burgess. Still innocent about sex – 'What do lovers do?' Leo eventually asks Ted – he finds himself employed as a messenger or go-between, a pleasant and exciting assignment, since he admires Ted and adores Marian. But storm clouds lie on the horizon and Leo increasingly finds himself subject to traumatic pressures, internal and external. The climax comes during a party given by Mrs Maudsley for Leo's thirteenth birthday. Marian's hysterical mother drags him off through the rain towards the outhouses . . . What they see, in L.P. Hartley's telling phrase, is 'a shadow on the wall that opened and closed like an umbrella'.

Ted will shoot himself, Marian will marry her fiancé, Ted's baby legitimized. Half a century passes. Leo Colston, a lifelong bachelor for whom this seminal experience had burned out all capacity for love, is pulled back to Brandham Hall by curiosity, nostalgia, a need for exorcism.

The Go-Between offered Pinter and Losey a perfect opportunity not only to penetrate the mysteries of time and memory – 'the past is a foreign country' – but also to re-explore the class codes of England, hypocrisy, masked desires, elliptical language, characters torn apart by acting out a role in conflict with concealed conduct: sex transgressing class boundaries, the shattering moment of discovering the forbidden coupling which incarnates betrayal. Soon after shooting *The Go-Between* Joe Losey married Patricia in Norfolk; within months he betrayed her when her back was turned, as Tony did Susan, as Stephen did Rosalind, as Marian did Hugh Trimingham.

Bringing *The Go-Between* to the screen had been an eight-year-long struggle. Alexander Korda had originally bought the rights but – Hartley later claimed – Korda 'never meant to make a film of the book . . . I was so annoyed when I learned this that I put a curse on him, and he died, almost the next morning.' In October 1963, before *The Servant* opened in London, Losey pressed Pinter whether he'd yet read Hartley's novel. Pinter replied: 'I think *The Go-Between* is superb . . . It's wonderful. But I can't write a film script of it. I can't touch it. It's too painful, too perfect, if you know what I mean.' Seven years later, Pinter recalled repeatedly bursting into tears while reading the novel. 'Really weeping. A few pages later I was off again.'[2] Although he did embark on a first-draft screenplay, the project was stalled by legal difficulties. Robert Velaise, of Liechtenstein and 28 Chester Square, London, had paid £8,000 for the rights and complained to Losey that everyone was trying to 'bypass' him and make a film without his participation. Losey's backer, Leslie Grade, would not proceed without legal assurances. By 31 March 1964, Losey had abandoned *The Go-Between* as his summer project, turning his energies to *King and Country*. Grade regretted that he could not be expected, 'in fairness', to remunerate Pinter for his work.

What Hartley initially thought of Losey's choice of Pinter as screenwriter is doubtful. J.W. Lambert recalled taking Hartley to see a Pinter double-bill, *The Room* and *The Dumb Waiter*. Later, 'As [Hartley] scoffed an omelette his head flopped not at all appreciatively . . . signalling that he didn't know what the theatre was coming to.'[3] However, having read the first sections of

Pinter's screenplay, he judged it 'splendid . . . so sensitive to the nuances of social condition and degree . . . Nor does he miss a dramatic development.' By 1967 Velaise was plaintiff and Hartley defendant in a law suit. A compromise formula was then reached, by which Velaise was to be the strictly nominal executive producer of the film. On 30 July 1968, Losey informed Hartley that all contracts were agreed and that Pinter was beginning work on his second draft.

Re-reading Pinter's first draft after a lapse of four years, neither Pinter nor Losey was happy with it. Pinter set to work again, with eight pages of notes from Losey – 'For God's sake throw them away if they begin to clutter your mind' – and in January 1969 completed a final revised screenplay which posed far more problems than his script of *Accident*. Precisely why, we shall see.

Losey's old friend (and script-reader) Baroness Moura Budberg helped with introductions to the local Norfolk gentry. A long trawl of possible houses was made with expert advice, most notably that of Lady Whilhelmine (Billa) Harrod, wife of the economist Sir Roy and author of the *Shell Guide to Norfolk*. Some years later, responding to an invitation from Angus Wilson, Losey wrote a vivid memoir of his first visit to Norfolk, stopping at the Red Lion in Swaffham – 'good old English plastic, neon and chrome – pinball and jukebox' – but friendly. The night was spent in a '"modern" little box hotel in East Dereham with walls of paper and sounds of parking lot and flushing toilets mingled with the giggles of amorous couples made more amorous by the freshness and surprise of a first snow-fall'. Morning brought smells of breakfast – 'stale fries of fish and sausage and bacon, watered coffee, tea-bag tea and lank toast grown cold, tinned fruit juice and thin marmalade sickeningly sweet'.

The first visit next day was to West Bradenham Hall, the scene of L.P. Hartley's summer of 1911. The owners were away and the caretakers were reluctant to open the doors. 'We peered through the window to see the famous double staircase – but the house and gardens had obviously been re-done and it was not the place my imagination searched for.' Several houses were visited. Losey's art director, Carmen Dillon, wrote to Lady Walpole: 'But as Mr Losey and I were driving around Aylsham I am afraid we trespassed and went up to the house hoping to see you. Unfortunately we arrived just as the shoot had returned; and you were having a very busy time with the young horses. So rather than be a bother we went away.' On, then, in driving snow, to Melton Constable, twenty-five miles north of Norwich, owned by a non-occupying farmer, G.W. Harrold. Built in the 1660s, with some spectacular ceilings, and set in several thousand acres, the empty house was in a state of some neglect but not suffering from wet or dry rot. Since the film would involve a crew of 120, arc lights and Mitchell cameras, the floors and roof were tested and found to be sound. Carmen Dillon was free to redecorate Melton Hall and furnish the desolate, stripped rooms.[4] 'Here we found . . . vast grounds, great if undistinguished halls, stables, dilapidated outhouses and kitchen gardens. Nature galore and opportunity to reconstruct our own reality. But, alas, no double staircase and no garden (that we did at Blickling). And no belladonna! The dower house and square and church were at Heydon . . . The cricket pitch was Thornage.' In 1991 parts of Melton Hall were for sale on a 250-year leasehold basis.

We turn now from sleepy Norfolk villages to film financiers far from sleepy. Lurking under the armpit of art is industry, the money men. Losey's favourite angel was John Heyman, the executive producer of *Boom!* and *Secret Ceremony*. Heyman was again willing, but financial support was required from a major distributor. On 6 February 1969, Losey sent Pinter's script to Bernard Delfont, head of ABC, with a flattering note: 'It would be marvellous if we could all combine to make another Grade/Pinter/Losey classic – which also makes money.' (Delfont was a brother of Leslie Grade.) In April Losey informed Hartley of terrible delays in raising the finance – the project had been turned down. On 13 May he wrote to Bogarde: '*The Go-Between* fell down owing to the predictable lack of courage of Bryan Forbes [vice-president in charge of production at ABC], so I have taken a film called *Figures in a Landscape* – very profitable.' On 19 May he sent his crew a standard letter, identically phrased, pledging that shooting would start on 1 July 1970.

But the crisis repeated itself the following year, during the run-up to shooting. ABC backed in then, in April 1970, backed out. Heyman took Delfont into the corridor and explained what cancellation would cost him: Losey's fee, Pinter's fee, Heyman's fee. Delfont went back into the room and told Bryan Forbes, 'We've got to make it work.' Despite Forbes's overt enthusiasm for the project, Losey rarely spoke well of him; as for Bernard Delfont, Losey's verdict was damning: 'The distributor who later claimed all credit for the film's success at Cannes (Grand Prix) and who used our work quite shamelessly to promote his own elevation to the peerage, had reneged on a signed contract . . .'[5]

Losey's fee (Launcelot Productions) was a modest £17,500, plus £2,500 'in lieu of all living expenses', plus a percentage of the gross receipts. The production budget reveals an unusual imbalance. Whereas only £30,000 (out of £532,841) was allocated to the entire supporting cast, £67,791 was spent on 'Story Rights' and 'Writers'. The agreement between World Film Services and Harold Pinter, Ltd was for the sterling equivalent of $75,000, plus 5 per cent of the profits – a reflection of Pinter's meteoric rise during the 1960s. An undated letter from Losey to Bryan Forbes reads: 'I am, as you know, sacrificing two thirds of my fee in order to get it made. Also of the remaining one third I risked nearly one half to part finance the writing of the script and to clear the rights.' Heyman confirms that Losey took a cut in his own fee.[6] Evidently L.P. Hartley himself was not, in the long run, happy about the financial share-out. On 11 October 1971, Losey reprimanded the author: 'I was somewhat shocked and disturbed to read in a recent clipping your complaint about the money received – or rather not received – on the picture.' The complaint was justified, but Losey found it 'unfortunate' since 'I and the actors worked for no cash at all'. Losey advised Hartley to rest content that the film had brought reissues of the novel in many languages.

Like other literary men, Hartley was a keen spotter of young acting talent and equally keen to bestow a little patronage. In December 1968 he suggested an actress to play Marian: she happened to be married to his godson. Fifteen months later he came up with another proposal. 'A few days ago I met Miss —, the daughter of —, who acts, I think, in films. She is on the large side! but blonde and beautiful, and might do for Marian ... She would like the part very much.' Alas. Yet Losey himself was in search of 'an unknown girl' for the part of Marian, 'because the girl ought to have been about eighteen, nineteen years old ...' Pinter describes a meeting with Delfont. Losey kept suggesting 'nineteen-year-old actresses' but as each name came up, Delfont snorted, 'She's nothing.' Losey, Pinter and John Heyman retired to a pub where (Pinter recalls) Heyman began to weep over all the work he'd put into the project. Heyman and Pinter told Losey that it would be Julie Christie or no film – and Losey surrendered. 'So,' Losey recalled, 'I conceded Julie Christie, whom I had tried to get initially in 1974 [sic: 1964], at which point she would have been almost the right age ... And I must say, to her very great credit ... she insisted that she was too old ...' Christie was hard to convince and her services had to be sought through Warren Beatty. As late as 21 May, Losey cabled Beatty desperately urging him to give a definite answer about her availability.[7] Gerry Fisher, director of photography, remembers being summoned to a meeting in the Dorchester Hotel. Could he make Christie look eighteen throughout the picture? Only by assuming total control over every shot and by imposing crippling restraints on everyone. Pinter then wondered whether Marian had to be eighteen.

Losey's attachment to 'unknowns' is open to question. Following their recent collaboration in Secret Ceremony, he had sent a script to Mia Farrow in February 1969, hoping to shoot that summer, and again in November 1969. Farrow gave birth to twins in late January 1970. On 2 March, Losey wrote to her: 'I gather you are not going to be in The Go-Between ... I can do no more to try and persuade you.' Even after the film was released, Losey, a lifelong subscriber to the star system, continued to complain that the film could have been made effectively without stars in the two main parts. Christie remembers receiving a public humiliation from Losey after arriving late on set; John Loring remembers Losey roaring, 'I would accept that from a star like Elizabeth Taylor, but not from you.' Christie regards him as a cinematographic rather than as an actor's director; pre-planning of movements and frames before each take rather than chats to actors about their roles. 'I was in a mess and in need of direction – directors were fathers. Later I would probably have had trouble with Joe, his paternalism.' In August 1973 Losey told Jerome Hellman that 'she has less talent than she is given credit for and she doesn't seem to me either willing or able to probe very deeply into herself'.[8]

Acting with children, Christie reflects, is difficult: 'You feed in but you don't get feed back. You have to make the best of what you have.' She remembers Losey as being infinitely patient with Dominic Guard. Fourteen years old, Dominic was the son of acting parents and suitably diminutive for his age, standing five feet, one inch high – essential that he look up to Marian, in every sense. But Dominic had a problem which had apparently escaped Losey during auditions. While shooting the smoking-room scene involving Trimingham, Mr Maudsley and young Leo, Losey became increasingly disturbed by Dominic's mannerism of 'sucking in his lips and swallowing'. Losey turned everyone except the three actors off the set for a rehearsal, warning Dominic that each time the mannerism occurred he would say 'lips' or 'tongue'. On the first run-through he said it thirty-five times; on the second try, only five. At lunch with the Loseys that afternoon Dominic was distant to the point of rudeness. Finally he explained that he had managed to conquer a stammer by these motions of mouth and tongue.

Alan Bates was not the original choice for the part of Ted Burgess, but his credentials as a fine actor were well established; he had played in Pinter's *The Caretaker*, and his partnership with Julie Christie in *Far from the Madding Crowd* made the pair a bankable package – although they are rarely seen together in *The Go-Between*. From Losey he received through the post a short essay on each character, which Bates found unprecedented, helpful and civilized. Edward Fox played Viscount Trimingham and deeply admired Losey's ability to convey what he wanted with scarcely a word or gesture, his intense concentration holding the actor in complete communion within a single take. But Losey fudged Hartley's physical description of the war hero whose bed Marian would be forced to share. Fox's facial scar is not grotesque and does not pull his eye down, 'exposing a tract of glistening red under-lid', or his mouth up, exposing the gum. The 'whole face' is not 'lop-sided'. If the cinema cannot risk deformity it cannot be taken seriously.

The part of Mrs Maudsley had been offered to Deborah Kerr. Tempers frayed. Kerr complained that she'd spent several hours in a hotel lobby waiting for Losey to call. She turned the part down because, she felt, Mrs Maudsley had only one good scene – at the end. 'I find her a one-dimensional figure, who is seen smiling enigmatically once or twice . . .' Evidently Losey was brusque when Kerr declined. 'I was left with "ashes" in my mouth,' she wrote, 'and distressed by your rather curt manner.' Margaret Leighton was brought in. By giving Mrs Maudsley a more volcanic inner life than the script suggested, Leighton won the SFTA award as 'Best Supporting Actress'. Richly talented as they were, the cast of *The Go-Between* performed for love (or kudos), not money. Edward Fox, who was about to fascinate cinema audiences across the world with his portrayal of a mercenary assassin in *The Day of the Jackal*, was paid £1,750 for six weeks' work. Margaret Leighton received £2,500, Dominic Guard £1,000. Julie Christie was guaranteed $50,000. She, Bates and Losey were granted an equal share of 10 per cent of EMI's gross receipts above the certified cost of production.[9]

And so to Norfolk. By 9 July the Loseys had installed themselves in Marsh Barn, Brancaster, an old barn on the edge of the marsh, tides ebbing and flowing in the creeks, sea birds and dunes, a walled garden of honeysuckle and roses. Often they had lunch or dinner in Holt, where the bar of the Feathers Hotel served as a rendezvous after work. It was to be a summer of flowering friendships, pleasure in life, artistic fulfilment. Pinter's play *Old Times* was conceived at Marsh Barn, which lies a full hour – a driver was laid on – from Melton Constable. Visitors included Sir Solly (Lord) Zuckerman and 'the gentle Lady Sylvia and her dear brother, Tommy Coke, Earl of Leicester' who 'loaned us many of those horse and carriages and carts . . .' and who lived in a beautiful, eighteenth-century, Palladian-style house, Holkham Hall (Losey wrote it as Hokum Hall) among Van Dykes and Poussins. Later the Earl asked why he hadn't been used in the cricket match scene.

Light is like gold, the summer sky is the presiding god. Losey's 30 July progress report reads: 'In sixteen days of shooting we have had two half-days of partial sun, one full day of intermittent sun, and today the first day of *consistent* sun . . . Approximately four minutes of film have been shot, involving nineteen set-ups.' A film is an industrial enterpise requiring the flexibility of a military campaign. The weather improved. The daily progress report for 24 August (the thirty-sixth of fifty-seven estimated shooting days), indicated that 88 minutes of film had now been shot, with 21 to follow. The average daily take was 2.28 minutes. By 15 September, 123 minutes of film had been shot and 59.125 feet printed. The unit returned to London on Friday 18 September.

In his pre-production notes, Losey conveyed his overall visual image of the film: 'The picture should look hot and like a slightly faded Renoir or Constable – the colours mostly gold and brown, the green minimized as much as possible under the circumstances. The skies and their clouds and the peculiar light of Norfolk . . . also the chiaroscuro of the corridors and secret passages . . . the present day sequences should stand out photographically. Whether this is done by filter or optically or with a change of raw stock . . . They should all at least be in dull weather . . . cold in tone . . . as against the dream-like quality of the 1900 story.'

Hartley's novel reminded Gerry Fisher of his own boyhood summers, lying in wheatfields, studying ladybirds and mice, finding old outbuildings, enjoying free run of an empty manor with a mulberry tree. In shooting *The Go-Between* he was able to evoke these memories, often with the aid of the zoom lens. But, Losey cautioned, the zoom should never be used for display; Fisher agrees that the zoom is 'a dangerous friend because it permits you to make things bigger but they're not really closer, you still feel at a distance'. Losey wrote to Fisher after shooting was done: 'In fact I think it is the best photographic work of its kind that I have ever seen. I only regret for your sake and mine that you haven't yet found an operator to equal you.' He also praised the work of the sound engineer, Peter Handford, a Norfolk man who wonderfully captured the lazy sounds of a dozing summer, swishing skirts, bees, harvesting, Ted Burgess loudly laying out tea cups.[10]

What was the developing relationship between Pinter's camera indications and Losey's implementation? Pinter explained: 'The only way I could actually write the script was to *see* it happening, *shot by shot* . . . Joe was not necessarily going to keep to them . . . but the point is that the feeling I was trying to suggest on those shots is something that Joe immediately understood, and would then translate into his own terms.' Losey's rough notes, scribbled on Pinter's final script, dated 27 January 1969, illustrate the process of translation from visual indications to a shooting script: 'Cut inside living room. Marcus is standing by Denys creating three groups'; 'V.L.S. from anti[*sic*]-room as Leo enters foreground goes towards Mrs Maudsley'; 'Straight down from above to large C.U. of Marian – very Renoir'. But such notes were merely provisional; the final framing was done 'through the lens'.

A striking example of Losey not adhering to Pinter's indicated shots comes during Leo's song at the village hall party. Pinter offered a complex sequence of camera movements:

During this song the camera gently cuts between STUBBS *and* BLUNT, *totally sober and attentive,* TRIMINGHAM, MR MAUDSLEY, MRS MAUDSLEY *and* TED. *Close-ups of* MARIAN *and* LEO. *Gleam of* MARIAN's *white arms and neck from* TED's *and* LEO's *point of view.*

But the finished film keeps the lens on Leo throughout his song, in a very simple manner. It's his moment, his ordeal – Hartley's novel, after all, was narrated by Leo alone. Subtle alterations in the written script result, also, from an actor's sense of character and idiom. Both Bogarde in *The Servant* and Bates in *The Go-Between* reworked certain phrases to convey the vernacular without altering the content. Bates, for example, changed 'pianist' to 'piano player', and:

TED: Would you like to have a shot with my gun?

to

TED: Want to have a shot with my gun?

Bates's Ted also says 'I'm sure she do' instead of Pinter's 'does'. He says, 'It's not what I want' instead of Pinter's 'It isn't what I want'.[11]

A distinctive feature of Pinter's screenplay is its unorthodox 'flash-forward' technique. *The Servant* had followed strict chronological sequence. *Accident* offers a quite simple time-pattern: present, past, present (the subtlety is achieved by delicately cross-referenced images). In dispensing with the prologue and epilogue of the *The Go-Between*, Pinter resorted to a bolder strategy; the summer of 1900 holds the main story, while interwoven flash-forwards to the late 1950s provide a subsidiary narrative whose cross-references are often achieved by a disjuncture between image and sound. Pinter's screenplay inter-cuts between young Leo in 1900 and a lonely Leo Colston being chauffeured *à la récherche du temps perdu* a lifetime later. The life-embracing boy has withered into a reserved, monkish celibate for whom a twisted sunlight flickers only in the distant memory. Pinter commented: 'I certainly feel more and more that the past is not past, that it never was past. It's present.'[12]

But would the flash-forwards work? Losey recalled that the fusing of past and present 'was fought to the death by everyone on the production side, from the beginning to the end, and it will still be fought even now'. Pinter's intercutting and the disjuncture between image and sound is complex: the boys' chat (1900) is heard as Norwich railway station (1950s) comes into view; the voices of the elderly Marian and Colston are heard against a slow shot of the cricket field in the evening sunlight fifty years earlier. As Young Leo hurries on his letter-carrying mission through the countryside, we hear:

MARIAN'S VOICE OLD (OVER): So you met my grandson?
COLSTON'S VOICE (OVER): Yes, I did.
MARIAN'S VOICE OLD (OVER): Does he remind you of anyone?
COLSTON'S VOICE (OVER): Of course. His grandfather.

On 18 March 1969, Losey's editor, Reginald Beck, severely criticized the script, 'albeit with humility and diffidence'. Beck found Colston's arrival at Norwich station confusing, 'since we don't know [who] Colston is . . . Also it breaks the spell of the growing relationship between Marian and Leo.' (Losey's comment: 'No.') 'Marian, I feel, one never wants to see in the present, except perhaps as a death mask.' (Here Losey commented 'Might be good?') Beck was also unhappy about the present-day appearances of Marian's grandson, the Young Man. The shooting of the film did not allay Beck's original doubts about the flash-forwards. On 28 October 1970, he wrote from the cutting-rooms at Elstree Studios: 'Regarding the present-day scenes my concern is that they are at present so much out of continuity that it is quite impossible to get any dramatic value from them.'[13] In its final form, the film reveals a quite drastic assault on Pinter's flash-forward sequences. Some examples follow (page numbers refer to the published screenplay):

On page 330, the evening after the cricket match, there is a shot of the cricket field.

> COLSTON'S VOICE (OVER): Isn't it dull for you to live here alone?
> MARIAN'S VOICE OLD (OVER): Alone? But people come in shoals. I'm quite a place of pilgrimage.

But in the film the voices are erased. Losey cut a scene in which the young Marian was pouring tea from a silver teapot, with voice-over of the old Marian:

> MARIAN'S VOICE (OLD): I rarely went to parties. People came to see me, of course, interesting people, artists and writers, not stuffy country neighbours . . . [etc.]

At another juncture the voice of Leo's mother (her letter to him) is heard while young Leo is seen in his pyjamas rather than against a present-day image, as Pinter intended. Near the end, two successive present-day scenes were removed, in both of which Pinter had laid the voices of the young Leo and the young Marian over an image of the old Marian talking to Leo Colston fifty years later. In these scenes the '1900' dialogue was a recapitulation of lines already spoken but of crucial significance in Colston's memory – for example, young Marian enjoining Leo not to tell anyone about the first letter he had brought her from Ted Burgess.

The inescapable conclusion is that whereas the screenplay of *Accident* was tailor-made for shooting, in *The Go-Between* Pinter's imagination tended to over-fly the boundaries of the feasible. For all its literary and directorial brilliance, the film bears the stamp of heavy editing. Whole scenes were cut, bringing protests from Reginald Beck, the guardian of logical continuity. Take, for example, the climax. After Marian has failed to attend Leo's birthday party, Mrs Maudsley drags the boy out into the rain and towards the outhouses in order to discover her daughter with Ted Burgess. But there is an apparent contradiction:

MRS MAUDSLEY: ... Leo, you know where she is. You shall show me
the way.

Yet Leo does not know where Marian is and it is palpably Mrs M. who leads
and pulls him towards the outhouses. Beck protested against Losey's decision
to cut a preceding scene showing the two boys approaching the fatal outhouses
and overhearing Marian with Burgess. This cut sequence – which would
explain 'Leo, you know where she is' – is found on pages 343–4 of the
screenplay. The scene ends with:

MARCUS: There are two of them. They're spooning. Let's go and rout
them out.
Quick close-up of LEO.
LEO: (*hushed whisper.*) No! It would be too boring!

Pinter's script stipulates: MRS MAUDSLEY *and* LEO *walk along the path, she
leading* – and later: *She pulls him after her.* Clearly, then, Pinter intended Mrs
M. to lead the way; therefore she did not need Leo as guide; therefore her
motivation was to punish him – all of which is psychologically plausible. But
Leo's awful, sublimated knowledge about the outhouses – evident in Dominic
Guard's performance as Mrs M. drags him along – really has no logical
foundation without the preceding scene.

Richard Rodney Bennett, composer for three of Losey's films, had been
commissioned to write the music, but Losey found his score too dramatic and
climactic. Could he come up with something 'non-descriptive and objective',
maybe electronic? Losey summoned Michel Legrand, showed him the rushes
in January 1971, then played him an album with a tenor saxophone and jazz
band. 'I said, "No, it won't work,"' Legrand recalls. 'He said, "Do what you
want".' Legrand steered clear of Elgar, basing his score on menacing
counterpoint, using strings, woodwind and two pianos scored to sound like
one. 'The boy almost hears the music. I wrote a theme on eleven variations.
When Joe heard the first cue he said, "No, No." He was like a bear in a cage.
It was the same on the second cue, and the third ... At the end he said, "It
doesn't work." I refused to re-do it. I advised him to dub and mix with the
music to see if it worked. I didn't hear from him again for months – not even
during Cannes. In June [1971] I received a huge telegram: "Best score – I'm
so proud."' Reginald Beck believes that Losey generally liked his music too
loud – and cites *The Go-Between*: portentous, obtrusive, and deliberately
anachronistic. Beck's French colleague, Marie Castro, agrees: 'It was an
invasion. Commercial. *Musique à la poubelle* [dustbin].'[14]

An unhappy sequel occurred in October 1974 when Losey fired off a
telegram to Legrand: 'Dear Michel, I don't know why we were worried about
The Go-Between music being used on cigarette commercials when last night

seeing a film on television, *The Thomas Crown Affair*, I heard the whole score. Are you ashamed of yourself. Joe.' Patricia concurred. Legrand replied with an indignant denial; he had written the *Thomas Crown* score before *The Go-Between*. Losey assured him that he had not intended 'to make any unpleasant accusations' but the fences were not rapidly mended. 'We didn't talk for years,' Legrand recalls. 'I was so furious.' (Losey could never come to terms with the work that his favoured artistes did for other directors; almost invariably he detected some element of imitation or theft from his own films.)

The rough cut shown on 16 October ran to a length of 1 hour 53 minutes. After the screening Heyman and Priggen arrived, in Patricia's words, 'like a couple of sleek businessmen, hands in their business pockets'. Both objected to Leo's exorcism ritual, which they found unclear and uncommercial. Losey was upset by their unholy alliance and wrongly ignored their advice. On 22 October, when the film was screened again, Patricia sat next to Pinter and behind Hartley; she was conscious of the young screenwriter observing the old novelist's reactions: Hartley was visibly moved. Losey asked him if the real 'Mrs Maudsley' had really dragged the boy like that to the outhouse in the year 1911. Hartley said, 'Well, no, really I was just made to follow.'

But the clouds over the film now equalled those over Norfolk during the summer. EMI had laid off half its production costs on MGM, whose boss was James Aubrey. John Heyman recalls: 'When Aubrey (the former head of CBS and creator of such TV shows as *Beverly Hills Billies* and *Green Acres*, known in the business as the Smiling Cobra) saw the film he hated it – boring – and said, "John, you've shafted me."' MGM wanted the première to be in Beverly Hills – ahead of Cannes. Losey rejected the idea; on 3 March, Forbes duly sent a cable to MGM stressing the importance of Cannes and the need to open in London and New York ahead of California. Aubrey finally blew his lid; from MGM Culver City came a fractious telegram to Delfont and Forbes. 'This film has been a constant problem and presents the first major breakdown in our association because of the complete lack of cooperation on the part of studio and producer. Strongly disagree with opinion that your film has any chance whatsoever at Cannes or that the reputation of Losey and Pinter enhance its potential success.' According to Aubrey, *The Go-Between* was 'the greatest still picture ever made'. MGM now backed out as American distributor, selling its half interest to Columbia for £250,000. In March Bryan Forbes resigned from his £40,000 a year job as EMI's production chief. Losey wrote regretting his resignation and expressing his sense of obligation to Forbes for his 'defence' of the film. EMI was not alone in its nervousness. John Heyman had taken flak from American distributors over *Secret Ceremony* and now he felt constrained to approach Losey and Pinter with a list of suggested cuts. Heyman wanted the flash-forwards removed.

Pinter responded frostily: 'How much your view of the film differs from ours is indicated, I think, by the range and nature of your proposals.' If distributors wanted cuts they would make them 'at their own risk', but 'for any of us to anticipate such action by taking the action ourselves I consider a wretched posture . . .' Twenty years later, Heyman remains convinced that the flash-forwards simply didn't work. 'I hate going to the cinema and not knowing where one is.'[15]

27

The Go-Between:
Deadly Nightshade

Losey launches the film at a faster pace than Pinter's initial, atmospheric, scene-painting. 'The past is a foreign country,' declares an elderly voice (which, it will transpire, is that of Leo Colston) over a long shot of an English country house, Brandham Hall. 'They do things differently there.' The voice is definitive, omniscient: it knows the story. Is the story of young Leo to be told by old Leo (as in Hartley) but by indirect narration? Pinter says no, he and Losey are doing the telling.[1]

Fast cheerful music as the trap carries the two schoolfriends, Leo (Dominic Guard) and Marcus (Richard Gibson), through the Norfolk countryside, Leo in his floppy brown hat gazing intently at the impressive world where he's to spend the summer holidays; Marcus − not a bad chap beneath his superior drawl − at ease in a boater. Cut to them mounting the stairs, a footman carrying their luggage; Losey establishes the main staircase, its walls coated in heavily framed paintings, as a symbol of the house's grandeur and Leo's well-masked awe. To further establish Marcus as the insider, Losey has him nip through a concealed door on the staircase to baffle his friend; soon Marcus will instruct Leo not to fold his used clothes but to toss them on the floor for servants to retrieve.

Rapidly Losey maps out the social and emotional geography of the great house, starting with Leo's glimpses of guests relaxing on the lawn − first a high shot from his attic window, then descending, floor by floor, until he gazes from the first floor balustrade at a beautiful young lady lying in a hammock, her face masked by a parasol, listening to a man reading from a book. We see Leo nervously descending the great staircase and pausing at the massive door of the drawing-room, where Mrs Maudsley is heard talking about him − a very nice boy who lives in rather a small house. He enters − a socially intimidating moment wonderfully choreographed. Mrs Maudsley (Margaret Leighton) takes his arm as she leads the procession into the dining-room. Across a vast table there are jokes about the young guest's powers as a magician; Marian speaks her first words to him, 'You're not going to bewitch us here?' It is he who is bewitched.

A very high shot from the roof of the adults strolling on the lawn on a summer night; the zoom swoops towards a couple; a high angle shot up the top floor of the house; the camera observes Marcus (asleep) and Leo (not) through their bedroom window. This is Leo's first experience of a grand house party, with its confident, baffling laughter drifting on grass-scented air. Next morning he finds Mr Maudsley (Michael Gough) examining his garden barometer. There are remarks about Leo's hot Norfolk jacket, and these continue during tea on the lawn. Poor Leo pretends that maybe his mother forgot to pack his summer suit. We first glimpse the tension between mother and daughter as Marian proposes to take the boy into Norwich to buy him a summer suit. Christie's brisk practicality and lemony sweetness well convey the calculations of a young lady engaged in massive subterfuge.

The carriage carrying Marian and Leo runs through beautiful fields of wheat, Marian's white parasol up – and we hear the old Leo Colston's voice over: 'You flew too near the sun and you were scorched.' This is the only occasion on which Pinter adopts the 'second-person' mode – and if the old man is addressing his boyhood self we might conclude, with Michael Riley and James Palmer, that the 1900 sequences are a form of indirect, 'mindscreen', first-person narration by old Colston. According to Pinter, old Leo is talking to himself. (Pinter and Losey were engaged in cross-referential 'stitching' of two distinct narratives by a variety of filmic devices more evocative than logical.)[2]

Now the first flashforward: Colston (Michael Redgrave), his back to us, is approaching a village cemetery on an overcast day. We do not yet know who the elderly gentleman is.

Marian treats Leo to a feast in a grand Norwich hotel, beginning with an angle reminiscent of Stephen's dinner with Francesca in *Accident*, a two-shot in profile, before zooming in. Marian inquires about Leo's dead father but she's not really interested; she is literally buttering the boy up. Invited to amuse himself for an hour while she does some 'shopping', Leo and the camera make the most of Norwich Cathedral. Later, at the horse-market, the camera zooms in from an immense distance on Marian found talking to Ted Burgess (Alan Bates), as observed by Leo – one of Losey's more meretricious motions.

Admiring Leo's new suit in the drawing-room of Brandham Hall – the social patronage inflicted on Leo is subtly conveyed – Mrs Maudsley asks Marian, 'You didn't see anyone in Norwich, I suppose?' Marian replies, 'Not a soul. We were hard at it all the time, weren't we, Leo?' After a moment's pause, Leo assents – his induction to the mendacity of devotion. Flashforward again: Colston is seen entering his sombre hired car outside Norwich station. As we are wondering who he is, and what time-frame we are in, Mrs Maudsley is heard, voice-over, telling Leo he can watch the others bathe. The bathing scene at the river establishes a number of factors with extraordinary rapidity: Ted Burgess, a somewhat 'Spanish' figure with a mop of black hair and sideburns running into small mutton-chop whiskers, is bathing out of bounds, but as an independent tenant farmer must be treated with courtesy; Marian's brother, Denys (Simon Hume-Kendall), a well-meaning ass, accepts Ted's apologies while Marian and her friend Kate take their swim at an appropriate distance. Leo, as the non-swimmer of the afternoon, is doubly the outsider as he observes all these movements through the rushes. Perhaps Dominic Guard's most conspicuous gift is his ability to convey spellbound curiosity. And how we empathize with his questing gaze! Walking home, Marian flirts and beguiles Leo by getting him to spread her wet hair over her shoulders. He's enchanted. A view of the house, then a rare fade, before a close-up of a gong sounding for breakfast.

Marian arrives late for breakfast prayers, while her father is already reading from the Bible, and gets a Leighton-look of thunder from her mother. At prayers we observe Viscount Hugh Trimingham (Edward Fox) for the first time; Leo is gazing at him; as the camera moves to Leo's view point we see Trimingham's Boer War facial scar. Trimingham and the scar smile at the boy. Each frame carries the quality of a still painting rich in depth, with the object of focus often viewed as the last layer in a series of carefully placed intervening objects: the back of a head, potted plants, a pair of figurines, a doorway or window, and finally the key person observed.

As the Maudsleys and their guests walk to church, we pass a cottage at which, in the next shot, the elderly Leo Colston, cobwebbed in memory, will be seen gazing, his back to the camera as he moves into the frame.

Leo is found gazing from a discreet distance at Marian lying voluptuously in the grass. He walks away; zoom as he is rapidly dwarfed against the vast and sumptuous countryside; deer are discovered in shimmering summer haze through the trees (Losey's use of trees and intervening undergrowth creates a tactile impression of threading through them). Leo reaches a farm, fondles a stabled horse, climbs a haystack, falls, cuts his knee on an axe, and is confronted by an angry Ted Burgess. But Ted is soon helping him hobble across the yard and patching his knee with a clean handkerchief in a comfortable kitchen furnished with good yeoman dressers. This hesitant, friendly encounter is in sharp contrast to the long, idle holiday of the upper-class, a ritual which sublimates human feeling beneath theatrical routines: meals, croquet, church-going, more meals, gossip. Ted works – a metaphor for his virility – whereas Trimingham can't produce a single suggestion for the day's activity when Mrs M. invites his opinion at breakfast, the Boer War hero turning feebly to the window for a sign of the weather. Only in Ted's presence does Leo feel able to go beyond the most confined of polite phrases.

Now a close-up of Marian removing Ted's blood-soaked handkerchief from Leo's knee as she tends his wound, Leo riveted by her administration. He gives her Ted's letter and the knot tightens. The grand picnic scene follows, with Pinter's ad libs murmured in the background as Leo stands diffidently near Trimingham and Marian, whose back is half-turned away from her suitor. The summer idyll continues as carriages run through open country and across a shallow ford. A skirt (Marian's) is seen flitting through the trees – then cut across half a century to old Leo Colston's legs descending from his car, the polished metal reflecting church, cottage and graveyard.

The blond Trimingham sends Leo on an errand; Marian is wanted for croquet but cannot be found. Leo encounters her emerging with tousled hair from the trees. More quick cutting as Leo delivers letters to and from Ted; a close-up of Marian kissing Leo for his valiant work; Leo watching Ted shoot a rabbit. The pace is extremely rapid here. Marian is surprised by Trimingham as she hands a letter to Leo and does not have time to seal it properly (but Trimingham is too much the gentleman to comment). Leo takes to the deer park again, running in long shot towards the sanctuary of a massive tree which occupies the foreground. The zoom swoops on him without a cut. Once beneath the tree – and his own guilty shadow on it – he cannot resist reading the unsealed letter. 'Darling, darling,' we see. Leo sinks to the ground in weeping misery, then smashes a stick against the tree in fury. (Pinter's script, by contrast, suggests a boy caught between genuine embarrassment and the standard posture of scoffing scorn at the approaching giant of sex.) Arriving at Ted's farm, Leo is in a deep sulk. He wants to stop carrying messages. The subject turns, by way of a mare about to foal, towards sex, 'spooning' and whatever it is animals and people do. Leo confesses his ignorance, turning to Ted for information, but the farmer is tongue-tied. Rather cynically, he promises to tell Leo what lovers do if he goes on carrying messages – a promise Ted will be reluctant to keep.

The long cricket match scene follows – possibly too long, although the 'social' set-up is amusing as Brandham Hall takes on the village in an annual ritual of social fraternization. Trimingham and Burgess – Marian's men – are the opposing captains; the Hall team strike the ball correctly, particularly Trimingham, while Ted lashes out with raw power, putting Marian and her mother in peril as the ball descends among the spectators – a nice stroke of sexual symbolism.

MRS MAUDSLEY: The ball didn't hurt you?
MARIAN: It didn't touch me, Mama.

But cricketers may be surprised when young Leo is seen fielding with intense concentration on the boundary, yet failing to 'walk in' as the bowler bowls. And when Ted Burgess's giant haymaker descends clean into his hands – sealing the match – Leo hasn't moved his feet an inch.

A convivial tea in the village hall puts Pinter's dialogue in to bat. After the match all return to the hall, in best clothes, for the sing-song, Marian accompanying Ted's 'Take a Pair of Sparkling Eyes' on the piano while Mrs

Maudsley's rolling eyes threaten to bring the roof in and Mr Maudsley pulls contentedly (perhaps) – Gough's performance is ambivalent – on his pipe. Leo is prevailed upon to sing his 'sacred' song – 'clad in robes of virgin white' – against a Union Jack (and against Leo Colston's hired car driving past the cricket field fifty years later). Needless to add, both the cricket match and the party afterwards are about Marian's relationship with Ted.

Through a window we see old Leo Colston handing his hat and umbrella to a maid. Later we will be shown the preceding moment, as the maid ushers him into the room. Why is this forgettable scene repeated, in reverse time sequence? Why, earlier, was Colston seen gazing at a village churchyard *before* he was shown arriving at Norwich station? The answer may reside in the hesitant nature of old Colston's return to Brandham; in visiting Marian after an interval of fifty years he is taking one step forward, two steps back. However, the burden on the audience is heavy: the dovetailing of the present-day scenes into the main story, with the many voice-overs overlapping half a century, is difficult enough to absorb without the time-scrambling of Colston's return to Brandham.

Leo and Marcus wrestle their way home from the village hall, Marcus revealing that Marian is to marry Trimingham. This is intercut with a long-shot of the silent Young Man, in fact Marian and Ted's grandson, stopping as Colston approaches him. Zoom in on his face – shades of Ted.

Marian finds Leo seated at a small table in a wooded part of the garden and gaily asks him to take another letter. Leo says he can't – 'Because of Hugh [Trimingham].' Marian loses her temper, moving around the table in rage, inflicting demeaning abuse on the boy: 'You come into this house, our guest, a poor nothing out of nowhere . . . we feed you, we clothe you, we make a great fuss of you . . .' When she demeaningly suggests that he wants payment, he snatches the letter from her and runs into the open countryside, pursued by the zoom lens, arriving at Ted's farm covered in sweat and tears. Ted comforts him with a clean handkerchief and distracts him by shooting at rooks. Leo demands that Ted live up to his promise to tell him what it is that lovers do; Ted tries to, searching for euphemisms but evading the details. Growing frantic, Leo keeps grabbing Ted's sleeve – a wonderful burst of acting from Dominic Guard as the boy's emotions finally break his self-control. Ted loses his temper and Leo runs from the farm as the crows screech.

Colston/Redgrave is gazing out of the old Marian's window, waiting to be received; over this we hear young Leo's voice as he writes to his mother, asking to be taken away from Brandham Hall. Furtively Leo posts his letter in the house box; Trimingham descends the stairs in evening dress, passes the boy with a silence which is friendly (there isn't always something useful to say), and enters the smoking-room. Drawn after him, Leo knocks and is cordially invited to join a masculine reserve of leather chairs and Teniers pictures. Without a hint of irony Trimingham offers him a cigar from a silver box. 'Nothing is ever a lady's fault,' Leo is firmly told when he ventures into the misty terrain of lovers and duels. Edward Fox effortlessly captures the public-school quality of being a fool yet no fool; of sheltering behind received formulas while knowing what really lurks in the undergrowth.

And what, Leo presses, does he think of Ted Burgess? Trimingham ponders this. Ted is a 'powerful hitter' and 'a bit of a lady-killer, that's all'. Mr Maudsley enters; Trimingham and Leo leap up respectfully. Both men courteously treat Leo like a man entitled to share the confidences of the smoking-room where the decanters brim with the true pleasures of life and where burial in a newspaper does not affront some woman's need of attention. The subject of conversation is still Ted Burgess (who enjoys the services of a daily housekeeper):

MR MAUDSLEY: They say he's got a woman up this way.
LEO: I know. But she doesn't come on Sundays.

Harvest time. Leo approaches Ted for a last visit, seen across cut corn. They stroll together like friends. Leo mentions Trimingham's engagement and Trimingham's thoughts of sending Ted into the army. Ted reacts: that's for Marian to decide. There are no longer any pretences between Ted and Leo – except about what lovers actually do. Cut. The Young Man (Ted's grandson, fifty years on) walks into close-up. Cut to deer. Cut to Leo asking Marian why she doesn't marry Ted. She cries; he puts his arm round her; she accepts his embrace. But what is the quality of those tears? Her 'I must' (marry Trimingham) signals her own sense of social hierarchy as much as her family's. Marian is not a tragic heroine of passion, but a self-willed young woman enjoying her terminal season of the senses.

Emerging imperiously from the house, Mrs Maudsley catches Marian in the act of forcing yet another letter upon her go-between, Leo. Marian stuffs it in the pocket of his summer suit and tells her mother it's for Nanny (Pinter) or Nannie (Hartley) Robson. Mrs Maudsley leads Leo away to inspect the garden, leaving Marian in worried profile. Now Margaret Leighton is offered a sustained scene in which she can display the range of her talent, alternating between tenderness and anger, maternal kindness and cunning – she places a flower in Leo's buttonhole – until she finally corners him in the greenhouse. Leo is driven to a series of desperate lies about the note in his pocket. The torment in this scene is almost unbearable – and the injustice. Pinter's writing and Losey's direction are merciless. (Yet Marian never subsequently asks Leo what happened to her love letter, a strange lapse.)

Alone in his room, the miserable Leo ponders his fate while we hear, over, the elderly Marian talking to Leo Colston. 'So you met my grandson . . .

Does he remind you of anyone?' For the first time we see the old woman's face. After hours of make-up, they shot Julie Christie in shadow and dubbed. Christie and Gerry Fisher agree that this didn't work. The elderly Marian should have been observed from a distant and oblique angle, perhaps walking on a stick in Colston's company – Colston uncertain whether to offer his arm. On the other hand, when she tells Colston, 'You're all dried up inside, I can tell that. Don't you feel any need of love?' we are moved by Redgrave's frozen features, his mute pain, a tragic contrast to the eager, ardent, life-embracing boy he had once been.

Leo's birthday party is Wagnerian: thunder and rain outside, Mrs Maudsley brooding over Marian's absence, Leo blowing out the candles on his cake (twelve plus one, Mrs M. being superstitious), crackers and paper hats, a carriage dispatched through sheeting rain to fetch Marian from Nanny Robson's, where she cannot be found, gloom descending on the party, Leo's birthday bicycle glimpsed in the hall, until finally Mrs Maudsley 'breaks' into dementia with rolling eyes and the famous line:

MRS MAUDSLEY: Leo, you know where she is. You shall show me the way.

Seizing him, she drags him out into the rain and on the last fateful journey to the outhouses. An anxious matron has become a throbbing Medea. The climax of *The Go-Between* poses a question which is Hartley's alone to answer. Mrs Maudsley is a woman of strict convention and propriety. If she expects to catch her daughter in the most dishonourable and scandalous of situations, why does this severe matron, in breach of every convention and decency, drag a boy, a guest in her house, to witness it? Would she want her family's shame to walk away from Brandham in Leo's head and on Leo's tongue? Has she, therefore, gone mad with anger? (Pinter comments, Yes.) But why take it out on Leo? Is it because the boy is an outsider of slightly inferior social status, who becomes in Mrs Maudsley's disturbed mind the scapegoat for the whole disaster?

At the moment of coming upon Marian and Ted *in flagrante delicto*, Mrs Maudsley instinctively draws Leo's head and eyes away – yet vengefully: her hands are claws in his hair. Whereas Marian, discovered, tenderly pulls Ted's head down to her breast. For the sexual climax in the outhouse, Pinter limited himself to Hartley's image of 'a shadow moving on the wall like an umbrella opening and closing'. Losey opted for heavy breathing, off, followed by a medium shot of a static coupling, head to toe, with Marian's bare thighs curled round her man – a simple and unevasive image.[3]

Ted Burgess responds to his shame, and Marian's, by killing himself with the shotgun that Leo has admired. Losey, who had pondered Hemingway's shotgun suicide, settled for a quiet, 'still-death' of a figure slumped across the kitchen table – death viewed as a version of sleep, and nothing splattered on the walls. The impact is nil. 'A bit of a lad' and 'a bit of a lady-killer', Ted would surely have reserved the wrong end of his shotgun for rabbits and rooks – or for the Boers. The last motions of the film belong to Redgrave, a bluebottle in his chauffeur-driven car, gazing through rain-flecked windows, and from a great distance, at the dull, lifeless façade of Brandham Hall.

Cannes was the target. As usual, Losey left nothing to chance, lobbying the film critic Dilys Powell, a personal friend and a member of the committee appointed to select the official British entry for the Festival. A greetings card from Dirk Bogarde was waiting for the Loseys at the Colombe d'Or: 'Welcome! – love and I hope you don't win!' Eric Kahane, who did the French subtitles, remembers much drinking with Pinter and the Loseys: 'Pissed every evening.' A supper party was held at the Carlton Hotel, attended by the Loseys, Pinter, Alan Bates, Norman Priggen, John Heyman – and, *frère-ennemi*, Dirk Bogarde. Five years had passed since they had worked together and now Bogarde was at Cannes as the star of a rival film, Luchino Visconti's *Death in Venice*. 'We had made a masterpiece and Joe had not,' Bogarde says. *The Go-Between* was 'all right but not that good'. Three years Losey's senior, the son of an immensely wealthy duke who was intimate with Puccini, Toscanini and the Queen of Italy, Luchino Visconti conducted himself like a monarch. When he filmed *The Leopard* or *The Damned* the great houses and flunkeys were his own habitat; Losey had to work his way into the décor. Visconti could hurl his money or his mother's jewels into a production, whereas Losey had to beg and scavenge. Yet by the late 1960s there was a striking convergence of style and aspiration. They employed the same actors. Losey felt acute rivalry. 'I met Visconti and his entourage a couple of times and we were not very compatible, even then [1951]. We took an instant dislike to each other . . .' Their directorial styles overlapped, they often used the same actors, and competition for literary properties was constant. Losey complained that Visconti arrogantly expected the rights of the modern literary masterpieces to land in his lap. This was true; with a princely disregard for dates, he declared himself a member of the same generation as Proust, Mahler and Mann.[4]

Bertrand Tavernier, Losey's publicist, claims that 'Visconti spent time and money to bribe the jury. He promised an actress on the jury a part in his Proust film.' According to Bogarde, it was Erich (*Love Story*) Segal who shed the casting jury vote against *Death in Venice*. Walter Alford wrote to Losey: 'The agonizing hung jury up to the midday last Thursday must have been hell for you – I said goodbye to Spike [Norman Priggen] and John Heyman, biting their fingernails in the Carlton lobby.' On 27 May, *The Go-Between* won the Grand Prix (Palme d'Or) for 'Best Film' at the Cannes International

Film Festival. It was also granted the INTER Film Award. Losey had reached the summit.[5]

Visconti (Bogarde reported) 'quietly packed all his Vuitton luggage, checked out of his hotel, and went to the airport where he sat on his suitcases awaiting the next flight to Rome'. But he was persuaded to return. On the evening of the presentation Heyman dined with Nat Cohen of EMI: 'We did not attend the ceremony because Lord Delfont had decided to pick up the award himself for a film he had almost managed to derail – something both Nat and I felt unjustifiable.' Heyman and Cohen took themselves to the Casino instead. Patricia Losey coaxed her husband into approaching Visconti; there was an awkward embrace in black ties; the cameras flashed; the impression (she says) was of Visconti holding court. When the Palme d'Or was presented by Jean-Louis Trintignant and Romy Schneider (who had been directed by Visconti but not yet by Losey), the director of *The Go-Between* walked 'rather uncomfortably' on to the stage. A roar went up for Visconti, who was awarded a special prize; he bowed graciously. According to Bogarde, Losey left soon after receiving his award, skipping the great supper party which ended the Festival. 'The only person not present was Joe who had gone earlier without any farewell.' Visconti asked Bogarde whether Losey was suffering from '*la grippe*'.[6]

After *Death in Venice* Visconti visited Normandy and Paris to prepare himself for Proust; a year later Losey embarked on the same mission. Pinter, whose Proust screenplay now held the field against that of Visconti and Suso Cecchi d'Amico, had conceived *Old Times* at Marsh Barn, while Losey was shooting *The Go-Between*. Was it out of revenge that Visconti directed the play in Rome in such a manner that Pinter held a press conference to denounce the indignities inflicted on his work – a pair of lesbians tirelessly caressing each other, a man masturbating his wife, and more? Visconti had to agree to changes.[7]

Two months after the 1971 Cannes Festival, Richard Roud travelled with the Loseys on the S S *France*, where Losey was treated like a king by staff and crew. On arrival in New York he let the gravel pour out: 'Sure Visconti acted like a prima donna at Cannes . . . kind of childish. It gets mild publicity and some attention for the film. I don't really believe in competitions and awards.' (The following year Losey served as chairman of the Cannes jury.) The Loseys were accommodated, courtesy of Columbia Pictures, at the St Regis Hotel. The première was held at the 68th Street Playhouse with a special screening at the Museum of Modern Art on 28 July: Julie Christie and Warren Beatty were the major exhibit for the trade press (Christie had not attended the Cannes Festival).[8] Losey pitched into a hectic publicity schedule: a 7.30 a.m. visit to ABC-TV's show, 'AM New York'; posing for *Harper's Bazaar*; lunch with Kathleen Carroll at Danny's Hide-A-Way; two cocktail sessions up to 7.30 p.m. On the Friday Losey was interviewed by Mel Gussow of the *New York Times* and by Mike Wallace on CBS Radio, had lunch with two people from *Life*, then met Rik Hertzberg of the *New Yorker*. Interviewed by *Time*, Losey agreed – 'in a voice full of measured flamboyance' – that he was a romantic, 'A middle-aged Marxist romantic'.

Columbia organized two press screenings on 1 July. Interviewed afterwards, the critics expressed admiration but warned that the film moved slowly; when they came, the reviews were even better than anticipated, many close to eulogistic. Vincent Canby, of the *New York Times* (30 July), greeted 'one of the loveliest, and one of the most perfectly formed, set and acted films we're likely to see this year . . . The production that Losey has designed for Pinter's screenplay is close to perfect, with never a single shot, a camera angle nor a cut from one scene to another that is not synchronized to the structure of the whole . . . It's one of the few new movies, in fact, that I can recommend without any real qualification.' Writing in *New York* (26 July), Judith Crist was ecstatic ('so deeply compassionate and wryly sophisticated . . . a brilliant film of classic stature'). Losey's attention turned to the American Oscars. He wrote to Delfont, 'I am also amazed that Columbia has done nothing to push *The Go-Between* for any of the Oscar nominations.' On 1 December Theo Cowan informed Losey that Jerry Pam had suggested a two-part campaign of publicity and advertising, at a fee of $5,000. Losey baulked. The film did win an Oscar nomination – but no award. Losey blamed the distributors – 'since Columbia chose to spend all the promotion on *Nicholas and Alexandra*'.[9]

The Go-Between was not released in Britain until September, opening at the ABC 1 and 2 in London. With a history of limerick-swapping behind them, Losey had no need to fear an 'X' certificate from John Trevelyan, on whose waistcoat the ash was pleasantly thickening. He had even proposed an actress for the part of Mrs Maudsley: '[Her] father is an ex-army man; she was educated at Girton College, Cambridge . . . I have not seen her work.' On official notepaper, Trevelyan wrote to Losey in March 1971: 'You will probably be hearing in the near future from my nephew . . . whose father is the Bishop of . . .' Apparently the nephew had written a screenplay. Trevelyan stood down as film censor that year, to Losey's regret.

David Robinson had earlier reported in the *Financial Times* that Losey's prestige stood 'as high as that of any director working in this country'. When the film opened in London, Robinson called it Losey's finest – 'a near-perfect example of adaptation'. This reflected the prevailing tone – almost unqualified celebration. In the *Observer* (26 September), George Melly announced that 'the Losey–Pinter partnership finally realizes its potential'; *The Times* concluded that the film surpassed both of the the partnership's predecessors.

The awards of the Society of Film and Television Arts were contested by a richly talented field in 1971. Nominated for 'Best Film' were *The Go-Between*,

Death in Venice, Taking Off (by Milos Forman), and the eventual winner, *Sunday Bloody Sunday* (by John Schlesinger, who was also named 'Best Director', a prize which always eluded Losey in Britain). The Best Screenplay award went to Pinter. Margaret Leighton emerged as 'Best Supporting Actress', Edward Fox as 'Best Supporting Actor', and Dominic Guard as 'Most Promising Newcomer'.

As with *Accident*, Losey rapidly concluded that the distributors were making a hash – promoting *The Go-Between* with neither the intelligence nor the enthusiasm merited by the Palme d'Or. His irritation surfaced in a cable sent to John Heyman on 24 September, proposing a public advertisement: 'We the creators of the film' dissociate ourselves from MGM/EMI's publicity campaign, which 'degrades our work'. Julie Christie and Harold Pinter supported the complaint. Heyman responded with a letter to Delfont: 'My dear Bernie, I would be most grateful if you could arrange to have my name and the name of this company removed from all posters and advertising material.' On 10 November, Delfont cabled Losey in Rome, assuring him that he was now giving his name additional prominence and that the Cannes award was 'fully publicized in main national advertising'. A 'Royal Première' took place in Norwich on 29 October, with the Queen Mother present. The Norfolk aristocracy turned out in force, led by Lady Harrod, chairman of the Norfolk branch of the Council for the Protection of Rural England. The smoothly groomed Delfont escorted the Queen Mother along a receiving line which included Patricia (Losey himself was working on *Trotsky* in Rome), Hartley and Pinter – the last two vividly contrasting physical types: Hartley over seventy, pale in the watery English way, stout and bald; Pinter, forty-one years old, in horn-rimmed glasses and big sideburns. Delfont sent Lady Harrod a bill for £535 for the hire of the cinema, which didn't seem fair at all. Losey responded by deploring 'Delfont's cupidity and stupidity . . . I believe that with your name and position, and with Delfont's desire for a knighthood, you can shame him into submission'. (Delfont became Sir Bernard, then Lord Delfont.)[10]

Four seconds of hayloft copulation were cut for the benefit of the American airlines. Losey complained to Heyman and Delfont: the hunt for the culprit was on. Stanley Schneider denied that Columbia had touched it – but Columbia was the guilty party. Columbia Pictures cut the same four copulatory seconds before KABC-TV received the print. Losey's admirer, Virginia Starr of KABC-TV, wrote to him in distress: ' . . . please, in future when a film of yours is released for distribution on television, do make the same stipulation as Mr Kubrick, Mr Stevens, and Mr Preminger. No editing! It is in our contracts – we cannot edit "Dr Strangelove", "A Place in the Sun", or "The Man with the Golden Arm".' Losey promptly declared Columbia's action to be 'totally illegal', urging Heyman to 'get enough out of Columbia to make it hurt' – but the only contractual protection was the standard one, 'removal of director credit if the editing is objectionable'. Mr Losey was not Mr Kubrick, Mr Stevens or Mr Preminger.[11]

The Go-Between cost £532,841 and eventually made a handsome profit. After eighteen months the net takings in the UK were £232,249 and total gross receipts for the 'Anglo-EMI Territories' were £453,632, including £88,883 from the British Film Fund. At 1 July 1972, Columbia's territories, including France ($621,071), showed a gross of $2,198,382. Of this, the USA and Canada accounted for takings of $1,581,972. Ten years after its première, the film had earned £290,888 from UK cinema and television, £204,566 from overseas sales (excluding the USA), and a boost grant of £96,599 from the British Film Fund. Columbia's gross receipts in the USA, France and Canada were the dollar equivalent of £1,375,300. Losey's personal percentage earnings amounted to £39,355.

PART VIII

Assassins

28

Politics and Film

From the agit-prop ultra-radicalism of the Living Newspaper, Losey retreated into the more cautious and coded messages – or 'issues' – grafted on to his 'genre' films in the 1940s and 1950s: in *The Boy with Green Hair*, war and racism; in *The Lawless*, bigotry; in *Time Without Pity*, capital punishment; in *Blind Date*, class-bound justice; in *The Criminal*, the prison system; in *The Damned*, atomic war and de-humanized science; in *King and Country*, the horror of war, the loss of justice. It was Losey's brilliant metaphorical imagery, his cinematic translation of the issues, rather than the scripts, which lent distinction to these early films. As Gilles Jacob commented, 'Racialism becomes incarnate in stone, imprisonment in cement, degradation and decay in mud ... the crumbling uncertainties of life in the atomic age, the decalcification of contaminated flesh and bone, is brilliantly translated by the chalky cliffs of *The Damned*.'

But in film-making, Losey pointed out, 'content is largely controlled by the group who are least interested in it – the producers and distributors ... I never once have been able to make a film that was wholly and exactly what I would freely have chosen. So ... one tries to inform relatively innocuous material with particular observation.' In June 1959, when *Films and Filming* invited him to contribute a piece on 'social content in films', he nailed his flag to 'progress' and to the struggle against hunger, fascism, race prejudice, inequities of class or nations, the H-bomb. (Of *The Damned* he said elsewhere: 'Nothing is done and people begin to give up. All those marches, all those sit-down strikes, Bertrand Russell ... and ultimately the governments have done nothing at all ...') He identified his own commitment with the 1930s, and with 'a positivism and conviction of over-all progress which is sadly lacking in the Fiddling Fifties'. The apparent hopelessness of the Beat generation didn't appeal to him. 'Surely the point of the artist's very functioning is to "say" something.' Without a point of view, art was sterile.

What was Losey's own point of view? At the risk of schematization, one observes four phases: the aestheticism and culture-hunger of his school and college days; the sharp left-turn into agit-prop from 1933 to 1945; the studio-formula-plus-message films from *The Boy* to *The Damned*; finally the embrace of modernist relativism which dominated his famous years. Losey noted Picasso's progress, his dramatic alterations and alternations of style; the extreme innovative hunger of the twentieth-century artist probably has no precedent. The film industry and the star system impose their own eclecticism; few directors have consistently enjoyed the privilege of total authorial integrity. The depth of Losey's early political conviction is impossible to measure; likewise his own sense of remaining true to that core of conviction. Dr Barrington Cooper, his physician for thirty years, suspects that his political commitment during the 1930s and 1940s reflected his need to be a hero in order to function; that the pretension was greater than the conviction: 'Joe was a polemicist rather than a thinker.'

Despite his credentials as a blacklistee, Losey remained an outsider to British left-wing theatre and cinema. Not only his realized films but his scores of projects demonstrated his continuing dependence on the studio-system formulas which the Hollywood blacklistees took for granted as being 'it'. By contrast, the new socialist literature and drama which began to flourish in the late 1950s was naturalistic, regional, the work of indigenous writers close to their roots in the vernacular. Losey took note of the work of the Angry Young Men and Women, whether novelists or Royal Court playwrights in the era of George Devine, but he was easily diverted by the rootless and rather glib Americo-international screenplays of film profession-als. Indeed Losey's most fatal error, with its underlying misreading of film professionalism, is found in the following comment to a colleague in February 1960: 'And then, of course, there are the Royal Court boys, though I wouldn't think any of them were very safe on a screenplay project.'[1]

Yet he came tantalizingly close to recognizing the obvious–that he must not only jettison the Hollywood professionals but grasp the new wave of British writing. In September 1960 he met Alan Sillitoe, who was working on a new novel. 'I'm not committed to [John] Osborne–[Tony] Richardson in any way for this one,' Sillitoe assured him.[2] David Storey's novels, *This Sporting Life* and *Flight into Camden*, and Willis Hall's *Billy Liar*, also led to discussions and feelers to producers, but these literary 'properties' rapidly passed into the hands of other directors. Losey was trading with the wrong producers. Daniel Angel, for example, sent *This Sporting Life* back to Losey with the news that sporting themes never worked well on the screen. In February 1961 Losey found himself sadly congratulating Lindsay Anderson on being about to film Storey's novel. Pinter apart, Losey missed the boat. Brian Phelan comments that he always wanted the big deal, the major names. Other directors – Reisz, Anderson, Richardson – were meanwhile pouncing on the new writing and discovering that 'kitchen sink' was not incompatible with box-office. New stars were born out of the vernacular – Finney, Courtenay – while Joe was meeting Gina Lollobrigida in Paris. The urgently contemporary films of the 1960s were not made by Losey.*

Highly illuminating is Losey's correspondence with the French critic and translator Louis Marcorelles. Commenting on *The Criminal* in August 1960, Marcorelles wrote: 'I only wish soon you can use your art to describe more important matters . . . I trust completely people like Lindsay [Anderson] and Karel [Reisz] to change the terrifying pattern of the middle-class cinema still prevailing on the British screens.' On 23 August, referring to Losey's disparaging view of the new 'Royal Court' theatre as compared with American productions of the 1930s, Marcorelles suspected that Losey was living off the nostalgia of an experience which was probably unique in its time. Describing Lindsay Anderson as the director who had best assimilated Brecht's teaching, he added: 'I don't think you did better in the great American era (*la belle époque americaine*).' Losey, he feared, was squandering his great gifts in a vacuum. On 28 November Losey struck back: 'You were right about my reaction to your review of *Billy Liar*, frankly I hated it and disagree with you on almost every single point.'

'I've always detested naturalism,' Losey declared in 1967. In art, images had to be larger than life. But naturalism was not really the issue; Anderson's masterly *If . . .*, a brilliant metaphor for youthful revolt, also offered images larger than life, blending 'realism' with fantasy. After *King and Country* Losey increasingly took refuge in the vestibule of allegory or fable and lush settings; talking to Ciment, he granted that this was an evasion. The Losey of the sixties and seventies resembled a two-eyed Cyclops – the left eye responded to the newspapers, the world, with radical instincts, while the right eye concluded that art and messages were like oil and water. 'I don't regard my films as social documents. I am not particularly interested by social reforms, and above all I don't believe in "message films".' *The Boy* and *The Lawless* he now categorized as 'message pictures' made by people who thought they knew the answers. 'I stopped somewhere along the line – I guess at *Eve* maybe – making that kind of picture, and have been much more interested in making pictures of provocation . . . it has to be a separate and individual reaction . . .' But when criticized for escapism or hedonism he was stung: in the course of a 1967 TV interview he complained that his recent

A Taste of Honey (Tony Richardson, 1961); *Saturday Night and Sunday Morning* (Reisz, 1961); *A Kind of Loving* (John Schlesinger, 1962); *The Loneliness of the Long-Distance Runner* (Richardson, 1963); *This Sporting Life* (Lindsay Anderson, 1963); *The Bomb* (Peter Watkins, 1965); *Morgan* (Reisz, 1966); *Charlie Bubbles* (Albert Finney, 1968); *If . . .* (Anderson, 1968); *Kes* (Ken Loach, 1969); *Family Life* (Loach, 1971).

films contained levels of political satire to which critics and public alike had
been blind. In *Modesty Blaise*, for example, Gabriel (Dirk Bogarde) stands
high above the sea in his sailor cap and dictates letters of condolence in the
sententious crocodile-tear tone – and indeed Texas accent – of Lyndon B.
Johnson. The public hadn't noticed.[3]

The swing of Losey's camera away from racism or capital punishment to
the beau monde excited as much private criticism as public – though it was
more diffidently expressed, or conveyed by silence. Numerous friends and
colleagues felt baffled, disillusioned. Basil Davidson, a champion of Black
Africa and anti-imperialism, tactfully expressed his disappointment in
December 1967, proposing that a work of art should be linked to an implicit
morality, 'a humanism that has a shape and points a meaning and makes a
judgement . . . your own art is not yet . . . completely fulfilled'. Writing again
in June 1968, Davidson strove to express his appreciation of *Boom!*'s qualities,
but had to concede his aversion to the star system, in which actors remain
essentially themselves. In January 1970 he gallantly strove to find pleasure in
Figures in a Landscape, but, 'I suppose it's inevitable that people like me
should long for your hand on some grand theme of what's nowadays belittingly
called "social significance".' Michel Ciment believes that Losey's radical
politics emerge more consistently if interpreted by (probably unconscious)
metaphor. The recurring Intruder-figure preys on the rich, who are relieved
to perish; their guilt mirrors that of the middle-class film director who had
joined the Communist Party – itself an Intruder.[4]

Towards the student–youth movement of the 1960s Losey underwent an
interesting change of attitude. Initially he displayed thirties chauvinism,
describing the young radicals as 'nihilistic, because they don't really know
what the fight was about, they don't really know what the hopes of the
Thirties were, and there are now no organizations that are organs of protest
. . . they are without theory and without orientation because they are
essentially anarchistic'.[5] However, interviewed by a French TV team late in
1968, he praised the May movement in France, the debates in the Odéon,
correctly pointing out that the original impetus came from West Germany. It
wasn't just a matter of smoking pot and taking LSD and growing long hair
and being promiscuous – but a real desire for learning, improvement,
communication. He felt 'deeply grateful' to the youth movement in Germany,
France, America, Spain, Italy and Chile. He had enjoyed *Hair* in New York;
when half the audience got up to dance at the end it was 'marvellous'. He also
criticized the failure of the French Communist Party to support the May
movement. 'I don't passionately desire the Soviet system as it has worked out.
I am also not anti-Soviet and not anti-Communist . . .'

He appropriated the philosophy of the New Left, when interviewed in
France, but vulgarized it on tourist terms: films should now be saying what

the students were saying – 'There is no comfort or any answers in preordained ideology or platforms and the only answer is in process . . . the biggest service I can render is to make people realize they are alive instead of dead years before they die . . .' The critic Jean Pierre Coursodon noted Losey's taste for inflated platitudes and pseudo-profundities. Talking to French television in December 1968, Losey declared; 'The most important word is dialectic, dynamic if you like, the inter-changing of opposites . . . let's go back to Hegel, the synthesis and antithesis, life has changed, life is growth, life is for living.'[6]

In May 1969 James Baldwin and Arnold Perl finished a first-draft screenplay, *Malcolm X*, which was initially offered to Columbia and in which Losey was interested. Nothing came of it and there is no discovered correspondence.* (Baldwin was a much loved friend, whose novel, *Another Country*, Losey ardently wished to film – 'the most important and successful *black* writer in the world'.)[7] In February 1970, while staying with his sister in New York, he wrote himself a memo: 'I should like to make a contemporary American subject that had originality and purpose . . . I would prefer such a subject to be original but it doesn't have to be. It could deal with color, commitment, the thirties, the fifties, second-class citizenship, youth or old-age or both, what it means to be an American now. Preferably through comedy or satire or even farce to the serious – as in STRANGELOVE . . . Secondly, I should like to tackle a "big" subject. Assassination in the modern world. The systematic, intentional extermination of the Indians. Automation and the human "soul". Third, just entertainment if it was new enough and had some purpose. Could be a musical. HAIR, for instance, if it isn't already out of date, as I think it is. Maybe its sequel, called LOVE?' Losey was stung when adversely compared to Stanley Kubrick: 'Sure I would love to make films about the parent–student–teacher union, city government strikes in New York, but this is not very easy.'

It was invariably in France that Losey's left eye opened widest. Talking to *Les Lettres françaises* (18 November 1970), he called *La Bataille d'Alger* a masterpiece because it had presented atrocity without sentimentality. He told *L'Express* (2–7 May 1972): 'I am – and this ought to be clear – a committed socialist. For me there is no alternative to Marxism, under one form or another. But I have forced myself all my life to preserve a certain independence of outlook . . . When you are on the Left . . . this poses certain problems vis-à-vis the world of business. And the cinema belongs to the world of business.' Mao's Cultural Revolution looked good: China was 'the only

*The *Malcolm X* screenplay was later published as *One Day When I Was Lost*. Baldwin was not credited in Spike Lee's 1992 film, *Malcolm X*.

country where democratic centralism has fulfilled its promises'. Mao had displayed the unique courage to destroy the bureaucracy which had been frustrating the revolution.

Back in England his emphasis was far from Maoist: 'I don't think that films necessarily have to indicate change, as long as they can produce the energies that will do it.' Interviewed on British television in 1967, he described 'class' as *the* major national problem. 'Almost everything stems from it.' As for the 'expensive monarchy, that serves no function except to perpetuate class nonsense'. Yet he sent Joshua to a public school, engaged in 'class nonsense' in Norfolk, and befriended titled people everywhere. Losey developed a snob-version of Marxism which rated the most interesting classes as 'the real aristocracy and the workers, and everything in the middle is a little bit confused . . . the people who can arrive at a truth, quickly and directly, tend to be really working class, not petit bourgeois or servants . . .' By 'real aristocracy' he meant 'people who have had many generations of education and never had to worry particularly about money. And they're most trust-worthy. And this is one of the reasons why Hollywood and the whole professional class in the United States was such a pushover for a McCarthy . . .' A year after he escaped from England into tax exile, he ruminated: '. . . but if one is a bourgeois making a working-class revolution, how can one help remaining bourgeois? So that, in the end, you have to have a revolution every few years because a lot of people involved in it have bourgeois habits they can't rid themselves of.' With an indulgent interviewer at his feet, he readily lapsed into ideological flatulence.[8]

His continuing loyalty to the Soviet Union was a form of loyalty to himself: no sell-out, he. In 1979 he claimed that Khrushchev's speech in 1956 had opened his eyes, but three months after the 20th Congress of the Soviet Party he and Ben Barzman, walking along a Paris street with Leon Clore, rose to Stalin's defence even with regard to the notorious 'Zionist doctors' plot'. 'You couldn't criticize the USSR to Joe,' Clore remarks. In September 1970, Baroness Moura Budberg told Losey how concerned she was, as a translator, not to do anything anti-Soviet. 'I don't want to die a renegade.' Losey responded that only someone of her generation – and his – could understand what this meant. Appeals from imprisoned Trotskyists got short shrift. In May 1969 John L. Davis, of IV International Publ (so written), wrote to Losey on behalf of imprisoned Trotskyists in Mexico, urging him to send a telegram to the authorities and asking for money to support the journey of a defence lawyer to Mexico. (The previous November Losey had visited the Acapulco Film Festival – lying on his back with his feet dangling in a swimming pool afloat with hibiscus – despite the recent massacres of Mexican students preceding the Olympic Games.) His response was to write angrily to his old and honoured friend Joris Ivens, who had given his address to Davis.

'What the hell is this ...? ... Please don't give my name and address to people sponsoring this kind of cause. And I haven't any more money at all. Love.' Ivens replied apologetically.[9]

Reluctantly, and under pressure, in April 1974 Losey agreed to lend his name as a patron of the Greater London Interdenominational Committee for the Release of Soviet Jury; but when the Committee solicited his signature on behalf of three Soviet Jews, a film director, a scriptwriter and a writer, who could no longer pursue their careers after applying to emigrate to Israel, Losey declined. The Committee also asked him to sign an open letter to the Soviet Government, to be published as an advertisement on behalf of over thirty Jews held in prison camps and seeking to emigrate. On this letter is written, 'Nothing done'. When the critic Dilys Powell indicated in print that Losey, Truffaut, Tati, Fellini and Visconti had agreed to sponsor a defence committee for a Soviet dissident, Sergei Paradjanov, Losey's secretary was instructed to inform the Committee that it was news to him – but he didn't wish to repudiate the initiative. In 1980 he was invited to direct *Boris Godunov* at the Paris Opéra after the intended director, Boris Lyubimov, had been refused permission to leave the USSR. Michel Ciment pointed out to Losey that an artist who had suffered persecution could do worse than pay public tribute to another – a few words on stage before the gala opening? Losey, recalls Ciment, took the thought seriously but did nothing.[10]

What really engaged his imagination was American politics and the Vietnam war. This may have affected his short-lived inclination, in 1968, to apply for British naturalization. Having drawn up a list of possible sponsors as if casting a film, he settled for Harold Pinter OBE, Dame Peggy Ashcroft OBE, John Trevelyan OBE, and Dilys Powell – all of whom endorsed his application. But he had second thoughts; on 10 December, he wrote to them in identical terms: 'My American advisers are now of two minds about my changing nationality ... It appears that if I give up the sanctuary of my blessed place of birth, I may very well not be allowed back in ...' In December 1972 he told an American comrade, F.V. Field: 'I had thought of renouncing [US] citizenship but to what purpose? I would then have to take more responsibility than I can for the Irish situation here, which differs only in degree.' (Presumably a reference to Vietnam.) In the mid-seventies he continued to renew his son Joshua's American passport.[11]

In June 1972 he visited New York and Los Angeles. 'All of my hopes are with McGovern,' he wrote to Carlos Fuentes, 'but I begin to expect that they are pretty forlorn ... I find the USA completely debilitating. ...' He told the *Saturday Review* that 'Nothing could be worse for the whole world than the re-election of Nixon. This espionage thing that's been uncovered [Watergate] is a perfect prototype [*sic*] of how the Nazis destroyed their opposition in Germany.' Having failed to interest Art Buchwald and Gore

Vidal in a project called 'Watergate Buggers', early in 1973 he enlisted Frantz Salieri, creator of La Grande Eugène cabaret in Paris, who devised a synopsis involving songs, dances, 'shocks', burlesque – Kissinger as the Merry Widow opening the ball of a new Congress of Vienna, etc. Losey envisaged a budget of only $640,000, but had no luck in raising money. To Salieri he denounced 'unimaginative flaccid mercenaries' and their 'stupid estimates of financial returns'. In 1976 he expressed enthusiasm for a highly political screenplay by Clayton Forhman, *Veteran's Day*. 'I have seldom been so completely and uncritically taken by a screenplay . . .' But, he wondered, 'Will anyone go any longer to see a film about Vietnam?' The agent George Litto advised him to forget it – the script was much too angry.[12]

He felt strongly about the right-wing coup which overthrew the Allende Government in Chile in 1973. The collusion of Henry Kissinger and the CIA guaranteed his anger – the enduring personal heritage of his post-war opposition to the Truman Doctrine, the *pax americana*. Rejecting the Chilean Critics' Award for 1973, he asked the British Council's Representative in Santiago to return it; he supported both the Arts Festival for Democracy in Chile and an underground paper.[13]

Art and industry: Losey fought against the industry but it was also lodged in his own bloodstream. He was no Bill Douglas. Losey's mode of operation (art, industry, politics, ambition, personal integrity) is vividly illuminated by an episode which might go by the working title, 'Gandhi with Caviar'.

Fereydoun Hoveyda, a francophone Iranian film critic, had flattered Losey's *Eva* in *Cahiers du Cinéma*: 'For the first time in cinema, the essence is found entirely in the form.' Brother of the Shah's Prime Minister, Hoveyda left Paris in the mid 1960s to take up a career as a diplomat and Iranian Ambassador to the United Nations – but found time to send Losey a flatteringly structuralist appreciation of *Secret Ceremony* under the letterhead of the Imperial Ministry of Foreign Affairs; he was also responsible for the Loseys' invitation to visit Iran in 1973. Hot from the Cannes Festival, they flew from Nice to Tehran via Rome, arriving at the Intercontinental Hotel on 13 May. 'Air Iran very comfortable, *caviar*, good food and drink, wonderful fresh vegetables and salad,' Patricia noted. At the airport they were met by Iradj Amini and an aide to the Prime Minister, Mr Shadri, plus a guide and TV cameras. Mr Shadri took them to lunch: yoghurt with cucumber; boiled rice with fresh butter and yolk of egg; lamb with tomatoes; spices; Persian ice cream full of lumps of curd. 'Adored the whole meal. For supper we had caviar,' Patricia noted again.

The key intermediary was Iradj Amini, an anglophone Foreign Ministry official recently transferred as a special adviser to Princess Ashraf and to the new Société Anonyme Cinématographique Iran. On 25 May the Loseys arrived at Ramsar after a five-hour drive across the desert and over the Elburz mountains. After eating sturgeon, they bathed in the Caspian and visited Amini's house, 'a perfect paradise'. That night the Loseys visited the casino. Joe was drunk and gambled heavily. 'In the end I was stealing his chips and changing them back into money. The croupiers were terribly upset,' Patricia noted. Back to Tehran, then Isfahan (19 May), sightseeing and the usual sunbathing. On 25 May Losey presented a prize at the Iran Academy awards, gave a press conference, and dined with Princess Ashraf.

Having accompanied the Princess to India and Nepal, Iradj Amini visited the Loseys while passing through London in September 1973. Losey subsequently sent him a copy of Robert Bolt's *Gandhi* script, although Bolt was tied in with Richard Attenborough as fundraiser and potential director. One third of the estimated $7 million budget had been promised by the Indian Government. Losey told John Heyman that Prime Minister Indira Gandhi knew of his interest in directing the film; as for Attenborough, he could remain 'associated' with the production. Losey advised Amini that 'this would be a tremendous prestige proposition for Iran . . .' Amini was polite about the Bolt script; Dr Mehdi Boushehri, the husband of Princess Ashraf and the official film mogul, had mentioned the project to 'His Majesty'. Maybe Iran would invest 1.5 million dollars . . . Maybe. At the end of the year, Bolt replied to an inquiry from Losey: 'Dickie [Attenborough] thinks that he has raised the money for himself to direct it . . .' Losey put Gandhi behind him, later describing Attenborough to Nadine Gordimer as 'a mediocre actor' and 'political idiot . . .' Fereydoun Hoveyda was still serving as UN Ambassador when the Iranian revolution occurred in 1979. His uncle was publicly humiliated and executed. Hoveyda received from Losey a cable of commiseration: 'Terrible for him and terrible for you.' For the Loseys there would be no more caviar dinners with Princess Ashraf.[14]

In September 1979 Losey met Nadine Gordimer in Paris; while keen to film one or other of her novels, he was worried about finding alternative locations to South Africa. Surely – he asked – the blacks differed in appearance 'from tribe to tribe and certainly from country to country?' Surely American or European blacks did not look like South African blacks? Gordimer, meanwhile, was urging him to visit the Cape Town and Durban Film Festivals, where *La Truite* and other Losey films were to be shown. He hesitated: 'I'm not sure, in terms of my position about racism, whether I should go at all . . .' He didn't.[15]

29

The Assassination of Trotsky

The great Bolshevik revolutionary Leon Trotsky was murdered in Mexico City in August 1940 by a Stalinist agent, Ramon Mercader, alias Jacques Mornard, alias Frank Jacson. The bloody fate of such a titanic protagonist might call for a Shakespearian vision of tragedy: the nemesis of the wordy Cassius or, more to the point, the fatal pride of Lear. The deadly choice in a political drama is to grab the bag of documentary history, a choice rendered more tempting in the age of the newsreel. Losey was sixty-two when he directed *The Assassination of Trotsky* – the deadly choice was the one he made. The genuinely political artist (Brecht, Resnais) sees beyond politics to metaphor or its cousin, satire; Losey had attempted this in his first feature film but could no longer raise the imaginative energy required by 'The Red with Green Hair'. Instead he took refuge in 'objectivity', 'facts', Mexican murals, and two male stars.

His first script, commissioned by the producer Josef Shaftel and written by Losey's fellow-blacklistee Ian McLellan Hunter, was grittily political in the Hollywood idiom, an awkward hybrid. The virtue of this screenplay was its rooting of Trotsky's assassination in the fierce political passions aroused by the Spanish Civil War; its vice was a thriller-genre packed with gunplay, police investigations and the occasional striptease show. Some (non-consecutive) examples of dialogue:

GITA: Yeah. I'm a Marxist . . .

FELIPE: He'll fall, Siqueiros. He'll fall.

FELIPE: The hardware has arrived. We're ready to go.

JACQUES: I know the score . . . History swept Trotsky into this backwater, and the tide carried me with him. Along with Gita . . . and Sheldon Harte . . . and Siqueiros . . . and a lot of other people. Helpless – all of us. If I'd known any way out, I'd probably have taken it.

Working with Losey in London, by December 1970 Hunter was suffering from depression and anxiety. Taken to the Harley Street Clinic, he discharged himself almost immediately, returned to New York, and wrote Losey a long, moving letter describing the history of his own (broadly medical) problems. In February Losey informed Hunter that he'd conferred in Paris with Alain

Delon, who disliked the script and its portrayal of Trotsky's assassin. In fact Delon had already signed on when Losey fired Hunter and turned to Nicholas Mosley, urging him to avoid Marxist jargon, which actors could never handle. The outcome was an assassin, Frank Jacson (alias Jacques Mornard), abstracted into a tormented, psychotic figure half-way into the underworld and lacking any clear motivation.

Invalided by a car accident, the author of *Accident* read Isaac Deutscher's biography of Trotsky while writing the script in bed. Director and writer headed straight for the cul-de-sac of History. Mosley's preferred solution was dramatized flashbacks filmed in old, grainy black-and-white. Losey agreed in principle to mock-ups but soon preferred documentary historical 'flashes' transferred to 35mm film: Trotsky storming the Winter Palace, Trotsky in Red Army uniform strutting in front of prisoners facing the firing squad; Lenin lying in state; Hitler's Stukas diving over Spain; Picasso's *Guernica*. In short, The March of Time. By 28 October 1971, with shooting almost completed, Losey had retreated to yet another documentary solution – a list of potential 'flash cuts' (still-shots, as in *King and Country*). For example, when the murdered bodyguard Sheldon Harte is identified by Trotsky in the hospital morgue, Losey planned to cut-in a flash of Lenin lying in state, 'with perhaps Stalin crying over the body'. Finally Losey scrapped the flash cut-ins as well. The burden of History was now purely literary – the loading of texts and quotations into Trotsky's mouth like herrings into a penguin. Losey had come full circle to the Marxist jargon he'd wished to avoid. Today Mosley still regrets the decision not to use historical flashes. 'I think all those people gathered around him bullied him into making a two-dimensional film: as politicians would!'[1]

This raises the question: was Losey inhibited politically or merely sterile dramatically? Recent claims by Frederic Raphael that he capitulated to subtle, or not-subtle, threats from old Stalinist colleagues can be discounted. Yet Losey was anxious to assure his old comrades that he wasn't betraying ancient loyalties. He promised Jeanette Pepper and Frederick V. Field, both residents of Mexico City, whose local help he might need, that the film would be 'neither anti-Soviet nor anti-socialist . . . Trotsky was and will be shown to be a boring, bullying egotist, nonetheless by virtue of the deed I presume he will acquire some sympathy.' This said, he headed straight for the Trotskyist camp in search of documentary validation. Trotsky's grandson, Esteban (Seva) Volkov, was fourteen years old and living with his grandparents when Trotsky was assassinated in his fortified house in the Calle Viena, Coyocan. Losey described his first visit to the Trotsky house in July 1971. 'First I was allowed to the door, then into the garden. Volkov was suspicious and, I think, also had been warned against me. He said, "Are you a Communist?" I said, "No, I'm not." He asked, "Were you a Communist?" I said, "Yes, I was."'[2]

Losey's designer, Richard Macdonald, who also visited Volkov, took a series of rapid Polaroid pictures, on the basis of which he would rebuild the Trotsky home in Rome's Cinecittà.

Volkov was both amenable and resolutely loyal to his grandfather. In May 1940, three months before his assassination, Trotsky had been attacked in his house by a Stalinist death squad led by the renowned mural painter David Alfaro Siqueiros. How had Trotsky and his wife Natalya survived in their bullet-ridden bedroom? Hunter's neo-Stalinist script indicated an advance tip-off; in short, Trotsky had allowed the attack to take place, without forewarning the police, in a bid for favourable publicity. On 18 August, Volkov wrote to Losey complaining that Mosley's script still adhered to this 'Stalinist' interpretation, making 'liars out of my grandfather and myself'. He also alleged that the script demeaned Trotsky – 'bragging and vulgarity'. Volkov was touchingly anxious to convince Losey about every detail, from bullet-holes to pet rabbits; from the size of Trotsky's brain to the exact angle at which the pickaxe entered it. On 26 November Volkov was back at it: 'The testimony of the victim's guards has been majestically ignored by you . . . your intent to falsify the facts of the murder . . .' He reminded Losey that his grandfather was 'the founder of the Red Army' – and 'you can in no way count on my silence'. On 14 December Losey rejected Volkov's complaints but further objections arrived from two experts to whom Losey sent the script, Marguerite Bonnet (a personal friend of Natalya Trotsky) and Jean Malaquais: the script was accused of 'blurred political perspective . . . like a mystery story motivated by a personal feud . . .' And 'Trotsky always wore gloves when tending to the rabbits.'

Losey's assistant, Bertrand Tavernier, identified the central problem, which had nothing to do with rabbits: '. . . the thing which is lacking in the script is your relationship with Trotsky. It is impossible to know if you are for him or against him . . . I would like the picture to be more involved.' Tavernier explains that he wanted the film to be 'about Joe' – about a victim of persecution whose perspectives had changed. Tavernier's instincts were sound but Losey was stuck in the history rut and Tavernier was dispatched to film libraries for documentary material; after a discussion with Trotsky's secretary, Pierre Frank, he suggested inclusion of Trotsky's attack on Marshal Pétain in his essay, 'Bonapartism and Fascism'. Losey's reply was weary: 'I can absorb no more and I believe practically nobody.' On 18 November, he wrote to Mosley from Rome with a virtual confession of defeat: 'What political statement will the picture make? . . . it may turn out to be no statement at all or, even worse, an essentially nihilistic one.' This long letter contains an astonishing personal attack on his chosen screenwriter. Hearing that the producer of the film, Josef Shaftel, had commissioned Mosley to write a quick book about Trotsky, Losey was furious: 'Firstly . . . Secondly . . .

thirdly, and forgive me for bringing this up, you have sprung from the loins of whom you have, and it is one thing for you to write a screenplay, and quite another to write a book about such a controversial figure; fourthly . . . fifthly . . .' Mosley replied with forebearance: 'I don't agree that a good book about Trotsky can't be written by the son of a fascist.'

Losey had buried himself in documentation but there was one private documentary film, of which he and Mosley were unaware, which might have animated their portrait of Trotsky. A hand-held film of Trotsky's domestic life in 1939–40, made by Alex Buchman, an electrical engineer who had been commissioned to instal a new security system, shows Trotsky as extremely expressive and lively in his gestures while talking or orating – he spat out ideas like a hail of cherry-stones, whereas Richard Burton's performance was grave and groomed. Burton dictated at 33rpm, Trotsky at 78. In Buchman's film we see him digging for new species of cacti (a passion), picnicking, his hair up in the air, fishing from a boat – echoing the younger Trotsky as hunter, soldier, buccaneer. The morbid stasis of the Burton/Trotsky compound in the film lacked this counterpoint.[3]

Officially a Franco-Italian production, *The Assassination of Trotsky* was financed by Josef Shaftel on the basis of territorial advances, including $440,000 in goods, services and cash facilities from Dino de Laurentiis Cinematografica. Losey's contract with Shaftel Productions was for $125,000, of which $25,000 was payable on signature, plus 20 per cent of the gross after deduction of 'negative cost'. Not believing in percentages, Losey would have settled for 10 per cent if the fee had been raised to $175,000 – his own original asking price being $200,000. As a producer, Shaftel was the automatic focus of Losey's animosity. On one occasion, visiting Royal Avenue, Nicholas Mosley listened to him abusing Shaftel over the telephone. 'I'm not fucking well coming to talk to you. If you want to talk to me you can shove your ass and come here.' Putting the receiver down, Losey smiled his 'demon king smile': 'If you want to see how the film industry works, stay around,' he advised Mosley. When Shaftel arrived – a quiet, self-contained man, as Mosley recalls – Losey continued the abuse, threatening to walk out. 'I'm finished with you!' Shaftel didn't turn a hair. When he departed, Joe gave Mosley another beaming smile: 'Now we can go and eat.'[4]

Alain Delon's participation held the key to the project, guaranteeing a big pre-sale in France. On 5 February 1971, Losey had expressed his gratitude: 'I was overwhelmed by your acceptance, considering your continuing uncertainty about the role [of Frank Jacson]. I hope I can join Visconti's company in proving to be worthy of your total confidence.' Italian actors, notably Valentina Cortese and Giorgio Albertazzi, guaranteed a pre-sale in Italy. On 18 March, Losey wrote to Romy Schneider about the part of Sylvia Ageloff

(Gita in the film): 'I think this is not a part for you. In the first place, the girl remains a big problem in the script and also she should be a Brooklyn American . . .' Schneider, too, expressed doubts about the script – Gita's attitude to Jacson, the sexual relationship, and Gita's position vis-à-vis Trotsky himself. Why Losey changed his mind is not clear – their tender yet fraught relationship is described elsewhere.[5]

The casting of Trotsky himself became a comedy. Bogarde declined; on 22 March Losey wrote to Marlon Brando, 'I hope to God you can do Trotsky.' Three days later he approached Burton. When Burton decided that he couldn't fit the film into his schedule, Losey approached Topol. In July, addressing a press conference in Brussels, he anticipated that George C. Scott would play Trotsky. Not until early August did Burton finally agree to take it on. On the 19th, Losey left Rome for Yugoslavia to talk to the actor, flying by Fokker to Dubrovnik, then by helicopter to Split. The following week he left for Mexico.

Should the film be shot mainly on location or in the studio? And in what country – Mexico, Spain (where censorship intervened), Italy? The great Marxist murals were of course in Mexico but the Mexican authorities were at first negative. Losey sent Richard Macdonald on a secret reconnaissance to Mexico City, where he was able to view the murals in situ, particularly those of the Communist painter who'd led the 24 May assault on the Trotsky compound, David Alfaro Siqueiros, 'by far the most visual and cinematic . . . also more integral to the script because of Siqueiros's personal involvement,' Macdonald recalls. Inspired by Macdonald's report, on 25 June Losey sent a new plea to Hiram Garcia Borja ('*Sufragio Efectivo, No Reelección El Director General*'), recalling Borja's previous hospitality. Borja's response was positive, and Losey himself arrived in Mexico on 9 July for the ten-day visit during which he met both Siqueiros and Volkov, and lunched at Luis Buñuel's home. On 31 July, the Mexican authorities set out their conditions. A number of changes in the script were demanded:

1. No drunken people after the parade of May 1st (Scene 13).
2. Eliminate the words where Felipe says to Jacson that Trotsky brings the President [of Mexico] 'By the ear' (Scene 38).
3. Felipe will not mention the President of Mexico when he speaks of Trotsky's power to persuade people (Scene 38).
4. Eliminate the words of Trotsky, saying that the Aztecs tore the hearts out of their still living victims to study the cardial affects of altitude (Scene 70).
5. No scenes of shanty towns in the outskirts of Mexico City.

If these conditions were met, the authorities agreed to the immediate release of the negative film shot and developed in Mexico 'after supervising "rushes" and verifying that you have complied with this office's dispositions'. By 16 August a deal had been struck.

But David Alfaro Siqueiros, whose murals Losey coveted for the film, was the hardest nut to crack. After the abortive attempt on Trotsky's life of 24 May 1940, Siqueiros had taken refuge among miners until betrayed to the police and put on trial in October 1940. At his trial he denied any link with the Soviet GPU, but insisted that Trotsky was a counter-revolutionary, 'hated by millions' and number one traitor to the proletariat since 1929. Siqueiros lied about the murder of Trotsky's secretary/bodyguard Sheldon Harte, when questioned by Judge Raul Carranca Trujillo, explaining his own evasions and silences in terms of revolutionary duty. Siqueiros denied ever having met Jacson/Mornard, the eventual assassin, despite evidence that Mornard had used his apartment as a safe house. When Siqueiros's wife, Angelica Arenal, testified in court she, too, answered some questions but ignored others. Siqueiros jumped bail and fled the country. When he returned twenty years later he was promptly imprisoned. At this juncture Losey himself has an unsolicited walk-on part. In December 1960 he received an appeal from Angelica Arenal de Siqueiros on behalf of her husband, then sixty-three. Addressing herself to fellow artists and intellectuals, she couched her appeal in the standard 'progressive' phraseology, evading the real reasons for Siqueiros's imprisonment: armed assault, attempted murder, murder, contempt of court in 1940. Ironically, Losey knew nothing of the case, nor did he want to; explaining his reluctance to intervene as 'a suspect alien' in England, Losey urged James Archibald, Kenneth Tynan and Sir Leslie (Dick) Plummer, MP, to write to the Press on Siqueiros's behalf.[6]

Now, eleven years later, Losey met Siqueiros and his wife in Mexico City. Whether his silence in 1960 counted against him we don't know – it cannot have helped. After the encounter, on 8 August, Losey wrote to the painter, addressing him as 'Maestro': 'I consider my afternoon with you one of the most memorable of an already fairly long life . . . Your work is dynamic dialectic – so at best I hope is mine . . . I believe we are both caught in the same corruptions and distortions of big business vs personal drive, purity, conviction, isolation . . . I could show a vast public what you are about . . . The script is not so superficial as you think.' Losey ended: 'I adore your tough angel, Angelica. Compared to her you and I are sentimentalists.' Clearly the tough angel had given Losey a hard time about the film and its capitalist backers. Losey's answer to her was remarkable: 'Yes, but to refuse to work in the system is to be Trotskyite, isn't it. And then we are all dead. To me as much as to you . . . Trotsky was in his effect an enemy . . .' Thus Losey spoke to Volkov and Siqueiros with a forked tongue. Siqueiros did not reply; his murals, *Revolt Against the Dictatorship of Profino* and *The Struggle of the Revolutionaries*, had to be eliminated from the film. In Mosley's revised scripts the Communist artist-commando is called Manuel Antonio Ruiz. A year after their meeting Losey learned that Siqueiros had recently been expelled from the Mexican CP after defending the Soviet invasion of Czechoslovakia, despite its condemnation by the Spanish CP in exile.

The Loseys left Rome for Mexico via New York on 25 August, visiting Cuernavaca, returning to Mexico City on 1 October, then travelling to Camino Real. The director of photography was Pasquale de Santis, operator on *Eve*, winner of the American Academy Cinematographic Award and, more recently, the top Cinematographic Award at Cannes in 1971 for *Death in Venice*. The unit began by focusing on murals, establishing shots of the Bell Tower and Lake, street scenes, Zocalo Square 'on May Day', the university, the cathedral. At a height of five feet, close to the first bulwark on the roof of the cathedral, at the back of the main cupola over the nave, Losey directed a 180-degree pan, using a zoom lens set at 35mm, then at the end of the pan slowly zooming in to 50mm through the windows of the dome to the crucifix at the front of the cathedral. Despite the presence of a representative of the censor on the set at all times, and her scrutiny of the rushes, Losey preferred the censor to the producer. When Shaftel insisted on visiting the rooftop where Losey was shooting, Losey stopped work at once. He had only had to hear Shaftel's name to explode.[7]

On 1 October Losey left Mexico for Rome, where Trotsky's house had been reconstructed by Macdonald with uncanny accuracy, bullet-holes and all. The interior scenes in the Montejo Hotel between Delon and Schneider were also shot in Rome. The Burton–Taylor yacht duly anchored in the port. On 28 October, Burton wrote in his diary: 'Joe is definitely not himself. He didn't seem to know the script as well as he usually does. Time and again I, or the continuity girl, have to remind him of things that are very obvious.' (In Norman Priggen's recall, Burton, having 'signed the pledge', almost, was so diligent in studying the script, so unrelaxed on set, and so controlled in his performance that Losey used to say, 'I wish he'd have a drink.') The Burtons were generous and concerned in their scribbled notes to Losey – get well notes, don't overdo it notes, take care notes, all suggesting that Losey was visibly under stress. In Burton's hand, on Grand Hotel Roma notepaper: 'Take it easy. Would you and Pat prefer to stay on the bateau.' On a torn-off piece of graph paper Taylor scribbled: 'Dear Joe, I think your film is absolutely marvelous and I think my Old Man is too! I wish that someday [one word] you could find a job for me – you are a great director and I just adore you.' Burton congratulated him more formally. 'Your brillance as a film-maker is self-evident . . . it compels one's imagination for many hours after the event.' He praised Delon's performance. 'I think that even I act with a most precise precision. There is a delicacy and frailty in my performance that I didn't think I was capable of and a reduction of my voice to its basic truth rather than a noble suggestion . . .' Burton also took a casual sideways swipe at 'Stanley Baker and Dirk Bogarde and others who chose with great taste the right man [Losey] to body forth their indifferent talents.' The Loseys travelled to Budapest for Taylor's birthday party at the Duna Intercontinental Hotel on 27 February. A note from Taylor thanked them for the gift: 'Darling Tapricia and Joe, My wee beastie is not only divine but I have the feeling it will bring me good luck and keep guard over me with his hideous wings.'[8]

What the film lacks is the guiding metaphor which inspires the singular sight-line that Losey achieved in *Accident*, in *The Go-Between* and in *Mr. Klein*. It was badly received by both audience and critics at the Edinburgh Festival in August 1972. Trotskyists protested outside and inside the Cameo cinema. Losey agreed to meet them after the screening, then decided to take the train to London instead. The film was reviewed in *Workers' Press* (28 August) by Roy Battersby, under the title 'The Second Assassination of Trotsky'. Battersby complained that the film reduced the assassin – and also Siqueiros – to psychologically motivated individuals rather than obedient agents of Stalin's GPU. The bullfight was mere titillating violence in the Hemingway tradition, and the whole thing reeked of the sterile philistinism that bewails man as half-devil, half-angel.

Anglo-EMI launched the film in October at the ABC Shaftesbury Avenue. The London reviews were the worst that a Losey film had ever received. Alexander Walker, of the *Evening Standard* (10 October), found Burton dull and the dialogue given to Trotsky unspeakable. Losey's comparison of Trotsky with the dying bull was 'crudely inept'. In *The Times* (13 October), Ronald Hayman roasted the film. Mosley's screenplay was 'naïve'. He and Losey had not found a way to convey Trotsky's career other than by dictating memoirs and such phrases as 'You should have been with us when we stormed the Winter Palace'. Talking to Michel Ciment five years later, Losey blamed the screenplay for the film's failings: '. . . the trouble is that he [Mosley] is not a real political theoretician, and he didn't know anything about Trotsky . . . all of the Trotsky political family mistrusted him . . . he was the wrong one for this.'[9]

The film was shown at the New York Film Festival on 13 October 1972 and opened at the Coronet Theatre later that week, distributed by Cinerama. Losey reported to Delon: 'I found the atmosphere much more welcome . . . than in England, where the reaction has been disastrous . . . subjective and stupid.' But Helen Dudar – who found Losey prowling his Algonquin suite, clutching his inhaler and longing for a drink early in the day – reported in the *Post* (21 October) that the Festival audience in Tully Hall had hissed and booed – she suspected the reason was aesthetic rather than political. A month later Vincent Canby lamented in the *Times* that the film had barely made it through a three-week first run at the Coronet. Canby urged his readers to hurry to the second-run theatres: 'Like great, primal figures in an Orozco work, the characters in the film are always moving towards a goal that fate has chosen for them.' Here was a film of style, intelligence and genuine historical interest. On 22 February 1973, Losey announced absolute destruction by the American critics. Arthur Manson, of Cinerama, reported on 7 March: 'It may seem that *Trotsky* sunk without trace, however, it did open in many key markets, but unfortunately with disastrous results.'

In July 1972 Losey wrote to Burton about his hopes of shooting *Under the Volcano*. The Burtons (both on the wagon) came to supper at Royal Avenue on 8 September, and were given Greek Shepherd's pie, which Burton likened to Welsh food. Hearing that the marriage was in difficulty, Losey wrote to Burton a year later: 'I will not talk about the present situation with Elizabeth . . . I will only say, tragically sad and mistaken.' Burton answered by telegram: 'You are too old a hand to believe what you read in the papers Elizabeth returns here to Rome next Friday . . . it was simply a burst of disillusioned inebriation am sober slim and beautiful . . .' By the summer of 1974 Losey's last major project with Burton, *Under the Volcano*, was dead and Burton was again failing to acknowledge his letters. A hint of Losey's asperity is found in a letter to Patrick White about the casting of a particular actor for *Voss*: 'He is as much of a drunk as Burton and even more difficult to handle.' The Burtons were divorced. On 28 March 1975, the Loseys were invited by Taylor to a sit-down dinner in the Dorchester. Taylor arrived two-and-a-half hours late. 'Naturally [Joe] was next to her and we left after main course served,' Patricia noted in her diary. And there it all evaporated. In October 1983 Losey wrote to Taylor proposing a Home Box Office TV film starring herself and Jeanne Moreau. Taylor expressed interest. Two months earlier he'd told a friend: 'I have known many child actors and worked with many and have seen how their lives, no matter how successful, are later ruined, like, for instance, the most extreme example, Elizabeth Taylor . . .'[10]

30

A Doll's House:
Filming Women

A Doll's House

In Ibsen's play *A Doll's House*, a loyal and dutiful wife and mother, Nora, finally walks out and slams the front door, her chauvinist husband's doll no longer. Losey's *A Doll's House* is a minor masterpiece and by no means a filmed play. David Mercer's script reworks the text into the shorter, cross-cutting scenes required by film; Hirsch calls it 'scrupulous cinematic realism'. Mercer opened up the text by introducing early scenes which take place before the action of Ibsen's play, thus removing the dramatic burden of contrived exposition and recall on which the dramatist relied. For this, Losey later took the credit: 'Having studied the play, I found . . . that Ibsen was terribly verbally expository . . . So I made a lot of very drastic cuts and tried to put all the expository elements into a kind of prelude at the skating rink.'[1] The action was extended to embrace Nora's children, Christmas revels, Krogstad playing with his children in the snow, Krogstad shadowing Nora while she's out walking – and we glimpse the shabby dwelling in which he fathers his motherless children.

Richard Macdonald being unavailable, the designer was Eileen Diss, with whom Losey travelled to Norway in search of 'an untouched town', Roros. 'And I used the townspeople, the bank manager, the school teacher, the doctor's wife, the grocer. All the townspeople participated as extras on the street, in the restaurant, at the party . . .'[2] From start to finish, Losey did not enter a studio. Co-financed by John Heyman's World Film Services and Tomorrow Entertainment, the film version cost a modest $912,850. His fee as producer/director was $40,000, plus $10,000 paid on account of a guaranteed 1.5 per cent of gross receipts.

The making of *A Doll's House* generated conflicts no less intense than those depicted within it. Whereas Nora and Torvald conducted their struggle within nineteenth-century domestic intimacy, Losey and Fonda eventually slugged it out under the searchlight of twentieth-century publicity. Losey did the lion's share of the slugging and slagging. Their working relationship began with friendly exchanges. Losey had been considering Fonda for *Under the Volcano* when he offered her the part of Nora on 31 May 1972: 'Congratulations and coraggio for your life and work.' Nervous about her politics, John

Heyman warned Fonda that Tomorrow Entertainment, which was co-financing the film, was a subsidiary of the major arms manufacturer, General Electric. According to Heyman, 'Fonda giggled and said she was a GE shareholder.' On 1 August, Losey reported to a colleague that although she could currently command $1.25 million a picture, she was prepared to play Nora for $30,000, plus 15 per cent of the producer's gross, four-and-a-half first-class round trip tickets, custom designs and a visit from the dress designer. But Fonda visited North Vietnam in July and was then heavily involved in the American elections. Uncertain of her and set on engaging a politically radical actress for the part, Losey also approached Shirley MacLaine, praising her performance at the Democratic Convention.[3]

Losey saw F. T. A. (i.e. 'Fuck the Army'), the movie of the Fonda/Donald Sutherland anti-war troupe, and wrote to congratulate her: '. . . a brilliant achievement . . . fresh, tough in the right way, vital, informed and informing . . . neither amateurish nor slick'. By the time Fonda travelled to Norway, Nixon had won a landslide victory over McGovern. Meeting Delphine Seyrig in Paris, her attention was diverted to the doll's-house war. A 'graduate' of Lee Strasberg's Actors' Studio, Seyrig had performed in New York, London and Paris theatres (twice in plays by Pinter), while also winning acclaim in films by Resnais (Marienbad and Muriel), Buñuel and Truffaut – and a cameo part in Accident, which she praised as beautiful, masterful and Chekhovian. In 1970 she stayed with the Loseys in Norfolk. 'What an extraordinary actress you are,' Losey told her on 1 June 1972, 'and . . . what a joy it is to work with you.' Both Fonda and Seyrig brought forceful aides on location: Seyrig's was Michele Richer, a co-activist on abortion platforms, while Fonda's team included Nancy Ellen Dowd, who had worked with her on F. T. A.. Richard Dalton, production manager on the film, recalls that when Fonda reached Oslo she immediately telephoned Losey and told him exactly what was wrong with David Mercer's script. Dalton put the anxious Mercer on 'a diet of milk' while Fonda publicly shredded his work.[4]

The Losey archive contains some of the critical script notes presented by the women. The general line was that Ibsen had been betrayed. A particular target was Mercer's (non-Ibsen) prologue: 'As for the opening scenes; they lack the tightness, the inevitability, the undercurrent of rigorousness and urgency which inform the play . . .' A curious feature of their critique was insistence that Mercer had minimized Krogstad's malevolence – curious because, however one reads Krogstad's motives, they have nothing to do with male chauvinism. Indeed, the women seem to have toppled into the naïve view that the individual virtues of Ibsen's female characters and the vices of his male characters should be banged home – 'naïve' because, if one adopts a close analogy, the evils of colonialism are best exposed when the oppressive attitudes belong to decent-enough whites rather than murderous bullies.

Thus one reads in these notes that the huge strains on Nora, the risks she took on behalf of her ill husband, should be highlighted, while, by contrast, Torvald should from the outset be shown as 'a terrible patient, a sniveler'. Torvald was too generously portrayed when one considered 'the unyielding imperiousness and pathological ambitiousness he exhibits all through the play'. But the feminist perspective is eroded, not enhanced, if one concludes that Nora had merely married the wrong man. A classic, early-1970s, note complained: 'In the original, Nora describes the presents she has bought for the children. "Here's a doll and a doll's bed for Emmy. They're rather plain but she'll soon smash them to bits anyway." Why omit such a great tribute to the disgust little girls have for their demeaning toys?' A long list of further challenges followed, of which a few:

'Why has Nora's "You're like all the rest – you think I can do nothing worthwhile" been omitted?'

'Why omit "wife can't borrow money without husband's consent"?'

'Why omit the fact that Nora has scrimped and saved to keep the payments up? . . . In the screenplay there is no reference to the sewing and copying jobs she has held for over seven years.'

'Why omit Nora's "dynamite line" in which she says that working to earn money was "almost like being a man"?'

Mercer's own feminism was invariably discounted as 'another clumsy sop to the women's movement', for example, 'Torvald's last remark about "tract-bemused women"'.

Clearly the British actors and film crew resented Fonda's behaviour, responding with condescension and male drinking covens. According to Gerry Fisher, the director of photography, each morning new pages would be issued by Fonda without consulting the other actors concerned. 'She arrived like a star pretending to be a non-star and occupied the biggest house in Roros in which all the lights burned day and night.' When she joined in conversation with the crew over a meal, she was crusading and obsessional – for example, acupuncture is the only way, the medical profession is a fraud. Fisher's clapper-loader, Steve Smith, found a book of quotations; each day a new one appeared on the side of the camera. The wit of these quotations is not obvious: 'Cleopatra needed liberation like a hole in the head.' Fonda was not amused. 'Steve was winding Fonda up,' Dalton comments, but concedes that Fonda was at least 'up-front' in pressing her point of view, whereas Seyrig was 'the snake in the grass, the smiling cobra, and all this French lib and all this crap'. An example of Seyrig's 'French lib' is found in a letter to Losey from the Bergstadens Turisthotel, Roros, enclosing a quotation from George Sand, written in 1837: 'Never choose . . . a woman who is strong-minded, disinterested and honest. The public will despise her and call her by opprobrious names . . . unfit. Unfit, yes, thank heaven! Unfit for servility or baseness, unfit for fawning on or fearing men.' And more.[5]

Tensions came to the boil during the final party for the cast, the crew and the residents of Roros. No sign of Fonda and Seyrig. Dinner proceeded without them – Losey thanked the town and its good citizens. He then sat in the lobby with Patricia, Carlo Lastricati and Richard Macdonald drinking schnapps, getting drunker and angrier by the minute. When Fonda's party finally arrived one-and-a-half hours late, Losey dressed her down. Patricia commented in her diary: 'I thought his behaviour even ruder, as we then left the party – although by then there was little choice as Joe could hardly walk.'

A remarkable series of Losey interviews appeared in the press from February to April 1973. The actresses themselves had given on-set interviews which had been held in the journalistic pipeline for several months. Anthony Haden-Guest's report in *Vogue* (1 March) quoted Fonda: 'The contradictions that exist in the production are exactly part of what the content of the play is about.' Seyrig was also quoted: 'I think women have no power at all in films. As actresses or anything else. I wish I weren't in this movie.' Interviewed by Keith Howes for *She*, Losey described Seyrig as 'a woman who uses every female wile and every seductress device while mouthing Women's Lib phrases. She wants it both ways.'[6] On 10 March the *Daily Mail* published a Losey interview with David Lewin. Fonda (he said) wasted time and energy on too many causes, she was unaware of other people, cruel even. 'The problem with Jane is that like most zealots she has little sense of humour.' He went further: 'In a love scene on set, she will kiss energetically, but off set she did not speak to any of the men and she shrinks from anyone touching her with the exception of her new husband Tom Hayden.' On 12 March, Jesse Birnbaum reported in *Time* – 'Oh, You Militant Doll' – that Fonda had wanted the film company to put Nancy Ellen Dowd – 'Fonda's full-time ideologue for women's rights' – on the payroll at $350 a week. Dowd was 'the sort of girl who goes around flashing her well-fingered copy of Ellen Frankfort's *Vaginal Politics*'. (She did or she didn't? – maybe she 'sort of' didn't.) Worse yet: 'So much did the two women [Fonda and Seyrig] kiss and touch each other before the camera that Director Losey had to complain about the unwarranted intrusion of lesbianism into the story.' Rex Reed published a full-page article in the New York *Sunday News* (1 April), which implied but did not explicitly claim an interview with Losey (to whom he later apologized): 'So Joseph Losey edits a *A Doll's House* in the London fog, Jane Fonda sulks in a Malibu beach house wondering what to do now that the war is over and there are no more soldiers to defect.' Losey began to panic, firing off letters of protest and slaughter to the journalists.

On 14 March Losey wrote to Fonda herself, 'with regret and shame and chagrin . . . I did indeed speak to all of the reporters concerned and some of the words are mine, but the intention is absolutely *not* mine . . . I feel that I have stupidly and unwittingly been used in what I can only regard as an

attempt to smear you for what I assume must be political reasons . . . If you
see Nancy Dowd would you apologize to her for me? I didn't ever say . . .
that we paid her $350 a week . . . I at no time referred to any warning about
lesbianism and a comment of mine that you were not a particularly tactile
person in my experience did not differentiate at all between male and female.'

Fonda's reply was not written until 27 July and came from the headquarters
of the Indo China Peace Campaign, Santa Monica. It was a demolition job:
'There were enough details within interviews . . . to make it quite clear that
you did say what you have been attributed as saying . . . I think what shocked
and angered me most is the irresponsibility on your part. The lack of
consciousness of the political ramifications of your attack (you who professed
sympathy with progressive movements) . . . The Right has used it against me
(is that what you intended?) and the Left is baffled, "wasn't he on the right
side back there in the '50s? What happened?" they ask. And so do I . . . your
inability to deal with, to countenance, strong women has done . . . irreparable
harm to the film . . . I was never able to penetrate your paranoia and
snobbery while we were working together . . .' A Fonda interview with Molly
Haskell appeared in the *Village Voice* eighteen months later, on 7 November
1974. Most of the men had been drunk all of the time. 'The strangest thing
happened: I found I had to become Nora with Losey, bat my eyelashes, and
make it seem as though it was his idea . . . a perfectly normal, well-disciplined
actress who had some ideas about the play, that he couldn't handle. And this
from a man who calls himself a progressive and a Marxist.'

The film was shown at Cannes in May 1973 as a '*sélection officielle*' outside
the competition, following some fretful messages from Losey to Favre le Bret,
Maurice Bessy and the Festival organizers. Distributed by British Lion, the
film opened at the Odeon, St Martin's Lane, on 4 July. Another filmed
version of the same play, directed by Patrick Garland and starring Claire
Bloom, had opened on 19 April. The British critics generally preferred Bloom
to Fonda, the American critics the reverse. In the *Evening Standard* (18
May), Alexander Walker dismissed Fonda as incapable of playing period
roles. 'The cheer-leader for Women's Lib would need a step ladder to get
within reaching distance of Ibsen's liberated woman . . . roles of Nora's
complexity are not mastered on the battlefields of Vietnam.' By 1 August the
film had been withdrawn. Delivering the John Player Lecture at the NFT,
Losey complained that he'd been promised six weeks in the West End, then
general release in fifty cinemas and twelve university cities. Instead, 'It was
dumped out at ten days notice. There are no fifty theatres and no bookings
until October.'

The US première took place at the New York Film Festival. Reporting
from Cannes in the *Village Voice* (24 May), Andrew Sarris preferred the
Losey version to Patrick Garland's because it was 'a real movie, with an acute

sense of period and place'. Nora Sayre, writing in the *New York Times* (2 October), embraced Fonda while hitting the film. 'The Losey script by David Mercer has been fattened with feeble lines and even short scenes that the old genius didn't write.' Sayre didn't like the intercutting either – 'the dramatic momentum is butchered'. By contrast, Fonda's performance achieved 'the ringing gaiety and the energy that the role demands . . . Gradually and subtly, we are given a portrait of a political prisoner – one who hasn't ever tasted the air outside the walls.'

The film never reached the West Coast art theatres and was diverted straight into ABC's *Sunday Night at the Movies* (Channel 7). John Heyman recalls that 'Tomorrow Entertainment made the decision to recoup $350,000 of their investment for a one showing license to ABC, despite my requests that the film be shown theatrically first'. What with the snow and the Christmas party in the film, Christmas week 1973 was it. As of 31 December, the world gross was $787,740, but this was based less on actual box-office returns than on rounded 'minimum guarantee' sums, including $500,000 from the USA and $100,000 from France.[7]

The Players

Losey's casting of actresses reflected a faintly perverse dialogue with other directors he admired. Casting Melina Mercouri as an English gypsy was a personal challenge to Jules Dassin. It was Antonioni's Jeanne Moreau, rather than Truffaut's, that aroused his predatory instinct. Directing Monica Vitti in *Modesty Blaise*, he was determined to subvert Antonioni's grip on her image. When he cast Delphine Seyrig in *Accident* he invited her to the George V, praised Resnais's *Muriel*, and informed her that he wanted to discover whether she was an independent actress or merely a creature of Resnais. Losey's rivalry with Visconti encouraged him to reach out for Romy Schneider (as well as Delon and Helmut Berger).

Gerry Fisher, who photographed eight of Losey's films, noted that he was never fully comfortable with actresses – an ambivalence which mirrored his attitude to women in general. John Heyman is of the same opinion. In Losey's case, sexual attraction was a source of resentment; indeed, any actress whose role involved erotic encounters within the film was likely to be put down on the spot or demeaned at a later date. Immunity from derogation was guaranteed to '*grandes dames*' or actresses portraying mother-figures like Karen Morley, Peggy Ashcroft, Brenda Bruce, Valentina Cortese, Margaret Leighton and Renée Houston.

He let it be known that Wendy Craig was miscast in *The Servant* and Monica Vitti miscast in *Modesty Blaise*. He publicized his early clashes with Elizabeth Taylor (*Boom!*). Losey told the *New York Times* (17 November

1968) that Joanna Shimkus (*Boom!*) 'will never be an actress in a million years'. He described his 'sadistic' treatment of Mia Farrow when he started shooting *Secret Ceremony*, how he broke her down to build her up. In reality his personal fascination with Farrow's half-woman, half-child allure caused tension at Royal Avenue – his public remarks about her can be read as a form of punishment. Sometimes the condemnation was decently delayed until later years. He depicted Gail Russell (*The Lawless*) as helplessly drunk and frightened. Of Shirley Ann Field, whose tight sweaters adorned the posters for *The Damned*, he said that she was 'an imposition . . . whom I didn't want at all but had to have because she had just played with Olivier . . . so they thought she was a star. She's a nice girl but, at least in my experience, with very little talent. In fact, she couldn't speak a line, let's be frank. It was just awful.' He aired his despair about Viveca Lindfors's 'method' acting (*The Damned*) and related his chastisement of Sarah Miles (*The Servant*). Virna Lisi (*Eve*) also won his hostility: 'much as I dislike her . . .,' he told his agent. Jacqueline Sassard, whom he reduced to paralysis while shooting *Accident*, he called 'just a catalyst . . . that poor unfortunate girl . . . she had these eyes which were windows, which you could look through . . .'[8]

Here we may observe Losey's interaction with four sexually charged actresses of spirit and temperament, Jeanne Moreau, Viveca Lindfors, Monica Vitti and Romy Schneider.

Eve was Jeanne Moreau's twenty-eighth film. She had recently worked with Roger Vadim (*Les Liaisons dangereuses*, 1959), with Louis Malle, with Antonioni (*La notte*, *1960*), with Peter Brook (*Moderato cantabile*, 1960) and most recently with François Truffaut (*Jules et Jim*, 1961). Personally attractive as Losey found her, the main source of magnetism was what she represented: a major star prized by the leading directors of Europe, the incarnation of 'modern woman', alluring, mysterious and ultimately unattainable.

Shooting *Eve* in Italy, Losey was in a state of personal exaltation and neurosis, sloughing off his third marriage, pursuing a variety of affairs, '*un homme couvert de femmes*'. Stories abounded, products of a sexually electric climate in which actors were eagerly identified with their roles in the film. Hugo Butler, the discarded scriptwriter of *Eve*, heard on the grapevine that Losey had been waiting in the hotel lobby for Moreau one evening after dinner. She arrived with Stanley Baker. Collecting their room keys from the reception desk, Baker had said, 'Your room or mine?' – and off they went. Florence Malraux, Moreau's companion on *Eve*, confirms Moreau's relationship with Baker – par for the film-course at that time, subject closed. According to Losey, he, Baker and Moreau were passing down the Grand Canal in a water-taxi when Baker mentioned a homosexual dream from the previous night. Dirk Bogarde blows this up: 'I was told that Stanley was broken when Moreau told him he was a suppressed queer. Joe said she had deballed Stanley.'⁹

Moreau's loyal devotion to her director is illuminated by François Truffaut's comments about her acting personality: 'On the set she's willing to give a fast or slow performance, to be funny or sad, serious or zany, to do whatever the director asks of her. And in case of disaster, she stands by the captain of the ship . . .' Losey told Ciment that he had been 'in love with her as an actress. I was very much attracted to her – but nothing happened between us, just in case that needs to be made clear . . . There was a period after *Eve* when I felt betrayed by her, but she and I have a kind of permanent love which came out of that film.'¹⁰

The 'betrayal' was squarely on Losey terrain; she did *Le Marin de Gibraltar* for Tony Richardson, not for Joe. There was little contact between Losey and Moreau for some years, despite mutual interest in two screenplays by Marguerite Duras. When Moreau wrote to him in March 1968, they were clearly not in touch: 'Not seeing you for so many months, why not say years . . . I felt very shy about writing to you.' Perhaps some day they would make another film together, 'as in our lives this is the only way to see one another long enough'. On 13 May, the great day of victory for the students of the Sorbonne, Losey wrote to Marguerite Duras that 'Jeanne Moreau has come to see me here, and looks younger than at the time of *Eve*.'

On 15 December 1969 he had lunch with Moreau; that evening they both attended the prize-giving dinner of the Académie du Cinéma – including the presentation of the Crystal Star for Losey's *Cérémonie sécrète*. A few days later Losey wrote to Moreau in appreciation of 'a marvellous day'; he wanted his young artist friend Jack Loring to look at her living-room in the rue de Cirque, 'as it is very much the sort of thing I would like for my house and he may be doing some of the buying for me'. The following May she wrote from the Beverly Wilshire Hotel: 'I am very serious about working with you on *Torrents of Spring* or even something else . . . You know how much confidence I have in you.' (Losey had recently discussed with Oscar Lewenstein and John Osborne a film of the Turgenev story, but it didn't come off.) The letter ends with regards to 'your wife'. In 1972 Losey was president of the Cannes Festival jury. His private notes on the film *Chère Louise* say of its star, Jeanne Moreau: 'I would not have believed she could be so bad.' In May 1975 it was her turn to be president of the Cannes jury; Patricia noted that Moreau 'gave us a terrific welcome . . .' Soon afterwards she agreed to play the chatelaine Florence in *Mr. Klein*.¹¹

Contacts were frequent and pleasant during the Loseys' years in Paris. On finishing *Don Giovanni* the Loseys headed for a cruelly cold winter break at Moreau's house at La Garde-Freinet, not far from Saint-Tropez – but without Moreau. He told Marie Castro that he'd loved Moreau and had been

1 Losey's father, Joseph Walton Losey II, pre 1909

2 Losey's mother, Ina (née Higbee), 1906

3 Losey with (left to right) his maternal great-grandfather, grandfather, mother 1911-12 outside Higbee House, La Crosse, Wisconsin

4   Losey (right) on the set of his animated puppet film, *Pete Roleum and His Cousins*

5   Losey with the playwright Francis Faragoh and child actors, including Sidney Lumet, rehearsing *Sun Up, Sun Down*

6 Losey's second wife, Louise Stuart,
   as a Paramount actress

7 Losey in the mid-1940s

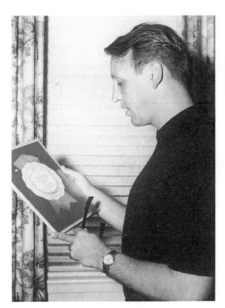

8 *The Boy with Green Hair*. Losey with
the Blue Ribbon Award for the Best Picture
   of the Month, February 1948

9 Portrait by Ernest A. Bachrach, 1948

10   Losey with the actor David Wayne on a street location for *M*

11   *The Big Night*. Losey with his assistant director, Robert Aldrich (centre) and the producer, Philip A. Waxman

12   Losey shooting *Time Without Pity*

13   Dorothy Bromiley at the time of her
marriage to Losey

14   Hardy Kruger and Micheline Presle in *Blind Date* (*Chance Meeting*)

15   *Blind Date* (*Chance Meeting*). Chelsea Embankment. Hardy Kruger (foreground), Micheline Presle beside Losey, David Deutsch behind Presle, Christopher Challis behind Deutsch

16   *Eve*. Losey, Jeanne Moreau and Stanley Baker

17   *The Criminal* (*The Concrete Jungle*) at
the Plaza Cinema, London

18   *The Servant*. Sarah Miles,
Dirk Bogarde

19 *The Servant*. Bogarde with Losey, haggard from pneumonia

20 *The Servant*. Losey with Richard Macdonald (centre)

21　*King & Country*. Tom Courtenay, Dirk Bogarde, Losey

22　*Modesty Blaise*. Losey and Monica
Vitti in Amsterdam

23   *Accident*. Delphine Seyrig, Dirk Bogarde

24   *Accident*. From left: Harold Pinter, Dirk Bogarde, Losey with Pamela Davies on platform, Norman Priggen seated below camera

25　Making *Secret Ceremony*. Left to right: Joshua Losey, Pamela Davies, Patricia Losey, Mia Farrow, Losey

26　*Secret Ceremony*. Losey with Elizabeth Taylor

27　*Secret Ceremony*. Losey with Mia Farrow

28   Losey with Michel Ciment in a Paris *lycée*, 1969

29   *The Go-Between*. Losey with Gerry Fisher

30　*The Go-Between*. Right to left: Losey, Richard Dalton (foreground), Gerry Fisher (behind camera), Mike Rutter (focus puller), Steve Smith (clapper, loader)

31　*The Go-Between*. Left to right: Losey, Pamela Davies, Gerry Fisher, Mike Rutter, Alan Bates, Dominic Guard

32 *The Assassination of Trotsky*. Losey with Alain Delon

33 *A Doll's House*. Losey with Jane Fonda

34 *The Romantic Englishwoman.* Helmut Berger, Losey, Gerry Fisher, Glenda Jackson

35 Losey on location for *Mr. Klein*

36　*Mr. Klein*. Left to right: Lucy Lichtig (script girl), Losey, Gerry Fisher,
Pierre-William Glenn (cameraman)

37　*Les Routes du Sud*. Losey with
Yves Montand

38　Losey at home in Royal Avenue, 1982

hurt by her interest in Stanley Baker. On 11 May 1981 he wrote to Moreau: 'I saw your program yesterday and thought you were beautiful. Your eyes haven't changed at all. The clip from Renoir was simply extraordinary. I wondered why you didn't run a little piece of uncut *Eva* . . .' What followed was Moreau's final role in a Losey film, the part of Lou in *La Truite*, a middle-aged woman utterly tormented by her husband's passion for a slip of a girl. Again Losey confided to Marie Castro: when the crew had left for Japan and Moreau was no longer to hand, he had lost interest. When the film was released in America, Moreau called at Losey's New York hotel; his young minder from Columbia Pictures, Mark Urman, observed they were very comfortable with one another, like old lovers. To Urman Losey conveyed his love of her.[12]

Viveca Lindfors, who played the emancipated sculptress in *The Damned*, had left Sweden when she was twenty-five and settled in America. In May 1961 her husband, George Tabori, wrote to Losey: 'The thing is to harness the enormous power and imagination she has . . . Don't let her become self-critical . . . She's done some fabulous work at the [Actors' Studio with Lee Strasberg] . . .' Lindfors (who describes Losey's visual sense as 'unbelievable') found herself playing opposite an actor, Alexander Knox, who worked off his lines – whereas she worked between them. Losey and Knox rapidly grew impatient with her – as Losey was happy to recall in print: 'At one point Alex [Knox] . . . said to me, "Is this woman going to speak her lines or isn't she? . . . I can't develop anything if she gives me a different cue each time."' Losey told Lindfors to stop being 'a bloody prima donna', to forget her 'psychological nonsense', and 'just play the scene'. Then 'she had a nervous breakdown and began to shiver and shake and I got blankets and wrapped her up and fed her with half a bottle of brandy. Then we had the scene [i.e. shot the scene].'[13]

Reading this passage for the first time almost thirty years later, Lindfors is shaken to tears. 'What difference does it make if you change your line? Improvisation is good. The other actor should listen to what you do. It may be a matter of changing pauses within or after a line. I call it brave acting and you have to protect your own inspiration.' She laughs charmingly. 'I was a turbulent actress. He wasn't open to one's ideas. It was his way or nothing.' While making *The Damned*, she stayed with Joe and Ruth Lipton in Ruth's flat. Lipton remembers Lindfors complaining to Joe, 'You're making me not be me . . . I can't feel it, can't feel it!' He would snap, 'Don't feel it, do it!'

'You had to be careful in conversation with Joe,' Lindfors continues. 'Say the right thing.' His ironic way of speaking made it difficult to have a straight, open conversation with him – particularly when Robin Fox was with them: 'The men of that generation knew in their heads that women were equals but they didn't feel it in the heart. It was easier for Joe to listen and take advice from a man. You often feel this with male directors.' Lindfors is pained that he never asked her to make another film. She sent him a telegram in 1971: 'I would love to play Mrs Trotsky . . . If you met all the men in my life you would see that I fitted in with all types . . . tall, short, fat, dark, blond . . . and there were divine times when I really was the perfect half to them all . . . so . . . I am your Mrs Trotsky . . .'

Originally a stage actress, Monica Vitti had achieved international fame in Antonioni's *L'avventura*, *La notte*, *L'eclisse* and *Deserto rosso* – drifting captivatingly through slow, sensuous, sunlit spaces of the heart, a darling of the European art cinemas. Her qualities as a comedienne convinced her she could do *Modesty Blaise*, but a letter dated 31 May 1965 shows that she was already disturbed by Losey's attitude before shooting began: 'Work together with somebody is for me live together with somebody, give him a period of my life.' On the same day Losey wrote to a friend that he wanted to get Vitti to London and away from her own 'particular entourage . . . the experience of the past week leads me to believe that nothing is going to be very simple or easy with her. Or at least, not until she understands my way of working . . .' Vitti was also stalling on provisions in the contract, darkening his temper further. 'I am so sorry,' she wrote, 'that you, two days ago, have lost your patience . . . If you had known me better, you would have trust me more . . . I need absolutely a period of rest.' She 'begged' and 'supplicated' for 'a free period to carry off the tiredness of nine months of emaciating managements'.

An undated letter by Vitti, written from Rome after shooting had begun, confessed: 'I simply don't yet feel at my ease neither with you nor in the picture. I feel you distrust me somehow . . .' The relationship had been further complicated by what she called 'the Bennett affair . . . a sad and awful thing that gaves to me many troubles.' This was a reference to the costume designer Beverly Bennett, whom Losey had dropped during shooting in Amsterdam. Janni recalls that Vitti brought in her own costume designer. Vitti was also unhappy about the script and the character of Modesty Blaise. 'I read the last changes in the script and I see Modesty weaking [sic] in strength and humor, in personality . . .' She reminded Losey that she had begun as a comedy actress. 'All I ask you is to give me a little more freedom, to believe more in me . . . I have a wild desire to explode, to act . . . please do let me try, to invent, let me have fun impersonating a role in which I believe.'

Evan Jones, who wrote the screenplay, recalls the mounting tension: 'He hated her. She wanted to go home from Amsterdam to Rome. Joe said to me, "You go over and teach her English and explain the next scene. I can't speak to that woman any more."' The executive producer, Joseph Janni, confirms that Losey and Vitti were no longer on speaking terms when they were staying in the Excelsior Hotel, Naples. Losey had convinced himself that she wasn't an actress – 'Antonioni just poses her against walls.' But for some reason Janni is adamant that Antonioni never appeared on the set. 'I knew him well. A month after we finished shooting he and Monica had split up.' According to Jones, Antonioni would call from Italy 'and then you sat there like a bump on a log while they held a long conversation in Italian'. Losey resented this relationship, which reminded him of the Melina Mercouri– Jules Dassin situation on *The Gypsy and the Gentleman*: 'pillow talk as to whether you agree with what I'm doing. Jules Dassin was so much in love with Melina that he couldn't be persuaded to stay away ... Finally, one day, I had to ask him not to stand behind the camera because it was destroying me and destroying her.' Likewise Antonioni: 'He was there, always in her dressing-room so she would come off the set and talk to him on every single set-up ... So it was not my idea of what Monica Vitti could be, but his idea of what he'd made her ... Plus his criticisms about what I was doing ... So eventually it became a set of terrifying frictions ...' Eve Arnold remembers Antonioni on location in Sardinia, standing beside or behind Losey, 'directing her with his eyes ... he had her on tenterhooks'. Arnold also recalls Vitti's obsessive avoidance of being filmed in profile.[14]

Bogarde adored Vitti and her imitations of Marilyn Monroe. (In 1966 she was to act the taxing role of Maggie in Franco Zeffirelli's Rome production of Arthur Miller's *After the Fall*.) 'Monica was wonderful. She was so funny. Joe respected her because she'd say get fucked in Italian. She screamed like a fishwife at him ... Joe kept photographing her in profile and she did everything she could to get herself in full-face, and she used to get terribly cross. She was at a terrible disadvantage because she couldn't speak English properly.'[15]

According to Janni, Vitti asked Losey to show her the finished film in London but 'he curtly refused'. On 18 March 1966 she wrote to the head of 20th Century-Fox, Darryl F. Zanuck: 'Now I know that I have suffered four months of bitterness and humiliation for nothing.' Zanuck soothed her fears but Losey's comments to Zanuck were belligerent: 'I had previously thought that she was at the worst frightened; now I must add the adjectives silly and malicious.' When Antonioni went on record that Losey made better films with men than with women, Losey told the *New York Times*, 'If Antonioni knew all about women, I wouldn't have the trouble I had with Monica Vitti on *Modesty Blaise*.'[16]

During the summer of 1971 Losey received several fond letters from the Austrian actress Romy Schneider, then thirty-two years old, whom Losey – after some hesitation – had chosen to play Gita in *The Assassination of Trotsky*. Ten years earlier she had partnered Alain Delon in Visconti's Paris stage production of John Ford's Elizabethan drama of sibling incest, *'Tis Pity She's a Whore*. At that time the couple were the 'fiancés of Europe' but by 1971 the relationship had ended. In a seven-page, handwritten letter to Losey, sent from Saint-Tropez, Schneider described 'a stone in my throat sitting next to you'. Losey had commented that she wasn't saying anything. 'I felt *terrible* when you said that.' Schneider's English was good but the following passage is not entirely clear: 'Then I might get on your nerves again . . . or we start hating each other (as you said, when you were here) because everything you hate are people – woman – like her . . .' – 'her' presumably being the intellectual film character Gita. Although they had not worked together previously, Schneider expressed touching faith in Losey's genius: 'I believe you know *more* and *better* about an actor's or actress's weakness . . . than they do themselves. *It's so rare!* – for me anyhow – to feel this – to have this confidence. . . .' Her small son David had learned from Joe how to lift his glass and say 'cheers'.

The experience of shooting was clearly a harrowing one for Schneider. Thomas Elsaesser remarked that, 'Delon and Romy Schneider in *The Assassination of Trotsky* bring their off-screen personal life into the film, as do the Burtons in *Boom!*' Patricia Losey noted (late September): 'She looked and talked like a sad, lost child.' She spoke about her feelings of inadequacy and insecurity, sitting on her bed in a nightgown and a sweater 'in a chaotic room'. On 23 October Losey sent her a sharp message: 'Your excesses alarm me . . . You are absurdly mistaken and self-indulgent in your judgements of your own work as it appears on the screen at this point. Naturally what you saw is . . . not even a first cut – it is hardly even an assembly . . . Such violent reactions from you . . . make work extremely difficult . . . As you know, much love and also respect . . .' After *Trotsky* Schneider immediately worked for Visconti again in *Ludwig of Bavaria*. When she died in 1982, Losey told an interviewer that the actress had been in constant need of attention and affection – if she wanted to talk to you in the middle of the night, so be it. She was haunted by the fear of not 'arriving' and was permanently dissatisfied with her performance. Her personal life had been difficult . . .[17]

The Parts

Losey was a better film director than philosopher, capable of saying more with a movement of the camera than in a hundred words, but a comment to Ciment about *The Big Night* does shed light on his general portrayal of the female of the species. 'As for the cruel reaction of the boy towards the women's affection, I think people who are searching for parents are also searching for a very particular kind of love . . . always with the expectation of being betrayed . . . Therefore, before you can be betrayed, you aggress and you test the person . . . by humiliation and cruelty and brutality . . . people who are searching that way – as I am – are to a certain extent incapable of either giving or receiving the thing they want.'[18]

Losey's doctor, Barrie Cooper, believes that he resented women as objects of desire, or as bonding agents creating dependency – but could not articulate his resentment. Perhaps so; but what is specific to Losey's work is the elimination of the heaven and purgatory which precede the hell. There is not a single depiction of mutual sexual passion or love in his films: in the *The Go-Between* it is there but off-screen, and the 'falling in love' of Macdonald Carey in *The Lawless* or *The Damned* is a peripheral box-office routine. The relationship of the sexes is a war without truce or Geneva Convention. One party is invariably exploiting or betraying the love of the other: Van Heflin in *The Prowler*; Bogarde in *The Sleeping Tiger*; Micheline Presle in *Blind Date* preying upon 'the purity of the young artist, who was a dedicated person, completely unaware of the guile and the involvement in evil and crime of his mistress . . .' Any relationship involving Stanley Baker (*The Criminal*, *Eve*, *Accident*) was flawed by his innate cynicism – even when, as in *Eve*, the greater sexual cynicism belongs to the woman.

Eve, of course, is a sexual nightmare. The emotional double-dealing in *The Servant* is blatant; *Modesty Blaise* is a pastiche of manipulative and transitory relationships. *Accident* transports us into the higher sphere of *mauvaise foi*. In *Boom!* the itinerant 'poet' incarnates duplicity and fortune-hunting; in *Secret Ceremony* sex is akin to madness (Cenci) or vulgar licentiousness (Albert); in *The Go-Between* sexual passion is doomed by class society and the necessity of an arranged marriage. In *Trotsky*, the assassin cruelly manipulates the woman's love to achieve his ends. In *The Romantic Englishwoman* Losey tortures Eros and once again poisons the flower by delivering his characters to mercenary opportunism and double dealing. In *Mr. Klein*, Robert keeps one mistress unhappily dangling while growing bored with the other. In *Les Routes du Sud* sex is subsumed by the mutual torture between the generations. Losey declared Don Giovanni to be a sick man – but added decorative touches to his emotional sadism. *La Truite* seems to summate all of Losey's cynicism about sexual relationships. In Losey, Eros is Eros denied. The beautiful body flaunts its wares before pulling the curtain.

But if, in Losey's cinema, mutual love is impossible, this emerges by result rather than explicit design. It's never stated, acknowledged. His statements pitch sideways from self-scrutiny into social refraction. When *The Servant* came out in France, he explained to *France-Observateur* (16 April 1964): 'Sexuality isn't essential in the film because it intervenes inevitably in the description of a total inversion of human relationships. It's this inversion which is the subject of the film, not the sexuality . . .' This is solid enough – except that in Losey's films 'inversion' usually takes a sexual form and sex an inverted form. Occasionally he took refuge in the broad church of Marxism: 'What interests me is not eroticism, but what provokes it in modern life, and of which it is the result: fear, solitude, boredom, lies, and above all hypocrisy.' But this is as obfuscatory as his claim that *Eve* was 'a fable about marriage, where money was involved'.[19]

Losey was apparently never asked why desire is invariably incompatible with respect and affection. The young delinquent in *The Sleeping Tiger* becomes the more attractive to the wife the more openly he mocks her desires. The nobleman in *The Gypsy* finds himself increasingly enslaved as Melina Mercouri no longer bothers to disguise her cynical manipulation of him. In *The Servant* Tony respects Susan and (therefore) cannot desire her; a sluttish smile from a common maid fires him with lust. Tyvian is not interested in his beautiful fiancée (Virna Lisi) because she is both loving and respectable. Eve, who is neither, sends him mad. What both Stephen and Charley obsessively pursue in *Accident* is not only Anna's beauty but their own loss of self-esteem. Chris Flanders's cynical gigoloism drives Mrs Goforth wild for him in *Boom!*; in *The Romantic Englishwoman*, the distinctly unromantic Elizabeth finally pursues not love but self-abandonment in the company of another phoney 'poet'. In *Secret Ceremony* the stepfather's lust feeds off contempt and cynicism. Robert's contemptuous treatment of his lawyer's wife in *Mr. Klein* merely inflames her passion. Isabelle Huppert's icy rejection of every suitor in *La Truite* drives them round the bend. A man murders his wife because of it.

Losey associated his mother with the predatory claims of beauty (sex) on wealth and blue blood (good breeding). Throughout his life he was haunted by his father's failure to be rich enough to command his wife's respect; responded to his father's defeat by populating his films with aggressive men prepared to treat women with casual contempt or ruthless exploitation: Van Heflin in *The Prowler*, Dirk Bogarde in *The Sleeping Tiger*, Leo McKern in *Time Without Pity*, Stanley Baker in *The Criminal*, *Eve* and *Accident*, Richard Burton in *Boom!*, Robert Mitchum in *Secret Ceremony*, Helmut Berger in *The Romantic Englishwoman*, Alain Delon in *Trotsky* and *Mr. Klein*, Jean-Pierre Cassel and Daniel Olbrychski in *La Truite* – and, supreme example, Don Giovanni.

Only in *Eve* did Losey open his art fully to the undeflected pain of childhood trauma. Here we are offered a full portrait of 'the Mother' as ruthless sexual predator (compared with her appearance in contemporary films by Antonioni and Truffaut, Jeanne Moreau's physical aspect in *Eve* is strikingly ravaged). As the lover, Stanley Baker displays the manly, athletic exterior which impressed Joe in his father; and the bleak, helpless jealousy of the loser. Tyvian's 'wealth' is fraudulent, just as Joseph Walton Losey II's was unreal, dissipated. Tossing down his banknotes at the feet of his fickle, scornful mistress, Tyvian is squandering what wealth he has – the story, in Joe's mind, of his own inheritance. Eve's dependence on her jazz records, the itching restlessness they convey, may well be Ina Losey at the piano; by contrast Tyvian, like Losey's father, is stranded in stone, a sculpted witness to an exotic force he can neither resist nor command. Yet Eve herself suffers social humiliation and stigma; she is 'that woman', an outcast, stung to the core as she watches Tyvian's grand Venetian wedding to Francesca. Losey takes care to punish Eve. At the end of the film, as she is led away by her latest sugar daddy, an unattractive Greek, her own humiliation is almost equal to Tyvian's – surely an echo of Joe's view of Ina Losey's men friends in New York and La Crosse after the death of his father.

Losey's animosity towards attractive women is clear. In *The Criminal*, a former girlfriend of Bannion's, played by Jill Bennett, a former girlfriend of Losey's, shows up unwanted at a party. Her drugged entrance is filmed through a kaleidoscope. 'I thought, this woman who is entering the room is not a whole woman – she's a fragmented woman. Let's make her enter in fragments.' Losey insisted that the crook-poet in *The Romantic Englishwoman* came off best and the wife, Elizabeth, worst, because 'her idea of romance was total bourgeois consumption of man'. The virulence of this comment certainly goes beyond the portrait of Elizabeth in the film and indicates that the idea of a woman – a kept wife, too! – picking her path between two men angered Losey.[20]

Losey perpetually humiliates women by making them mere adjuncts of man. Leaving aside Sarah's purely notional identity as a lawyer in *Steaming*, his thirty-one films do not yield more than one educated woman whose career figures prominently in her life and in the film: the sculptress Freya in *The Damned* possesses talent, independence and vision; she is happy to live alone. To depict Frédérique in *La Truite* as a businesswoman, an expert in trout-farming, is to miss the point; her true career is sexual, arousing but not answering male desire. Losey's women are predominantly mere annexes to man and sex. Susan in *The Servant* clearly has a career 'off stage' but we never hear a word about it. She and Tony talk about his dreamily absurd business plans, but he never asks, 'How are things with you?' and she never says. Her ambition is to 'settle down', though the phrase is not used.

In his production notes for *Galileo* (1974) Losey called Brecht an 'MCP' but noted that he wrote some remarkable female parts: in *St Joan of the Stockyards*; Gruscha in *The Caucasian Chalk Circle*; in *The Good Woman of Setzuan*; in *The Mother* (Gorky); and in *Mother Courage*. Had he directed any of these plays, Losey would have confronted a truly formidable portrait of female strength and intelligence operating outside the confinement of a strictly sexual or domestic field of action.[21]

31

A Directory of Directors

In Hollywood Losey was never more than a second-league director with an insecure future. The release of his first feature film coincided with his fortieth birthday. He was no Kazan and no Welles. Kazan and Losey were born the same year; the radical New York theatre of the 1930s had briefly crossed their paths, but then Kazan raced ahead into the big league while Losey floundered in low-level agit-prop. Indeed, when he arrived at MGM in 1945 he cited Kazan in his private notes as an example of a director who knew how to lay down his terms to the studios. By 1960 Kazan had directed *A Streetcar Named Desire, Viva Zapata, On the Waterfront, East of Eden, Wild River, Baby Doll.* He had also testified to HUAC, naming names; it is hard not to regard Losey's protracted scorn for Kazan's capitulation as a blend of genuine disgust and of revenge against the more precocious talent. Arthur Miller, who risked imprisonment rather than 'name names' when subpoenaed by HUAC, felt it would be intolerant to boycott Kazan for ever: 'As for morals perhaps it was just as well not to cast too wide a net; for one thing, how many who knew by now that they had been supporting a paranoid and murderous Stalinist regime had really confronted their abetting of it?'[1]

Losey was not inclined to such reflectiveness. Before the first showing of *The Servant* in Paris, wooden seats had to be placed in the aisles; hearing that Kazan was among the overspill, Losey grabbed Bogarde, dragged him out of the cinema and straight to Fouquet's in the Champs-Elysées, a showbiz restaurant with wickerwork chairs whose red velvet seats match the curtains, yellow tablecloths, mirrored pillars and sofa-seats along the panelled walls where brass plaques indicate '*la table préférée*' of stars such as Jean Gabin. Shaking with rage, Losey ate his meal in reverse order: cheese, then raspberries, ending up with soup.[2]

The film industry succumbed to the blacklist out of fear. Losey's real condemnation by the American studios and film financiers is a subterranean torment running from the end of the blacklist to his death, twenty-five years later. American money financed *Modesty Blaise*, *Boom!*, *Secret Ceremony* and *Figures in a Landscape*, but these were regarded as European films. In December 1970 he accepted an invitation, sent by George Cukor and Robert Wise, to become a member of the Academy of Motion Picture Arts and Sciences (which awards Oscars), but from 1951 to 1984 Losey shot not a foot of film on American soil. During the 1970s every project involving American themes and settings was turned down. Producers and agents failed time and again to raise money against his name. Hollywood, Jerome Hellman sadly explained, regarded Losey 'in the category of darkly conceived and esoteric character studies which "ain't box office"'. This was his second exile and it was more humiliating than the first. 'I have a feeling there is a generation or two in Hollywood who don't really know who I am,' he remarked in 1982.[3]

To pour scorn on what rejects you is a common human reaction, and Losey possessed it, uncommonly. Time and again he insisted that nothing good could come out of West Coast studios, and he simply ignored (in public at least) the work of the younger Hollywood directors. He didn't want to know about young talent. Mark Urman comments: 'In understanding Losey's waning reputation in the 1970s, it has to be remembered that this was the decade of breakthrough for American directors in terms of techniques and innovations.' Losey's pink note-cards for his Dartmouth College lectures in 1975 display the confusions of a good mind racked by animosity: 'America is still a province, thanks to Hollywood. Even now very few Americans have been innovators – without the Godards, Antonionis, Fellinis, Bressons, Welleses – even the Robbe-Grillets and Duras – who have by and large reached you only lately – you wouldn't be here. You would still be "enjoying" second-rate Capras. And there would be no film art (American i.e.) excepting perhaps Welles and Chaplin – and Von Stroheim and Sternberg – and look at what happened to them – and how many of them were Americans? One. Of course there are exceptions like Ford . . .'

The influence of European cinema on the young Losey was patchy. He met Sergei Eisenstein in New York but, contrary to a legend propagated by *Cahiers du Cinéma*, he did not, in 1935, attend Eisenstein's courses in Moscow. Eisenstein's work was 'heavy and academic'; he found 'Ivan terribly boring' rather than Terrible. The post-war years brought him into contact with Italian neo-realism. 'Of course I was influenced. Enormously by Rossellini, and also by De Sica. I thought *Open City* and *Paisa* were absolute revelations . . . I was also very much influenced by Giuseppe De Santis's *Bitter Rice* . . .' Paul Mayersberg recalls that Losey picked things up fast. 'He was like a magpie. What he discovered in Europe is that film can work by analogy, rather than by cause-and-effect.' Contingent scenes make their impact. While shooting *Blind Date*, the lighting cameraman, Christopher Challis, discovered Losey's admiration for the new European cinema, 'rather like Van Gogh when he moved from Holland to France, and he wanted to make films like that'. Mayersberg regards Antonioni as the crucial influence on Losey, although he found Fellini's flamboyance as seductive as Antonioni's reticence. As for Visconti, his style was fundamentally operatic, whereas Losey's musical appreciation was mainly confined to jazz.[4]

As he gained in confidence and stature, Losey began to air his opinions about other directors. *Sight and Sound* (Autumn 1961) reported: 'The first demand Losey makes of a film-maker is that he should have a "signature", an immediately recognizable style ... *Saturday Night and Sunday Morning*, for all its qualities, misses out on this point, as does *Room at the Top* ("Clayton's signature here is Signoret"). He finds the indispensable signature in *L'avventura*, which he greatly admires, and, with reservations, in [Godard's] *Breathless* ("But aren't they more kids playing at film-making than kids in the film playing at life?") There's a craftsman's impatience with some of the nouvelle vague "innovations" which, he says, are only old tricks refurbished.'[5]

Losey wrestled with the new boys. 'Godard is talented but I don't like his work very much. He makes a virtue of his unprofessionalism and parades his ignorance and self-indulgence, which is very annoying, even though bits of his films have power.' According to Andrew Sarris, he was unprintable about Godard. *Weekend* was dismissed as affected amateurism, a pile of junk. However, in the late 1960s favourable references to Godard became more frequent. In 1966 Losey expressed a desire to move 'into less directly realistic and less literal aspects of film. A few people are beginning to do this, but the explorations have hardly started ... Resnais is a pioneer, and in quite a different way, Godard also.' But: 'If it seems a trick, then it's a failure; if it doesn't, then I think it's as much an extension of the medium as Picasso ... when he began to paint three or more aspects of the same face in one portrait.' Richard Lester's *Yellow Submarine* he called 'an absolutely marvellous film, full of surrealist ideas, well ahead of their time, very Magritte'. In April 1974 he congratulated Robert Bresson on *Quatre Nuits de Rêveurs* – 'an absolute masterpiece'. He wrote to Bernardo Bertolucci praising *Last Tango in Paris* as 'extraordinarily beautiful' and 'a total piece of work'. They met in London in December 1973; as Richard Roud recalled, Bertolucci sat at Losey's feet. 'He warmed to any young director who did that.'[6]

Among the directors whose work Losey screened for his Dartmouth College students in 1975 were Bergman, Resnais, Antonioni and Fellini. Asked in 1976 what interested him in contemporary cinema, he replied: 'Nothing. I don't have the time to go to the cinema. Fellini, Bergman, Bertolucci, always interest me. But I don't go to see films. I've done with it (*j'en fait*).' In 1980 the Festa Nazionale de *l'Unità* (the Communist daily) asked him to suggest the most 'significant film' he would like to be present at. Losey replied that he wouldn't be present, but nominated *Salvatore Giuliano*

(Francesco Rosi), *8½* (Fellini), his own *Mr. Klein*, *Le Chagrin et la pitié* (Marcel Ophuls), *Asphalt Jungle* (Huston), and *Lacombe Lucien* (Malle).[7]

Michelangelo Antonioni

Losey wrote to Louis Marcorelles in February 1961, praising *L'avventura* as 'one of the most complete and satisfying films of its kind I have ever seen'. As for the same director's *Il grido*, although 'beautifully patterned . . . I found it false, the performances stilted, the opticals abhorrent . . . and the whole thing a bore'. Interviewed by *Le Figaro littéraire* (13 October 1962), he said: '*L'avventura* is a very full film, but Antonioni does not believe in life, nor in the spirit, nor in the senses. He sees beauty through black spectacles: for him the world is perhaps beautiful, but it's not worth the trouble to live there.' Later he commented that *L'avventura* was 'the most perfectly realized film, the strongest director's statement of the past decade'. *La notte* was masterful but colder. Losey hated *The Eclipse*. 'As for his new one, *The Red Desert*, the first few reels are overwhelming . . . Each sequence is perfect but they don't add up.'

The last phrase is significant: even in his new, modernist incarnation, Losey remained anchored to a tradition of strict storyline relevance – an action must reveal human motive. Losey's digressions were strictly decorative and atmospheric; he came to town (Venice, Paris, Oxford) like a businessman with a loaded itinerary, whereas Antonioni had the tourist's habit of wandering down alleyways just in case. In *L'avventura* the male population of a town noisily crowds around a famous actress; later, in a subtle echo of this scene, Monica Vitti finds herself surrounded in total silence by the men of a poor southern village. In *Blow-Up*, David Hemmings fights furiously with other fans for a pop star's broken guitar and then, having won it, tosses it away. Antonioni's sensibility expands laterally to make a public observation which also touches the private predicament of his characters. Losey's remarkable eye avoided the contingent crowds and the alleyways.

Eve's debt to *La notte* (1960) is clear. Losey used the same director of photography; his fashionable party-scene echoes Antonioni's; exotic black dancers writhe in both films. Losey turned Moreau's earlier performance inside out; the idle, unloved, dignified Lidia becomes the vulgar predator Eve. Like Lidia, Eve embarks on solitary walks through the city, but Lidia's peramulations incarnate her predicament whereas Eve's are narcissistic. Losey's film lost the humanistic trail in the alleys: whereas *La notte* is a tragedy in which alienation overcomes good will, in *Eve* virtue is banished. After Monica Vitti's fraught participation in *Modesty Blaise*, personal relations between Losey and Antonioni inevitably suffered. When Losey told Andrew Sarris, 'I don't treat actors like raw clay to be moulded to my specification,' Antonioni was his target.[8]

Ingmar Bergman

Losey lauded Bergman's *Virgin Spring* (1960) as an example of Brechtian form adapted to film language. Although the film depicted the world at its 'most cruel and barbarous', he had left the cinema purified: life – pitiless and marvellous – was affirmed. 'I hate the religious mysticism of Bergman but I approve of his positions.' He also envied him his crew of twelve, the cameraman doing the operating, the grips working with the electricians, the budget only $200,000. Bergman, he said, was 'unique in the world' in being able to move from theatre to film with real freedom of subject and the same team of actors and technicians. Bergman's *Cries and Whispers* was 'one of the best films ever made'.[9]

Federico Fellini

In October 1969, Losey congratulated Fellini on his *Satyricon*, 'one of the great pictures I have seen in my life and an achievement to justify a whole lifetime . . . awful, horrible, terrifying and yet ceaselessly of extreme beauty; it is in fact this use of life and of the cinema that I hope one day to approach myself . . . I envy you!' Fellini replied cordially. In March 1972 he sent Losey a warm and enthusiastic letter about 'your picture' – possibly *Trotsky* viewed at a private screening in Rome. Fellini had been chained to his chair from the first frame to the last, possessed by anguish while admiring 'this virile feeling this civil courage *expressed* with artistic health and powerful talent'. In April 1974 Losey wrote to Fellini about *Roma*: 'The memory of the fleeing camera car with its flapping tarpaulins in the wind like a witch from *Macbeth* haunts me as many other images . . .'

Yet Fellini's work has little in common with Losey's. Fellini is an 'Elizabethan', whereas Losey's stricter style is 'Restoration'. He took no joy in the hot tides of popular life, the garlic embrace of hot bodies (although the carnival scene in *Galileo* is well done). Losey's crowd scenes are few and rigidly orchestrated: the May Day parade in *Trotsky*, the herding of the Jews in *Mr. Klein*, the chorus in *Don Giovanni*. Fellini's eye is naturalistic: In *Roma* a boy is found sitting on a lavatory: 'I've done it!' he calls. Building a new metro line, the drillers pierce a buried Roman chamber – the frescoes cloud over as the air enters. Such spontaneity of invention was not Losey's.

In March 1975 he went to see Fellini's *Amacord* while drunk (according to Patricia), fell asleep during the film, and disliked what he remembered, although the images were beautiful. In November he sent Fellini a telegram offering him a small part (Old Klein) in *Mr. Klein*, if he could spend one or two days in Strasbourg. 'Possibility journey by private plane stop Possibility production offers you Cadillac or girl or young man stop.' If this was humour it did not seduce: Fellini cordially declined (Losey's gesture had in fact been prompted by a sudden need to bring in an additional actor of Italian nationality, to meet co-production requirements). An undated cable from Fellini congratulated Losey on *Don Giovanni*. Patricia Losey translated from the Italian: 'I didn't believe it possible to realize so happily an opera so complex and delicate. You have managed to make it in the most evocative aristocratic and spectacular way.'

Alain Resnais

L'Année dernière à Marienbad struck Losey as technically masterful but largely incomprehensible and self-indulgent. According to Paul Mayersberg, he concluded that the technique employed by Resnais was semi-detached from the film. Losey was impressed by *Hiroshima mon amour*: the analogies, the cutting, the opening of doors of perception. *Muriel*, which explores the indirect impact of the Algerian war, the break between the pre-war and post-war generations, by means of subtle time-shifts, he greatly admired – likewise *La Guerre est finie*. 'Resnais has no side, no pretentiousness, only craft and thought.' Resnais's postcards address him as 'Joë'. 'Joë' wrote to Florence Malraux Resnais in January 1968: 'Please tell Resnais how very moved and pleased I was by what you told me of his reaction to *Accident* . . . Resnais's work seems to be the only work I learn from, and he is almost the only director with whom I have been able to have long stimulating sustained conversations.' The overlap between the two directors revived in 1977 when Losey 'borrowed' Montand and Semprun from Resnais, while the latter 'borrowed' Bogarde and Mercer for *Providence*. In January 1978 it was Losey's privilege to present the 'Best Director' César to Resnais. Later that year Resnais visited Vicenza, where Losey was shooting *Don Giovanni*, and expressed admiration for the rushes.[10]

Stanley Kubrick

In 1960 Losey told Dwight Macdonald that Stanley Kubrick was 'almost the only director of the youngest generation whose work I respect'. In November 1964 he explained to Francis Wyndham that the directors he most admired adopted different styles from film to film yet always displayed a distinctive signature: for example Kubrick's *Doctor Strangelove*, *Paths of Glory* and *The Killing*. In Losey's estimation, more than any other director Kubrick had beaten commercial pressures by facing them on a practical level, buying his own equipment and controlling every aspect of production and distribution. In October 1969 he wrote to Kubrick, who lived near Elstree Studios, 'It is ridiculous that we have never really met.' In July 1972, he again approached

Kubrick, alluding vaguely to post-production difficulties with *The Assassination of Trotsky*. After a sequence of attempts to reach each other by telephone, Kubrick wrote: 'I have admired all of your films, and *Trotsky* was no exception.' But it was not, he said, one of Losey's best. Stung, Losey threw a counter-punch. He had seen *A Clockwork Orange* in Paris: 'I am afraid I would say about it much the same as you say about *Trotsky*: I found it technically brilliant and much more exciting as to images than any of your previous pictures, but I also found it unclear, dangerous and, I regret to say, gratuitous . . .'[11]

As a director Losey aspired to autonomy, supervising and controlling every aspect of his art – casting, design and locations through to processing, editing, mixing and music. But in the English-speaking countries the film industry granted directors neither copyright nor '*droit d'auteur*' in their work – merely the right to remove their name from a 'final cut' which they decided to disown. In 1963–4 Losey published several letters complaining about the mutilation of films by producers and distributors while retaining the director's name on the print. He quoted France's Minister of Culture, André Malraux – even if the cinema must make money, not every film must. Yet Losey's own attitude was not consistently idealistic. During the contract negotiations for *Secret Ceremony* in 1967, he wrote to his agent Robin Fox: 'I am far more deeply concerned with how the money is handled than with the artistic provisions which have always proved to be meaningless anyway.'[12]

By the early 1970s he was publicly expressing his rising frustration with 'the ugly world of films. The commercial monsters don't really care about the films themselves, they just care about their money . . .' He told the *Observer*: 'I am dealing with a bastard that is bred of the most blatant commercialism and cynicism. It's an industry that's basically about the exploitation of bad taste.' The only rational solution was the European '*droit d'auteur*': 'to vest the copyright to a completed film in perpetuity in the director, the writer and the composer, as is the case in France and Italy.' Could the directors break free of the stranglehold? At the 1972 Cannes Festival, Losey discussed with Louis Malle, Agnès Varda and Jacques Demy a possible '*Quinzaine*' of directors, an international directors' guild association. In September 1974 he informed the critic Pauline Kael that he and six other directors – Richard Lester, Tony Richardson, Ken Russell, Karel Reisz, Lindsay Anderson and John Boorman – had spent nearly a year endeavouring to launch a system by which each would make one film a year, of the individual's choice, on a fixed minimum salary, with a guaranteed budget not in excess of $1 million. 'Each director would have to raise his own completion bond. We also agreed to cross-collaterize each other . . . a large percentage of the profits of a successful picture would go towards the recoupment of losses on any [commercially unsuccessful] picture . . . We planned to start on a small revolving fund of five million dollars . . . In return for this we asked for total freedom – from scripts to completion . . . We succeeded in getting half of the five million dollars from a merchant bank on condition that we got the other half from a distribution company. We were turned down for one reason or another by every single American and English distributor. The reasons varied from one group wanting veto of the scripts . . . another wanting exclusion of one of the members of the initiating group [not me] . . . We finally had to abandon the plan.'[13]

Protectionist trade-union rules were another source of frustration: Losey called them 'syndicalist suffocations'. He envied Bergman's freedom to work constantly with the same company and technicians, with only the crew he needed, and coveted Godard's privilege of working with 'a crew of seven . . . I have to work with sixty'. Two thirds of his own crew might never know what the film was about – went into the building trades during a recession, 'and then they come out for a little more gravy . . .'[14]

By the early 1970s the film industry faced a crisis rooted in changing consumer patterns; between 1960 and 1973 the number of operating cinemas in Britain declined from 3,034 to 1,530. Annual ticket sales slumped from £500 million in 1960 to £244 million in 1968 to £111 million in 1973. Investment in British films fell from £12 million in 1972 to £4.7 million in 1973. The major studios at Shepperton and Elstree were being used to only half their capacity, and the ACTT reported that 60 per cent of its members were out of work. In a letter to *The Times* (10 December 1973), Losey, Pinter and Peter Hall called for a regenerative investment fund. On 5 November 1974, Losey wrote to Lady Falkender, Personal and Political Secretary to Prime Minister Wilson, offering to 'put together' a group to discuss the future of the film industry. This received a guarded response. In April of the following year he again wrote – he'd heard that the PM was calling a gathering which included no director, actor, writer or technician, only the 'wreckers' of the industry. Falkender responded that the proposed dinner was to be 'representative of all areas in the film industry' – but clearly Losey was not invited.

Lecturing at Dartmouth College in 1975, he released his full rage in a lecture called 'Art and Industry'. Here the normal short headings on his pink cards give way to the prose of passion as he considered what producers did to directors:

'Mr Mayer, Mr Ponti, Mr Albery [Aubrey] and the other strippers never lost anything – in the long run they always win.'

'Neither Antonioni, nor Resnais, nor Bergman, nor Robbe-Grillet, nor Bergman, nor Buñuel are rich. They have paid heavily for embattled position in "wasted creative life", energy, anxt [*sic*: angst], lack of continuity, compromise, and private life . . .'

'I cannot think of a rich director excepting perhaps Kubrick or Visconti who inherited it; Kubrick – 6 or 7 films no personal life; Bresson – obscure and poor; Buñuel – nearly 20 years wasted; Lester – five years without work; Welles – one step ahead of the debt collectors – how many unrealized films – a ruin – a magnificent one . . .; Resnais – five or six years without work – only 5 or 6 features – really poor. All more or less happy. Like me.'

32

By Design

Losey's most important and enduring collaboration with a production designer began with his first British film, *The Sleeping Tiger*. He had met Richard Macdonald 'at a big spaghetti and Chianti party in London in Campden Hill when I first arrived in England'. Ten years the younger, Macdonald had served in the Royal Navy during the war and was now teaching drawing and painting at Camberwell School of Art – but he later left the art school world 'in disgust', moving into film design and advertising. His commercial television studio became what he calls his personal 'film school'. Because of union rules, Macdonald's crucial role as Losey's 'eye' was for many years unofficial, even clandestine, resulting in numerous clashes between Losey and his officially credited art directors.* As Losey expressed it, he and John Hubley in Hollywood, then he and Macdonald, formed 'a kind of island – and I did my talking to them because I didn't dare to talk to anybody else . . . because of studio politics, because of unfeelingness . . .' It was like a marriage.[1]

Macdonald remained contemptuous of script writers' detailed descriptions of décor: usually incompetent. 'My first task is to read the screenplay and then to forget it, or at least detach myself from it as much as possible . . .' Bored, also, by actors, Macdonald grants that their specific qualities have to be taken into account for design purposes. His method was to draw a series of cartoons charting the progress of a scene, indicating the dominant gestures; then work on the angles, then turn to the essential design, starting with the walls, later the furniture. Subsequently he enlarged his drawings, or frames, to accommodate more detail. Lighting was always a crucial factor.[2]

For Losey, Macdonald was simply a 'genius' who understood his own need to find a particular painter or artist who was a frame of reference for the picture: Rowlandson in the case of *The Gypsy*, Goya in the case of *Time Without Pity*, Joszef Herman, whom Macdonald imitated in *Blind Date*. Macdonald had few inhibitions about enlisting the work of other artists. Elisabeth Frink's sculptures thrust into the sky along the clifftop in *The Damned*; op and pop art were ransacked for *Modesty Blaise*, virtually turning

*Macdonald worked with Losey as design consultant, production designer or art director on seventeen films: *The Sleeping Tiger*, *A Man on the Beach*, *The Intimate Stranger*, *Time Without Pity*, *The Gypsy and the Gentleman*, *Blind Date*, *The Criminal*, *The Damned*, *Eve*, *The Servant*, *King and Country*, *Modesty Blaise*, *Boom!*, *Secret Ceremony*, *The Assassination of Trotsky*, *Galileo* and *The Romantic Englishwoman*.

Bridget Riley's style into wallpaper. 'It was "anything goes",' Macdonald comments. 'Splash it on. Use anything. The white and blue squares in the communications room. Vasari and Bridget Riley. I knew her and talked to her at the time – an example of what you can steal – but the ones I took were from Vasari not Bridget Riley because I knew her. You can't be a pirate.' (Safe passage for friends?) Despite the abstraction of design in these Sicilian scenes, Macdonald characterizes Losey as a 'Renaissance person'. 'He didn't like abstract art or thoughts, he was much too flesh and blood for abstraction. He wanted to know what makes people tick.'[3]

On *Blind Date* Macdonald again worked with Losey in anonymity, Losey regularly passing off his sketches as his own, although, David Deutsch recalls, 'No one was fooled.' The officially credited art director was Edward Carrick, described by Deutsch as a very nice person, rather easy-going, who 'got up Joe's nose'. The script had specified a wall of a particular height, and all the pre-production drawings by Macdonald were based on that height. Carrick shrugged: 'I don't bother about reading scripts, old boy.' Losey blew his top and Carrick was replaced by a draughtsman, Harry Pottle.[4]

With *Eve*, let loose in Venice, Losey and Macdonald embarked on a visual riot of mirrors, water, reflections, reverse images, paintings. They scoured Rome and Venice for furniture and *objets d'art*, constructing the bedrooms in a Rome studio. Losey felt himself to have been reared 'in a kind of documentary tradition', instinctively preferring location interiors to studio sets. 'Most of the interiors of *Mr. Klein*, all of the interiors of *The Go-Between*, and of *The Romantic Englishwoman* were done on location,' he stressed. In *Boom!* Mrs Goforth's villa was erected by Macdonald on a Sardinian rock; *Secret Ceremony* was a mix of location and studio interiors. However, studio set interiors prevailed not only in his British films of the 1950s, but also in *The Servant* and *Accident*. Of necessity, *King and Country* had to be filmed entirely in a studio – Macdonald designing what he calls 'a compact set held together by outside references to the war, photo inserts . . . holes in the walls, mud, a stark feeling, impermanence . . .'[5]

Macdonald is a cavalier, unpopular with production roundheads. Norman Priggen, for example, calls him 'a very annoying character. Uncontrolled.' On the set of *King and Country* he would join in digging mud – 'Do it like this, twice the speed' – putting up the backs of NATKE members as well as the Association of Cinematograph Television and Allied Technicians (ACTT). The animosity towards Macdonald is reflected in a letter written on 25 November 1963 by W. Alexander, who described himself as assistant art director on *The Servant*: '. . . in fact Richard Macdonald's major contribution to this film was the cessation of shooting for a week while he decided how to dress the sets! . . . I would like to see the ACTT enabled to take action to stop this man being allowed to enter any British studios . . .' (In reality Macdonald's artistic contribution to the film was spectacular.)

Macdonald pleaded his case to George Elvin, General Secretary of the ACTT: 'There was in fact some misunderstanding . . . as Joe had been given to understand in conversation between himself and Mr Leach [*sic*: Leech] . . . that my transfer to the category of Art Director had been passed and agreed.' (ACTT Senior Organizer Paddy Leech had merely informed Losey that there was no objection to Macdonald being credited as production designer on *The Servant*.) Macdonald now wished to re-apply for 'transfer to the ACTT category which would permit me full functioning as Art Director'. Losey and he had worked together 'as he worked with John Hubley in Hollywood to try to produce a cohesive over-all image and visual style. This function is one which is peculiar to Losey and which apparently is not understood . . . or used . . . by most other British directors or art directors.' Macdonald regretted that ACTT had limited his accreditation to title design on *Blind Date*, production design on *The Criminal* and *The Servant*. His full creditation as art director on *Eve* had come about only because it was a Franco-Italian co-production. The ACTT's executive committee regarded Macdonald with suspicion; he was too versatile and wore too many hats. As head of the television department of the advertising agency Batten, Barton, Durtsine and Osborn, he was hiring ACTT members: the craft unions were particularly sensitive about individuals who operated on both sides of the employment line. *Modesty Blaise* was a British quota production subject to ACTT staffing regulations. Despite pleas to George Elvin from Losey, Macdonald was once again confined to off-set production designer rather than art director. Macdonald's conduct on *Modesty Blaise* clearly angered ACTT. On 19 October, Elvin wrote to Losey: 'The Executive Committee agreed that the application for lifting the 10(C) restriction should be rejected and in addition that Joseph Losey and Richard Macdonald should be asked to show reason why they should not be dealt with under Rule for not carrying out the policy of the Union.' Losey protested his innocence. On 25 October, Elvin wrote refusing Macdonald's application for transfer from 10(C) to ordinary membership. Losey declared himself 'shocked and disturbed' by this 'flagrantly unjust and unjustified decision'. Macdonald recalls being called with Losey to the union's 'Star Chamber'. Carmen Dillon (later Losey's designer on *Accident* and *The Go-Between*) sat on the other side of the table. The ACTT later granted Macdonald art director status for John Schlesinger's *Far from the Madding Crowd*.[6]

Both awed and angered by Macdonald's relaxed attitude to sexual relations, Losey expressed almost boyish admiration: 'I mean when he gets drunk he does pornographic, obscene drawings, some of which are very, very funny.' Once on an aeroplane 'we were discussing how many men a woman could take on at one time; and in about five minutes he did a quite beautiful drawing of a woman having five men at the same time'. Then came *Boom!* 'And hc had a whole string of girls down there . . . and some that were not his girls but other people's girls. And he would put them to work on waxing the plaster to make it look like marble.' In Macdonald's view, *Boom!* was 'the last glimpse of super-film making'. Everyone played king, including Losey. 'It was the Rolls-Royce era, a court. No one asked who was down what corridor.' But they did ask. Norman Priggen, the production manager, recalls Macdonald's affair with Losey's secretary on location – then a Macdonald affair from *Modesty Blaise* arrived and furiously wrote 'Bastard!' in lipstick over the guilty bed sheet. Macdonald's wife of the time arrived that evening, by which time he was in hiding.[7]

'In one sense Macdonald is the star of that film,' Losey commented. 'He did a most fabulous job and made it very easy for camera movement.' Macdonald recalls that Tennessee Williams hated the set: he wanted a Palladian villa with cypress trees, mazes and stables. 'I built it with Italians who spoke no English, six Roman foremen and some Sicilian workmen all of whom understood proportions. That wouldn't be possible now.' Losey told Richard Roud how, after *Boom!*, Elizabeth Taylor invited Macdonald to decorate the private jet that she and Burton had bought. 'But in what style?' he inquired. She said Regency. 'Where am I going to put the fireplace?' he asked. Macdonald regards the ornate bedroom in *Secret Ceremony* as a more elaborate development of the ones in *Eve*, and it's clear that his passion for bravura display was increasingly responsible for Losey's reputation as a director of style-without-substance.[8]

Losey's new home at Royal Avenue reflected the decorative tastes of his films – especially the mirrors: a large mantel mirror in a pine bolection moulded frame; an antique oval convex mirror in an ebonized frame; an oblong bevelled mirror, the frame inset with mirror segments with designs made from metal washers, screws, bolts, rivets. The drawing-room at Royal Avenue had been designed by David Hicks and carried a Hicks fitted carpet with a black trellis pattern on a white ground. There were beautiful chairs: a bamboo and brass-bound easy chair, the seat and back cushions in blue velvet; a pair of ebonized frame chairs, the seats and backs in ponyskin; a settee on chromium-plated supports, with back cushions upholstered in brown fabric; an Italian easy chair upholstered in black leather on polished metal supports (valued at £200 in 1969). The *pièce de résistance* was a pre-classical (1500-500 BC) Mexican pottery globular shaped vase, its neck formed with a head and a double-headed mask fragment, valued at £750.

Losey tended to expect his most valued collaborators to make themselves available, regardless of other commitments. Macdonald was not available for *Accident* or *The Go-Between*: 'Oh yes, he expected people to be available.

When he wanted me for *The Go-Between* I had been working for three months on *Far from the Madding Crowd*. He said, "Fuck you". Next day he had to ring me up to ask where his viewfinder was. [Laughs] He'd love to say "Fuck you for ever" – and hit you ...' For *Trotsky*, Losey dispatched Macdonald to Mexico on a secret reconnaissance mission, the purpose of his visit concealed. Macdonald tracked down David Alfaro Siqueiros, who was at that time painting three-dimensional murals in the auditorium of the Europa Hotel. Assuming that Macdonald was merely an artist interested in murals, Siqueiros invited him to visit his studio. From Mexico he flew to Italy and advised Losey that the Mexican face, with its blend of Spanish and Indian, its Aztec mask, could not be imitated. Sicilians were not the same.[9] He was unavailable to design *A Doll's House* because currently working in Israel on *Jesus Christ Superstar*. (Losey, misunderstanding a cable alerting him to this, signed 'Richard', promptly sent off a cable to Richard Burton berating him for taking part in such a project.)

Losey generally preferred his colleagues to remain faithful to their wives or partners. Macdonald became involved with Ruth Myers, the costume designer of *Galileo* and *The Romantic Englishwoman*. 'How do we stop this?' Losey asked Richard Dalton, the associate producer (who'd never imagined that his job-description involved stopping such things). Losey instructed Dalton to send Macdonald back to England from Baden-Baden, 'to prepare the titles' – which struck Dalton as a pretty pathetic excuse, but he duly procured a ticket for Macdonald, who then vanished with Ruth Myers and later sent a postcard to Dalton from Baden-Baden (not England): 'Escaped'. Losey and Macdonald never again worked together. Macdonald settled in California. 'There was no enmity,' he says. 'I don't know why Joe chose to have a period of no-speak. It was all on his side.' After moving to France and beginning his collaboration with Alexandre Trauner, Losey began to voice criticisms of Macdonald. With Trauner, he told Ciment, he never had to worry that he might have made 'some hideous mistake or gone off on a complete tangent ...' Macdonald was 'rather flamboyant and selfish and emotional ...' Losey's own 'baroque' style was more in evidence when he worked with Macdonald because they both had 'a tendency towards excesses ... On the other hand, Macdonald's imagination is visually richer – often wrong, but richer to draw on and a wider palette than anybody else I've ever worked with.' Macdonald calls it a 'love-hate' relationship. Losey invariably found people's weak spots and 'made an absolute art of finding where it would hurt most. He would save it up in his mind. He almost resented that I knew who I was and didn't care – I was happy to be a brokendown Scot in exile, whereas he wanted to be a Red Indian – all that guilt about being American.' Like Dirk Bogarde, Macdonald despises politics as a sequence of fashionable postures and clichés. Losey constantly chided him for being right-wing. 'He was always getting at me. If

I said anything remotely liberal, he'd say, "Ah, so you're beginning to change your colours."' Macdonald offers a final, barbed observation: In *Don Giovanni*, Losey and Gerry Fisher 'stood Palladian architecture on its ears'.[10]

33

Metteur Oui, Auteur Non

Journalists and visitors to the set were usually struck by Losey's calm manner – the flare-ups with producers, designers and actresses were rarely observed. An unidentified journalist reported a day's shooting of *Accident* in 1966: 'Losey on the set directs with infinite calm and patience, always guiding gently; when something goes wrong that should have been avoided – a carpenter noisily dismantling a set which is no longer needed during a tense rehearsal of a crucial sequence, for example – his expression is one of weariness rather than of anger. He seems both relaxed and alert at the same time, never overlooking a detail. Sometimes his face lights up in a boyish smile, perhaps when he is making a quiet joke at his own expense, and he momentarily seems to shed forty years.'

Donald Wiedenman described Losey at work on *Secret Ceremony*: 'He stands nervously behind the cameraman, running his hand through his grey hair, shifting his huge frame from one foot to the other. He is tall and continuously adjusts his trousers, pulling them up around his waist, only to have them slide down again a moment later.' Between takes he confers endlessly with technicians. He sighs heavily: 'I sigh as a way of living, as a way of breathing and not because I am mentally burdened.' Geoffrey Moorhouse observed him shooting *Figures in a Landscape* on a hillside behind Málaga. 'Losey does not roar or mutter. He crouches by his cameraman and murmurs . . . When a shot has to be taken for the fifth time because of spectator noise he just sighs and says, "I think that was one of our people this time." He doesn't even glower in the coughing extra's direction.' Businesslike about keeping within a budget and meeting a schedule, he was at his best when shooting a film. Patricia laughs: 'If he'd been like that all the time . . .' Observing Losey on location while shooting *Mr. Klein*, his former Dartmouth student Jonathan Sa'adah was struck by the respect and affection he inspired, and 'the energy that taxed the endurance of people thirty years his junior . . .'[1]

Those on the inside track witnessed another side. Ben Barzman recalled the production director of *The Boy with Green Hair*, Ruby Rosenberg, gripping his arm on the set. 'It seems that Losey is one of your friends? You'll do well to talk to your pal and without delay. He pokes his nose into everything. He checks the cables, the projectors, the lights – he's driving us mad! He wanted to make sure that the lamps were properly screwed! This shit doesn't let anyone get on with it. The technicians are mad. What does he think he's making – a fucking masterpiece?' At that moment they saw Losey standing beside a technician, gesturing and explaining how to alter the lighting with the aid of an oil cloth. A heavy silence lay across the set as actors and crew watched. Finally the outraged technician shoved the oil cloth in Losey's hand and walked away with an angry gesture. Losey then rubbed the lights with the oil cloth. Barzman approached him but before he could get in a word Losey growled, 'This Ruby Rosenberg is one of your friends? Tell him to leave me alone. He's driving me mad.'[2]

In his early days as a director Losey embarked on each day's shooting with a detailed schedule. But later, as he told *Cahiers du Cinéma*, he might work out a *pré-découpage* of thirteen '*plans*' (scenes) – then not give it a glance all day. He liked to improvise and to respond to actors, cameraman, editor. 'At the moment of shooting . . . there is an enormous nervous tension and excitement . . . which gives the film form and style.' Asked in later life, 'Do you storyboard all the shots?' he replied: 'Storyboard? Certainly not. I used to, when I was frightened . . . Occasionally I still do, if there's a very complicated shot that has to be laid out for the cameraman and the actors . . . the preparation for me is something to fall back on if I panic, if I suddenly don't know what to do.'[3]

Gerry Fisher observed that Losey's method was to build up a complex mosaic of visual and sound elements; Richard Macdonald emphasizes his capacity to integrate various skills in order to achieve an overall style. Gilles Jacob noticed 'his precision of gesture and expression, his mastery of décor, framing, angle, lighting and the line and property of objects. Out of this control a style is born . . .' Freddy Buache singled out his capacity to create a perfect framework where décor, light and movement are orchestrated in a dynamic theatricality – he had conquered the authentic modern cinematographic language of which Visconti was the master. From their first collaboration, Dirk Bogarde was struck by Losey's professionalism, his attention to detail. Confronted by plaster banisters on a set, he had them stripped, repainted and polished several times. 'English directors always said, "It won't show, old boy", and took you off for a pink gin. For Joe everything showed. Visconti was the same.'[4]

After thirty-two years on the production side, Richard Dalton regards Losey as the only director he totally respected. 'I loved the man out of respect.' No other director prepared in such detail: each department was notified, in advance, of variations and alternatives (sunshine, bad weather) for the coming sequence. One might panic when spending all day lining up a long and complicated shot: five p.m. arrived and the camera had not turned. But when the cameras rolled the next morning, two days' work was in the can. Losey would look for suggestions on the initial line-ups, his manner not dictatorial. 'Everyone was dying to help. We were all involved.' John Heyman agrees: 'He was a sublime professional, a fine technician, easy to work with, prepared to listen, always on budget, on schedule, and he always made the film he'd been contracted to make . . . He was a warm and emotional person who found it hard to translate those qualities on to the screen. You got an antiseptic feeling from his films – he was the king of involved disinvolvement. He always wanted to angle the hammer blow, which is why he was the darling of the French cinema and incomprehensible to mainstream America.' (But his work was often incomprehensible to Heyman, too.)[5]

Shooting *The Prowler*, for the first time Losey used long, sustained takes to achieve greater continuity of acting. Robert Aldrich, his first assistant, built an electrically operated crane almost as small as the hand-operated model. This was the beginning of a characteristic technique. 'In many films I had set-ups as long as five or six minutes without interruption.' Norman Priggen recalls Losey's mastery of movement, his fabulous track shots, a single take instead of ten cuts. 'That's why we could shoot *King and Country* in three weeks.' But more was involved: Losey is celebrated for his divination of a locale – he described Venice, Oxford and particular houses as 'characters for me'. The camera arrives at a location ahead of the human characters and lingers after their departure. (In *L'avventura*, Antonioni broods on the deserted, sleepy square of a small Italian town in a long stasis before the couple's car edges into view.) Losey's camera slowly pans round empty rooms, capturing the expectation of a curtain rising on an empty stage.[6]

An equivalent Losey 'signature' is the vacated space. The camera dwells for a moment on the scene of a dramatic action vacated by the protagonists. In *Accident*, the kitchen; in *Secret Ceremony*, a patch of lawn. In *The Go-Between* the camera holds a spot on a village lane after the two boys have departed, the cheerful Leo having been traumatized by the news that Marian is engaged to Trimingham. The empty lane not only 'marks the spot' of young Leo's desolation, it also absorbs and incarnates the vital moment. This is one of the minor miracles of cinema; in *L'avventura*, the static camera lingers on an empty alley and a fragment of the little town after the couple have driven away, then glimpses their car as it briefly reappears round a bend. In *Trotsky*, Losey observes the now-empty hall where the painter Ruiz and his comrades have cheerfully dressed up in police uniforms; in *Mr. Klein*, he scans Robert's walls stripped of their paintings by the police. A recurrent feature of Losey's later films is the separation of the image from the overheard conversation.[7]

He believed in clean cuts, rather than fades or dissolves. His retrospective notes on *Blind Date* stress that he used very few 'opticals' (effects devised in the laboratory, instead of with the camera, such as overlapping images and fades). 'I abhor opticals.' This attitude was encouraged by his first British editor (cutter) Reginald Mills. During the following thirty years Losey felt entirely confident in only two editors, Mills and Reginald Beck.*

A Cambridge graduate who had entered the industry in 1935, then directed the Army Kinema Service during the war, Reginald (Reggie) Mills accumulated credits which included *The Tales of Hoffman* and *Red Shoes*. After *The Servant*, Losey wrote to him: 'Your work is superb, unequalled and absolutely indispensable to me.' But it wasn't; *King and Country* was to be their last collaboration. Explanations differ: Mills is no longer alive; Reginald Beck, who supplanted him as Losey's regular editor, says: 'Mills decided he wanted to direct. He left *King and Country* to attend some function, leaving Joe high and dry. He never came back.' Ruth Lipton adopts Mills's side of the argument: 'He gave up a lot of jobs waiting for Joe . . . But Joe concluded that Reggie had betrayed him when he finally did take another film and wasn't available for Joe's.' An additional factor, quite probably, was Norman Priggen's personal dislike of Mills. Either way, Losey brought in Beck for *Modesty Blaise*. On 1 April 1966 he wrote to Mills, referring to his next project (without naming *Accident*) 'which I had expected and should be offering to you' but which would be 'unattractive and impossible for you'. Why? Losey listed a number of factors, including a conflict of dates with Mills's (assumed) annual itinerary and the necessary presence of some of the *Modesty Blaise* crew. Mills found Losey's letter 'hard to believe . . . I resent the way in which twice during the last nine months you have taken the trouble to tell me the reasons why I am unable to work with you.' If Losey no longer needed his services, why not say so? 'People change I know but I never thought that you would.'[8]

Losey's attention to detail extended from pre-design to the projection of the film in the cinema. Constantly he harassed producers about the quality of laboratory work, printing, colour, light, spots, subtitles. Priggen would work with Losey checking the prints. 'We'd go and check every copy going out for distribution.' While shooting *Galileo*, Losey warned Ely Landau: '. . . quality depends entirely on personal supervision and taste because of the fluctuation of standards in photographic stock, baths, lab lights, etc.'[9]

Losey was a 'writer's director' in several senses: he didn't write his own screenplays; he invariably worked from a fully elaborated shooting script; and he valued writers, normally welcoming their presence during shooting, an attitude less than universal among directors. His editorial skills were remarkable, his capacity to trap the essence of a script. He would telephone Ben

*Mills: *The Sleeping Tiger, Blind Date, The Criminal, The Damned, Eve, The Servant* and *King and Country*. Beck: *The Gypsy, Modesty Blaise, Accident, Boom!, Secret Ceremony, Figures, The Go-Between, Trotsky, A Doll's House, Galileo, The Romantic Englishwoman, Mr. Klein* (with Henri Lanöe), *Les Routes du sud, Don Giovanni* (with Emma Menenti and Marie Castro-Vazquez) and *Steaming*.

Barzman in France to ask if he accepted the smallest changes, which the majority of directors would have made without consultation. Their collaboration was 'stimulating, enriching, marvellous . . .' – yet Losey always somehow made him feel guilty, as if Barzman hadn't quite given his best.[10] The basic problem with Losey's films from 1948 to 1960 can be summed up in two words: 'bad scripts'. During the first thirty years of his career he was offered only one play (*Galileo*) and not a single screenplay of genuine literary distinction. Ten out of eleven of the pictures he directed before *The Criminal* were written by left-wing Hollywood writers cast in the Movie Mould and content to smuggle a few 'social' points into blatant melodrama. Their dialogue was terse, tough and sometimes crass. Ben Barzman wrote six scripts for Losey (four of which were made) before Losey finally saw the light and made the break. But then four of his next five films were written by Evan Jones, despite striking variations in the material and required sensibility. In 1968, interviewed by Andrew Sarris for the *New York Times*, he disparaged Hollywood screenplays loaded with explanations and precise camera directives. The Hollywood style – Losey here cited *The Prowler* – was to spell everything out. 'A writer like Dalton Trumbo would hand me a script with all the camera angles indicated. I would never accept such a detailed script today.' By contrast, Pinter's explanations (in particular, motives, states of mind) were implicit in the dialogue and the suggested scenarios. The *Times* published Trumbo's indignant rejoinder. Having worked with thirty directors, 'some almost as distinguished as Mr Losey', he insisted that not one of them had objected to his custom of breaking a script into shots. Besides, a script written in master scenes might be allotted ten or twelve days, less shooting time.

After 1961 Losey never again worked with a writer from the stable of Hollywood blacklistees. Writing to a novelist in 1975 about potential screenwriters, Losey warned against 'Hollywood hacks' and 'one also has to avoid any of the Hollywood 19' – except perhaps Waldo Salt.[11] Paul Jarrico was a friend but Losey was still saying no to his projects in 1980: 'Not for me.' Alvah Bessie's jaunty, sometimes facetious, letters to Losey make painful reading – erstwhile political persecution becomes a permanent alibi.

'Auteur' means 'author' but the notion of an 'auteur' film director is sometimes employed as an accolade, sometimes to indicate stylistic distinction, a personal signature on the film which would be identifiable without a credit. The young critics of *Cahiers du Cinéma* in the mid-1950s, Truffaut among them, were keen to elevate their preferred directors as 'auteurs', and when the *Cahiers* and *Positif* turned their spotlight on Losey it was discovered that the derivative quality of the screenplays he worked from was either irrelevant to his status or a bonus; Losey 'wrote' with his camera.

The influential critic André Bazin had embraced the notion of an 'auteur' director as one who 'speaks in the first person'. The fundamental question is whether this 'speaking' is mainly cinematic and stylistic, or whether it should also embrace the genesis of the film: choice of subject and scenario. Crude though the distinction between originating and interpretative talent may be, that is the uncertain frontier we are mapping.[12]

In Losey's view, writing and directing were 'separate functions', although a creative director would contribute to the writing, just as a creative designer or cameraman must contribute to the mise-en-scène. (This was also Resnais's attitude.) Yet many directors either wrote or supervised their own scripts: for example, Welles, Bresson, Satyajit Ray, Buñuel. In January 1969 Dalton Trumbo and Losey held a private debate on the subject, Trumbo insisting that many writers became directors of their own screenplays, mentioning Dryer, Rossellini, Antonioni, Bergman, Cocteau, De Sica, Fellini, Pasolini, Kurosawa, Preston Sturges and Truffaut among others. Losey replied: '. . . surely you cannot be serious about Antonioni, De Sica and Fellini as writers? They are no more writers than I am and always work with half a dozen or more stooges on their scripts.' Interviewed for Thames Television's *Moviemen* series in October 1969, Losey remarked: '. . . whether a director is an auteur or not, this must obviously relate to whether people create their own material.' But, having agreed to this definition, he went on to doubt whether it ever applied in practice: 'Fellini, Antonioni and others whom I admire, and I'm not meaning to derogate them, may have as many as seven writers on the film. I prefer to have one writer . . . who is strong rather than many who are manipulated.'[13]

Even Losey's French admirers drew the distinction that he himself declined to recognize. Michel Ciment noted that unlike many other major directors, Losey was not his own screenwriter; he injected his own visualization and obsessions into the work of others. Gilles Jacob went further: 'But Losey's real problem is that he has nothing much to say, nothing original at least, and has been content to work for the most part on subjects of some banality . . . one does jib when he heaps his films with pseudo-philosophical ideas borrowed from psychopathology and half-digested Freud, or existential messages whose ambiguity can mean either everything or nothing.'[14]

Losey was inclined to state the available choices in quite different terms: 'A director doesn't choose his subjects, they're offered to him.' (But in January 1967 he told *Le Républicain Lorrain*: 'Bergman and Fellini are the great punters (*pontes*) of the cinema because everything is permitted them, they can undertake everything.') In 1980 he remarked: 'I don't know of any director who really chooses his subjects because . . . you are always bound by the people who think you're commercial, or you're not commercial . . .'[15] This raises two distinct questions: how many of Losey's films were imposed on him by producers and by the need to work; and how eclectic was his own choice of 'material' (a word he used frequently)? Michel Lambert of *Télémoustique* (15 February 1980) commented critically: 'Losey has made many films to order (*films de commande*). At first sight, it's quite astonishing.' Michel Ciment came to Losey's defence: 'Among all these things proposed to him, he retains only one out of ten.' (But this is a myth, unless one equates 'things proposed' with mere scripts flooding in. A real film proposal is one backed by a producer with development funds; if we adopt this definition, Losey's refusals are hard to find.) Ciment then moved back into the 'auteur' debate: 'Let's say the commission becomes a kind of choice . . . Losey directs it in such and such a way, he appropriates the commission.' This indeed takes us to the heart of the debate, though Ciment pressed his point beyond credibility when stressing that Mozart also worked to commissions, and was asked to write operas on given subjects.

Losey displayed striking eclecticism in his search for subjects. In October 1958 he wrote to the producer Daniel Angel: 'I am still looking for a subject to star Jeremy Spenser, who I think will be the next British star, and if we could team him with a sexy Bardot or someone like that I think that's what we are looking for.' This is not the voice of the 'auteur' browbeaten by commerce. 'There's no conscious pattern at all in my work,' he said in 1963. 'If I get a subject that is interesting to me I think I can do something with, I do it . . .' Writing to his agent in 1963 he expressed a desire to pursue David Storey's new novel 'or any other new material that may come along that seems exploitable . . .' 'Exploitable' meant, mainly, star casting. Three years later he informed Angel: 'Of John Galsworthy's social plays, *Loyalties* was in many ways the most interesting and lends itself to star casting'. A Somerset Maugham story could star Ingrid Bergman. James M. Barrie's *Dear Brutus* and *Mary Rose* both carried star-casting potential – as did D.H. Lawrence's *The Woman Who Rode Away* and two plays by Pirandello. Lermontov's *Hero of Our Time* was 'a possible vehicle for James Fox'. Pinero's *Dandy Dick* was right for Gladys Cooper. And so on. In May 1972 he wrote to Delon: 'I am in search of a bitter love story or an original suspense story . . . I will be going through my file of abandoned ideas tomorrow.' In October 1973 he approached Jack Lemmon: 'How would you like to play an ageing member of The Three Musketeers in a straight version of their last venture – in a sense a re-make of *The Iron Mask*?' Two months later he wrote to John Heyman: 'I would be prepared to do practically anything that could make money for us both.' 'He wanted to do everything,' Patricia Losey recalls. 'He was eclectic.' She discussed with him the advantage of starting from an idea, but 'it was never going to happen'.[16]

Losey became a 'modernist' in the stylistic sense: 'I think more and more the film medium is . . . not a novel, not the conventional Hollywood screenwriter's form, it is a visual form . . . In other words a form in which images take the place of words . . . And you never use a word when you can use an image.' (By contrast, the standard Hollywood formula might be defined as functionally expressive camerawork at the service of stage dialogue, plus visual spectaculars.) But Losey's modernism should not be confused with the authorial instincts of an 'auteur' director. The 'auteur''s line of vision is an insider's – he 'paints' the world he knows best. Losey was semi-detached from the new society of the 1960s. Comparing Godard's *Pierrot le fou* to *Modesty Blaise*, Penelope Houston commented: 'And one reason why Godard is so modern an artist is that he takes the whole disposable civilization for granted, works from inside its values and sorts them out on his own terms. A director of an older generation is simply unequipped to do this . . . the defence of middle age is to take up the fashion, but to protect itself through a whole series of barricades against any risk of genuine involvement.'[17]

An outstanding example of a contemporary 'auteur' director is Woody Allen. Losey dismissed his work as 'unvisual'. Allen not only writes and performs in his own films, pulling the rabbits from an apparently bottomless hat, but each film is a report on the world he knows best. Like Godard, Allen sometimes gives the impression of almost frivolous spontaneity as he plants new flags across a familiar landscape. By contrast, Losey avoided contingent contemporary references and the 'sideways glance'. There are no wayward incidents, no demos crossing the street (as in *Blow-Up*). Nothing intrudes which is peripheral to the action, the script, the *découpage*. While filming *The Servant*, Losey thought of injecting a short extract from the popular political satire *That Was the Week That Was* [*TW3*], to be screened on Tony's television set, but dropped the idea. There are no accidents within the seamless stitching of *Accident*. While shooting in Oxford Losey was irritated by World Cup football commentaries on the crew's transistor radios; an 'auteur' director might well have slipped this national fever into the soundscape of the film as a counterpoint to the tolling bells. Unlike the *nouvelle vague* directors (and Buñuel), Losey was too fastidious and manicured to respond to the surrealist heritage. The neo-Brechtian farce of Godard's *Weekend*, the insolent vomiting up of the here and now, remained alien to him.

Although he dreamed heavily – his father being alive, after all, was a recurrent dream – unlike Buñuel and Bergman he never surrendered to the world of sleep. No hearse runs out of control into a lamppost, dumping the coffin and disgorging the corpse (as in Bergman's *Wild Strawberries*); no character steps up the camera and offers his dreams (*The Discreet Charm of the Bourgeoisie*). A relative lack of humour accompanies the lack of spontaneity. 'It's very hard for me to get any kind of comedy material. I mean get any backing for it.'[18]

Losey's brilliant observation of British society remained strictly classical and sculpted. *Accident* is a better film than John Schlesinger's *Darling* or *Sunday Bloody Sunday* (which took the British 'Best Film' award from *The Go-Between*), but what distinguishes both of Schlesinger's films, and particularly *Sunday*, from Losey's is a relaxed immersion in topicality – telephone answering services, kinetic and op art, permissive parenthood, sexual confusion, push-button gadgetry, snatches of radio, children passing in the street and casually ruining the paintwork of cars. The heroine (Glenda Jackson) is a kind of everywoman, mobilizing generosity and good sense to the point where the script lies down under life – she's us. Really no one in Losey's three Pinter films is 'us'; the contingent 'ad libs' that Losey loved were heavily stylized rather than 'hand-held camera'. As a high priest of style, Losey became a 'director's director'. *Sunday* attracted large audiences because it radiated a warmth, a fondness for the here-and-now, that Losey's work lacked. Arriving at *The Romantic Englishwoman*, Glenda Jackson brought with her the same 'everywoman' combination of common sense, emotional longing, domestic acuity – but the script and Losey's direction cornered her in elaborate levels of non-reality and substituted artificially contrived confrontations for *Sunday*'s deadly innocence. The story becomes a 'found object' which is worked upon with elaborate pre-design into an 'international film' starring Helmut Berger and some fancy locations. Ten years earlier Gilles Jacob had observed 'a sort of abstraction, a laboratory chill in the construction of [Losey's] characters. Because they have been carefully and patiently cultivated in the test-tube, they lack the freedom of life, that tiny seed of incalculability . . .' which registers real emotion. In life Losey's eagle eye pounced on a hole in a garment, an abandoned supermarket trolley, a dog's turd, but in his art he exacted a kind of 'revenge' (Monique Lange's term) on the mundane world by passing over it into pre-design and a classical rigour.[19]

PART X

The Years of Frustration

34

Proust and Pinter Found and Lost

Could Marcel Proust's *A la recherche du temps perdu* (*Remembrance of Things Past* or, more correctly, *In Search of Lost Time*), an eight-volume novel of supremely literary texture, be adapted as a film worthy of its literary provenance? The parabolic work of a master craftsman, Pinter's screenplay traps Proust's seminal experiences, his longings, inhibitions and uncertainties – and most spectacularly, the physical fabric of his life – but this inevitably involves stripping away his genius, which resides in layers of elaborate prose and discursive reflection without which he is anybody who might not have written the novel. A film of *A la recherche* can only provide an X-ray of the skeleton beneath the flesh, with Marcel himself emerging as a wide-eyed owl whose ultimate powers lie beyond the frame of the film. (Claude Chabrol's film *Madame Bovary* (1990) offers voice-over passages from Flaubert's novel, but they merely remind us of divorce rather than of marriage.)

In August 1964 François Truffaut had declined an offer from Nicole Stéphane to direct a film of *Un amour de Swann*, listing the major literary works he had felt constrained to resist. In Michel Ciment's view, '*snobisme*' (and Visconti) led Losey to Proust – great films have not, on the whole, emerged from great books. The producer Daniel Toscan du Plantier, who became involved in the project, agrees: 'You don't make great literature into films – *Gone with the Wind*, yes.'[1] For Losey the great trap was obvious: milieu and mannerism, decadence and ennui, pursued as ends in themselves. An atmosphere of idle arrogance and sexual ambiguity alternating with the splendours of the opera, great dinners and soirées would inevitably raise the twin problems of 'style' and 'integrity' in acute form.

A group of French production companies had already spent nearly one million dollars on acquiring the rights and investing in previous directors and their writers. Losey later told Patrick White: 'Also a great deal of money was wasted . . . by Visconti and Réné Clément, who were attempting to make films of limited parts of the book: ("Swann's Way" – Clément; "Sodom and

Gomorrah" – Visconti).' During the 1971 Cannes Festival, Visconti informed Nicole Stéphane and her colleague Robert Dorfmann that he had decided to film *Ludwig II of Bavaria* at dates conflicting with his commitment to make his Proust film; he was accordingly informed on 30 July that all obligations to him were ended. Nicole Stéphane (once the marvellous girl in *Les Enfants du paradis*), had been devoted to the project for years and felt cruelly let down by Visconti. She now approached Losey and Pinter. Barbara Bray, a Proustian authority and distinguished translator, was enlisted as literary adviser; she takes credit for having persuaded Losey and Pinter that Proust's novel should be adapted in its entirety. But Stéphane was always sceptical; although she expressed admiration for Pinter's screenplay, in retrospect she calls it '*immontable*' (unproducable): 'I told Joe it could not be done. His version was four times more expensive than Visconti's.'[2]

A draft agreement dated 18 January 1972 involved a large number of parties and partners, but principally Nepenthe Productions (Nicole Stéphane), her esteemed colleague Robert Dorfmann (Les Films Corona), and La Société Ancinex. A draft contract dated 28 April, between Nepenthe and Launcelot, indicated a directorial fee of $145,000, but it remained a draft. An agreement dated 5 October guaranteed Losey payments totalling $30,000 up to delivery of the final script. By June bad-tempered correspondence was passing between agents and lawyers; Florence Malraux called it a 'wasps' nest'.

Losey and Pinter embarked on summer location trips to Illiers, Cabourg (Proust's 'Balbec'), and Paris. Losey enjoyed a 'Rabelaisian meal' as guest of Bruno Coquatrix, Mayor of Cabourg (Calvados). Diligent and passionate in his research, Losey compiled three folders containing: Hôtels Particuliers Divers; Rues-Ruelles; Jardins, Halles et Escaliers, Opéra-Théâtre; Restaurants Pâtisséries; Appartements Bourgeois; Châteaux. The Baron Charlus was to occupy the Hôtel du Baron Élie de Rothschild and the Hôtel du Baron Alain de Rothschild. As with *The Go-Between* and – later – *Don Giovanni*, Losey's depiction of a bygone aristocracy brought him into negotiation with a contemporary aristocracy devoted to beautiful films: various Rothschilds, the Duc de Gramont, the Comtesse de Cossette among them.

Pinter's first screenplay (October 1972) was revised in April 1973, and re-revised in October. The first words of dialogue are not heard until the thirty-sixth shot. The revisions were partly influenced by Losey's own annotations. Here and there he put lines through or beside the dialogue, with the comment 'Pinter' or 'H.P.', indicating that the word-patterns owed more to Pinter than to Proust – particularly where words were deliberately repeated in conversation.*Losey began by writing occasional camera directives in the margin – 'Her eyes to Marcel's to L.S. to zoom to men' – but did not persevere. One or two annotations were dyslexic. Against shot 22, the yellow wall in Vermeer's *View of Delft*, Losey wrote, 'The Haig' [Hague].

Recognizing another Pinter masterpiece, he pursued potential collaborators in a mood of euphoria. Writing to his intended art director, Carmen Dillon, he predicted pre-production shooting from May 1973 in Normandy, Venice and Holland, with principal shooting in France to follow in August, lasting sixteen to twenty weeks. He told Leonard Bernstein that his recent Stravinsky concert at the Albert Hall demonstrated that he was 'the unique person' to collaborate on the Proust project. 'Music, obviously, will play an enormous part ... the violin sonata and the septet ... the character of Vinteuil ... based on a combination of Saint-Saëns and César Franck ...' Bernstein pleaded lack of time. Losey persisted. Pierre Boulez also declined.

On 4 February 1973 a crisis conference was held at Royal Avenue; the required financial backing had not emerged. Losey told the *New York Times* (14 February): 'I think it will be commercially profitable, but it's not a short-term tax shelter. We need investors ... who will look for their rewards like they would in buying a Renoir' (perhaps a doubtful analogy). In desperation he turned to Sam Goldwyn Jr, Dino de Laurentiis, Sam Spiegel, Goddard Lieberson of CBS, United Artists and other movie moguls. American reactions gave little cause for optimism. The agent Robert Lantz described the project as 'special, remote, lacking in action, very interior, and generally "literary". This is the age of Gene Hackman and Barbra Streisand. There are no roles for them here ...' Losey consulted Gregory Peck, who also drew a blank. The Vietnam war notwithstanding, Losey sent a four-page appeal to his Dartmouth College classmate of '29, Governor Nelson Rockefeller: 'Dear Nelson, Forgive my intrusion on your obviously very busy life with a query, which seems to me to fit in with what I know about your tastes and your sponsorship of the arts.' Would Rockefeller read the Proust script? Could Losey call on him? The reply from the Executive Chamber, State of New York, was polite but brief and negative.[3]

Losey's temper shortened. Monique Montivier, director general of Ancinex, ran into a storm when she dared to send him a detailed timing of the screenplay at three hours fifty-eight minutes. Nicole Stéphane concedes that this letter was '*maladroit*', its purpose being to pressurize Losey and Pinter into shortening the screenplay. Following a meeting with him and Pinter in London, Nicole Stéphane (herself a Rothschild with access to their houses) lamented certain remarks by Losey belittling her professional competence, 'unexpected mistrust and this bad temper'. In reply, he suggested that his

*For example, Losey crossed out three speeches in Scene 87 (Swann – Odette – Swann) and seven speeches in Scene 128 (Charlus and the page); an entire Charlus monologue went from Scene 129, ten lines from Saint-Loup in Scene 187, in which the word 'idiot' is repeated and 'rumours' repeated twice. Pinter's 'Revised 1973' script is largely responsive to Losey's points and reduces 468 shots to 455.

much distorted clashes with Jane Fonda had given him an unjustified reputation as a male chauvinist. 'You have been reacting more subjectively than you might have done because of the particular conditions which affect a woman in the Hollywood commercial film world . . .' Stéphane becomes emotional at the memory of Losey's treatment of her: 'He was more than a misogynist – for him women didn't exist.'[4]

On 8 December 1973, Losey, Pinter, Bray and Stéphane gathered for the day at Royal Avenue. It was a grim meeting, Pinter tight-lipped and holding his anger in check. Sceptical voices were in the ascendant. Graham Greene sent Losey commiserations but thought it 'a rather impossible project unless you took Swann's story and his relations with Odette by itself. But then as I seem to remember that would have left out Charlus.' Nastier was a letter Pinter received from Laurence Olivier, to whom he'd sent a script: 'May I offer a little word of warning about kaleidoscopic insert-cum-flash technique? . . . I am terribly sorry that I don't want to play Charlus. I can't see anything to him beyond his characteristic values, and a plain dilettante decadent or plainer still a *filthy* old snob and nothing else somehow doesn't feel interesting.' It was a long letter, mocking in tone – 'Thank you for thinking of me. It is a masterly achievement' – with a savage ending: 'Perhaps you could get somebody of normal intelligence who's never read Proust to tell every time they are foxed as they read it through . . . someone like a good continuity girl, I mean.' Another actor whom Losey had approached, Dirk Bogarde, applied a stiletto: 'I cannot see the Yank Orientals on the Coast having an idea as to what it is all about . . . the audiences too . . . will *they* know who is who? Or what is what? . . . Luchino [Visconti] said, years ago, that there was nothing for me in Proust . . .' (But Bogarde also says that Visconti had him earmarked for Swann.)[5]

In June 1974, Losey turned to the recently elected President of the Republic, Valéry Giscard d'Estaing, whom he had met once, briefly, inviting him to discuss a Franco-British production. 'I consider this is the major project of my life.' A copy of the screenplay was delivered by hand to the President on 29 July; early in October Losey was granted half an hour and found the encounter awe-inspiring. Politely but with undisguised scepticism Giscard inquired whether he was sure he was the person to tackle so French a subject; Losey assured him that there were many parallels with other worlds. Giscard had not read the script, and Pinter's name didn't seem to mean anything to him. A year later Georges Beaume's office informed Losey 'the interview with Michel Guy, the Minister of Culture, was very positive and you can count on the government to aid the project . . . Giscard d'Estaing wants the film to be done.' But there was a snag: 'The French government, as a condition for their help, would want the film shot in French.'

Was language a problem? None of those involved seems sure. Barbara Bray speaks of bilingual casting plus some dubbing; the master production would be in English. Nicolas Seydoux, President of Gaumont, comments that his company could not have produced Proust in English – 'But that wasn't the problem.' Losey was now looking to Gaumont for the money. On 3 October, Losey reported to Pinter an hour-long meeting with Seydoux and Daniel Toscan du Plantier, Vice-President in charge of production, about a possible collaboration with one of the French television channels. Here Losey was virtually pleading for Pinter's consent: 'I can easily give them five 50-minute episodes ... There would be no changes of text ...' By now Losey's scapegoat was the woman who had first brought him the project. 'The other problem, always of course, is the mindless intervention of Nicole Stéphane who has again raised her price by $250,000 ... I feel that if anybody can ever handle this mad woman it will be these people' (i.e. Gaumont, to whom Stéphane had in fact introduced him). Barbara Bray recalls that more than once Stéphane proposed limiting the film to 'Un amour de Swann': 'She asked me to write it but I said no. She also suggested that we drop Joe – we didn't tell him.' Bray points out that Stéphane's devotion to the project had brought her into real financial difficulties, 'but Joe could never see other people'. Stéphane agrees.

On 19 October Losey's agent Georges Beaume gave a dinner at the Ritz – at Proust's table, with the same *maître d'hotel*. Pinter arrived from London with Antonia Fraser, whom the Loseys met for the first time. Toscan du Plantier was present but the empty chair belonged to Nicolas Seydoux, who was hunting in Spain. Patricia nevertheless adjudged the dinner 'exquisite': 'The sommelier, as he poured each glass of champagne or wine or port or *fine*, announced name and year, sweetly, like a sacrament.' The '*chaise vide*' remained a memory that neither Losey nor Pinter could forget; Stéphane was also 'enraged' and claims that Seydoux sank ('*a fait sauter*') the project. 'He's a shit' (she uses the English word). Seated atop the Gaumont building in the Avenue Charles de Gaulle, Seydoux does not remember the occasion: 'If I was absent it was not deliberate.'[6]

Losey clung to hope: 'Gaumont will buy all the rights from Nicole and Dorfmann for $350,000 ... Gaumont is then committed to roughly $2,250,000 with French TV O.R.T.F. contributing in equal amounts ... the rest of the [$10,000,000] to be raised through Italian, English and German co-productions with financing to be completed eventually by John Heyman through one of his tax-shelter deals ... I will work on the film and the TV simultaneously with two cutting crews and no doubt two post-sync crews to produce five TV episodes ... and one film of three to three-and-a-half hours. French TV will refrain from showing their series for two years after the film distribution.[7] But the estimated budget continued to climb: by May 1976 production costs were predicted at $11,083,407 (55,417,036 francs). Losey would receive

$400,000, while the literary rights and the Pinter screenplay would cost $425,000.

That month a distressing incident occurred during the Cannes Film Festival. With great embarrassment Bernardo Bertolucci told Patricia that he'd been offered *Un amour de Swann*. Apparently the offer came from Gaumont but it could only be with Nicole Stéphane's consent. The Italian director left it to Patricia to tell Joe. 'Joe had more vodka. I felt full of special hurt for him as he was wearing his Proust T-shirt.'

While Losey was clearly prepared to bend to television, Pinter wasn't. He would not contemplate a shorter version. 'I have no confidence in Nicole's television idea and can have nothing to do with it. However . . . I think we should let her go ahead with it . . . she has had a terrible ten years . . . We must still hold on to the right to do the film, should the money turn up. I know you never give up and you might even consider what I've said here as a betrayal. I hope not.' In 1978 Pinter published the screenplay, dedicating it 'To Joe and Barbara' (and according each a sixth of the royalties). In his despondency Losey turned to merchandizing Proust, negotiating with Paris *Vogue* to use either actors (risky, he thought) or people from '*le haute monde*' who could dress up as six characters 'from the Pinter–Losey screenplay'. Each couturier would take one character but make three different costumes. The feature would occupy fourteen colour-pages and would use either extracts from the English screenplay or from Proust himself. Losey spelled out his own required, non-taxable, remuneration in detail, but nothing came of it.[8]

Visiting New York in 1982 for the opening of *La Truite*, Losey was angered to hear that Volker Schlondorff was to film a section of Proust's novel, *Swann in Love*, for Nicole Stéphane, starring Jeremy Irons and (cruel cut) Alain Delon. Such anger had become a familiar condition: *Under the Volcano*, *Voss*, *Ibn Saud* and *Dr Fischer of Geneva* had all passed into the hands of other directors. He later conceded in conversation that he had admired Delon's performance.[9]

During the 'Proust years' the relationship between Losey and Pinter subtly altered as the one struggled against decline while the younger man fended off hubris by force of character. On 19 October 1972 *The Times* published an indignant letter from Pinter in which he described as 'impertinent' the opening paragraph of Ronald Hayman's savage review of *The Assassination of Trotsky*. Hayman had written: 'Three of Pinter's screen plays and a chronic need for film directors who can be elevated to the rank of culture-hero have helped to inflate Joseph Losey's reputation beyond all reasonable bounds.' In March 1973, Losey mentioned to Bogarde that he had sounded Pinter out about a film of *Old Times* 'and got a non-response – neither negative nor positive . . . I suggested that he himself should write the script.' Pinter had pondered a film version but explained that he hadn't discovered a cinematic solution. In July, Losey lectured at the National Film Theatre on the occasion of a major retrospective of his work. Afterwards the Pinters came home for dinner and sat in the garden on a balmy night until two or three in the morning, drinking the whole of an 1842 brandy, previously decanted, without any subsequent hang-over.[10]

Deep into one of the most frustrating interludes of his career – fourteen months since turning a camera and nothing in sight – Losey was invited to a succession of Pinter first nights, with their attendant choreography of celebrities, receptions, photographers, handshakes and smiles. It was hard; Proust was stymied, but Pinter's disappointment was relieved by a storm of work, his reputation soaring. For Losey, the Proust project was only one of many bogged down in frustration. Joe was never at ease on social occasions, even when held in honour of his own work, but 'Pinter's Progress' was like a summer to his own autumn. Twenty-one years separated them in age, but here was Pinter directing a film (*Butley*), here was Losey not directing the film of a famous Pinter play (Peter Hall's *The Homecoming*), here was a double-bill (*Landscape* and *Silence*) at the Aldwych; it was like being a middle-row guest at the many weddings of your own best man.

On top of this Losey had never really liked – and perhaps not always understood – Pinter's plays. He celebrated and took some reflected glory from Pinter's reputation, but he didn't like Pinter's plays in his gut. Although himself a 'modernist' by conversion, Losey's modernism was always grafted on to an inalienable socialist-realist root culture. Russian avant-garde theatre, yes, the Living Newspaper, yes, Brecht, certainly – but that line of consequential inconsequence known as 'the absurd' was never Losey. Each of Pinter's four screen adaptations for Losey remained basically faithful to the original author's traditional psychology of human behaviour. Losey thrived on Pinter's highly individual twists and turns and tones, but always on the basis of Maugham, Mosley, Hartley, Proust. With these screenplays Losey knew where he was; with Pinter's plays he didn't (always, or fully).

It was bound to show, during these flash-bulb first nights of the early 1970s. Losey would leave the theatre or cinema hurriedly, without staying for the drinks, without a word, just a wave on the way out if Pinter's anxious eye crossed with his own. Losey's temperament left little space for good manners or false smiles, or the murmured 'Wonderful'. Pinter cared about Losey's good opinion, always sent him his new play as soon as it was typed up, and registered these cumulative rebuffs with a pain that led to anger. The verbal violence in Pinter's plays – often swathed in ice – doesn't come from nowhere. The skin is thin. Losey's too: as Patricia recalls, he 'did wish Harold would let him direct one of his plays'.

Casting *The Romantic Englishwoman* in February 1974, before it was definitely 'on', Losey offered minor parts to both Vivien Merchant and Harold Pinter. Patrick White named Pinter as the screenwriter he most wanted for *Voss*, 'if it could possibly be fitted in to his writing schedule . . .' Losey immediately advised White that Pinter was extremely busy with a new play and re-writes for *The Last Tycoon*. 'I wouldn't hold forth a great deal of hope in relation to Harold because I really don't think it's his kind of subject . . . Harold, of course, is considerably more expensive than any of the other writers we have been talking about . . .' But also the best: 'In all my years of working on films I have never found any writer at all who even approached Harold's contribution . . .' Despite his warnings to White, Losey urged Pinter to give serious consideration to *Voss*, but Pinter was preoccupied with finishing *No Man's Land*.

On 17 August Losey asked Pinter whether he still wanted to play the Doctor in *The Romantic Englishwoman*. Evidently his reply was negative; four days later Losey informed Tom Stoppard that Pinter 'no longer finds the Doctor scene interesting for himself . . . Previously found it amusing – now finds it factual and not amusing for himself . . . Vivien likes the script but wants to discuss it . . .' Things then turned very bad indeed. Losey withdrew his offer to Vivien Merchant to play Isabel; the role went to Kate Nelligan. Pinter was evidently furious and Losey expressed his regret and grief, explaining that the producer held a right of veto on casting: 'I got him down to EMI yesterday to see Vivien in a reel of *The Maids*. Danny [Angel] insists that "she is the wrong generation". This of course is nonsense . . . but I am very sorry about the humiliation and disappointment – if any – and I hope there isn't any.'

'I've read your play [*No Man's Land*] twice,' Losey wrote on 4 October, 'and think it quite staggering – brilliant and lyrical. If asked what it was about I would be at a loss to say excepting something like myths and reality.' (Three weeks later Patricia read the play: 'Very brilliant. Very funny . . . Joe says he doesn't understand it . . .') Meanwhile, Vivien Merchant conveyed her own tensions by sending Losey some glossy stills from Genet's *The Maids* (the film that Losey had shown Daniel Angel) with a barbed comment about 'Miss Jackson' – Glenda Jackson having played in both *The Maids* and *The Romantic Englishwoman*.

On 6 January 1975, Pinter had lunch at Royal Avenue the day after the Loseys came off one of their periodic liquid diets, and drinking began again in earnest. Unknown to the Loseys, the Pinters' marriage was approaching a terminal crisis. It was at dinner on Easter Sunday that (Patricia recalls) 'Harold dropped his bombshell over coffee.' He had fallen in love and had told his wife. The conversation ran deep into the night. What followed became the delight and delirium of the gossip columnists – with the inevitable corollary for the Pinters: persecution and pain. Such affairs command the absorbed attention of every class and estate, from cardinal to char. The victims' friends feign shock, disgust, at the vulgar clamour, while privately passing on the gossip, one to another. In July, one such friend, David Mercer, reported to the Loseys that he'd heard from 'a reliable source' that ... The point is not what Mercer had heard, but the fact that, having passed on the details, he lamented: 'Show business London is agog with it all, and I find people's sneaky, cockahoop-morbid interest both tiresome and distasteful ... as far as my friends are concerned I am deaf to rumour. Since Harold hasn't been in touch, I have mentally sealed off and quarantined the whole area.' But he hadn't, had he? In his published interviews Losey's tongue often wagged about the private lives of colleagues, yet when challenged about his published remarks he never hesitated to declare himself misquoted or disparage the journalist. In September Mercer reported again: 'I had lunch with Harold a few days ago, and he seemed much recovered. The attentions of the slimy press are dwindling.' Mercer again went into the personal details, ending with the quotable reflection that, 'One hardly creates hell in life in order to have something to write about, but I never yet met a writer who didn't sooner or later profit from even the most painful of experiences.'[11] Well put. Losey, meanwhile, had to live with the knowledge that his *bête noire*, Elia Kazan, had directed Pinter's screenplay *The Last Tycoon*, written for another figure from Losey's past, Sam Spiegel.

The Loseys first met Pinter's future wife, Antonia Fraser, when she and Pinter flew to Paris in October 1975 for the grand 'Proust' dinner at the Ritz. The Loseys drove to Roissy to meet them but the flight was two hours late and they left their chauffeur with a physical briefing: 'Antonia blonde, tall, English, beautiful. Harold, dark and elegant. Antonia very flattered as she said Boda [the chauffeur] went to her first.'[12]

Early in 1970 Pinter had gone to the country and written a first draft of *Langrishe, Go Down*, based on a novel by Aidan Higgins. This screenplay, as Losey immediately realized, matches Pinter's best film work: clear, sensitive, finely crafted, moving effortlessly between two time sequences, that of 1932 and that of 1937–8. The story concerns the impact of a visiting German scholar, Otto, on two Anglo-Irish sisters whose family has declined economically since their father's time. Sexual repression and tension pervades the script, which makes frequent use of voice-over and rapid alternations to convey remembered exchanges – an ideal script for Losey but one which Pinter had aspired to direct himself. But *Langrishe* had languished and in April 1976 Losey made a proposal to John Heyman: 'It would be a German-French co-production, possibly with some English and Irish money, in the three to five million category. Plan is to start shooting September 15th.' Heyman sent his regrets.

In February 1978 Pinter again heard from Losey: 'Perhaps I must let myself go and have a total nervous breakdown so that I might stand a chance of getting your next play away from Peter Hall.' Two weeks later, having read *Betrayal*, Losey added: 'It's clean, tough, funny, cruel and I think destined for considerable success. I will confess to a slight disappointment, because I had hoped for a beginning – a new style – a new point of view. For me, instead of that, it's the absolute culmination of one aspect of your work.' As Losey's career again faltered he increasingly reached out to Pinter to validate his projects. He told Graham Greene, whose *The Human Factor* he wanted to film: 'It is the first time in five years I've heard Harold so enthusiastic on any project I proposed to him.' In June Losey inquired whether Pinter would be inclined to collaborate on (the thoroughly defunct) *Under the Volcano* if the rights reverted to Margerie Lowry in May 1979. With notable disloyalty to Guillermo Cabrera Infante – given the history of the project – Losey referred to the previous scriptwriter as 'somebody else. I don't know what has become of that somebody else. The script that he wrote was not very good and more than four years have gone by.' Pinter's initial response was encouraging but he changed his mind after rereading the novel. 'I find the language dense and convoluted, to the point of being impenetrable.' To Margerie Lowry Losey described Pinter's reversal as a 'blow to me'.

Another was to follow – on 28 September, Pinter sent Losey 'a very quick note on a rainy morning' to inform him that he was now scheduled to write *The French Lieutenant's Woman*, with Karel Reisz directing. (Reisz had been a rival contender to film *The Go-Between*; as for *The French Lieutenant's Woman*, it was classic Pinter–Losey terrain, the perfect subject.) Losey responded on 15 November: 'I cannot think of anyone as good as you' for this project. He didn't mention Reisz. Plaintively, he inquired whether Pinter knew John Fowles's last novel, *Daniel Martin*. 'It might be interesting for us. I think it is quite remarkable, but very difficult.' But one Fowles filled any hunter's bag. The desperate nominating of Pinter continued. Bidding for the right to film the late James Kennaway's *Silence*, Losey fired the idea at Pinter while admitting that the writer should ideally be an American. Pinter had not liked an earlier Kennaway novel that Losey sent him, and his response was clearly not encouraging. 'I do not share your doubts about this project,' Losey told him. 'Your comments on *Silence* I don't understand.' In April Mrs Kennaway reinforced the 'Pinter-factor' by advising Losey that if Pinter were to accept, 'then this would go a long way towards convincing us that we should be happy to let you tackle it'.

Visiting Hollywood, Losey bought a ticket to see *The French Lieutenant's Woman*, in which Pinter had found a cinematic parallel for John Fowles's inescapably literary alienation effects. 'A superb film,' Losey told him, and enormously successful in its first run – but the pain surfaced. 'It has its antecedents which I find disturbing but do not resent. The leading man

[Jeremy Irons] is very much like James Fox, but better. I find bits of [*The*] *Servant, Accident,* [*The*] *Go-Between.* I kept feeling I had directed a lot of it, but a lot I wouldn't have directed that way.' When Pinter's play *Betrayal* was filmed, the director eventually chosen by Sam Spiegel was not Losey – and there is no evidence that he was Pinter's choice, either. The chemistry between Spiegel and Losey would result in instant combustion.

Pinter had become more directly involved in politics and human rights issues and Losey, he recalls, was pleased about this. 'In England,' Pinter reflects, 'well-off people who are indignant are laughed at.' Losey's own indignation about the state of the world was, he believes, strong and genuine.

In the last months of his life, the dream of making a film with Pinter was plausibly revived when Losey's collaborator Moira Williams drew his attention to William Trevor's novel *Fools of Fortune.* Writing to Trevor in August 1983, Losey stressed the enthusiasm shared by Pinter and himself. '*Fools of Fortune* is particularly interesting from the cinematic use of time and place, which Pinter and I have worked on through three films and four scripts, getting nearer and nearer the mark.' In September Patricia Losey bought an option on behalf of Launcelot Productions for £5,000 against a minimum purchase price of £60,000. On 11 October, Losey and Pinter signed an agreement dependent on financing. In November Trevor's novel won the Whitbread Prize. Losey turned again to John Heyman for financial support: 'Harold and I have agreed to take equal sums of $200,000 and 5 per cent each of the net profits.' He added: 'I know you didn't like the subject [of the novel] but . . . I think with Harold I could make a masterpiece – not a blockbuster, but a very longtime earner . . .' Heyman refused outright to buy the rights or put up development money, yet, anxious to retain Pinter's interest, Losey assured him on 30 March 1984 that EMI/ World Film Services would buy the rights and that Heyman 'is very positive and pleased and has done a deal . . . to set a ceiling on William Trevor's fee'. Heyman denies any deal: 'I did not guarantee or finance any part of the development on William Trevor's *Fools of Fortune.*' Losey died. *Fools of Fortune* was later directed by Pat O'Connor and released in 1990, starring Julie Christie.[13]

Clearly, Losey's relationship with Pinter had been, during more than twenty years, a close, loyal and exceptionally fertile one, indeed it was arguably the most important artistic collaboration of his life. Yet there were limits to their intimacy. Losey's fifty years prior to 1960 remained, more or less, *terra incognita* to Pinter. The past was, literally, a foreign country. Pinter is not sure, when asked, how many times Losey had married. He detects an element of father-and-son in the relationship from Losey's end, but not, he adds, from his own. Life with Losey invariably involved an element of challenge, even combat – a sardonic man, relishing sudden rapier thrusts, eyes alight, intent on one's reaction. 'You had to keep on your feet with Joe.' A pause. 'In fact you had to keep on your toes.' Then no pause at all: 'I loved him.'[14]

35

Save Me the Waltz

Losey's constant disappointments with unfulfilled projects assumed tragic proportions in the 1970s. This narrative cannot do justice to their malign and cancerous role in Losey's life.* Resnais commented that his own still-born projects were far outnumbered by Losey's; by outlook and temperament Resnais was also the more philosophical.[1]

'I have had a passion for Scott Fitzgerald's work for as long as I can remember,' Losey told Jerome Hellman in August 1971. In 1931 he had read *Tender is the Night* on the beach at Le Lavandou. 'The works of both Zelda and Scott are for me enormously visually evocative . . .' He therefore wanted to make a film of Zelda Fitzgerald's novel, *Save Me the Waltz* – ideally with Mia Farrow. It was 'touching, tender and horrifying', with remarkable dialogue. He hoped that Hellman, the producer of *Midnight Cowboy*, could set it up with a script by Waldo Salt, but in August 1973 Hellman reported 'depressingly negative reactions' in Hollywood to Losey's participation – and asked to be released from his commitment to Losey. But this was a relatively minor disappointment compared with the years of frustrated labour, emotional investment and erased expectations he committed to *Under the Volcano* and *Voss*.

Under the Volcano

May 1968: John Heyman reports correspondence with his New York attorneys about the rights to Malcolm Lowry's *Under the Volcano*. 'Stick to the book,' Losey advises Guillermo Cabrera Infante (also known as Guillermo Cain) when he begins writing a screenplay early in 1972. Despite his traumatic experiences with Robert and Raymond Hakim on *Eve*, and despite having publicly denounced their philistine depradations for a decade, Losey is (perhaps) now prepared to do business with them again, as holders of an option on Lowry's novel. Before flying to Rome to discuss the script with

---

*To mention a few: Nicholas Mosley's *Impossible Object*, which might have starred Catherine Deneuve, and which Losey abandoned at the close of 1971; Conrad's *The Secret Agent*, for which Losey had asked Graham Greene to write the screenplay, but which was dead by 1973; *Courier to Peking*, by a leading biologist, Dr June Goodfield, which went down in the same year; *In the Hours of the Night*, a novel by Bradford Huie – Losey backed out in 1975 after the author assured him, 'I can finance it in any form you want to make it. And you can have Paul Newman or Marlon Brando or any other actor you want . . .'

Losey, Cabrera Infante is waylaid by the Hakims at Orly airport. Warning him that Losey is 'mad', they claim that before signing the *Volcano* contract he is demanding that they reinsert 'twelve feet' into the print of *Eve*. 'He's mad,' they repeat. Losey arranges for Jane Fonda to meet the Hakims in Rome; she describes them as 'rug merchants' for whom she won't work. Losey later informs Cabrera Infante that at Cannes 'I had two painfully (surface) agreeable encounters with the Hakims . . .'[2]

Cabrera Infante produces a first draft, 247 pages long, on May Day 1972. On the back of it Losey writes: 'Too long by 50%. Inconsistent style. Inconsistent conventions. Obscure references. Hugh talks too much . . . Inserts don't all work . . . End hard to follow, confused as to style.' The usual actors are scribbled on the page: Burton, Brando, Scofield, Mia, Julie, Sarah (the latter crossed out), plus John Voight and Donald Sutherland.

April 1973: the Hakims' option expires and is not renewed. Losey tells Carlo Ponti that the Hakims 'did not accept delivery [of Cabrera Infante's script] and they did not pay the money owed . . . and the author returned the money he had received so they have no rights in this particular version whatsoever.' The story's setting is Mexico. Losey is in correspondence with Rudolfo Echeverria, director general, Banco Nacional Cinematografica (not to be confused with President Echeverria). Carlos Fuentes is also helpful. July 1973: Losey sends Burton 'the same script which nearly a year ago I didn't want to give you . . . bounteous material faithful to the novel but incompletely resolved as a screenplay.' It would be a great picture, and 'I'm sure you are the only person to play it.' The Mexican Government will finance 'all below the line costs' and he has the support of Margerie Lowry. Burton replies by return: he has just read a script 'sent by an agent on behalf of Mrs Lowry' and it doesn't seem the same as Cabrera Infante's. He suggests accepting Mrs Lowry's script 'and play it by your perfect ear in order to get it done'. About to make a film with Sophia Loren, Burton declares himself available thereafter. August 1973: Margerie Lowry sends Losey her own 'treatment', in non-dramatic form, insisting that Lowry himself had always sworn he'd never allow a film unless she worked on the script – because she knew more about his novel than he did. This is bad news. Given Mrs Lowry's pivotal position as executrix, Losey replies circumspectly on 21 August, enclosing a long list of comments concerning her 'treatment'.[3]

January 1974: Losey and Buñuel meet for a drink in Paris. Buñuel writes down alternative Mexican locations for *Volcano* on a slip of paper. Losey loses it and writes asking Buñuel to do it again. Buñuel obliges and mentions Tepotzlán and Cuantla, about fifteen and thirty kilometres from Cuernavaca respectively. Also friendly advice on hotels. (Buñuel's kindness is not diminished by the fact that he had read 'eight' *Volcano* screenplays and been converted by none – the essential action, he was convinced, remained *within* the Consul.)[4] Losey visits Mexico, his trip financed by Jacques Perrin, of Reggane Films: his travel expenses come to $2,504. In Los Angeles he talks to Donald Sutherland and Marlon Brando – who declares himself unavailable for two years.

Jacques Perrin and Giorgio Silvagni have reached an agreement with Mrs Lowry; they will pay $150,000 cash, plus 10 per cent of the profits, subject to assistance from the Mexican Government and subject to a satisfactory 'quit claim' from the Hakims. Perrin is a well-known actor and businessman, the producer of *Z* and *State of Siege*. Silvagni is president of the production and distribution firm which put up $600,000 for the French rights to *Trotsky*. Losey has in mind a budget not exceeding $2 million. 15 March, Losey to Burton: 'Financing and rights settled. Proceeding full speed.'

Mia Farrow agrees to play Yvonne. Now Mrs André Previn, she is living at Leigh, in Surrey: 'HOWEVER, it is a long trudge on rainy nights to the local pub to try to call fucking Rome.' Losey cables Donald Sutherland, profoundly hoping that he will play the part of Hugh. He hopes that John Heyman will sort out the Burton contract in New York, but a week later the negotiations are bogged down. Tempers fray. On 20 June, Perrin informs Losey that he won't be coming to London since Heyman doesn't wish to meet him (*ne m'attend pas*). The rights pass to a Mexican film group. 'Perrin blames John Heyman for reneging,' Losey reports. 'John Heyman blames the French for dilatory tactics.' Heyman comments, in retrospect: 'I had nothing whatsoever to do with *Under the Volcano*. I asked not to be involved. My assumption is that Joe in his bid to pull people in said, "I'll get Heyman".'[5]

1976: Losey receives a new script, uncommissioned, from Tom Ropelewski, his most brilliant Dartmouth student of 1975. He sends it to Margerie Lowry with the comment that it's far superior to the one by Guillermo Cabrera Infante (whom he calls 'Infanta Cabrerra'.) August: Mrs Lowry rejects Ropelewski's script.

May 1982: 'I had lunch with ex-President Echeverria and others in Mexico. Echeverria is interested in doing *Under the Volcano* with me. I had practically given it up.' The same month Losey meets Volker Schlondorff in Cannes, and is told of the German director's plans to direct *Volcano*. The following year the film is at long last made by John Huston.[6]

Voss

October 1973: Losey cables Patrick White congratulating him on his Nobel Prize. March 1974: they meet in London. Losey is now White's preferred director for a film of *Voss* to be produced by the Australian tycoon Harry Miller. With White's approval, Losey approaches several writers – including Graham Greene, Waldo Salt, Pinter and Mercer. Mercer is commissioned. February 1974: Losey flies from Rome by Alitalia – VIP treatment, paid for by Harry Miller under a deal with the Italian airline – to Australia. With Donald Sutherland in mind for the German explorer Voss, Losey describes the actor in eulogistic terms when writing to White. White's response is negative: 'Before anything else, that flabby wet mouth is entirely wrong. Voss was dry and ascetic – he had a thin mouth like a piece of fence-wire.' Losey conveys the veto as tactfully as he can to Sutherland, who responds with a dignified expression of hurt.

Losey's contract – between Miller's Attractions (Aust.) Pty Ltd and Citel Films [Losey] – is signed in the late summer of 1976. Michael Linnit, of Chatto & Linnit, informs Miller in April 1976 that during the previous two years Losey's 'international situation had changed quite radically. I am currently quoting and being offered deals around $300,000 for pictures of far less duration and effort than yours.' There is no evidence for this claim but Miller buys it, his offer climbing from 125,000 Australian dollars to 150,000, then in July 1976 to 275,000 (£170,000) plus 10 per cent of the net profit. Losey is also guaranteed overall creative control, including choice of key technicians, actors, composer and locations – plus the sole right to edit the film except in the case of official censorship. November: Losey is pressing Linnit to get the US$18,000 non-returnable advance out of Miller.[7]

December: during a field trip to Australia with Mercer, financed by Klaus Hellwig's Frankfurt company, Janus, Losey becomes ill and runs into every kind of local production problem. February 1977: he cables Gerry Fisher and other potential colleagues: 'As of tonight Australian project killed by unions . . .' The unions have accepted Fisher but insist on local technicians; Losey wants his own. To Hellwig he writes: '. . . my life is being bogged down by epics that after years of work, lumberingly come to a halt, with great disappointment and waste.' September 1977: 20th Century-Fox say no: '. . . too long before the real adventure and drama begin to unfold,' explains Sandy Lieberson, Vice-President of International Production.[8]

Harry Miller goes to jail and the film rights pass to new proprietors, who approach David Mercer late in 1979 – but not Losey. Losey's alleged remarks about Stuart Cooper as the potential director lead to exchanges of letters with Cooper's solicitors. White's own comments about Cooper are not flattering, particularly as the great novelist deems him to be in cahoots with his old friend and new 'enemy', the painter Sidney Nolan.

Losey and White remain close, affecting correspondents to the end ('friends' would be a large word), each urging the other not to kick the bucket first. Losey's nine-year correspondence with White is a monument of mutual esteem, wit, intelligence, openness – and a fascinating commentary on the cultural life of the age. They meet from time to time and the Loseys' dog Tyger takes a friendly bite out of the Nobel laureate's overcoat in Paris – no doubt Tyger would have done the same for William Blake.

David Mercer: a Friendship

The bitter *Doll's House* war of the sexes in Roros did not undermine Losey's confidence in his screenwriter, but every subsequent venture with David Mercer was destined to founder. Whereas three out of four projects with Pinter reached the screen, Mercer's work for Losey was constantly bedevilled.

As with Pinter, Losey's enthusiasm for Mercer's stage plays was less than total. Stated perhaps too simply, whereas Pinter's work was too modernist for Losey's taste, Mercer's was too preoccupied with professed politics, too angst-ridden about utopia betrayed. It was Mercer's flexibility and skill in adapting the work of others for the screen that Losey valued – his adaptations of Joyce, Ibsen, Roy Fuller, Thomas Mann, Patrick White. When Mercer made a trip to Sardinia in the last quarter of 1967, during the shooting of *Boom!*, to discuss his adaptation of Wyndham Lewis's *Snooty Baronet*, his career was in fine shape. On 12 February 1968 he reported: 'I had my German première of *Ride A Cock Horse* in Zurich ... then Pierre Brasseur contacted me to say he wants to do *Belcher's Luck* in Paris ...' By July Losey had drawn up a cast list for *Snooty Baronet* – Burton again – but nothing more is heard of it.

On 11 December Mercer wrote congratulating Losey on *Secret Ceremony* as 'one of the best films for years ... hardly room for criticism. It's haunting.' Did he mean it? Ten days later he arrived at Patricia's birthday party 'paralytically drunk'. A year later it happened again. Paradoxically Mercer, who suffered from a pitiless lucidity about his own personal failings, gradually discovered a kind of father-figure in Joe – a man whose self-scrutiny was not obvious, and who normally found fault in others.

On 15 May 1972, Mercer tactfully conveyed his reactions to *The Assassination of Trotsky*, concluding that 'the long historical antecedents to the film maybe just can't be compressed adequately into an account of [Trotsky's] assassination ... Unless seen in action and conflict, his *meaning* takes on a reported air rather than something we can experience.' Losey had been talking to Mercer about writing a screenplay of *Galileo*; Mercer's response was cautious. What he did write was a fourteen-page adaptation of James Joyce's Dublin story, *Counterparts*, which BBC Television wanted Losey to direct for its 'Full House' series. But Losey apparently hesitated and by January 1973 the project had been offered to Peter Gill.

On 8 September 1972 Mercer and Losey had their first work session on *A Doll's House*. On 29 October Patricia noted that Mercer had been on the bottle for a week. On 10 November Mercer and his friend Lynn Horsford reached Roros by train from Oslo – she remembers the grim comedy of extracting the sozzled writer from the train. For her part Patricia 'felt angry with him for the first time – I've usually felt more indulgent'. With Jane

Fonda about to savage his screenplay for alleged male chauvinism, Mercer had to sober up; as he and Losey went through Fonda's pages of notes, he was stunned and saddened by the almost militarized intensity of the attack. After it was all over Patricia joined the Mercers and David Warners for lunch at Bergstaden. Mercer was now drunk again 'and encouraging male backlash . . .' She and Joe 'both lashed out at David about his indulgent and arrogant drunkenness. Lynn left the table . . .' Back in London Mercer confessed his sins to Pinter – would Joe ever forgive him? In April 1973 Losey expressed his gratitude to Mercer for accepting £1,750 to write a screenplay of Roy Fuller's *The Carnal Island* – this being Granada Television's top fee for a 16mm television film. But Granada eventually cancelled the project as too expensive.

In June 1974, the Loseys attended a publishing party. Mercer was drunk and upset that Losey had offered *Voss* to Pinter. Patricia described him bending over the bar threatening to kill Lynn (a whore) and Joe (who'd betrayed him). 'I've heard I'm second to Harold Pinter,' Mercer greeted Patricia. He was right, Losey wanted Pinter – but Pinter declined. On 17 June Losey reminded White that Mercer's films included *Family Life* and *Morgan*, and mentioned such plays as *After Haggerty*, *Flint* and the recent *Duck Song*, 'which I think is his best, although badly reviewed, and now gone from the West End'. Despite Mercer's 'heavy drink problem he has never let me down in work and I have worked with him four times . . .'

Losey began rehearsals for *Galileo* on 1 July 1974. The cast list dated 18 June shows the name of David Warner (Torvald in *A Doll's House*) against the part of Sagredo, but on the night of 2 July Losey received a long and alarming call from Warner, who was clearly in no state to play Sagredo or anyone. Sir John Gielgud's only memory of the production, on which he spent half a day, was of the actor's temporary flip from sanity into ECT. Losey turned for help to Mercer who, confronted by his friend's condition, rapidly became the father-figure himself. What ensued plunged Mercer into a protracted nightmare, including the destruction by Warner of his production stills, his plaque for *Morgan*, in which Warner had played, and Mercer's notes on *The Magic Mountain* (which Losey was scheduled to direct). Warner himself, who was passing through a period of deep personal disturbance, cannot explain why he did it to his closest friend, apart from the obvious point that, 'It was through friendship that one could allow one's vulnerabilities to show.' The character that he played three years later in Resnais's *Providence* (written by Mercer) conveyed a version of rational lucidity within a thin envelope of derangement. Mercer, meanwhile, had plunged from doctor to patient; arriving in London, Alain and Florence Resnais found him blind drunk: 'I've burned it!' He meant the screenplay of *Providence*. They got him to hospital in a taxi.[9]

Thomas Mann's *The Magic Mountain* was within Losey's sights from 1973. Rondo Productions had acquired the rights in conjunction with Induna Film GmbH of Munich. Luchino Visconti, meanwhile, was announcing that the film was his; Induna issued a public correction on 22 August 1974. Losey read and greatly admired – 'superb' – David Mercer's script, commissioned by Rondo. The project was aimed at television, a seven-part serial, each episode to be shot in twenty days. In 1975 a contract was agreed, worth $205,000 to Losey, plus $1,500 a week expenses for twenty-six weeks, but in May Rondo informed him that shooting could not now take place before the autumn of 1976, at the earliest. The project then faded from his horizon.

Mercer asked Losey to be his best man when he married his Israeli fiancée, Dafna. The wedding feast was held at Neal's and Mercer stayed sober. Three months later it was Losey's turn to get drunk after the top financier behind the *Magic Mountain* project flew from Germany to double his fee and grant his demand for complete artistic control. Losey vomited all night and in the morning had a temperature of 102°. In May he congratulated Mercer on seven hour-long television episodes of *The Magic Mountain* – 'brilliant, profound, moving'. In November Mercer wrote to Losey about *The Romantic Englishwoman*. Despite Losey's 'tremendous mastery', he hoped it would not be impertinent if he suggested that the film had 'very nearly' had an 'artistic and moral vacuum at the centre'.

On 5 January 1976 he sent Losey a breakdown of *Voss* and mentioned tactfully that he'd only now learned that his own contract with the Australian impresario Harry Miller, which he had signed, was conditional on Losey's contract, as yet unsigned. On 6 March, Mercer sent a letter of apology for 'my débâcle last Thursday' – they had met in Paris – and for 'the infantile . . . ravings of my boozy daemon . . . The material disgorged when I am badly drunk is aboriginal, all of it.' Losey replied, very kindly: 'As for your getting over your boozy daemon, I doubt if you ever will. Just as I never will, more than twenty years later . . . get over my weaknesses.' Losey had responded to Mercer's gentle reproach about the *Voss* contract by sending him a copy of his own letter of agreement with Miller. But on 22 March, with Mercer working flat-out on the script, Losey wrote to his agent, Michael Linnit, rejecting Miller's more detailed terms. At this juncture Mercer's famous agent, Margaret Ramsay, intervened to protect her client's interests – which angered Losey. Mercer fled into a hole, disparaging Ramsay and abjectly lying: 'And of course I had not meant for her to get a copy of your letter . . .'[10]

On 6 June 1976, Harry Miller wrote to Losey: 'I personally am ecstatic about David Mercer's script. It is *exactly* what *Voss* is all about . . .' Patrick White was also delighted with it, and Mercer was paid the first half of his £20,000 fee. But, writing from Haifa in July, his anxieties again surfaced –

apparently Miller wanted to pay for the second draft script only after Losey signed his own contract – Mercer concluded ruefully that he'd been working 'on spec'. Israel, meanwhile, had been rejoicing over Entebbe: 'I'm as popeyed with admiration as the next untermensch, but I would have thought the toll of dead might have toned down the popular jubilance and crowing a little.' There was no concern in Israel for 'the grave issues behind the PLO's grim terrorism'. Losey could not respond; his several visits to Israel in connection with John Heyman's TV epic on the Bible (which continued until 1979) incurred the risk of forfeiting the Arab-financed *Ibn Saud* project if revealed.

David and Dafna Mercer resumed residence in London, he unwell and gloomy at the age of forty-eight: 'One tends to brood all day and have nightmares all night. Top of all else, some kind neighbour stopped Dafna in the street and asked if she knew I had "had an affaire" with someone while she was in Israel. *For once* – it is characteristic of this marriage, it is my committment [*sic*] – it is not true. But she believes this gossip rather than me.' By the end of September his letters conveyed domestic torment. 'I get to feel I am drowning.' Could Losey possibly come to the rescue by inviting him to Paris on some professional pretext? Mercer arrived in Paris sober and drank nothing during dinner. On his return to London he wrote defending his wife: 'I do believe that in many ways you are wrong about her. She is certainly far more than an [. . .].'

On 3 November Mercer registered despair about himself and the wider scene: 'At least in the twenties artists were violent and rebellious, iconoclastic.' Here he took a swipe at some colleagues, including his friend Pinter: 'Now . . . we have the remote murmurings of *No Man's Land*, the arid wit of Simon Grey [*sic*: Gray], the squalid entertainments of the boulevard. No man's land indeed. It would be splendid to be cheerful for a day or two, a lifting of the spirit. IT'S GOT TO BE THERE *SOMEWHERE*.'

There followed a visit to Australia by both men – Mercer described it as an amazing and rich experience, for which he was grateful, though he felt like 'superfluous baggage' in the eyes of Harry Miller, for whom 'I haven't the slightest respect . . . as a human being'. He hoped to tackle Brecht's *Puntila* for Losey, who confided to a friend that Mercer 'went off his rockers rather badly in Australia a couple of times, although as long as he is under the arm or wing of Dafna he seems to be okay and to write well.' But the *Voss* project foundered. At Losey's suggestion Mercer wrote a screenplay of Muriel Spark's *The Takeover*, for Klaus Hellwig's Janus Productions, but without much enthusiasm for the novel. Losey didn't like Mercer's screenplay and told him so when Mercer visited him at Fregene, near Rome: Mercer also got badly drunk after overhearing some disparaging comments by Losey while talking to Hellwig on the telephone. Later Losey sent him a devastating critique of his script, point by point: 'a kind of vulgarity of the wrong kind ... sometimes it is cheap and inelegant ... there is hardly a situation or a character in the script with credibility'. On 28 August Mercer received a telephone call from Hellwig terminating his work. His rage flowed over – he found it 'more than faintly disgusting' that Losey had not told him so himself. Was Losey joining Sam Goldwyn and Walter Wagner, who used to hire and fire reputable writers like secretaries? 'Over the years plenty of people warned me that you would act something like this one day.' And no to Brecht's *Puntila*. Replying on 7 September, Losey described Mercer's letter as 'more than "faintly" disgusting'.

Happily the rift was mended and they embraced in Paris when Mercer received a French Film Academy César for *Providence*. Losey sent Mercer his dearest new project, James Kennaway's novel *Silence* (having tried, and failed, to interest Pinter). In March 1979 Mercer responded, admiring the novel but regretting that 'The Chicago vernacular would elude me ... the whole frame of American society is one I have little true experience of.' Losey sent him Nadine Gordimer's novel *Burger's Daughter*. Again Mercer expressed misgivings: 'Why am I talking myself out of an honourable and attractive job of work? I'd certainly use the money. But intuition says "no" ...' His own news was bad – his play, *Then and Now*, hadn't done well at the Hampstead Theatre; and no one wanted to produce his *The Monster of Karlovy Vary*. Was his talent faltering (he wondered) or had he merely become 'unfashionable'? Either way, 'virtual eclipse of Mercer in the theatre ... Trevor Nunn and the Royal Shakespeare Company – for whom I was virtually the house-writer for ten years – seem loath to touch my work ...' Losey replied very briefly: 'I'll take a deck chair beside you on the Titanic. I'm already aboard, only mine is stuck in a sea of shit.'

Then a new storm. In December 1979 Losey was outraged to learn that Stuart Cooper had bought the film rights to *Voss*, including Mercer's script. He wrote to Patrick White: 'I was astonished a few days ago to receive a phone call from David Mercer in Adelaide. He was, I'm afraid, at his most maudlin ... he's an alcoholic schizophrenic ...' Losey felt deeply betrayed. White replied that Mercer had come as an emissary on Cooper's behalf and had 'confessed before leaving that he would like to see his own work on the screen – this after I told him I thought the whole project imbecile: to hand *Voss* over to a nonentity who proposes to cast Jack Nicholson'.

The *Voss* project was in any case defunct and Mercer wrote no more. He died on 8 August 1980, at the age of fifty-two. Harold Pinter organized a memorial meeting in London on 26 October. Losey, still living in Paris, composed a generous tribute, wishing that he himself had been more patient 'with his idiosyncrasies ... I was often too grudging, too late, and too little, in thanking David for himself and his work.' On 29 October Pinter reported:

'Your statement about David arrived the day after the memorial evening. Pity. I have sent it to Dafna.' On 4 November Losey wrote to Dafna Mercer, regretting that 'no one' had let him know the date for the memorial meeting . . .

36

Parting the Traffic

Interviewed on television in 1967, at the age of fifty-eight, Losey was still a lean figure, dressed in a sombre tweed suit, a white shirt and a tie. His hands were nervously active and frequently fingered his short hair. His appearance changed drastically from about 1969. He ballooned, lost his physique, and turned dandy. Michael Wale described 'the craggy face, the sad eyes, the slightly stooped shoulders and the shuffling feet . . .' David Gow reported 'a rather large, rotund belly that makes him sort of flop backwards into chairs . . . The face is almost as lined as that of Auden . . .' 'He was like a tree,' says Nicole Stéphan.[1]

He was now dressed by Marc Bohan of Dior, among others. David Sterritt interviewed him in July 1973: 'Losey paces to a chair, looking like a bulky rhapsody in blue – blue jacket, blue shirt, blue slacks, blue kerchief tied around the neck.' In 1975, posing in the Luxembourg Gardens for a glossy magazine, he wore a cape by Yves Saint-Laurent and a *'foulard en crêpe'* by Cerruti, while carrying a cane from La Bagagerie. In another shot he posed in a black cotton kimono braided (*gansé*) by La Factorerie. He favoured extravagant cravats and satin jackets and longish hair rigorously styled – visits to the barber figure prominently in his Paris desk diaries. David Mercer wrote: 'I now think of you in Norfolk smock and green cloak, parting the Parisian traffic as if dealing summarily with the Red Sea. Oscar Wilde it is said parted Parisian traffic in similar fashion . . .' Ciment noted that despite asthma and varicose veins, Losey gave off an extraordinary impression of power, a wild beast in repose with a sculpted lion's head. In 1982 Losey was refused entry to Doubles, a private club in the lobby of the Sheraton-Netherlands Hotel, because wearing a large cravat in place of a tie.

Dr Barrie Cooper nursed him through film after film: bronchitis, irregular heartbeat, high blood pressure, alcohol abuse. From about 1970 on, it never rained but it poured. For example, in January 1972 he contracted virus pneumonia while living in Rome; the congestion had not fully cleared in early February. In April he had the stitches of hand and chest wounds removed. On the 13th he'd told Pinter: 'There has been a general fuck-up about my "operation", so, having gone into hospital, I came out still with the lumps and protruberances untouched . . .' He had a septic tooth. A chest infection which began in Norway continued to harass him through the first half of 1973. In February he had pneumonia. By early March he had developed fluid in his right eye. In July he had a chest infection. Periodically he put himself on a liquid diet, yoghurts, fruit, eggs, all liquidized, something to eat every hour. Alcohol contributed to obesity. Leaving the Quiberon health farm in October 1975, he weighed 97 kilos (15 stone 6 pounds).

His health during the shooting of *Mr. Klein* was in bad shape. By December he was taking anti-coagulants to clear up thrombosis in the eye. In January he contracted a chest virus: 'I am drugged to the gills with anti-biotics and anti-histamines, etc.' In the lobby of a Strasbourg hotel he suffered the worst asthma attack in fifteen years: 'Everyone clustered around to be helpful which of course was worse than nothing.' 22 January: 'I don't sleep well, am not well. The asthma is controlled, the bronchitis still under antibiotics but going, the eye bad, probably all of it fatigue and nerves.' In January 1977, while in the Australian bush (*Voss*), he contracted a serious virus pneumonia.[2]

During the last decade of his life, drink was the main factor in his decline in Dr Cooper's opinion: 'Drink affected his decisions. Faced with an impasse – a contract, a story, a friend – he would drink his way through to an impulsive outcome and then try to rationalize that outcome.' On 18 January 1974 Losey wrote to Tom and Miriam Stoppard to express 'my shame at my boorish conduct at your New Year's Eve party . . .' On 4 April Patricia noted: '. . . we are both drinking excessively but at this moment it seems the best solution'. 29 June: 'Joe having given up drink for four days has gone back on again . . . Hope ebbs and flows with Joe's alcoholic intake . . .'

Rehearsals for *Galileo* began at EMI Elstree on 1 July 1974. On 6 August Patricia was 'reduced almost to tears . . . at the sight of him raising his cup with such tremors in him that it clattered and rattled and shook and only with difficulty could he get it to his lips. Nothing to be said.' On 9 August she watched Joe shooting the scene of Galileo's disciples awaiting his non-recantation. Joe kept falling asleep. Reginald Beck asked her what was the matter with Joe – he was sleeping through rushes and on the set. Richard Dalton was also worried – Joe had sent for a vodka at 11 a.m. Tom Conti told Patricia that Joe had fallen asleep after saying 'Action!' 14 August: 'He was up at 6.15 still drunk and hopping about complaining about Topol. Losey missed a private showing of *Galileo* on Sunday, 20 October 'because I had a minor, but extremely painful accident to my knee over the weekend'. This was the result of a drunken fall. At film festivals, Barbara Bray comments, there were acolytes who would pour him a vodka for breakfast. While living in the rue du Dragon, he would drink himself into incoherence until, frequently, he passed out, cold. Lifting him from the floor was not easy. Monique Lange asked him why he drank so much vodka. 'From despair. From fear of everything.' Late in the evening he could be monstrously aggressive on the telephone, 'undoing', as Barbara Bray puts it, 'the good work of the day'.

Drunkenness contributed towards his unemployability; he was likely to 'blow' a conference or social gathering connected with work he desperately wanted.

He dictated hundreds of letters. Typically he would begin an adversarial letter by professing himself 'amazed' at the conduct of the other party. Some letters start mildly enough, then change gear into stricture – but often ending on a mellower, friendly note. He drew up endless lists: names, tasks, what he owed people and what they owed him, whether financially or morally. Suffering from lingering bronchitis, on the wagon and beset by acute melancholia, he dictated twelve business letters on 2 March 1973; eleven on 16 March; ten on 19 March; eleven the following day; thirteen on 22–3 March; thirteen on the 26th; nine the following day. He demanded to know from the editor of *New York Magazine* why all four January issues had been posted to him on the same day. He berated his telephone answering service, Air Call Ltd, for 'error, inefficiency and just plain stupidity'. 'I want no more misunderstandings or "damaged feelings",' he told Simone Signoret. He fired off angry letters to British Airways and the Peter Jones department store, and moved his money out of Barclays Bank. He informed the Hon. Secretary, Club du Château, Castellaras, that 'no apology is owed for [my own] black humour, but rather, apologies are in order for the abysmal service as well as its tone'.[3]

Most of these letters were dictated to Celia Parker. She arrived in May 1970, after a succession of secretaries had gone with the wind. Joe appeared in a dressing-gown, told her she'd probably be no good either, then commanded, 'Get me Gerry Burke and Harold Pinter.' Who was Gerry Burke? One dare not ask. In 1971 she was summoned to Rome. On her first evening the Loseys and the Richard Macdonalds took her out on the town, drinking stupendously. (Nicholas Mosley also describes Losey 'holding court' in Rome, with great dinners in the Piazza Novona, often starting as late as 10 p.m., after which the tired writer was cross-examined about the motivation of Trotsky's assassin.) But back to Celia Parker: Fellini was shooting *Roma* that night, on the streets, and Joe went over to greet him. Next morning she awoke in the grand Losey apartment with the worst hangover of her life. 'Joe growled that I should learn to hold my liquor. His temper remained vile. He kept talking about the actor "Jason Robard" and when I finally corrected him he went into such a sustained fury that I decided to quit. Patricia tried to dissuade me but I left for Tangier.' Later Losey rang to make amends and their relationship flowered into one of mutual respect and affection. 'Joe respected only those who stood up to him or bounced back.' She grew to love him: a big man in every sense, baronial, bullying, warm, helpless when alone, often seeking her company in Don Luigi's, frequently disappointed and pessimistic, constantly writing people off.[4]

'Joe behaved like a Medici prince,' says Pieter Rogers. He knew – and cared – where people stood on the several ladders of life – fame, talent, wealth, breeding, title, influence. Time and again his correspondence mentions a person's connections; or the paintings he owned. John Loring recalls: 'He had a strong sense of who was who. If someone like Delon arrived, people like Patricia and I were just flicked away, dismissed.' As Pieter Rogers puts it, 'He liked to be closeted. If Harold was with him I would never be invited in.'

Loring met the Loseys in Venice in 1969 through a mutual friend, the formidable Peggy Guggenheim, who combined great wealth, a palace in Venice, an unrivalled personal collection of modern art (and not a few modern artists – she had been married to Max Ernst). A beautiful bedhead by Alexander Calder guarded her sleep; the Grand Canal greeted her at dawn. The Palazzo Venier dei Leoni, where the Loseys were made welcome, is in fact a kind of folly. Begun in 1749, it had failed to grow more than one floor above the waterline by the time Peggy Guggenheim made it a home for herself and her Collection in 1949. But she found the perfect statue to place before her wrought-iron gates, Marino Marini's *Angel of the City*: a squat figure, with arms and cock outstretched, astride a horse whose humble, horizontal, mulish neck seems to challenge the elegant, Byzantine horses and the winged lions of San Marco to face the facts. A decade after meeting Peggy Guggenheim, Losey made a short speech at a party honouring her for bequeathing her Collection to the city; she did not, he stressed, suffer fools gladly.

'We were all stifled, moneyed Americans who'd escaped into the arts and Europe,' Loring reflects. A handsome young artist turning thirty, Loring was running the Yves Saint Laurent shop in the Cipriani hotel (where the Loseys often stayed) when they invited him to spend some months in Royal Avenue. He admired the David Hicks drawing-room which the Loseys had inherited but deplored the dining-room they'd commissioned from Tony Cloughley, who was designing shops in the swinging-sixties King's Road; on the corner of Royal Avenue he turned a pub into the notorious Drug Store that was to plague Losey's nights and tax his temper. Loring sighs over the 'Albrizzi flexiglass table, very chi chi, an extravaganza of pink perspex and geometric carpets, quite out of character in an early nineteenth-century dining-room.' A fellow-director felt sceptical about the Losey décor: 'Did the furniture Joe used in his later films represent his characters' tastes or his own? That leather winged chair he used for the seduction scene in *The Servant* . . . Directors like Visconti, Losey and Bertolucci set up a bogus Marxist mockery of upper-class opulence while, frankly, not only capitalizing on it but adoring it.'

Monique Lange reported a dream Losey suffered while living in Paris. He finds himself in an American hospital, receiving a visit from Mrs Samuel Goldwyn Mayer, who offers him $25,000 for the sumptuous Oriental sofa, re-upholstered in pieces of beautiful carpet, which graces 29 Royal Avenue. An electric torch is beamed into his eye and he knows he's going to sell the sofa.[5]

Defending the habitat had become a major preoccupation. The Royal Avenue Residents' Association had been founded in 1968 and Losey elected chairman despite overwhelming opposition from the local pigeons, for whom he had a Nixon-Cambodia solution: 'When, O Lord, is the carnage going to start? Soon, I pray. Car, garden, roofs are covered with excrement and the creatures grow fatter and more numerous by the second ...' *Chelsea News* was tipped-off about 'the titled woman who makes a habit of four times daily pigeon-feeding in all accessible squares in the area ...' He informed Bogarde that chauffered ladies were turning Royal Avenue into a canine lavatory. 'Between the dog-shit and pigeons we are vermin-ridden, which seems to concern nobody excepting me.' Human vermin infested the area after nightfall, though mad ladies didn't feed them. In July 1971 he wrote to the director of Charrington & Co., owners of the Drug Store, complaining about 'incessant noise of cars parking and the shouting and general rowdyism ... the breaking of glass from beer mugs being carried into the Avenue and then being smashed by your customers ... the constant blare of loud pop music ... filth in the Avenue'. All this and more was conveyed to the Borough Planning Officer – the deterioriation and degradation creeping through Chelsea. To Chelsea Police Station he reported: 'This morning I counted six Sainsbury and three Safeway trolleys, abandoned at random points on the pavement ... hooliganism has been ever-increasing particularly ... on football match occasions ... shouting "Fuck ... cunt ... piss ... cock-sucker ... etc."' A major crime ensued – someone had unscrewed the front registration plate of his Lancia and laid it, with the screws, on the bonnet. (No doubt the owner of the MGB in No. 31 whose illegal parking he'd reported to the police – or a guerrilla pigeon.)[6]

Clearly Losey was not an 'auteur' director – the films of the 1970s portray not a single pigeon, hooligan, dog's mess or abandoned supermarket trolley.

Although he often derided the British class system, he was impressed by its higher reaches – Dames, CBEs, Knights; genuine aristocrats, he told Ciment, were 'classless'. Depicting the moribund stasis of the Norfolk landowners of 1900 in *The Go-Between*, he befriended the Norfolk landowners of 1970: Lady Sylvia Combe, the Earl of Leicester, Lady Cory-Wright, Lord and Lady Zuckerman, Lady Harrod ... Poised to depict the decadence of Proust's aristocracy, he cultivated the Rothschilds of the 1970s. While portraying the collapse of the 'old order' of 1790 in *Don Giovanni*, he cultivated the aristocrats who occupied the same villas, notably Conte Lodovico di Valmarana and Marchese Giuseppe Roi. The Loseys rented the Villa Emo, described by Patricia as 'full of pop art and oriental statues and objects'. Contessa Barbara Emo, a Canadian, wrote: '... you added so much glamour

and excitement to all our lives ...' Losey thanked Contessa Maria Teresa
Ceschi, of the Villa San Bastiano, 'for your great help and unflagging
graciousness – not to mention your performance in *Don Giovanni*'.[7]

Soon after Joe's legal marriage to Patricia in September 1970 Bruno Tolusso
relented: Patricia could see her children for two weeks if she visited Brazil.
On 2 January 1971, she spoke to her son for the first time by telephone,
asking him in Italian whether he knew she was English. He did. Joe sat and
listened and said it was very touching, but his attitude towards her imminent
departure was 'ambivalent'. 4 January: 'Joe being terrible all day both to me
and to Joshua.' Later he told her that all of his important films had been
made since her time. 'But then too much to drink and so we're back to the
nastiness.' She left for Brazil on 9 January.

Joe was promptly unfaithful (by intention if perhaps not quite in deed)
with the wife of a younger colleague. The lady does not wish to be identified.
The Loseys knew the couple well and had passed an evening in their
company at the Cannes Festival, smoking grass; Patricia noted in her diary
that as Joe couldn't smoke, they ate it mixed with *petits suisses*. Disaster
ensued – too much to drink. Joe lay under the glass coffee table making rude
gestures then pottered about all night, trying to rouse Patricia, who lay poleaxed. *La dolce vita* continued later that year at a party, Patricia's diary
complaining of the husband's drunken remarks to her about wife-swapping.
It had happened before and in her view had 'nothing to do with me and
nothing to do with sex'. Envy was the motivation: 'It feels violent.'

Joe hated to be alone, he was shaken by the death of Robin Fox, and
Patricia's independent journey to Brazil was her first since their relationship
began, a reminder that he alone did not occupy her emotional landscape. She
returned to London, innocent of what had transpired. Joe was away when the
husband telephoned her at three in the morning, asking to come round. On
arrival, less than sober and clearly badly twisted, he told her that he and his
wife had a bet that he could seduce her: 'He then tried to throttle me and I
screamed. His hands closed round my neck.' Joshua, who was sleeping in the
house, appeared. (Joshua says she summoned him.) Next day the husband
turned up with flowers – she tossed them back into the street. Even then
Patricia didn't know what it was all about – only later did she open a sarcastic
letter from this man to Joe saying how honoured he was that his wife should
have enjoyed the attentions of so distinguished a person – or words to that
effect. On Joe's return Patricia confronted him. Alarmed, he confided his
worries to friends such as Richard Macdonald and Nicholas Mosley, who
tried to console Patricia: 'I told her it was typical show business behaviour
and didn't mean anything.' (According to Jack Loring, Joe told him that the
adultery had not actually taken place, be that as it may.)

The effect of this incident on Patricia was devastating. One sign of her depression was the complete lapse of her diary – normally vividly responsive to events – for eighteen months. Even the Palme d'Or at Cannes went unrecorded. Of this period of her life she recalls that she 'wanted to be doing something of my own. I never gave up that struggle but Joe was completely ambivalent about all that. On the one hand he would say how much he would like it if I got work and became separate. On the other hand when anything did come up he was brilliant at sabotaging things for me.' What came up is not clear.[8]

The little Tolussos are coming! In December 1971, Patricia's children flew to Rome from Brazil. Joe found it difficult to cope with their presence or her own divided attention. During the holiday at Flaine in Haute-Savoie, he was in a dire mood. To Burton and Taylor he reported: '. . . we have the Brazilians for a couple of weeks'. Then to Budapest for Elizabeth Taylor's birthday party. In October the Loseys gave a big champagne party in Venice – apparently the conversation was dominated not by death in but death-duties in. In June 1973 they returned, stretched out on the Excelsior beach of the Lido. Peggy Guggenheim joined them for dinner. 'I wish,' Patricia wrote, 'we had a house here.'

On 9 August she noted that Joe 'seems to have nothing to do but hate me and I begin to be afraid of coming to hate him'. But despair was subsumed in restless movement. She was a survivor for whom tomorrow was another day. Borrowing heavily and lamenting his poverty, Joe rented a beautiful house full of interior pools with water-lilies, with a view of Málaga and the mountains and a Spanish woman to cook lunch. Patricia was involved in a serious car smash and returned to London in a plaster cast and considerable pain. On 11 October she rebuked the Fish Department, Harrods, about a piece of smoked salmon 'about one foot long' which had been ordered by 'my secretary' at an estimated cost of £7 but which, when collected by 'my chauffeur', cost £12.35.

In January 1974 Flavia arrived from Brazil. Joe was 'unrelentingly critical' of the girl, 'so unable to approach her kindly and gently – and therefore getting nothing from her'. Each time her children departed from London Patricia wept. In March the Loseys were in Paris and Saint-Paul-de-Vence; two months later it was the Cannes Festival, returning via Paris, a whirl of expense. Patricia was now campaigning for SLAG, Save London Action Group. Standing for Kensington and Chelsea Council, she achieved 118 votes in Royal Hospital Ward – solid Tory terrain. The Loseys acquired a long-haired dachshund, Tyger (after Blake). Patricia took the train from Calais to Monaco with Joe and Richard Macdonald, to film *The Romantic English-woman*. She cried at leaving Tyger: '. . . my tears flow for them [the children], for me . . . for little Tyger . . . My eyes are swollen. I'm just a raw piece of meat and emotion . . .' On 11 January 1975, Joe departed for New York after a 'monumental row' with her – evidently her children had no manners and no respect for the house, she didn't discipline them, she didn't accept criticism, etc. She often felt like quitting: 'People used to take it out on me because they didn't have the guts to attack Joe.'

Joe's first wife, Lizzie Hawes, died alone, in the Chelsea Hotel, New York, on 6 September 1971. Joe wrote to Gavrik from Rome: 'There is a great deal to be said about Lizzie, but probably – almost certainly – not by you, or at least not at this time.' But his own views were a different matter: 'She was an almost classic example of a wasted talent and exceptional person. A bright mind and soul with no guiding star or purpose, shy to the point of self-extinction and covering it with an often intolerable aggression or violence of opinion, sometimes, or even most times, just returning into a schizoid shell or an alcoholic haze, or both.' Joe hoped that Gavrik had inherited from her the Lurçats, the Mirós and the Noguchi – a pity that the Alexander ('Sandy') Calder wood sculpture of a reclining male nude had gone astray. He himself would love to keep the Lurçat screen until Gavrik could find a place large enough for it. A year later he wrote to Gavrik: 'Are you aware that I have only once been to your house to dinner and that you have never ever asked Patricia there for a meal . . .' Joe's contact with his elder son and his two grandchildren remained minimal.

How many sons had Joe: two, or three? At the turn of 1974–5 he visited California and stayed with Robert and Marsha Presnell in Sherman Oaks. He telephoned Louise for the first time in over twenty years. His call began (Louise recalls): 'This is Joe. I did not call to discuss our marriage.' He wanted to meet his adopted son, Mike – but without Louise. Marsha Hunt Presnell recalls the meeting. 'And Joe, feeling guilty over such long neglect of a lad he'd given his name, driven by curiosity, and acutely uncomfortable, did keep tossing down scotches . . . And Mike, having grown up with a succession of "fathers", none of them interested in him (however much they loved his mother) . . . was self-protectively reticent . . . So Joe, to cover his own unease, kept up a stream of words, more off-hand and callous than actually cruel, but surely enlarging the wall between them . . . And it's to Joe's credit that he was disappointed with himself and haunted by it afterward . . .' On 24 January, after his return to London, he wrote to the Presnells: 'I find myself constantly plagued by the memory of that night . . . I was absolutely shitty to Mike. I should never have seen him and I should never have permitted myself to drink so much under those circumstances . . . I shall always regret it . . . How ghastly the whole thing was.' On 5 September, a statement of account from the Georges Beaume agency in Paris recorded a payment on Joe's behalf of $5,076 to 'M. Losey, Junior'.[9]

At the time of Joshua's birth, Joe had put his name down for one of London's major public schools, Westminster, probably influenced by his friend and agent, Robin Fox. By November 1970 Joe had arranged private tuition in an effort to get him into Westminster but Joshua (as he puts it) 'managed to fail effortlessly'. Joe rallied energetically, shooting off letters to progressive, co-educational boarding schools; Bedales offered a place. Impatient with Joshua's defensive listlessness, Joe regularly reminded him that he had gone to Dartmouth at the age of sixteen and had done well in his exams, despite suffering from the dyslexia which also afflicted both his sons. He advised Dorothy that their son was capable of intense suffering, especially from counter-pressures. In November, Joshua spent half-term in Norway, where Joe was shooting *A Doll's House*.

To Dorothy Joe became increasingly short-tempered. In March 1974 he wrote to her about 'a tax rebate of several hundred pounds in connection with Joshua's school fees. Considering that the fees have gone up steadily every year and the annuity depreciated and that I am broke and that I pay all of his practical expenses, don't you think it reasonable that the rebate should apply to me?' Dorothy's response can be gleaned from his next letter: 'Naturally, being broke is "relative", but perhaps it would be salutary for you to know that I have . . . paid all carrying charges, including Joshua and you, on borrowed money which it now seems impossible to repay. Naturally there are privileges of achievement (also "relative") such as a certain amount of free travel . . .'

Holiday time: on 26 July he sent Dorothy a telegram: 'We want Joshua to do what he wishes but cannot accommodate to vagueness caprice and opportunism stop . . . Relationships are disappearing and expenses are becoming grotesque stop Ring specifics and decisions which can be adhered to Joe stop.' When it came to it everyone, including Joshua himself, consulted Dr Barrie Cooper as to whether he should accompany Joe and Patricia on holiday. He did. Clearly Joe couldn't cope with anything he had not planned and controlled. Family life was grafted in like business. 'To see him one had to make an appointment through his secretary. You had to go through a committee,' Joshua recalls. Fortunately, Joe's secretary during these years, Celia Parker, was Joshua's friend and ally. Between father and son there were also passages of tenderness. When Joshua visited Paris while Joe was shooting *Mr. Klein*, Joe reported to Patricia: 'Joshua . . . is desperately in need of love which he doesn't know how to take or get. But good. Not envious or jealous or contemptuous.'

Patricia's son Ghigo remembers how Joe regularly insisted that the children should remain at the table while the adults continued to talk – or 'drank cognac for a further hour'. Ghigo would rebel – by crawling under the table or whatever. As he grew older, the rebellion assumed other forms. Dining at

Fregene with Franco Solinas and his children, Ghigo became irritated by young Francesco Solinas's eulogies concerning Joe's films, and intervened with the comment that one viewing of any film was enough. Joe threw a glass of brandy in his face. And yet Joe took him through books of photographs and paintings – particularly Magritte – inspiring Ghigo to study art, though this didn't happen.

Now a student at Warwick University, Joshua took lodgings in Coventry. 'Why d'ya want to live in a backwater town like that?' Joe asked. When *Les Routes du sud* came out Joshua was fascinated by the depiction of a father–son relationship which paralleled his own – material generosity instead of real contact, a father whose work and mission were everything, impatient with his son's lack of purpose and commitment. In July 1982 a depressed Joe, writing to a friend, mentioned that Ghigo and Flavia had returned to Brazil: 'He's great but not doing anything. She's not doing anything but isn't great. Joshua is Joshua and not doing anything.'

Joe gave his son an insurance policy before his death. Joshua spent the money – 'a lifetime's lesson', he reflects. Joe had been a devoted, troubled, conscientious father but the exotic film sets which had delighted the boy were a paradigm: for Joe everyone was an extension of himself; camera angles, lighting, scenario were always his, the Godfather. Looking back, Joshua now emphasizes Joe's generosity, his mischievious (but not malicious) humour – 'the lovable rogue' – and his own enduring love for his father.

37

Fading Loyalties

Losey first met Dr Barrington Cooper in 1954 through a mutual friend, Olwen Vaughan, who ran Le Petit Club Français in St James's. Himself involved in a variety of show-business projects as investor or co-producer, Cooper's medical clients have included Elizabeth Taylor and Nicholas Ray; he successfully treated Melina Mercouri for a skin complaint before Losey shot *The Gypsy and the Gentleman*.

A family friendship developed. In June and July 1959 the Loseys and the Coopers shared a holiday house near Cannes. When Patricia Tolusso arrived in a battered condition from Italy in 1963, Cooper took care of her (his bill, submitted in March 1965, was for £315). He and his daughter accompanied Joe and Patricia on holiday to Cabourg.

Cooper's financial relationship with Losey was not confined to medical fees; he also took a percentage on Losey's film profits under the formula of 'script consultancy'. Cooper does not recall (or choose to recall) which of them first proposed it and is resolutely reticent on the details. The Leon Clore files show that an arrangement applied as early as May 1956, to *Time Without Pity*. A letter from Cooper to Losey's solicitor indicates that he was due '1 per cent of the gross profits less agent's fees' in respect of script consultancy for *Blind Date*, *The Criminal* and *The Servant*. The same arrangement had been agreed for *King and Country* and all future Losey films. And more, too: he was due 2 per cent of profits on *The Damned* and 25 per cent of Losey's deferment of £2,000 (minus agency commission) on *Time Without Pity*. 'Additionally there is an amount of £210 for consultancy on various scripts for which Joe was billed in December 1959.' Dr Cooper was now (1 October 1964) asking for payment of all these debts.

Cooper's letters to Losey's solicitor sometimes came under the letterhead of Fabyan (1964) Ltd, the address being that of his surgery at 19 Wimpole Street. He was informed that the amount now due to Fabyan in respect of *The Servant* was £317. Losey reacted by wishing to impose a 'ceiling' on future liabilities, but after discussion with Cooper the idea was dropped. A new agreement between them was signed in March 1965 but has not been traced. The arrangement appears to have covered all of Losey's subsequent films, although Cooper is adamant that the only film from which he ever received a penny was *The Servant*. With a stake in the profits, he joined with Losey in establishing a fund to fight for American revenue allegedly owed by the Landau organization in respect of *The Servant* and *King and Country*. As late as 1979 Cooper was taking initiatives to enlist the help of American lawyers on a 'results' basis.[1]

Cooper flew out to Sardinia for both *Modesty Blaise* and *Boom!* He remarks that Losey could be enormously generous – off the back of a production. Patricia Losey recalls that when Cooper visited Joe on location there was often an argument as to whether the main purpose was medical and the journey could be put down to expenses.

Wearing his producer hat, Cooper was keen to set up *The Man from Nowhere* for Losey, with Topol in the lead. Cooper denies this: 'I wore no producer's hat in relationship to Joe. I may well have suggested Topol for a role in conversation although I have no memory of it.'[2] In fact Cooper wrote to Losey on 28 July 1967: 'If you are being serious about it, will you let me know what the script costs are at present; who actually owns it; whom you would want to produce it; what fees should be sought in the first instance; what your rough budget estimate is; and an authority to act for you in this matter . . .' Whatever hat fits this, nothing came of it.

Cooper deplores Losey's ornate period – *Modesty Blaise*, *Boom!* and *Secret Ceremony* – and believes Losey was fooling nobody but himself when he argued that these films would provide him with the financial means to make the films he really wanted to make, yet on each occasion deluding himself that he had made a good and worthy film even though the casting of Elizabeth Taylor was (says Cooper) a 'corruption' and the films a betrayal of talent which 'did him nothing but damage'. Dr Cooper lights another cigarette. 'I also betrayed myself,' he adds. Why? He pauses for reflection. 'Because of the position I put myself in.'

Losey adored Cooper's young daughter Vicky and gave her a small part (later cut) in *Secret Ceremony*. Often they all lunched together in Don Luigi's or Alvaro's. Patricia consistently leaned on his advice, his kindness, his dependability. Yet Losey became obsessively aggrieved about his financial relationship with his friend, both as doctor and script consultant. On 15 June 1971, he urged Cooper to be 'more explicit' about his fees – 'I presume it's drugs. The amount seems to be exorbitant in the light of percentages and retainers. If they are accurate we will simply have to make some other arrangement, because I cannot possibly afford this. It is better, as I have said before, to die.' (But it wasn't.) On 20 August, Losey turned to their profit/percentage arrangement, after pausing to wish business and friendship could be separated and to acknowledge 'my extreme personal and moral indebtedness to you . . .' Then back to his 1964 proposition of a 'ceiling' on his 'improbably astronomical profits . . .'

Clearly money brought Losey close to mania. His calculations and new schemes were complex, pathological. On 15 January 1972 he asked his solicitor, Laurence Harbottle, to look again at the agreements with Cooper dating from 1965: 'If he continues to be as evasive and avaricious as he has seemed to be lately, I will want to eliminate the whole thing categorically and find another doctor.' To Cooper he wrote: 'Both money and generosity are obsessional with both you and me to a point where neither of us can consider either between us rationally.' This flash, however, did not illuminate the night; again he urged Harbottle to cancel or renegotiate the agreement with Cooper, while drawing his accountant Gerry Burke into the fray: 'The limits should be 1% after $50,000 – up to $150,000 . . . there should be a top amount, or a reduction of percentage, on any picture or period in which we are out of the country for a total of six months or more of the year'. Further thoughts rained in: '. . . supposing he [Cooper] suffers a stroke and lives for another twenty years; or supposing, as this year when I had pneumonia, he was absolutely unreachable for nearly a month'.

A new agreement with Barrie Cooper was drawn up and dated 26 October 1972. If Losey was out of the UK for more than six months in a fiscal year, the retainer would be reduced to £150. But it would revert to £250 if Dr Cooper's services were used.[3]

Dr Cooper is not keen to discuss these financial details and moves readily into abstraction, employing psychoanalytical terms and words ending in 'ity' – he speaks of Losey's 'perversity' and 'obsessivity'. This vocabulary does not conceal a sub-text of deep personal resentment towards Losey. Apart from *The Servant*, he says, 'none of the other agreements was honoured'. By the end of his life Losey had received about £40,000 in percentage-profits on *The Go-Between*, but Cooper is adamant that he received not a penny. Gerry Burke, re-examining the files, confirms this in general: however, £350 was paid to Fabyan Ltd in 1975 on behalf of Launcelot – which *may* indicate a percentage on *The Go-Between*.

But the money, Cooper says, was never the point. Losey and he were engaged in 'a rather perverse game'. Cooper's annual retainer 'barely covered postage', but it was 'symbolic'. Only when geography allowed could Cooper respond 'clinically' to Losey's 'pathological behaviour'. On reflection he regards the film-percentage arrangement as 'absurd and counter-productive', provoking Joe's hostility for no realistic reward to himself. 'With hindsight, I was trying to bring about a measure of self-esteem.'

Cooper's role as healer, counsellor and guru in the Losey household is well mirrored in passages from Patricia's diary for August 1974, in which she describes Joe as 'the alcoholic'. Cooper dropped in and administered a vitamin shot to Joe, who was 'hostile, evasive'. The following day, 16 August, Cooper came again, had a long talk with Joe, and told Patricia it had been 'Lecture No. 1'. Joe had been more receptive than expected. 'I cried. Barrie left. He said my hand visibly worse. Operation necessary.' Sunday 18 August: 'Joe drank brandy. Barrie came in . . .'

After moving to France in 1975, Joe and Patricia continued to depend on Barrie Cooper for the care of his and her children – and Losey continued to grumble about the retainer and the bills. Cooper visited the Loseys in Paris several times. At this time (1977–8) he was involved in expensive legal proceedings; asked whether Losey contributed to his defence costs, Cooper does not reply. On 23 September 1981, Cooper wrote to Losey's accountant, Gerry Burke: 'I do know that things are not rosy financially for Joe at the moment, but it does seem to me most extraordinarily unfair that my small fees, diminutive retainer and significant disbursements should go unpaid.' The outstanding medical bill was then £880.15. Losey was becoming more than ever amnesiac when charged for services rendered. In December 1982 he grumbled to Burke: 'Also he [Cooper] has not seen me in a professional capacity in three or four years.' (In fact: one year previously, if not more recently than that.) After the Loseys returned to London both were unwell, Joe mortally, and acutely dependent on their doctor. His skills and friendship became, once again, indispensable. A Cooper bill dated February 1984 showed £1,250 unpaid.

The agent Rosalind Chatto comments that Joe was the only one of her film-director clients who employed his own solicitor on contracts; he wrote more business letters than any other director she knew, circulating them widely; and he was her only client likely to distribute a written agenda before a meeting. 'Don't forget he was American.'[4]

Losey's frustrations were increased by constantly involving a constellation of advisers – lawyers, agents, accountants, and of course, their in-house stand-ins. A letter to Robin Fox makes the point: 'I am not prepared to sign a contract which I question, sent to me by John Stutter, which has not been vetted by you, [by] Gerry and [by] Harbottle in spite of a message from Ros that you think "from the agency point of view" it is OK . . .' (John Stutter was a solicitor at the firm of Harbottle & Lewis; 'Gerry' was Losey's accountant, Gerry Burke.) Whenever one adviser was deemed to be at fault, Losey circulated copies of his strictures to the others. On 21 September 1967, he wrote to Fox from Sardinia. 'It is simply that I have become rather disaffected with Laurence Harbottle . . . I think his life has become so successful that my affairs are small fry . . . Too much of the work in his office – at least in my case is turned over to people who do a half-assed and uninformed job. I do not want any more word of this to be conveyed to Laurence than has already been stated in my previous letters . . .'[5]

At a later date Losey became incensed about his contract for *The Go-Between* – the definition of profits and percentages. Losey complained to John Heyman, who disclaimed personal responsibility and hinted that Losey might have been better represented. Harbottle took umbrage at these 'defamatory' remarks: '. . . it sounds to me as if he [Heyman] is simply stirring things up for some purpose of his own.'[6] Gavrik Losey believes that his father was absurdly loyal to his advisers, who either missed important contractual protections or failed to track down and collect moneys due. Neither point is proven. Losey was usually desperate for work, and the advisers were employing common sense, the norm, on his behalf. But clearly the constellation of professionals resulted in a plethora of percentages and fees – a seigneurial posture he needed but could not afford.

Losey's most harmonious relationship was with Robin Fox, who represented him throughout the 1960s, following his acrimonious rupture with the Christopher Mann agency during negotiations for *The Servant* (Mann was granted a small percentage on the profits). Fox became his beloved friend and mentor in things English; his son James (Willy) played in *The Servant*, his son Edward in *The Go-Between*, *A Doll's House* and *Galileo*. In 1966 Robin Fox moved from the Grade Organization to London International, a company formed out of various Grade affiliates. By February 1970 Fox was with International Famous Agency in Hanover Street, but that year he became fatally ill with a widespread lung cancer. Losey was distraught. Fox was sent to Dr Issels's clinic in Bavaria. When Angela Fox brought him home after nine weeks Losey travelled down to Cuckfield. She recalls: 'The only person who behaved foolishly, even cruelly, towards me was Joe Losey, who made a hysterical scene and beat his fists against the wall, screaming at me that I was a murderer and that if it had not been for me and my inability to be disciplined enough to stay in Germany, Robin would have been cured.' She regarded Joe as 'a dull, groaning humourless American' – though a brilliant director, utterly transformed when working with devoted colleagues, and uncannily perceptive about the acting talents of her sons. Although Fox was the last father-figure in Joe's life, he did not attend the funeral and wake at Holy Trinity Church, Cuckfield. The clue, perhaps, lies in the attendance of 'all the big names that worked in the English entertainment business'.[7]

After Robin Fox's death, his colleague Rosalind Chatto set up her own agency which became, a year later, Chatto & Linnit. Losey and Bogarde went with her – but Bogarde not for long. Her love for Joe was no secret. Her letters begin 'Darling Jo'. She describes him as an attentive listener and a wise counsellor, invariably kind to her mother, friendly to her husband, the actor Tom Chatto, and great with her children. It was a short walk to the Chatto home from Royal Avenue. 'They came,' she recalls, 'to lunch every Sunday – and Christmas. Perhaps he liked my cooking.' This comment masks the wound. One evening John Loring heard her say to Patricia, 'You're so lucky, I'd love to be in your place.' Patricia told her coolly she couldn't have handled it. For Boxing Day 1974 the Loseys had made a provisional lunch date with Dr Barrie Cooper's family, but subject to Cooper's confirmation.

Patricia's diary reads: 'Joe got restless by twelve, called Barrie, called it off as Barrie was only just awake. So we had another Chatto lunch – delicious but there's always that gush.'[8]

Ros Chatto has her detractors, Bogarde among them. Celia Parker, Losey's secretary, feels that she failed to forage for work on Joe's behalf – but emphasizes that this was her own view, not Losey's. Losey's script consultant, Moira Williams, admired Chatto's skills and in 1974 Losey himself noted that Chatto & Linnit had devoted much work to his unrealized projects with little or no remuneration. In June 1979 Losey wrote to her: 'I am embarrassed by the amount of personal attention you give me and the demands I make on you on the few occasions I come to England.' But the multiple disappointments in his working life were pulling him into restless discontent. In March 1981 he wrote to his press agent, Theo Cowan: 'I'm also very discouraged about Ros and [Michael] Linnit. Can you recommend any other agents?' On 5 August 1982 he sent Ros Chatto a telegram: 'No word from you. No proposal television or otherwise. No script or contacts in fact. Under the circumstances think we should end our professional association. Considering my role in starting the agency trust this change will not affect our personal relations. Losey.' But he hesitated and partially retracted; not until early 1983 did he commit himself to a new London agent, Judy Daish Associates. Ros Chatto comments that finding work for him was very difficult and she wanted him to do what he wanted to do. 'It was sort of a relief.' Their friendship, she says, was not impaired – but the wound is evident.[9]

Losey's 'second exile' from Hollywood was commercial, not political. He would dearly have liked to work in America again. But Losey offered a fatal combination: irascibility and box-office failure. Bertrand Tavernier comments that the blacklist became, in the end, his alibi: 'John Berry never played on his past, nor did Trumbo.' A succession of West Coast agents failed to deliver a single deal. Correspondence with the George Litto Agency, 90000 Sunset Boulevard, begins in December 1969. In October 1972 Losey wrote to Litto: 'I have come to the reluctant conclusion that our professional relationship is unproductive for us both and should be terminated forthwith.' However, he faced a dilemma; this termination, he added, need not apply to such projects as *Save Me the Waltz*, *The Secret Agent*, *Under the Volcano* or *Galileo*. Clearly not gratified, Litto intimated that he did not wish to represent Losey on anything. In April 1977 Losey humbly inquired whether Litto knew of any 'routine work out of which I can make money'.[10]

Robert Lantz (for whom Louise Losey had briefly worked twenty years earlier) wrote from his New York office in August 1973, promising to keep his eyes open for any Hollywood project 'worthy of attention, that you could compromise on . . .' In April 1974, Losey complained that he'd received nothing from Lantz. Passing through Hollywood in February of that year, he made an arrangement with Evarts Ziegler, of Ziegler-Ross, but within a couple of months Losey was complaining: 'Dear Zig, Whatever happened to Zig? And to Evarts?' The last correspondence on file is dated March 1975.

A few demeaning scripts arrived from Hollywood agencies. In March 1978 the William Morris Agency, Beverly Hills, dispatched the synopsis of a forthcoming novel: 'Neither her husband nor any other man has ever given slim, bosomy, pretty Martha Sullivan an orgasm . . .' Losey duly conveyed his chagrin. In 1978 he angrily broke off with another agent, Paul Kohner, after barely a year. In March 1980 he made an arrangement with Leon Kaplan, a member of a Beverly Hills law firm; although not a talent agent, Kaplan agreed to find him work on a commission basis, tried hard, did not succeed. No luck with Stan Kamen, of the William Morris agency. In November 1981 Losey lamented: 'I have been to Hollywood, seen Kamen . . . seen Kaplan – nothing develops out of Hollywood. I suppose I might as well forget it.'[11]

Losey's close friends and colleagues in France knew where his heart lay and understood that the French films he made after *Mr. Klein* were '*faute de mieux*', the by-product of a succession of collapsed projects in *le monde anglophone*.[12]

The Romantic Englishwoman

A pulp writer (Lewis) confronts a compulsion to stage-prompt his own life into the triangular relationship which will generate the fiction he writes. Sexual jealousy is both real and imaginatively necessary to him. Masochistically, he tests his wife's love by pushing her towards an affair with a gigolo-poet – which genuinely wounds Lewis when it materializes.

'I have no reason to be ashamed of it as a craftsman, although I think it is a piece of junk.' This was Losey's private verdict on his film *The Romantic Englishwoman* (based on the novel by Thomas Wiseman). In July 1972 Losey had informed Alain Delon that the material was 'not of sufficient quality for you and me'. Two years later, when setting up the film, he sadly reflected that he'd twice refused the project but, deprived of work, had finally succumbed.[13] On 1 February 1974, Launcelot Productions signed a contract with Film and General Investments, who would acquire the film rights to the novel and engage Wiseman himself to write a screenplay on the basis of Losey's services as director and producer, subject to production finance being found. Losey sent Wiseman's script to several producers. In April John Heyman turned it down. 'I am a great loather of impressionistic flashbacks and imagined scenes . . . I think it is cold, analytical and a little tricky from the film angle.' Finally Major Daniel Angel's Dial Films came in as the production company. Losey's contract with Dial was for £60,000. Total production costs were £582,776.

Star casting was the film's best hope. Bogarde and Julie Christie both declined. From Belle Mead, New Jersey, Christie sent Losey an arresting portrait of cold weather and beautiful countryside, plus anti-Tory sentiments and reflections on the male leads whom Losey had proposed: she admired Alan Bates but there had never been any 'electricity' between them; Alain Delon was 'unpredictable' – why not the 'exciting' Jean-Louis Trintignant? Her noncommittal letter irritated Losey and he admonished her to 'change your habits' regarding communications. Eventually he was able to sign up Glenda Jackson, Michael Caine and Helmut Berger for the principal roles. Berger was a coveted star from the Visconti firmament; Yves Saint-Laurent designed his clothes. On 17 August Losey wrote impatiently to his agent in Rome: 'Caravans and other trappings are, in these days, bullshit. An actor gets what he needs.' Glenda Jackson found Berger irritating and achieved no rapport with Losey. While shooting in the south of France she asked the actress Doris Nolan if she could explain Joe's melancholia; Nolan replied that it was normal, but when Losey lectured at Dartmouth he classified Jackson as a 'JL failure', alongside Robert Mitchum.[14]

Losey had worked closely with Wiseman on his script; the writer was invited to accompany a small production unit to the south of France. 'The mood was one of fidelity to the author,' Wiseman recalls. But by June Losey was evidently dissatisfied and decided to bring in a new writer – a step so uncharacteristic that *The Damned* is the only precedent (leaving aside unrealized projects). The new writer was to be Tom Stoppard, a wizard in the field of comic cross-references between life and literature, the real and the imagined. Wiseman – who complained to the producer that this was 'a crazy Hollywood way of going about things' – was not appraised of Stoppard's identity until the beginning of September. 'No doubt Tom Wiseman will be infuriated – maybe even destroyed. I'm afraid I don't care,' Losey wrote to Stoppard.[15]

Daniel Angel hated Stoppard's first draft. Having described his efforts to make notes on it as 'a complete waste of time', Angel sent Losey notes on the first fifteen pages: 'Why know he is a poet so soon . . . Enid Blyton – very unfunny . . . The policeman scene means nothing . . . What is the scene all about on the roof . . . There is not a gigolo market in Baden . . . I have never read such balls!! . . . A film producer in any script is death!! . . . Lewis – "I want to see Elizabeth!" She has only *just gone* . . . She never went to Baden-Baden to be screwed standing up in a lift and it was a good scene in Wiseman's script . . . Lewis's head noises all through the script is rubbish . . . Sequence 48 MIND STUFF – stupid and unbelievable.'

Clearly shaken, Losey counter-attacked, describing Angel's annotations as 'quite unworthy of both your intellect and your character', as 'insipidities', as 'quibbling and querulous and even childish to a degree which amazes me and I must try and look for the motivation ... You seem to treat Tom Stoppard as if he were some upstart ... He happens to be one of the major playwrights in the world and also an important screenwriter ... I could not ... go back to the Wiseman script, which I never wanted to do in the first place.' (Wiseman comments that Losey 'was suddenly acting like a Hollywood rat straight out of *What Makes Sammy Run*'.) On 17 August, writing to Helmut Berger's agent, Losey described the new script as 'more interesting, more fun, much less pretentious, more commercial, but at the same time more complex'.[16]

Losey's intention was to distinguish what had really happened between the wife, Elizabeth, and the lover, Thomas, in Baden-Baden from what the husband/writer (Lewis) imagined had happened, by filming the imagined scenes in a manner he described as 'all very Hollywood, very glittery and lush and overblown'. But with Stoppard at work it became considerably more complicated than a simple dichotomy of the real and the imagined. A memo from the writer, dated 29 September, shows him wrestling with three categories of scene: (1) The writer's imagination (silent); (2) The wife's memory (real sound); (3) The writer's own fictional script (special musical score and audible dialogue where necessary). Losey, meanwhile, invoked his own (all too familiar) categories of 'real' and 'unreal', which were ideological rather than empirical: in the home of Lewis and Elizabeth, he commented, every sequence involved a mirror or reflection, 'because I want to convey that their reality was totally unreal. Richard Macdonald and I set out to make the ugliest house that we could possibly make. There's not one thing in that excepting the Magritte reproductions that either Richard or I or anybody I know would want, and they're all immensely expensive.' In the same spirit Macdonald adds: 'The glass gazebo by the swimming pool was meant to be ridiculous, but people said, "That's just what I want."' However! Of the furniture that Macdonald obtained from Waring & Gillow, Losey put in a private bid to Angel for two metal black leather chairs, one with a footstool and the other a rocking chair. He did not bid for the gazebo.[17]

Shooting began on 21 October at the green-belt house in Weybridge, Surrey, chosen for the Fielding home. Denham Laboratories would process the Gevacolor stock. Losey recommended to Angel that Richard Hartley's score be recorded at Olympic Sound Studios in Barnes, 'where I have done my last four pictures'. And so to Monte Carlo (the Hotel de Paris) and Baden-Baden (Brenners Park Hotel); from 24 November to 18 December, the Loseys ran up bills for £2,174 (about £9,000 in current terms). Patricia's diary provides a vivid description: 3 December, Monaco: drinks with Princess Grace at the Palace 'and liked her much better'; on the Sunday Helmut Berger took them to lunch with Hélène Rochas and Kim d'Estainville. On 4

December Losey shot some car chase scenes up in the mountains while Patricia packed and bought a black outfit from Saint-Laurent. 'We had large lunch – oysters and spaghetti Grimaldi.' By 5 December they were in Baden-Baden. It rained and Helmut Berger was being 'impossible'. 'He couldn't be found this morning and told everyone he went to commit suicide at 2 a.m. but water too cold.' On Friday the 13th Losey and Angel had a row and there was 'bickering and discontent' in the unit. At close of play, the Loseys took the Rheingold Express from Karlsruhe to Hook of Holland, where Joe became enraged over a cabin mix-up on the boat. On 15 April 1975, he commented on the screening of the first answer print at the '2nd Salle de la Projection Lincoln': 'Note particularly that Glenda, when too light, as she frequently is, shows skin defects and bad teeth . . .'

In May the film was shown, out of competition, at Cannes under the title *L'Anglaise romantique*. Losey was photographed arm-in-arm with Berger and Caine – Jackson's absence conspicuous. Glenda Jackson comments: 'My experiences of working with Losey were not particularly happy, or from my point of view, fruitful.'[18] Whatever his feelings, Thomas Wiseman, the abandoned screenwriter, congratulated Losey on the film. Wiseman recalls that Losey's press agent, Theo Cowan, invited him to join Losey, Caine and Berger on the platform at the press conference. When Losey was asked why the couples in the film were always interrupted when making love, he turned to Wiseman for an answer. Wiseman declared that 'the director must have had reasons of his own for interrupting the love scenes, because in my novel and in my film script they had not been interrupted'. Losey fumed. According to Patricia's diary, Wiseman 'turned up, to our surprise, at the press conference . . . it entirely prevented Joe from talking about Tom Stoppard'.

Later that day, after visiting a TV studio with Michael Caine, Losey and a group of friends were having lunch on the beach when Wiseman, who happened to be passing, greeted him. By now Losey was badly drunk: 'He stood up and said angrily to me, "You really fucked me up this morning." I replied: "No, Joe, on the contrary you fucked yourself up." He was furious and made a violent gesture, in my recollection knocking over a bottle of wine.' Patricia took to the sea when (she says) Wiseman accused Losey of being anti-Semitic (though Wiseman has no recollection of doing so). Returning to the table from her swim, Patricia saw Losey throw down his glass in Wiseman's direction and walk away in a state of violent anger. Wiseman then threw a glass at Losey's head, 'hitting him rather well', leaving Losey's hair full of glass. Barbara Bray, who observed the incident, was disgusted by Losey's performance and later that day congratulated Wiseman for having stood up to him. The writer himself was left feeling 'pretty furious that Joe had fucked-up my book, and saddened that I should have joined the long line of those who had fallen out with him'.[19]

The film opened in London at the Plaza in October 1975. The critics slaughtered it. In the *Evening Standard* (16 October), Alexander Walker found only smart décor and nudges. 'His film looks like too many pieces broken off his previous films and even, in the Marienbad-like prologue, other people's films.' By 14 October, every major American distributor had turned the film down. Rescue came in the shape of Roger Corman, head of New World Pictures. Corman wanted a few cuts. Losey summoned Reginald Beck to Paris. The film opened in New York on 26 November, at two theatres, the Embassy and the UA Eastside Theatre.

Of a New York woman film critic – unnamed in print – Losey had remarked in 1972: 'She gets almost sexual pleasure out of venom.' Pauline Kael or Judith Crist? In September 1974 he had written to Kael pointing out that they had never met because 'you seemed so consistently to dislike all of my work'. However, 'I admire your capacity as a writer and your ability to get things over.' This laurel branch was destined not to be hugely rewarded. Writing in the *New Yorker* (8 December), Kael disposed of *The Romantic Englishwoman* as 'mystification melodrama with leftish overtones . . . As usual he empties everything definite out of the characters, as if that would make them richly suggestive . . .'

The Romantic Englishwoman did not make money. Years later, in October 1980, Daniel Angel wrote to Losey: 'I have had a dreadful time with the Rank Organization – and in self-defence I had to buy the film back for £55,000!!!' On 27 January 1982, Angel reported that the film was still not in profit; box-office returns in Britain, America and Australia had been 'disastrous'. 'It was not a good film,' he added.

Lefty at Dartmouth

Losey returned to Dartmouth in 1973 to receive an honorary doctorate; in 1975 to teach again; and in 1979 for a showing of *Don Giovanni*. On 8 March 1973 he wrote to Peter Smith: 'Naturally I wish to avoid the '29 Forty-Fifth reunion and at the same time I want to see old friends . . .' Losey kept changing his mind – 'the usual stop/go' in Patricia's phrase – whether to buy an extra ticket for his wife. He did. The visit lasted four days.

During his 1975 term of residence, which was organized by Maurice Rapf, director of film studies, Losey used Pinter's volume of five screenplays. Among the available films that he chose for screening were *Juliet and the Spirits*, *8½*, *Hiroshima mon amour*, *L'Année dernière à Marienbad*, *Battle of Algiers*, and *The Seventh Seal*. His teaching schedule, beginning 23 June, included:

'History of *The Go-Between* – from inception to now.'

Screening of Antonioni's *Zabriskie Point*.

'Art vs Industry (Pauline Kael article of 5 August 1974, *New Yorker* magazine, required reading).'

'Reel by reel analysis of *The Go-Between*.'

'Discussion of *Accident* and *Marienbad*.'

'Role of the actor in films.'

Losey's lectures in the Fairbanks Theatre embraced such topics as: editing, camera, colour, labs – producers and finance – the role of the writer with special reference to Pinter, Mercer, Solinas, Tennessee Williams 'and the Hollywood phenomenon' – pre-design and the role of the art director – music in films – professionalism – rehearsals and preparation vs improvisation, spontaneity, openness; finally, 'social responsibility'.

He also directed the Dartmouth Players Repertory Company in *Waiting for Lefty*, using student actors – his first stage production for twenty years. The eighteen-day rehearsal period began on 28 July, Losey describing himself to the cast as a 'romantic Marxist'. The setting of the play is a 1935 meeting of the New York cab drivers' union called to vote on a strike. In a programme note Losey commented: 'Now the play . . . may seem naïve and grotesquely over-simplified . . . [but] its freeing influence on the theatre was enormous. It was the beginning of the political theatre of the Thirties which the war, reaction and the atomic bomb, not to mention prosperity, aborted . . . Of course *Lefty* is sentimental, but it is truth . . . honest and passionate . . . We fancy ourselves safer and immune . . . Perhaps similar times are creeping up on us.'

There were nine performances, starting 14 August. The *Dartmouth* magazine carried a review by Howard Fielding which noted that Losey had placed 'union men' in the audience. Scenes from private life were always performed front-stage, with comments from union members. In 1935 the audience had jumped up and yelled 'Strike!' along with the actors; now, forty years later, there had been prolonged applause from the opening-night audience in the Warner Bentley Theatre. Losey's old teacher Stearns Morse wrote to him: '. . . perhaps it was nostalgia working in you, or another sort of challenge: to see if you were as good now as you were at twenty-six [a reference to Losey's Moscow production]. Anyway . . . you brought it off beautifully.' Morse marvelled at Losey's ability as director 'to take recalcitrant humans and make them do what you want them to do . . .' Nicholas Biel Jacobson, playwright, wrote: 'And the performances you got out of the cast! You must be the 1975 equivalent of Max Reinhardt in 1925 . . .' Peter Smith calls it a galvanizing production.

38

The Baron's Business

In 1975 Joe Losey took himself into a second 'exile', explaining publicly and privately that he could no longer afford to pay British income tax. This decision affected not only his personal life but his work as a director – henceforward he was making films in a language, French, which he did not speak. With Losey, money was a lifelong preoccupation which grew into an obsession and (like vodka) a source of alienation. He could be profoundly generous – for example his will of 1970, at the time of his marriage to Patricia – but his generosity functioned only when it brought him pleasure and self-esteem.

From his earliest years he had been impressed by what money can buy – not only servants but style, not only comfort but Culture capitalized. Living in New York, Paris, London, Rome, his postal codes were always good. Invariably he lived beyond his means by borrowing and by not paying bills. The evidence is clear by the time he left America in 1952: a string of debts to friends, running to more than $4,000. The devastating blow of the blacklist and exile reinforced his tendency to maximize his needs and minimize his obligations; his debts were subsumed by a moral credit balance. He proposed to his business manager, Henry Bamberger, that his creditors be invited to write off his debts against tax, and then be reimbursed for the amount of their actual loss. Bamberger found the idea not very acceptable because hazardous for the creditors.[1]

Losey benefited from exceptional loyalty and devotion from his professional advisers, but he hated to pay their bills. Sidney Cohn, a staunch friend and fixer throughout the blacklist era, had travelled from New York to London to win him a new passport, the lynchpin document in Joe's life. Cohn's bill of 13 June 1956 was for $2,500. Losey answered with bad news about his career and earnings; Cohn promptly reduced his fee to $1,675 (£625). But still Losey didn't pay – he was 'out of work' and his mother was completely dependent on him after an eye operation. Yet Losey's own financial records reveal that between March 1956 and 1958 his gross earnings amounted to £27,320 ($73,164) minus commission – with a lifestyle to match. Confronted by dollar debts to remoter figures, Exchange Control regulations were his alibi: his Swiss dollar bank account was forgotten. Pursued by a Los Angeles psychiatrist who was owed $835, he replied: 'I have no dollars and have not earned any dollars since 1951 . . .' In September 1964 a New York dentist called up a very old debt of $690; 'I have no funds in America,' Losey replied.[2]

By 1965, thirteen years after he left America, he had still not paid off his debts to close friends like Edward Hambleton, Jules Dassin and Bob Aldrich. By now he owed money to the leading actors he had worked with in Britain, including Bogarde, Kruger and Baker. Unable to plead poverty as he entered his 'time of the Burtons', he offered his creditors payment in the Bahamas – 'Dear Edie [Somer], the usual batter of bloody authorities are trying to work out ways and means of setting up a little nest-egg for you in the Caribbean.' But a letter from Harbottle to Stanley Baker indicates that Losey's solicitor was as uncomfortable about his strategy of debt repayment as Bamberger had been in 1952. Baker fell in with the scheme but Hardy Kruger was distempered; he wanted his money sent to Tanzania for his farm and pointed out that the £500 he had loaned was now worth £600.[3]

When buying a leasehold on 29 Royal Avenue in 1966, Losey borrowed £16,000 from Leslie Grade. In May 1970 Grade, whose company had financed *The Servant*, begged him for repayment of the £11,700 still owed: '... if you could do this for me, Joe, I would be eternally in your debt'. Losey replied on 21 May: 'I have no cash sum or anything like it available now or in the near future ...' The loan was repaid in June 1971 by incurring a new debt of £25,000 to Trident Insurance Company. In February 1972, confronted by debts twenty years old or more from Doctors Cogan and Prince, he told Henry Bamberger: '... if all the people who now thought I was rich billed me ... for how long would I be paying off and how far back would it go? ... while I have made more money in the last few years than ever before, it is very hard either to keep or have any.'[4]

In 1969 Losey's accountant estimated that he needed £25,000 a year to meet his UK expenditure (equivalent to £160,000 in contemporary values). In April 1971 a bill arrived from Morgan Furze, wine merchants, for £549 (£3,000 at contemporary values). At that time a bottle of Cinzano cost £1.07, Martell brandy £3.45 and vodka £2.75. But the bottles were now empty; angrily he instructed his accountant to demand a 20 per cent discount or he would take his business elsewhere. In November Losey asked John Heyman for a loan of £25,000 against future earnings, pleading tax demands in the pipeline. In April 1974 he privately noted that Heyman had 'advanced me money against nothing ... without strings and without interest and without dates'. To Alain Delon, from whom he borrowed large sums, he expressed gratitude in identical terms.[5]

Only first-class travel and first-class hotels would do. A film director is a prince and Joe was at ease with the role; on a train to Norwich he asked the guard to fetch him a vodka with such assurance that the guard transformed himself into a steward. 'He didn't know there was any more airplane behind those six seats in first class,' comments Ruth Lipton. His secretary of the

1970s, Celia Parker, made inquiries about flights to Morocco at Joe's request: on hearing that no first-class seating was available from Paris to Tangier, he called it off. He used to chide Gerry Fisher for not demanding first-class travel in his contracts. 'He thought of himself as a socialist,' Fisher laughs. Reginald Beck offers an identical comment. Both recall Losey's taste for style and luxury when on location: a big house in Málaga; one in Normandy with a racing stud in the grounds; a splendid Vicenza residence; a grand Rome apartment at 10 via Della Vetrina, obtained through the good offices of 'our old friend Princess Soldatenkov'. Losey bitterly complained about the unavailable cook and cleaning woman, but when the writer Guillermo Cabrera Infante stayed there he was struck by antique works of art and the kind of luxury which had always pleased American exiles such as Henry James.[6]

Expense accounts further eroded Losey's sense of reality – defined as what he could afford and what he spent. Despite his pleasure in winning the Palme d'Or at Cannes, he complained to his press agent: 'When I came to pay the Colombe d'Or bill which amounted to something over 3,100 francs, I found that the Festival had contributed less than 500 . . . nothing to drinks or entertainment of Dirk, Alan Bates, Harold Pinter etc . . . I told the [hotel] to submit the rest to Columbia . . .' The following year, when he served as chairman of the jury, the Festival paid for his accommodation and meals, but this left an outstanding debt to the hotel of 3,699 francs mainly on account of drinks (2,580 francs – £1,500 in 1990 values). Again he challenged his own liability; according to Bogarde, the Roux family, proprietors of the hotel, became aggrieved; Patricia denies it. Friction over expenses was constant. In July 1971 the Loseys sailed for New York on the SS *France* to attend the première of *The Go-Between*. Columbia (British) Productions Ltd pointed out that the boat passage from Southampton with double de luxe cabin and single cabin (for Joshua) had cost £928.50, compared to the price of two first-class air tickets (£325.90).[7]

Thoughtful and sensitive to the tastes of friends and employees, he often bought them beautiful and appropriate gifts. Ros Chatto remembers his generosity to her family. He gave his secretary, Celia Parker, a silver link chain bracelet watch. Dr Barrie Cooper recalls that for many years his family and the Loseys had Saturday lunch at Don Luigi's in the King's Road. Joe would insist on paying 'to the point of embarrassment'. In Cooper's view, this generosity operated when not demanded (a bill) or morally claimed (a loan); it required the spontaneous combustion of *noblesse oblige* or love. 'Joe resented everyone's fees,' adds Cooper.

Periodically serious calls were made on his generosity from friends in dire need. When Hugo Butler (*The Prowler, The Big Night, Eve*) fell ill, recalls his widow, 'Trumbo, Aldrich and Ingo Preminger helped the family survive.' Butler died in January 1968, a victim of Alzheimer's disease, leaving six children. The same month Losey signed a contract to direct *Secret Ceremony* for £93,750 ($225,000). On 14 February he wrote to Robert Aldrich: 'It may well be that there are others closer and better able to help [Jean Butler] than I. Please tell me.' Aldrich did: 'Jean is in debt. They don't have "what to eat" and anything you can do would be most appreciated. At present Trumbo, Maltz, Ingo Preminger and I are continuing monthly sustenance contributions. Mine, for example, is $166 a month . . .' Losey then wrote to Ingo Preminger: '. . . it is almost impossible for me to get money out of the sterling area without express currency exchange permission . . .' A year later, in February 1969, he explained to Jean Butler (who had not asked him for help): 'Anything that is brought into England or out of the sterling area . . . would be subject to tax, which would make it practically disappear.'

In June 1969, the documentary film director Paul Rotha begged him for help – he'd suffered 'a particularly rough few years through which I am still struggling'. He added: 'You may remember way back in 1953 when I was at the BBC-TV how I tried to help you? . . . If you want to know the truth, I am living on National Assistance and that does not cover food and rent.' Losey responded with a letter to a mutual friend, John Trevelyan: '. . . would he [Rotha] be insulted by an amount such as £50? . . . I cannot bear to get re-involved in the neuroticisms of his personal life.' It was at about this time that Losey gave one of his professional advisers a Christmas present in the form of a bespoke Douglas Hayward suit at a cost of £90. Impulse-gifts were life, pleas and demands were death.

But he did, on occasion, lend. In February 1969 he lent $4,000 to Capi-Films, Paris, so that Joris Ivens, then seventy years old, could complete his film, *La Grotte Sacrée*, about an underground village in Laos under American aerial bombardment. When lending small sums Losey could be restlessly insistent about early repayment. He repeatedly badgered an indigent writer, who'd been working on a screenplay for him, to repay £250, which 'was only extended . . . out of compassion for you and your family'.

Losey went into tax exile in 1975. Who would pay taxes if they could be avoided or evaded? Few.

He had sought refuge in London, strangers helped him, and he was allowed to work at a time when the unions were acutely protective of their own personnel. He used the National Health Service (though it was too slow, he complained). From the moment of his arrival his British earnings were far above the national average; by 1956–8 fifteenfold; but he felt no obligation to contribute to the local amenities. By his own cheerful admission, the Danziger brothers paid him £100 a week 'under the table so that I didn't have to pay tax on it . . .'[8]

Despite official residence status with the Bank of England, he maintained his Swiss bank account – strictly illegal. By law his worldwide earnings were taxable in the UK; however, when setting up *The Sleeping Tiger* in January 1954, he entered into a clandestine arrangement with Dorast Pictures, Inc., of New York. In October Carl Foreman dispatched a cheque for $1,380 to Losey's Crédit Suisse account in Zurich; Dorast would reimburse Foreman on the basis of a $4,000 loan to Losey against his share of the profits in the film.[9]

In December 1955 Losey was contemplating further tax evasion. Expecting to receive 'some dollars' from Sol Lesser's film company, he asked Henry Bamberger to receive the whole amount in California and pass $500 to his Swiss account. Bamberger's response was encouraging, although he did not mention the Swiss account: 'It will be a pleasure to see you receiving some dollars again.' He would deposit such funds in his own account then transfer them to Losey's: 'your name may never have to appear anywhere'.[10] (But the deal fell through.)

Legal tax avoidance is another matter. Losey was normally well served by the professionals. On 30 January 1954 he informed Sidney Cohn that his London accountants, Gorrie, Whitson and Birkett (where May Davey looked after his affairs), had justifiably claimed expenses in excess of his earnings, 'and I have so far paid no British tax'. His secretary, Pieter Rogers, who reckoned his annual income to be less than a tenth of his employer's, regularly noticed that Losey had paid less income tax than he. Stanley Gorrie later summarized Losey's first ten years in Britain (1952–62): 'He certainly paid very little tax at all . . .'[11]

He had ceased to pay tax in America. His last US tax return was filed for 1953. 'For tax purposes,' explained Gorrie, 'he is resident and ordinarily resident in the UK, and not resident in the US. For UK tax purposes he still retains domicile in the US.' This status entitled him to a reduced UK tax liability as compared to the natives. From 1959 to 1964 he paid only £5,090 by way of income tax and national insurance on an income of £18,597 from Yetell, Ltd. This modest declared 'salary' in fact represented total earnings of £60,344, from which had been deducted such expenses as £7,564 for 'Travel & Hotel', £3,729 for 'Entertaining', £2,368 for 'Car Hire & Taxi'. In short, meals and foreign holidays could be written off against tax if they involved a chat with a producer or a location search. If this was a sin it was a universal one – and legal – and therefore none.[12]

Entering, with *Boom!*, his 'time of the Burtons', Losey was advised by Robin Fox and Laurence Harbottle to set up an 'offshore' financial arrangement to minimize his tax liability: Glenbrook Investments, Ltd, located in the Bahamas. This was fairly typical of the time – elaborate schemes which often came to grief. Losey's solicitor explained that henceforward 'all rights would be vested in Glenbrook which would be a company held for your benefit' and jointly controlled by the Royal Bank of Canada and the Westminster Bank. Income would be taxed only when remitted to the UK; on the other hand money could not be remitted outside the sterling area without permission of the Bank of England or Exchange Control in the Bahamas.[13]

In December 1967 the Loseys took an extended family holiday. This, too, was written off against tax: 'I am enclosing a bank receipt for money I drew from Glenbrook in Vienna on the trip which was for the purposes of location search and professional discussions with Dirk Bogarde in Vienna, Alain Delon in Venice, and Brigitte Bardot, George Tabori and John Heyman in Paris.' Payments in respect of *Secret Ceremony* and *Figures in a Landscape* were remitted to the Bahamas, either in dollars or in Swiss francs. In December 1969 his new accountants, Norden & Company, requested Glenbrook to transfer 'hard currency' to the amount of $10,000 to Losey in Los Angeles (where he was beginning a three-month visit to America). If allowed, this money would never have been taxed. But the reply was discouraging: Bahamas Exchange Control would not permit it.[14]

In May 1970 Glenbrook decided that his recent films had not proved sufficiently profitable and that it wished to end the contract. The severance settlement offered was £25,000 plus '£32,500 due to be received by us from third parties in respect of sums already agreed . . .' It was this settlement into which British Inland Revenue slowly fastened its teeth – as we shall see. On 27 July 1970 Glenbrook Investments was put into liquidation, and Losey's 'offshore' life was transferred in June 1971 to Bookers Artists Services (Bahamas), for 20 per cent of net profits.[15]

In April 1974, Inland Revenue argued that the payment of £57,500 to Mr Losey from Glenbrook was assessable by reference to Section 480(4) of the Income and Corporation Taxes Act 1970. 'Alternatively, it is arguable that this payment was a capital sum within the meaning of [the 1970 Act] . . .' Norden & Company put up appropriately complex arguments to fend off the tax claim, but when Gerry Burke flew to Paris in September 1975 to spend the day with the Loseys, Patricia noted 'awful news' of back taxes owned: 'We seem to be paupers. Hell . . . J[oe] had dinner at Le Chamarée.' It dragged on. A letter from Losey to Burke dated 30 January 1978 indicates that the British tax claim was £22,000. The final settlement is not known but according to Burke it was an 'amicable' one, much smaller than the original demand. Losey had in effect substantially avoided the normal British tax liability thanks to skilled professional representation but he complained about the lawyers' bills, 'considering the demands were a direct result of . . . a set-up which [Harbottle] made and advised and was paid for at that time'.[16]

Gerry Burke served him well, not merely as an accountant but as a financial psychiatrist. Losey's long letters to Burke – with their 'items 1, 2, 3, 4' – served a therapeutic purpose not unlike the comic count who asks his man, 'Do my shoes feel too tight?' Burke recalls that Losey never directed a cross word at him in nineteen years – perhaps a unique distinction, though it may also have applied to the adored agent, Robin Fox. Invariably Burke used the telephone when responding to Losey's denunciations of others. Norden's annual fees ranged between £1,000 and £2,000; without doubt Losey was often not billed for time-consuming services rendered.[17]

He had of course paid tax on moneys remitted from the Bahamas to Britain. Taxation in Britain was high and severely escalatory until Margaret Thatcher came to power. Writing to friends, Losey invented punitive rates unknown to Inland Revenue; thanking Alain Delon in July 1973 for a generous loan, he explained: 'My chief problem . . . is taxes. Anything I bring in I pay 90 per cent plus on.' This was not so. In fact Losey's non-British citizenship entitled him to special privileges. During the three years preceding his tax exile, his UK tax returns were as follows (the multiplier for contemporary values being 5):

1972–3 £27,847 (£7,476 tax paid)
1973–4 £13,877 (£4,571)
1974–5 £25,824 (£11,137)

Losey's tax status was 'resident but not domiciled in the UK'. 'We fought for that,' recalls Burke. Indeed, Losey's brief flirtation with British citizenship in 1968 had foundered precisely on the increased tax liability it would have entailed. This 'resident but not domiciled' status, peculiar to British tax law, limited his liability to 50 per cent of the top rate of tax for which a resident British citizen was liable. Now came the great shock. The Labour Government's Finance Act of March 1974 modified – but did not eliminate – this immunity for foreign residents who had lived in the UK for nine out of the previous ten years. Losey decided to leave Britain. In April 1975 he wrote to Lady Falkender, Personal and Political Secretary to the Prime Minister, informing her that 'for three out of the past four years, I had virtually no income and lived on money borrowed from colleagues'. (His aggregate income for the past three years had been £67,575 gross, £44,391 net.) A month after drawing Downing Street's attention to his poverty, the Loseys travelled from Genoa to New York on the *Leonardo da Vinci* but were much discontented about the accommodation. Prior to the return journey Celia Parker was instructed to confirm the size and privacy of their suite on the *Queen Elizabeth* – a sunbathing balcony was required and the dachshund Tyger's needs must be met.[18]

Briefed by Losey in a Paris bar, Sam White reported in the *Evening Standard* (14 May 1976) that Losey had been faced with 83–7 per cent British tax on worldwide earnings over £20,000. In November, while in London for the Film Festival, he publicly complained: 'All of this suddenly descended on me . . . there is no way I can make enough money to pay my English taxes.' But he and his advisers knew the real position, which was lucidly set out by Robert Sheldon, MP, the Financial Secretary to the Treasury, in a letter to Losey's union, the ACTT. After nine years' residence in the United Kingdom foreign artists were now liable to pay a maximum rate of tax on their earnings of 62.25 per cent (i.e. 75 per cent of 83 per cent). The Government wished to eliminate 'glaring inequalities in tax treatment between different groups of people living and working in this country . . . a special tax category for foreign artists would be very hard to justify, especially to the British writers and actors who pay tax on the whole of their income in the ordinary way'. Mr Sheldon believed that 'for most people questions of language and environment played a much larger part in such decisions than the level of taxation is likely to'. Joe wasn't most people.[19]

Money had become a source of alienation and self-alienation, divorcing his personal practice from his professed beliefs. On the other hand, if it's time for a smile, Losey had nicely stolen the Marxian adage, 'to each according to his needs'. He embraced the acquisitive individualism of bourgeois society while remaining aggressively contemptuous of the bourgeoisie. It should not be assumed that pillows and pockets in the lives of artists are private matters (not that 'private' is a valid biographical category). Leaving England, Losey took his knife to the empty materialism of the bourgeoisie in *The Romantic Englishwoman*. Arriving in France, he embarked on a devastating collective portrait of the parasites in *Mr. Klein*, a withering one in *La Truite*. The artist, on the other hand, bestows on himself a special exemption; how can he flourish unless comfortable and surrounded by beautiful objects? Robert Klein, by contrast, buys paintings for profit. The true bourgeois – and worse, the uncultivated petty-bourgeois – is also a Collaborator: he deals on the black market while abandoning Jews to their fate. The lifestyle of the progressive intelligentsia, by contrast, is irrelevant. To be on the Left is enough; *ça suffit*. The happy image of the artist as sceptically observant anarchist is mirrored in Losey's fleeting, bar-stool appearances in *Eve* (Harry's Bar, on the most expensive strip of the Venice waterfront) and in *La Truite*. But Losey also quoted Hegel: 'Truth is concrete'.

39

The Losey Cult:
A Chevalier on Tour

It was in Paris that Losey had first achieved international status; with the arrival of *The Servant* he entered the pantheon and by the late 1960s he was regarded by the French ciné-clubs as incontestably a great artist.

The cult began with the rediscovery of his Hollywood films by a small group of enthusiasts passionately involved in the relatively new art of cinema. Situated on the Avenue Mac Mahon, near the Etoile, the Mac Mahon cinema showed original, subtitled versions of foreign films; some projections were private – *'une salle d'exclusivité'* – at the wish of its director, Emile Villion. Here Joe Losey was praised, received, honoured. As one of his critical devotees, employing the hyperbole typical of the time, put it: 'Losey's reign on the French screen established itself from June 1960 to September 1962. Outside of any fashion, he imposed himself as one of the ten best directors in the world . . . It was also the beginning of a new fashion, dethroning Bergman from the podium of critical snobbery.'[1]

The cult of 'genre' movies owed much to the writing and teaching of André Bazin, co-founder of *Cahiers du Cinéma* in 1951, and an active force at the Institut des Hautes Etudes Cinématographiques. Bazin was fascinated by the stylistic rules applicable to each genre (not least the Western); his disciples were taught to admire the long take and depth-of-field camerawork rather than inter-cutting (montage); depth-of-field enabled the action to develop over a period of time on several spatial planes, each an aspect of the 'realism' so essential to the cinema (as distinct from the theatre). It was in critical terms such as these that Losey became a cult-figure in France.[2]

Paul Mayersberg comments that what the *Cahiers* and Mac Mahon critics admired in Losey's American films was their 'limpid, exact expression of the story'. The ultimate hero was Fritz Lang, whose American films revealed a residue of expressionism, but without obvious stylization. Lang would strip down a hotel room and then stick one object on the wall to signal, undeniably, 'cheap hotel room'. Losey, too, conveyed essences with occasional expressionist-Brechtian touches – but when he began to play up the expressionist or baroque 'in the image', some admirers were disenchanted. Their written texts had excavated the half-buried motifs in his films; now he seemed to be filming their texts.[3]

At Pierre Rissient's instigation, *Temps sans pitié* (*Time Without Pity*) was first shown in Paris on June 1960, and distributed by Télécinex. Losey described it as a major turning point 'because through the French it reached many other people, many other nations, and particularly colleagues and artists. I owe all this to the Mac Mahon group . . .' Michel Ciment describes this 'big psychological turning point' as Losey's late revenge on life. Bertrand Tavernier discovered *Temps sans pitié* at the Caumartin experimental cinema (*cinéma d'éssai*), on the advice of Pierre Rissient, and was so '*bouleversé*' that he wrote two reviews, in *Cinéma* and *Positif*, lauding this 'obscure director of B movies'. The film's screenwriter, Ben Barzman, lived in Paris and was much liked in film circles: 'He brought warmth, affection, social values, kindness to his work,' Tavernier recalls. But admiration was far from universal. In *Figaro* Louis Chauvet called the film 'badly thought out, badly written, badly made. It rambles through successive moments of clumsiness in each sequence. Removed from the snobbery of the chapel, such a film would be regarded as a mediocre B movie complicated by absurd pretensions.'[4]

Although Losey's Hollywood films had been faintly noticed when released in Paris in the early 1950s (*The Prowler* and *The Lawless* were screened at the Mac Mahon in 1955–6), it was only now, in 1960–61, that they became the object of a full-scale cult. Having viewed *Le Rodeur* (*The Prowler*) and *La Grande Nuit* (*The Big Night*), Jacques Serguine did not trust words to describe them – Losey's art had revealed, 'like a blow from a fist, a being greater than the hero, more beautiful than the gods – a man. The cinema is the art of the true: Joseph Losey is the first artist.' The September 1960 issue of *Cahiers du Cinéma* was devoted to Losey – an accolade engineered by Rissient and Michel Fabre from *Présence du Cinéma*, with the slightly sceptical consent of Eric Rohmer, editor of the *Cahiers*. But Losey's admirers were soon to divide between those who enthusiastically embraced *both* the Hollywood period and the British films post-1956 (the three intervening films remained virtually unknown), and those who, by contrast, regarded his British films as an over-elaborate degeneration from the cinematic purity of the Hollywood phase.[5]

Writing about *Le Garçon aux cheveux verts* (*The Boy with Green Hair*), Pierre Rissient described the first scene between Dean Stockwell and the police doctor as achieving an unforgettable simplicity: 'completely unpolished – and today still unapproached'. Tavernier recalls with fervour the same scene, largely because the depiction of the police doctor and the schoolteacher represented 'respect for intellectuals'. A fable, he adds, was an original concept in Hollywood: 'The Cold War was on. Here was ingenuity and a sense of happiness, trust in human beings. And when they cut off the boy's green hair!' Rissient was equally ecstatic about Losey's second film, *Haines* (*The Lawless*): 'Never have we encountered such nakedness, such truth, such a human presence on the screen . . . at root this is the greatest Western and the only Western ever made.' Marc Bernard called *Haines* the most beautiful of films: 'The fervour which filled me at each viewing is like feeling one's breathing become easier, the certainty of purer blood . . . The tone is that of youth, irresistible, equal to the ardour of a national swimming champion . . . the clarity of the editing give an adorable sparkle to a free and vivacious body. Example: a tree is a tree, a young girl is a young girl, the young girl climbs the tree (and bruises her knee).'[6]

The Prowler had first been shown in France in 1952 and scarcely noticed, just another '*film policier*' – although Truffaut had regarded it as '*un film d'auteur*'. Ten years later it became a masterpiece. 'With *Le Rodeur*,' wrote Raymond Lefevre, 'Joseph Losey becomes Joseph Losey. That is to say, the director of alienation (*distanciation*), the lucid painter of a progressive disintegration within a décor which becomes the action.' His use of depth of field allowed 'the total expression of behaviour which was often ambiguous', while 'the extreme subtlety of his camera movements' enriched interior psychology. *M. le maudit* (*M*) had come out in Paris in December 1951 and been poorly received. But now Rissient wrote: 'The action is inscribed in the décor. For the first time a city exists on the screen. It's this expansion of the action in the world which allows us to call Joseph Losey a cosmic director.' According to Raymond Lefevre, in *La Grande Nuit* Losey portrayed 'purity, violence, perversity, justice, lies, force, virility, absurd fatality, lucidity, culpability . . .' – and from these qualities found his décors. Writing in *Combat*, Henri Chapier argued that the film followed the expressionist tradition while applying the principles of Greek tragedy to the cinema. Rissient added his own commendation: 'And also certain movements are achieved with a force, a brutality, unknown to the cinema before Losey . . .' In 1961 the Paris Cinémathèque presented a week-long Losey retrospective – standing room only.[7]

L'Enquête de l'Inspecteur Morgan (*Blind Date*) was distributed in France by Paramount. Jean Douchet interviewed Losey in *Cahiers du Cinéma* (March 1961) then launched himself into hyperbole: 'Losey is above all a researcher, his direction a method [of research]. The avowed aim is knowledge, as with a scholar . . . Losey restores to the camera its original function as a scientific instrument. In this resides the novelty of his contribution.' And more: 'Losey is the first director to take as his subject the truth about human beings, regardless of moral, metaphysical and religious reference.' While saying nothing about the script, dialogue, flimsy plot or performances, Douchet poured forth abstractions. *Cahiers du Cinéma* reported that *L'Enquête de l'Inspecteur Morgan* was the most critically praised foreign film of the month; Louis Marcorelles informed Losey that the film had broken previous records during its first week at five cinemas.[8]

Les Criminels (*The Criminal*) opened at the Marbeuf on 22 March; rated the best film of the year by the 'young French critics', it won the *Nouvelle Critique* prize for the 'Best Foreign Film' of 1961. Marcorelles, who was working on the French subtitles for new films by Anderson and Reisz, reviewed it enthusiastically in *France-Observateur* (23 March). 'Now I disagree entirely with Lindsay [Anderson] and Karel [Reisz],' he wrote to Losey, 'I think your film is, within the limitations of the gangster story, the most genuine and revealing I have seen for a long time.' *Les Criminels* was enjoying an extraordinary success at the Marbeuf, which had refused to show John Huston's *The Misfits* '*pour poursuivre l'exclusivité de votre film*'. In *Combat* (27 March) Pierre Marcabru wrote: 'The creation is purely visual; the words don't count. The conflicting images, the brusque cuts . . . give *Les Criminels* a shattering dimension.' Karel Reisz himself saw the film in Paris in the company of the director Claude Berry, who adored it. 'Of course the French loved all that heightening,' recalls Reisz, who wonders whether Losey was seduced into pursuing his own French mirror-image.[9]

Losey never forgot what he owed to his French admirers – though with time he became irritated by the exclusive personal relationships which his young admirers demanded. It 'turned bitter in many instances and it was often exploited. However, I am very conscious of the debt that I owe.'[10]

Of all his films, none was so scorned, denigrated and detested by French critics as *Eva*. Two examples must suffice. For Robert Benayoun, writing in *France Observateur* (11 October 1962), *Eva* was 'the apotheosis of the chichi'. Moreau was a great actress led astray ('*fourvoyée*'). Benayoun noted that at the Venice Festival publicity handouts declared Losey to be the world's leading director, in the opinion of French critics. 'Flattered by an international coterie of snobs, he seems to have let himself be entangled in a skein of embellishments and affectations. He tirelessly blows up bizarre detail, the pretentious soliloquy and the gratuitous ellipse.' Panic reigned among his admirers. In *Le Figaro littéraire* (13 October), Claude Mauriac wrote: 'With *Eva*, he tries to be equal not to himself, but to that part of himself he desires to promote. As a result the voice is stridently raised, the posture derivative, beautiful ambitions lose themselves in pretensions.' The denigratory terms piled up: 'fakery and pretence', 'pretentious', 'confectionery shop', 'virtuosity in a human void' – and why was the radiant Moreau of *Jules et Jim* here rendered so ugly?

In vain Losey sought to offset the devastating bombardment. *Les Lettres françaises* (17 October) published his remarks about 'a clinical study', 'clarity of expression' and 'that simplicity which is the best of modern art' (retranslated). Interviewed in *Le Figaro littéraire* (13 October), he played the Puritan card: 'As I predicted, it's difficult for *you* to understand Tyvian: one must be Anglo-Saxon to understand that he's precisely a Puritan.'

On 12 July 1963, Losey reported that a private screening of *The Servant* in Paris, organized by Claude Makovski and Florence Malraux, and attended by about fifty people, had gone well. The costs were about £250. Publicity was entrusted to the critic Pierre Rissient. Losey dispatched a version of the script in which dialogue was reduced 'to practically the bare minimum' as a basis for the French subtitles. Tony and Barrett should address one another as '*vous*' until the moment that Barrett is fired from employment; after his return, '*tu*'. (In fact '*tu*' begins during the rough game on the stairs.) Expressing delight that Nissim Calef was supervising the subtitling, Losey advised: 'An absolutely imperative rule is that . . . the same sentence [subtitle] should never ride across a cut.' The rule is right but the result was to excise some good lines:

'You can't have it on a plate for ever, can you?'
'I'm not staying in a place where they just chuck balls in your face.'

A young audience not yet born at this time, viewing the film near the Sorbonne in 1991, took off at Barrett's orgasmic reversal of the master–servant role: 'Well go and pour me a glass of brandy . . . Don't just stand there! Go and do it!' But translation (as distinct from subtitling) must miss the untranslatable idioms: 'Better safe than sorry' is '*Mieux y preventir*'. Susan's killing remark at the dinner table, in response to Barrett's claim that white gloves are 'used in Italy' is badly translated: 'Who by? – *En quel milieu?*'

The première took place on 9 April 1964, when the Association Française de la Critique staged its annual gala at the Raimu cinema in the Champs-Elysées. Although the French title still seemed in doubt – some critics tried *Le Valet*, some *Le Domestique* – there was no division of opinion about Losey's triumph. 'I was pinned down by journalists the moment I arrived,' he reported. The guest list he proposed included 'key members of Mac Mahon', 'key members of *Cahiers du Cinéma*', 'key directors at discretion such as Resnais, Chabrol, Godard, Truffaut, Rivette, Dassin . . . André Malraux and entourage . . . Picasso and guest'. During the first four weeks of its run, the film was shown at three cinemas simultaneously. The predominant tone of the press was ecstatic. *Eva* was forgiven, almost forgotten. Rissient reported 99 per cent capacity for the second weekend at the Saint-Sévérin, 94 per cent at the Raimu, but the total audience figure, 41,755, did not contractually oblige AZ Distribution to dub the film; it would be released in the provinces with subtitles, like *Les Criminels*.[11]

Cahiers du Cinéma (December 1963–January 1964) carried an interview and a eulogy by Richard Roud. In *France-Observateur* (6 April), Robert Benayoun called it the most 'scabrous' film of the season, a film of perfect ambiguity, a libertine nightmare of perverse decadence, done in a classical style with reserve, rigour and glacial humour. In *Le Figaro littéraire* (9–15 April), Claude Mauriac broke free from the prevailing obsession with Losey as 'auteur' by emphasizing Pinter's reputation for boldness and originality in the theatre, his handling of daily language through repetitions, interruptions, broken rhythms – and what Nathalie Sarraute called '*sous-conversation*'.

Michel Mourlet achieved the deification of Losey by way of Valéry, Nietzsche, Hegel, not to mention Bach and Stendhal. Losey had achieved pure art, 'that perfect limpidity of consciousness at the bottom of which the true forms of the world manifest themselves . . . The genius of Losey recovers perfect objectivity: the most deliberate art ends up with the birth of spontaneous sentiment. Losey is the only prophet . . . The clearest eye selects the noblest form. We know of no other definition of art.'[12]

King and Country was released in France in January 1965 by Ucinex under the title *Pour l'exemple* – possibly better would have been '*Pour encourager les autres*'. Arriving for the Paris première, Losey, Bogarde and Priggen caught sight of Losey's name in huge letters above Bogarde's. 'I think I'll stay in the car,' Bogarde remarked. Tom Courtenay remembers walking up the Champs-Elysées with Losey; a passing posse of beautiful English girls recognized Courtenay and 'after that I had a very happy time'. The critical reception equalled that of *The Servant*. In *Nouvel Observateur* (7 January), Michel Cournot was ecstatic: 'He is one of the princes of his art, one of the twenty living directors who know how to launch, directly, physically, the heart and mind into the image. A movement of the camera, a short and simple movement, reveals his unique generosity and grandeur of spirit.' Yet the film was not a commercial success in France. Philippe Modica of Ucinex informed British Home Entertainment that in Marseille the box office returns were poor because the ladies stayed away.[13]

20th Century-Fox, distributors of *Modesty Blaise*, booked Losey into the Prince de Galles Hotel on 14 April 1966, the purpose being collaboration with M. Perdrix of Cinétitres. Losey returned to Paris with Patricia a week later, to check the subtitled print, Theo Cowan notifying him that Air France had been alerted to give them 'special prestige treatment'. The Paris première was held on 18 October, at the Marbeuf. All the guests were born under the zodiacal sign of the Scorpion, with twenty-three cover girls forming a guard of honour on the staircase, their thighs decorated with – a scorpion. *L'Express* was one of several French magazines to give the film huge publicity – Vitti as Superwoman. But *Modesty Blaise* was poorly received.

Losey's rift with his ardent young admirer Pierre Rissient no doubt reflects on both men but also sheds light on the heated French film culture which brought Losey international prominence. Rissient had been nineteen when his passionate admiration for Losey began. He was, and remains, a cine-fanatic: critic, agent, publicist and distributor. 'Pierre fought alone for *Temps sans pitié*,' stresses his friend and colleague, Bertrand Tavernier. But the price of Rissient's admiration was high: his hero was always in danger of 'betraying himself' by, for example, working with Jeanne Moreau (who had let down Fritz Lang, etc.); or with Monica Vitti; or with Marguerite Duras (who belonged to literature not cinema, etc.). Losey appointed him to handle French publicity for *The Servant*, but his mounting impatience with him was obvious. Writing under the letterhead of the Cercle du Mac Mahon on 28 March 1964, Rissient complained in clear but fractured English: 'Just before receiving your letter which, I believe, wants to be an unpleasant one, I obtained from Wasserman the expenses for Patricia. People here look to be more happy of my work, and more confident in me than you.' Rissient was lobbying on behalf of his 'new-born company' to handle the publicity and distribution of *King and Country* – but Losey was evasive. The distribution contact was eventually made with Aurelia Films, and sub-contracted to Ucinex. In the event, Rissient hated *King and Country*: a *'film du salon'*. Having introduced Losey to the author of *La Truite*, Roger Vailland, he complained when Losey pursued Bardot for the lead part: 'You want to work with a star.' As he puts it today, 'For me, very young at that time, a great director should be a saint.' Patricia Losey agrees that he was very useful, but remembers him as a bore, a nuisance, whose arrival depressed the spirit.[14]

The crunch came with the publication in 1966 of his book, *Losey*. He now turned against his idol with a vengeance worthy of Balzac's *Les Illusions perdues*. Of *Temps sans pitié* he had once written: 'The décor . . . leaves no doubt that the word "cosmic" is appropriate to Losey . . . dialogue, interpretation, décor, light, sound, music . . . all are integrated in the art of direction, and thus is projected a prodigious force that we have never felt before.' But now he condemned the same film as melodramatic, the character portrayed by Michael Redgrave as improbable, and more. As for *Gypsy*, it was cold and empty, complex and sterile – symptomatic of Losey's recent tendency to make *'films de salon'* for his entourage. (But what entourage did Losey possess in 1957?) Clearly evident is the muddy water in a film culture where critics become publicists and entrepreneurs.[15]

They fell out. In December 1972, when Losey had been nominated president of the Cannes jury and Rissient was representing the Festival in some of its preliminary selections, Losey expressed a desire 'not to deal with him'. Rissient's comments on Losey's performance as jury president are scathing. Two years later Rissient was dining in a restaurant and felt two hands on his shoulders. 'Losey said, "How are you?" I made a cutting remark – I was wrong. He wanted to heal the situation.'

Following the banning of Jacques Rivette's film *La Réligieuse* (original title), Jean-Luc Godard published an open letter to André Malraux, the Minister of Culture, in April 1966. On 22 July Losey wrote to Godard: 'I wanted to congratulate you on your most extraordinarily courageous and valuable statement . . . unique in its force and content.'

The Loseys arrived at Pontarlier on 29 October 1966, for the fifth Rencontres Internationales de Cinéma, where eight Losey films were screened. Thanking the Sous-Préfet and Maire, Losey wrote also to Pierre Blondeau, a local *lycée* professor and secretary of the *ciné-club*: 'The fondue was marvellous, although difficult to manage for a novice . . .' A 'royal tour', organized by the Fédération Française des Ciné-Clubs (representing 250,000 film enthusiasts) followed – full houses all the way, a storm of hospitality, big dinners in the best restaurants, summed up by Jacques Robert, national president of FFCC, as '*la "tournée" idéale*'. Losey's aversion to formal lectures was understood, participation in discussions would suffice. All expenses would be paid and the presentations would engender revenue of about 5,500 francs to cover the cost of making new prints of four or five 16mm copies of the films shown. To cap it, a few days' vacation at the Colombe d'Or, Saint-Paul-de-Vence, would be laid on. Coordination was in the hands of Albert Cervoni, the Marxist film critic of *France-Nouvelle*, whose reviews of Losey's films were unfailingly adulatory. What more could Losey want? We shall see.

At Nancy, *Haines* (*The Lawless*) was shown at the Caméo cinema to the Ciné-Club 35 de l'AGEN. Losey was hugely praised in the *Journal de Nancy*, then interviewed in *Le Républicain Lorrain*. Invited to Lyon by the Association Lyonnaise des Amis de la Cinémathèque, he was present when *Les Criminels* was shown at the Salle des Carmes. There followed two days in Grenoble at the invitation of the Ciné-Club Universitaire. On to Bordeaux: five films were screened at the Capitole cinema and in the auditorium of ORTF by the Ciné-Club Bordelais. A local newspaper photograph showed Losey smiling as he left his plane, flanked by two Air France hostesses at the head of the steps. *Liberté*, of Lille, announced that 'the great director' would be on hand when the student ciné-club showed *Haines* and *Les Damnés*.

The Loseys moved to the Colombe d'Or for a rest, as guests of the FFCC. Losey told Titine, Francis and Yvonne Roux that their hotel was becoming 'a second home and refuge' – he thanked them for the beautiful roses. On 27 January, Jacques Robert reported that the tour had earned 9,405 francs from the *ciné-clubs*. He now proposed yet another visit, involving a flight to Marseille and a visit to Pezenas '*pour le weekend des animateurs de Ciné-Clubs*'. Meanwhile, the Mac Mahon and Panthéon cinemas in Paris were showing *Le Garçon aux cheveux verts*, greeted with enthusiastic reviews in *l'Humanité* and *Le Monde*. The Loseys flew to Marseille, where the ciné-club showed ten of his films: '*Hommage à Joseph Losey*'. On 13 February he was present for a debate at which he remarked that he was too fond of human beings to allow them to close their eyes. Local newspaper coverage in Provence was massive, including a long piece in *Le Soir* – one of the most important directors of his generation. *Provençal-Dimanche*, styling him as 'half-cowboy, half-Viking', went on to describe his personal décor: blouson, velvet trousers, checked shirt, pepper-and-salt haircut, blue eyes, very lively and changeable. The conclusion was an odd one: the Viking cowboy could also be mistaken for a 'gentleman farmer' (using the English phrase).

Exhausted, the Loseys stayed at the Colombe d'Or from 21 to 27 February, at a tarif of 190 francs a day. The bill for drinks during the six days – mainly vodka, cognac and cocktails – came to 554 francs. But Losey's nature was ill at ease with lasting contentment. Patricia Losey wrote to Jacques Robert, conveying acute vexation about their visit to Marseille. Four towns had been visited in four days – too much. Unscheduled press conferences and radio interviews had been foisted on them. The journey from Marseille to Béziers had amounted to a ninety-minute round trip, the audience being interested neither in cinema nor in ideas. On the Wednesday the train from Marseille to Narbonne had arrived three hours late. The press had gone home. In short the whole timetable was an impossible burden, on top of which they had ended up on a plane without first-class tickets. Jacques Robert duly expressed his regrets that the Loseys' hosts had not respected the rules (*consignes formelles*) which he had sent them; in future he would endeavour to accompany them everywhere. Following print, transport and Customs expenses for *King and Country*, Losey's account now stood at 1,924 francs in his benefit.

Next stop, Algeria – a Losey retrospective presented by the Cinémathèque Algérienne from 19 March until 1 April. Losey had laid down his conditions: the visit must be mainly a holiday, with not too many interviews. At the end of the first week he would bring nine-year-old Joshua and his French tutor Mlle Delphine Bouvet at the expense of the FFCC: 'First class travel for all concerned. I will not fly on any plane that has tourist class only.' Approaching the coast of Africa at dusk, Losey marvelled at the first-class view of the 'brilliant orange, violent violet sky'. He and Patricia were greeted by 'a cordial and anxious group of men and women, of whom only Jean-Michel Arnold and François Roulet were previously known to me'. Following libations, they were driven to the Cinémathèque theatre where they were shown 'the most extraordinarily skilful exhibition of . . . my life and work . . . huge blow-ups, brilliant juxtapositions of design and montages of stills from quite different films of mine . . . I have had retrospective shows . . . in Paris, in London, in New York, in Stockholm, in Amsterdam, in Italy, even in Spain . . . and I have *never* seen a job of this sort done with such understanding and precise effect.'

The Cinémathèque, 26 rue Ben m'Hidi Larbi, charged two dinars for tickets to see each film. In both Algiers and Oran Losey was impressed by 'the incredible hunger of the audiences for films and . . . contact with film makers. All the showings were sold out . . . there were as many standees as the place could cram in . . . The audiences were predominantly young . . . students and workers of one sort or another. (By which I mean to say *not* bourgeois.) . . . In Oran there was such a demand . . . that the house was oversold by 100 per cent and had to be emptied and the audience readmitted . . .' Meetings were arranged with the young writer Kateb Yacine and officials from the Ministry of Culture. 'We travelled to the border of the desert, we travelled over the mountains to Constantine.' After three weeks they departed loaded down with gifts. Another four-day holiday at the Colombe d'Or followed on 4 April, at a cost of 1,310 francs (which Losey referred to the FFCC). A month later they were back again for the Cannes Festival, an eleven-day visit costing 5,009 francs.

With such a life, and such adulation, most of us would be content; but most of us wouldn't get it. On 11 April Losey dispatched a long indictment of the luckless Jacques Robert, president of the FFCC: 'I am very sorry about the recent performance in Paris . . . you were spoken to well in advance about the arrival of our plane from Algiers, not once but three times . . .' Yet the Loseys had found themselves with 'no tickets, no money, no reimbursement, and a great deal of our badly needed rest in St Paul was squandered in attempting to reach you by telephone'. The complaints piled up: 'I should like to have a full accounting . . . I should also like to know what happened regarding the 16mm prints of [three films] . . . what indeed did the peculiar sum of 1,721 francs which you sent me at the Colombe d'Or represent? . . . I hope very much that all this is a misunderstanding because if it is not it is excessively ugly.'

Ignoring the tone of Losey's letter, Robert promised a full accounting soon. But Losey kept digging up further grievances: despite paying a $7.60 supplement for first class on a flight from Marseille to Paris, they had finally been compelled to fly tourist. About this crime Robert remained silent. Losey raged: Robert's 'discourtesy' was hard to 'reconcile with your graciousness when you wanted something from me'. Not surprisingly, the FFCC never again invited him to undertake 'une "tournée" idéale'.

40

President of the Jury

Against Losey's wishes, London Independent Productions commissioned a firm of their own choice, Cocinor, to subtitle *Accident*, but capitulated when he tore into the result: in low-key sections the subtitles were so bright that they stole the eye, and most of them were set too high in the frame. He got his way: Nissim Calef would again do the subtitles for Cinétitres, with Pinter's translator, Eric Kahane, working in an advisory capacity. On 27 April, Losey wrote to Monsieur Perdrix, of Cinétitres, asking for certain amendments, including the elimination of two 'fucks':

'Who made this fucking soup? – *Qui a fait cette foutue soupe?*'

'How the fuck would I know – *Comment foutre le saurais-je.*'

For *Accident*, Losey engaged Bertrand Tavernier for the first time as his personal representative in France. Tavernier's colleague-in-disgrace, Pierre Rissient, recalls: 'The French distributors grew [were driven] crazy about Losey's expenses. It was a kind of revenge. If he wasn't highly paid, he could get it in expenses.'[1] Patricia Losey denies this.

The critical reactions showed Losey's star still in the ascendant. For Yvonne Baby, of *Le Monde* (7–8 May), this was as brilliant and intelligent a film as could be imagined; the 'two authors', Losey and Pinter, had achieved an implacable precision, worthy of goldsmiths. Claude Mauriac, also writing in *Le Monde* (16 June), praised Losey's superb '*écriture*', his art of seeing and making others see. Writing in *Cinéma 67*, Gilles Jacob went wide of the mark in detecting signs of Stephen's latent homosexuality and a past pederastic relationship between Stephen and Charley. This, explained Jacob, was a whispering England – a different face of England from Antonioni's *Blow-Up*; this was the world of '*pique-niques*', of 'high teas', of improvised '*après-soupers*' at which 'long drinks' were served. The critic was on safer ground when he observed that in *Accident* every smile is a dagger-thrust, every meeting of friends a settling of accounts. He applauded Losey's brilliant touches: a cat on a tennis court, a fallen umbrella, a cigarette smouldering on a deserted dinner-table, a morsel of food on Stephen's lip. Like Bresson, Losey had an extraordinary appreciation of sound, the breath of tennis players, eggs cracking, a policeman's cough, the whistling of a kettle, the wings of a swan, a lawn-mower.[2]

By early October, 157,000 tickets had been sold in Paris's '*salles de premières visions*', making it 'the second most successful British picture that has been released in Paris during the past twelve months'. The film won the *Spéctateur* award.[3]

In July 1967 Losey heard that he was one of two nominees by the Centre National de la Cinématographie for the title Chevalier de l'Ordre des Arts et des Lettres. Ironically, he was to be honoured by the Gaullist regime in the revolutionary year, 1968, which began with the dismissal of the director of the Cinémathèque Française, Henri Langlois, by the Minister for Cultural Affairs, André Malraux – who in turn bestowed the title of Chevalier on Losey. Losey, who had known Langlois since the early 1950s, was among the directors who withdrew permission to have their films shown by the CF unless Langlois was restored to his post. Victorious, on 15 May Langlois confirmed to '*Cher Jo*' that his precious documents had at last arrived from Algeria; he also spared a thought (in English) for the revolution: 'I am so proud that the beginning was the Cinémathèque.' Malraux's case against Langlois rested on the valid complaint that it was impossible to find out the number of titles held by the CF, where they were kept, and on what legal basis they had been deposited. Langlois was an authoritarian and paranoid genius; the inventory he kept was a shambles. By a second irony, it was precisely on this issue that Losey and Langlois were later to fall out.[4]

Losey first met Marguerite Duras through Sonia Orwell. In December 1963, he was hoping to direct Duras's *Le Marin de Gibraltar*, on which Jeanne Moreau had an option. 'Our communications are never exactly fluent and certainly not at their best on the telephone . . .' Losey commented to Duras in February 1966. 'I am now deluged with propositions and flattering money – the major number of which I have no interest in . . .' Keen to shoot her script, *Le Ravissement de Lol Von Stein*, he was rebuffed in a handwritten note (March 1967). In February 1968 she sent him her script, *La Chaise longue*, a slow-moving, dreamy, sophisticated confessional about love, set in a country hotel. Bertrand Tavernier recalls Duras visiting Losey in May 1968 and leaving with an enormous tin of caviar, which she took to the striking Renault workers. The original plan was to make the film in French, as a Franco-Italian co-production, but after Losey read a translation by Barbara Bray he insisted on making it in English. While Duras wanted Marcello Mastroianni and Silvana Mangano, Losey preferred Peter Finch, Mia Farrow or Delphine Seyrig. He had met a will as strong as his own: '*Te prie dire oui ou non définitivement*,' she cabled. In November 1969 *La Chaise longue* emerged as *Détruire, dit-elle*, at the London Film Festival, directed by the author herself. *The Times* quoted her: 'Money ruins everything. Joseph Losey wanted the rights. I refused. He would have made the film for too much money.' His budget had been close to £320,000 – hers about £20,000. In a vigorous riposte, Losey argued that her conditions 'would have made it quite impossible to do justice to one of the best film scripts I have ever read. I do resent, however, the implication that money "ruined" *me* along with "everything else". What money?' Duras summed up her view of him in an observation to Lee Langley: Joe, arriving at a station and finding his train ready to depart, would curse because he'd missed the one before.[5]

Losey wrote to Marc Bohan (designer of Elizabeth Taylor's clothes for *Secret Ceremony*) on 13 May 1968: 'I was distressed about the brutality in Paris and personally concerned to see that you yourself were badly beaten ... The clothes [for *Secret Ceremony*] are marvellous ... The Burtons, and perhaps I too, will be in Paris for the world opening of *Boom!*' The film opened on 20 June, as France underwent a wave of repression in the wake of the May uprisings. Losey visited Paris briefly, to open the new '*super-salle d'exclusivité*' furnished in fawn leather, the Paramount-Elysées. Bertrand Tavernier, his *attaché de presse*, described the betrayal of the revolution by the PCF and CGT, then turned to *Boom!*: 'All the reviews we have controlled were very good. The only bad ones were the very right (*Minute, Carrefour*).' But more could have been done (Tavernier complained) if Losey had made himself available for previews and private screenings. Jean-Louis Bory's review in *Nouvel Observateur* (3 July) fastened on the incongruity of the film's arrival in post-revolutionary Paris. France, he wrote, had seen the sea turned into a storm by an incredible wind, yet Losey had provided a second grinding (*mouture*) of *Modesty Blaise*, this time with aspirations towards Shakespearian tragedy, the result being 'facile and frenetic'.

On 9 July, Losey told Noël Coward that the French notices for *Boom!* 'are the best I have ever had for a picture ... and the business is "smash"'. The Communist critics seemed content with the arrival of a lavish bourgeois décor-drama in the midst of the political crisis, but the Communist press censored his remarks about the May events. Discussing this a year later with Michel Ciment and Bertrand Tavernier, he commented: 'It doesn't surprise me. I wanted an explanation of the failure of the leaders of the Left, their slowness and opportunism ... the desire to destroy outdated forms of expression was so powerful and healthy that it was criminal not to grasp them immediately, setting aside egotistical considerations ...'[6]

In November 1968 Losey paid Tavernier £40 for his work as public relations liaison in Paris; Tavernier arrived in London to represent Losey and interview him for a ninety-minute French TV programme in the *Cinéastes de Notre Temps* series. Previous subjects included Renoir, Ford, Buñuel, Lang, Samuel Fuller, Godard, and Ophuls. Seven hours of interview were recorded, covering sixty-four typed pages translated into English for Losey's perusal, but the *cinéaste* himself was in a black mood. On 13 February 1969, he wrote to ORTF complaining about mistakes, confusions, misspellings. On 25 March a cable followed: 'Understand first cut [of the film] is disastrous parody ... unless drastic changes are made under the supervision of ...

someone acceptable to me must withdraw consent . . . will not advise distributors to give you clips.' No more was heard.

How – with regard to Tavernier – can one be simultaneously Losey's paid public relations agent and his interviewer? This is Tavernier's answer: 'We never considered ourselves as press agents. We took on only those films and directors we liked. We fostered interviews, emphasis, history, selection. We fought for unknown directors like Losey. We brought them over [in person].' (But by 1968 Losey was scarcely unknown?) 'No, you can't call *Boom!* commercial,' Tavernier continues. 'And I liked *Secret Ceremony*. I was fascinated by the [camera] movements in empty rooms and by the colour in *Boom!*. In *Figures* there were brilliant things, terrific moments. I told him [Joe] what I didn't like. I took these films because I liked them. Joe needed comfort and assurance after what he'd taken. In France people fought for Jules Dassin, his permission to work, no one helped Joe in London. He got no interviews [in the 1950s]. He was warm and lovely. He was faithful to me. He wrote a good review of [my] *Le Juge et l'assassin* in *Le Point*.' Eric Kahane comments: 'Tavernier was very assertive. He used to protect Joe by answering for him. Joe would say to journalists, "What did he say?"' Tavernier himself recalls that he used to beg Losey not to slag his producer at press conferences.[7]

Under the title *Cérémonie sécrète*, the film opened in Paris in March 1969, subtitled at the Vendôme, Publicis-Saint-Germain, and Publicis-Champs-Elysées, and in a dubbed version at the Paramount-Gobelins and Paramount-Montparnasse. Despite the common belief that Losey's films were more rapturously received in Paris than in London or New York, the evidence of the reviews themselves indicates a surprising convergence of judgement. The major exception was *Secret Ceremony*, which brought forth a flood of hyperbole in Paris. In *Combat* (25 March), an admiring Henry Chapier likened the film to *The Servant* – both dealt with the 'victim–executioner' relationship which, according to Sade, Baudelaire and Freud, characterizes the erotic life of two beings devoured by desire: 'an oppressive poetry of psychosis . . . a theatre of cruelty . . . a sort of fable on the futility of all human relationships'. Losey's characters, gripped by crises, resembled Racine's. The film won Les Etoiles Award of the Académie du Cinéma for the 'Best Foreign Film' of 1969 and the 'Best Two Performances by Foreign Actresses' (Taylor and Farrow). On 14 December the Loseys arrived in Paris for the award, presented at the Laserre restaurant. Henri Alekan met them at the airport in his 'rather ramshackle car' and took them home to dinner. 'He is very sweet but the dinner interminable,' Patricia noted in her diary.

Losey arrived alone for a retrospective at the Théâtre de la Commune, in the Paris suburb of Aubervilliers. Patricia was ill with hepatitis; deprived of her company – and her invaluable role as interpreter – he sank into despondency. His visit coincided with the national strike of 11 March 1969 – all audiences fell by at least 50 per cent – and he became exhausted by a succession of publicity interviews, including four television programmes and eight discussion groups. Following his visit, his hostess at the Théâtre de la Commune, Claudine Boriès, admitted that it had been '*un semi-échec*' – a more mature (*formé*) public was required to appreciate his work. Conversations between them in the streets '*et au petit bistrot du coin*' had not been easy, since neither spoke the same language: '*Mais vous avez "l'oeil". Et celà, c'est irremplacable.*' Losey replied, in the kindest way, that he had enjoyed the discussion with the *lycéens* of Suresnes – whereas 'too often at your Theatre it required a star performance to galvanize an otherwise preponderantly unresponsive audience'.

Figures in a Landscape opened in France in November 1970 under a title that Losey objected to, *Deux hommes en fuite*. The subtitled version was shown at the Publicis-Elysées, the Publicis-Saint-Germain, and the Studio Jean-Cocteau; the dubbed version at the Paramount-Montparnasse and the Capri. Losey complained to Goddard Lieberson, of CBS, about the poor posters and publicity. *Paris Match* accorded the film a spread. Jean de Baroncelli, of *Le Monde*, hailed it as a supreme cinematographic feat, the work of a real stylist, a commentary on freedom. *Les Lettres Françaises* displayed a photo of Losey bare-chested, in sun goggles: 'We do not know all our enemies,' was the quotation. Tavernier reported: 'The film did not make much money. Several reasons: the title, the story . . .'

By 1971 Nissim Calef was dead and Losey invited Eric Kahane to undertake the translation and subtitling of *The Go-Between*, the technical work to be performed, as before, by Cinétitres and LTC. Losey anticipated problems with the boyish slang of 1900 – 'spooning', 'hard cheese' – and with the main title. At Cannes the film went by the provisional title of *L'Intermédiaire*, but it was *Le Messager* by the time of its French release. Kahane put the voice-overs, whether flashforwards or flashbacks, in italicized subtitles. The white dresses and tablecloths presented a problem for subtitling; Kahane recalls that Cinétitres worked a chemical miracle to produce yellow subtitles: 'Joe sneered at it . . . It should be more this or that.'[8]

The critical reception from Cannes was ecstatic. Writing in *Les Lettres Françaises* (2 June), Michel Capdenac likened the film to a bolt of lightning which not merely dazzled (like *Death in Venice*) but transfixed, subjugated and invaded the spectator. Harmony, originality, perfection and polyphonic sensuality – here was the language both of the intelligence and of the soul. *Le Messager* opened in Paris – four months before it was first shown in London – on 18 June, at the France Elysées, the Madeleine Gaumont, Studio Raspail and Saint-Germain-Huchette. Advertisements announced that '*le chef-d'œuvre de Joseph Losey*' – had won '*la plus haute/suprême récompense du XXVe Festival de Cannes*'. *Paris Match* awarded its maximum three stars while *L'Express* (21–7 June) displayed a photograph of Losey with Pinter, captioned 'For the third time'. Claude Mauriac delivered a eulogy in *Le Figaro* (18 June). The Loseys attended the Brussels première on 22 June and were royally entertained with Bols gin and other gifts, the photographers pursuing the couple to the Grand-Place. Losey was asked how he could know the English so well and how a great creator like himself supported the boredom of incessant interviews. *Le Soir* (25 June) called him 'a giant of world cinema'; with his red-brick skin, long silvery hair, engraved features, and luminous blue poet's eyes, he was simultaneously 'inspired and simple'. By 19 September, Columbia Pictures International was able to report 341,329 tickets sold in Paris, with good business for the dubbed version in the provinces. At 1 July 1972, *Le Messager* had earned $621,071 in France. Losey's pleasure was converted into a tribute: 'Concerning the attitudes of distributors ... I'd say that the United States is the most commercial, England is the most provincial, and France is the most cultivated.'[9]

Disaster followed: *L'Assassinat de Trotski*. On 20 March 1972, the Loseys arrived at the Hotel Raphael to work on the dubbed and subtitled versions with Eric Kahane. Nine days later Valoria Films launched the film in two subtitled versions at Cluny-Ecoles and the Elysées, while the dubbed version opened at eight cinemas in the Paris region. A minority of reviews were admiring but Losey informed Alain Delon on 20 April 1972 that the Paris reviews were 'pretty devastating and the business is more so ...' The first run in Paris grossed only $151,500 in seven weeks. Nevertheless, Losey offered rare praise to the distributors, Valoria Films, as 'the most intelligent and accommodating that I have had to deal with ... the only distributors who ever gave me a present'. On 5 June, Valoria's Hercule Mucchielli lamented that it was the greatest commercial disappointment of his life: having paid Shaftel a high price for French rights (the Delon factor), Valoria faced a disaster. Losey later forgot his praise of Valoria: 'They put it on in the ten biggest ... theatres in Paris, which was very silly. Because this is a film that had to grow over the years ... It is being handled in that way in the United States and England.' (In reality it was a critical and commercial flop in both countries.)[10]

In 1971 Losey was invited by Favre le Bret to preside over the 1972 Cannes Festival jury.* He agreed to attend every screening and every jury session, and to appear at three ceremonies: the opening, the closing and the official

*Losey served on the Festival jury alongside Bibi Andersson, Georges Auric, Erskine Caldwell, Milos Forman, Marc Donskoi, Alain Tanner, Georges Papi, Jean Rochereau and Naoki Togawa.

lunch – to which was later added 'The French Night'. On 17 April, he wrote to Robin Livio, Directeur Adjoint du Bulletin du Festival de Cannes, resisting further commitments: he would not give interviews or press conferences. 'I was . . . dumbfounded by your questionnaire . . . it seems astonishing . . . that you should ask me a number of questions, the answers to which you can find in . . . Who's Who editions etc. It also astonishes me that quite a number of your questions are of the genre which I normally refuse to answer even to the Press.'

The French press was alert to Losey's enmity towards Elia Kazan, whose new film, *The Visitors*, was the first to be shown to the jury. With slapdash inaccuracy, *Paris Match* (May 1972) claimed that it was because of Kazan ('*à cause de lui*') that Losey, Jules Dassin and Dalton Trumbo had been forced into exile. (Trumbo had remained in America.) The magazine reported no handshake between Losey and Kazan when *The Visitors* was screened; Losey had departed without a word. Kazan has since claimed that although he had discounted the possibility of Losey allowing personal bias to influence his judgement, Bibi Andersson later told him that *The Visitors* enjoyed the support of herself and 'a couple' of other jurors during the final deliberations, but Losey argued so vehemently against the film that it was set aside. Kazan's film, shot in Super-8 and skilfully blown up, depicted the havoc wrought by returning Vietnam veterans on a comfortable and complacent group of American liberals. Losey's private notes, to be found on his '*fiche du jury*', praised this aspect of the film as 'good and terrifying'. But the main theme raised, as in *On the Waterfront*, was whether it was right to inform (or betray) a criminal comrade, in this case a rapist. Patricia Losey had been allowed to attend jury meetings because of Losey's difficulties with writing by hand. Joe, she recalls, explained to the jury his own history with Kazan – though in fact there was none; Losey was appropriating the world's wound as his own – and insisted that he could not be objective enough to speak about the film or vote. 'I persuaded him that the film was dastardly and that he should speak, particularly as there was an old Soviet director on the jury in favour of Kazan.' Losey's jury *fiche* tells the rest: 'He [Kazan] is professional, skilful and much more restrained than in any previous film. Actors and all things relating brilliantly handled.' However! Important political statements were 'mixed up with confused metaphor and personal whine. Kazan's problem has nothing to do with the peg on which he seeks to hang it. Enfin, a dirty, disgusting, meretricious film by a near master of everything but himself.'[11]

A Belgian film, *Malpetius*, with Losey's other *bête noire*, Orson Welles, earned similar opprobrium, extending to the photography by his colleague Gerry Fisher ('also *enmerde*'). In the 'Appreciations' column of his *fiche* Losey wrote 'shit' five times. A French film, *Les Feux de la Chandeleur*, he described as 'Imitative, pretentious, self-conscious'. By contrast a Canadian film, *A Fan's Notes*, directed by Eric Till, was 'original and passionate', 'brilliant', 'good', 'superb'. *Trotta*, the German entry, directed by Johannes Schaaf, was 'superb skillful stylist moving excellent'. Roy Hill's *Slaughterhouse Five* was: 'Ugly, unfunny, pretentious. It is all the things it pretends to mock. Also no style. It is a dying gasp of Hollywood . . .' Philippe de Broca's *Chère Louise* he put down as 'cute, provincial, tourist, sentimental, dishonest, not even competent'. He slaughtered the British entry, *The Ruling Class*, though he credited Peter Medak's direction as 'competent and hard working'. 'The whole thing is thirty years too late, déjà vu, opportunist. *Modesty* but even less good. No style. Should go down great with the "Carry On" audience in England . . .' Clearly Losey's preference was for Robert Altman's Irish entry, *Images*: '. . . absolutely first-rate in all ways . . . very much the film of an auteur. Great style. Extraordinary movement. Main contender.' *Il caso Mattei*, directed by Francesco Rosi, won his admiration – 'A master of his genre' – but he found the script hard to follow and damned the music as 'electronic cliché'. This film won the Palme d'Or.[12]

In 1972–3 he became involved in a project to film the Chicago sequence of Simone de Beauvoir's novel, *Les Mandarins*. Annie Girardot was to play the fictional de Beauvoir. At a meeting with the author in June 1973 Losey told her he didn't think much of the existing script and had asked the critic and screenwriter Penelope Gilliatt to concentrate on de Beauvoir's relationship with Nelson Algren, perhaps working in her past sexual and political relationships in flashes, 'in somewhat the same way, only in reverse, as Resnais did in *Hiroshima mon amour*'. In an undated letter scribbled on cross-lined paper, de Beauvoir agreed that her fictional *alter ego*, Anne, needed 'deep links with her country lest one fails to understand why she does not remain in Chicago . . .' In November Losey informed her that he'd heard no more from Mitchell Brower, of Columbia Pictures, concerning finance. One must regret the demise of this project, given the absence in his films of a female character of intellectual calibre; whether Losey understood de Beauvoir's stature is not clear. The political literature of modern Europe was not noticeably his thing – throughout 1974 he regularly referred in correspondence to Malraux's novel, *La Condition humaine*, as the *Comédie humaine*. Like almost every other leading director, he wanted to film it.

Malraux's daughter Florence, wife and first assistant to Alain Resnais, was a close friend; '. . . has your Father put his foot in everyone's mouth again?' he asked her on 2 May 1968, after Malraux dismissed Henri Langlois as director of the Cinémathèque Française. When visiting Paris in June of that year he offended her by failing to call, but he was quick to counter-attack: 'Dirk [Bogarde] tells me that you found I have "gone Hollywood". How would you know? Or is that what you think of *Boom*!?' Florence Malraux's letters to Losey would suffer in translation due to her playful juxtaposition of

French and English. 'I never said that you have gone Hollywood. Je ne l'ai pas dit parce que je ne le pense pas ... However, il est vrai que j'ai moins aimé *Boom*! que tes autres films.' (In short, she had liked it less than his other films.)[13] She pulled strings on his behalf while he attempted to set up Proust; and she was always ready to bring him together with their mutual friend, Jeanne Moreau. In April 1975 the Loseys visited the Resnais' new flat, their first guests. In 1976 she introduced Losey to Jorge Semprun, the future screenwriter of *Les Routes du sud*, and in November 1980 to Monique Lange, the future screenwriter of *La Truite*.

Maison de poupée (*A Doll's House*) – subtitled and dubbed by Eric Kahane – opened on 17 May 1973 at eight Paris cinemas including the Bonaparte and the Biarritz. During the first week 25,341 seats were sold in Paris and five '*périphérie*' cinemas. Mireille Amiel (*Cinéma*, June 1973) declared that the feminist issues raised by Ibsen were alive and '*brulant*' in the film.

At the turn of 1973–4 Losey was talking to Jack Lang about a possible Losey production at the Théâtre Nationale Populaire – probably a kind of Living Newspaper presentation about children's rights, but in May 1974, with two films on his immediate horizon, he made his apologies to Lang: 'I am deeply sorry and even humiliated to have kept you waiting so long ...' (The previous November he had hinted to Lang that he didn't regard theatre work as sufficiently lucrative.)

Fantasizing, Losey assured Ely Landau that in Paris he could guarantee a preview audience of 250–500 people for *Galileo*, including President Giscard, André Malraux, Pierre Boulez, the leading writers 'and the whole of the top art world, not to mention socialites'. The reality was rather different. An undated memo from Georges Beaume lists those he had invited by telephone to attend the showing of an un-subtitled version at the Palais de Chaillot: Louis Malle had left town, Milos Forman had checked out of the Hotel Lancaster, Werner Herzog was leaving, Alain Resnais had not replied, and Simone Signoret was a case of 'message left'. Not even reviewed in *Positif*, the film journal which most ardently cultivated Losey, *Galileo* was never released in France. As Eric Kahane recalls, subtitles worthy of Brecht's text would have covered the entire screen.[14]

In April 1975 the Loseys became tax exiles and moved from London to Paris. Their intention had been to buy Jack Loring's flat in the rue du Dragon, but on inspection they found it too cramped and remained there only a few months. Françoise Sagan, meanwhile, published a brief eulogy in a glossy magazine, *l'Egoïste*, which carried photos of Losey posing in the Luxembourg Gardens, wearing a stylish cloak and raffish cravat.

Sagan commented that discomfort (*inconfort*) was the obsessive source of his work – here was a man profoundly of the Left but 'with the means of a rich bourgeois' who was deeply troubled by the plight of the world's unprivileged and who sometimes felt his own privileges to be 'truly odious'. A vulnerable man: 'I know few women who wouldn't want to look after him if he was injured one day . . . and few, also, who would not like to wound him in order to look after him.'

Losey's relationship with Sagan had its ups and downs. In March 1961 he had hoped to direct a play of hers in London, but her brother (Losey complained) had changed the terms too often. When compiling a guest list for the Paris opening of *The Servant*, he noted: 'I would not ban, but would prefer not to see there: any of the Hakims; Françoise Sagan.' In April 1977 Sagan sent him *Le Lit défait*, which Patricia greeted with the highest praise. Losey sent it to Pinter and Romy Schneider. When Sagan invited the Loseys to dinner, Joe sent his regrets from Normandy in the name of his dog, 'Monsieur le Tyger'. By February 1978 he had read an English translation of *The Unmade Bed*: 'It presents immense difficulties,' he told Sagan. In January 1979 he returned to her an English version of her new play: 'I'm not quite sure why you wanted me to read it at this point. If it is for a film, I'm afraid it won't make one – too discoursive and static.' Somewhat stung by this, Sagan responded (7 February) in her graceful English: '. . . it was after dinner at my home, when you recalled, as you sometimes do – the uneffaceable scandal of my behaviour in London, some twenty years ago . . . and you kindly showed your interest in reading [my play] . . .' Sardonically she expressed guilt at having 'encumbered your luggage, all summer, with my paperwork'.

L'Anglaise romantique opened in June 1975, subtitled, at the Concorde-Pathé, Cluny-Palace, and Saint-Germain-Huchette cinemas; the dubbed version was shown in four cinemas of the Paris *périphérie*. From Cannes the ever-supportive Robert Chazal (*France Soir*, 21 May 1975) announced: 'From a completely banal story, he has made a brilliant film, corrosive, unusual and insolent.' In *L'Express* (19–25 May), 'G.J.' recorded pleasure at seeing a great artist at the summit of his form; despite the film's modish and unoriginal theme, 'Losey's stiletto chisels gold . . .' *Le Point*'s Pierre Brillard (19 May) greeted a supremely intelligent comedy of manners.

Returning to Paris from a summer's teaching at Dartmouth, the Loseys found themselves involved in protests against Franco's execution of five young anti-fascists. Faced with a widespread boycott of the San Sebastian Film Festival, at which *The Romantic Englishwoman* was to be shown, Losey reluctantly decided not to attend. On 26 September the Loseys found themselves caught up in anti-Franco demonstrations in Paris (later echoed, though not reproduced, in *Les Routes du sud*). After attending a dinner for some seventy-five guests, including three government ministers, they were driving home in the car of the Mexican ambassador, the writer Carlos Fuentes, when they encountered a riot in the Champs-Elysées. Fuentes assured them, 'We'll be all right. Mexico has protested.' The following day they hit the same problem: 'It was very violent by then and totally indiscriminate,' Patricia noted.

Searching for alternative accommodation, in November they rented a flat at 86 rue de Bac at a rent of 4,500 francs. Patricia described it: 'Tiny loo off hall, but pleasant, comfortable and easy (we think). Only drawback: no balcony, courtyard or anything for Tyger.' Losey was now filming *Mr. Klein*, in which the rue de Bac is constantly referred to; indeed the many buses grinding along it had not greatly altered from 1942 to 1975. The dog Tyger also went into tax exile: Patricia's diaries for 1975 are much preoccupied by Tyger, walking Tyger, feeding Tyger, shampooing Tyger's 'undercarriage', mopping up Tyger's diarrhoea, complete with X-certificate descriptions of the dog latrine deck on board the *Leonardo da Vinci* while crossing the Atlantic. In December Patricia took Tyger to Brazil. Joe wrote regularly to the dog: 'Dear Mr Tyger I hear you were very careless and got an infection in your arm pit.' Tyger had to be in the films: Patricia called him 'a good luck piece' and took him to the auction-room scene in *Mr. Klein* at the Hotel Intercontinentale. 'Annalia gave me a revolting little hat with a veil and a heavy mink coat de l'époque . . . I rather enjoyed it except . . . I couldn't bear to be thinking I was a collaborator or something.' But a friend assured her that Resistance people had attended such functions 'to see who was collaborating and who had money – so then I felt completely okay including for Tyger'.[15] Croix de Guerre for him.

The present writer's view of the tax exile, one should add, is not Patricia Losey's: 'We were never rich. Any money that came in went to pay off debts. It's your Anglo-Saxon puritanism. On the Continent they're not hung up about these things. The intelligentsia are valued and admired on the Left and no one worries about their standard of living. They're so much more relaxed and self-confident than we are.' But observant, too: as Bertrand Tavernier remarks, 'Joe would complain about Lipp and yet go to Lipp, the place to be seen.'

41

Collaboration and Resistance:
Losey's French Films

Mr. Klein

Losey made three films in the French language, the first of which, *Mr. Klein*, is universally regarded as among the finest of his career.

On 22 November 1975, Losey wrote to President Giscard d'Estaing, informing him that he was about to make a feature film about the Grande Rafle (Round-Up) of 1942, 'which has the theme of indifference'. *Le Monde* (16 July 1975) had quoted Giscard's own recall: '. . . I saw them leave. On the morning of 16 July 1942, we were awakened by the unusual noise of buses passing along the avenues of Paris before the start of the day. Several hours later, one learned that it involved Jews who had been arrested at dawn and assembled in the Vélodrome d'Hiver. I observed among them children of our age, tense and still, faces crushed against the window during the crossing of this icy city, at the hour made for the sweetness of sleep. I think of their eyes, black kernels, which became millions of stars in the night.' Congratulating the President, Losey asked for permission to use this quotation after the credits: 'This would set the whole tone and intention of the picture.' Giscard replied on 18 December, giving his permission in most gracious terms, but the screenwriter, Franco Solinas, was sceptical and the President was dropped.

Losey had been offered the project after Konstantin Costa-Gavras stepped down. The production company was Lira Films, the star Alain Delon. In September 1975 Losey and Solinas settled down to work on the script in the fishing village of Fregene. Solinas had worked with Gillo Pontecorvo (*Battle of Algiers*) and with Costa-Gavras. As he spoke no English, and Losey no French, Patricia Losey served as English–Italian translator. Losey and Solinas then worked through the script with Alain Delon who, as star and 'minority co-producer', held the key to the entire project. Losey began viewing films about the Occupation, including *Lacombe Lucien* and *Chantons sous l'occupation*.[1]

In December 1974, Losey had assigned to Citel Films of Geneva his services as director and co-adaptor. In July 1975 Citel signed a contract with Lira Films. Losey was to enjoy total artistic control and his montage would be '*définitif*'. The fee for his services was 400,000 francs. The total budget was 18,050,781 francs; Delon received 1.25 million. Losey cabled Jeanne Moreau at the Beverly Wilshire Hotel, California, hoping that she would accept $20,000 for one week's work – 'and if you possibly can please accept for my own sake love Joe'. The Loseys had recently crossed the Atlantic on the *Leonardo da Vinci* with Francine Berge and her husband Donald Sutherland – Berge was cast as Robert Klein's ex-mistress, Nicole.

Some members of the crew had direct experience of the Holocaust. 'The extraordinary casting director, Margot Chapellier, lost many of her family in the camps. The head of the laboratory at LTC, Claude Lyon, lost his mother this way too.' Joris Ivens's wife, Marcelline, described to the Loseys how she had been arrested in the south of France at the age of fourteen and taken to Paris. When the train stopped at her home town, she saw a neighbour on the platform and dropped a note out of the window – this may be the origin of the note tossed out of the bus by Robert Klein as he is driven to the stadium. After a year in Auschwitz Marcelline caught sight of her father: they both broke ranks and embraced; her father was later machine-gunned in another camp.[2]

Losey got on well with Raymond Danon, of Lira Films, and with Ralph Baum, the director of production, despite 'all of his bad jokes'. Of his assistant directors, Losey declared Philippe Monnier to be a 'gem'. This was his first film with Alexandre Trauner, the *directeur artistique*, a Hungarian Jewish emigré who had survived the war in Paris. Trauner built the interior of Klein's apartment in the Billancourt studio, 'a bit flashy (*tape-à-l'oeil*)'. Losey brought in Gerry Fisher as photographer (*chef operateur*); it was Fisher's first job with a French crew but the much-admired cameraman (*cadreur*), Pierre-William Glenn, spoke good English. For quota reasons, Losey's perennial editor, Reginald Beck, was hired for uncredited 'collaboration'; the official editor (*chef monteur*) was Henri Lanoë, assisted by Marie Castro-Vasquez (*monteuse*). She was to work with Losey again, and admired his 'extraordinary personality – beauty, force, generosity, aura. Even seeing the rushes one felt emotional at the way he inscribed his image: the black machine of the Occupation, implacable, infernal.'[3]

On 6 November, Danon wrote to Losey expressing disquiet about the budget and the chosen locations. Would he consider dropping Saint-Sulpice and using 'a less imposing church, which would occasion fewer material problems'? Secondly, could he find a château nearer Paris, rather than sixty-seven kilometres away? Thirdly, regarding the house of the second Klein, could an old building be found which would not require expensive repairs prior to shooting? Danon was worried about the cabaret scene, for the creation of which Frantz Salieri wanted 40,000 francs (nearly $10,000).

Shooting began on 1 December 1975 and continued until mid February. Forty-two days were spent on location, fourteen days in the studio; actors and technicians addressed Losey in English; Patricia had translated the screenplay for his benefit.* The crew gave Losey a birthday party, 'complete with five foot cake and 21 candles, a life-size clapper board in chocolate and six velvet and corduroy blouses . . . A speech was called for but I didn't make it. Danon sent several bottles of Shivas and a card.' Alain Delon blew out the candles for the asthmatic Joe. On 16 January they travelled to Strasbourg (Losey by plane) and shot sequence 36 in the Public Gardens. Losey reported to his wife: 'Strasbourg was a near disaster. Much too large a crew, horrible journey . . . It pissed rain from the moment we arrived . . . We got the interior . . . very well, thanks to the miracle of Gerry [Fisher] . . .'

To Patricia he also reported his gloom about Delon, from whom he had received nothing but kindness and support since *The Assassination of Trotsky*. Delon had dined off roast beef with the Loseys and been presented with a turn-of-the-century silver tipstaff. In July 1973 Losey had asked Delon for a loan: 'If you could make me an official pound sterling loan however informal and on whatever terms you wish – I could deposit it in my account in France which would make that account negotiable outside of France and make it possible for me to repay you in francs . . . This would enormously relieve my worries about current carrying charges and taxes.' Losey was in bed with a clogged lung. Delon phoned anxiously. The next day his friend Mireille Darc (heroine of Godard's *Weekend*) arrived in London and withdrew so much cash from Delon's account that the bank staff protested on security grounds. She arrived at Royal Avenue carrying a Vuitton bag the contents of which she tipped on to the bed: £10,000 for Joe. Losey wrote thanking Delon for his 'unique' generosity, an 'extraordinarily warm demonstration of love'. The following year, in a memo to himself, Losey reflected that Delon had lent him money, 'without strings, without interest and without dates' in order 'to permit me to live and make choices'.[4]

Delon remained loyal, sending Losey warmly supportive letters and telegrams, some of them signed 'Robert Klein'. *Screen International* (23 May 1976) quoted him: '. . . if all the actors are panicked at the beginning of their work with Losey, it's finally he himself, by his humility, his reserve and his understanding of others, who admits to being the most embarrassed'. But

---

*The elegant auction sequence was shot on 8 December at the Hotel Intercontinental, 3 rue de Castiglione; the theatre and stage door scenes at the Cabaret La Nouvelle Eve; the communion service for children at the Eglise Saint-Eustache; needed were choir-boys and girls, priests, and technical advice from M. l'Abbé Galines of the Office Catholique du Cinéma. The apartment house of the other Klein (Klein II), at 42 rue des Panoyaux in the 28th *arrondissement*; the factory sequence at the Usine Citroën in Paris XV (involving 150 women workers, forty bicycles and four German lorries with camouflage nets). La Coupole restaurant – at La Coupole. Extras required included German officers, 150 men and women, plus twenty-five '*jeunes femmes élégantes*', their costumes to include local fascist badges ('*insignes Doriotistes ou franquistes*').

Losey's letters to Patricia in Brazil conveyed profound misgivings. Delon – he reported – was experiencing violent swings of mood, alternately brooding and euphoric, resistant and cooperative. 22 January: 'Completely psychopathic behaviour from Alain that delayed us half a day and nearly stopped the picture. I haven't a clue what it's all about, except, perhaps, ego.'

At the Gare d'Austerlitz Delon had arrived late, despite elaborate negotiations with SNCF,* dashed to change his clothes, then streaked away, half-dressed, when Losey commented that he'd cut it a bit fine. Raymond Danon, who has made eleven films with Delon, begged the actor to resume shooting. 'It was the clash of two strong personalities.' Losey wrote: 'Alain ... is perfectly okay with me and rather a broken and sad man. No one now speaks to him excepting me, including [Mme Ludmilla Goulian, the production manager] ... and all of the people more or less imposed by Alain are disasters ... He is a lost soul, and I think I have to give up about him ... If he took it out on me, it would be much easier to deal with directly, but he is on one knee all the time. It is sort of the worst Burton behaviour but not excused by alcohol and more vulnerable and more sick.' By May there was talk of removing Delon's name as co-producer; on the 14th Patricia Losey noted that Delon had telephoned, very emotional, declining to have his producer credit removed.[5]

The climactic stadium scene (shot at the Vélodrome Municipal de Vincennes with the help of Jewish organizations) involved 2,000 extras, including 400 children, plus *service d'ordre*, Croix-Rouge, firemen and caravans, eighteen extra make-up artists and twenty-three additional *coiffeures*. Losey wrote to his wife: 'We have four cameras and twenty first assistants. God knows how it will turn out.' Losey later recalled survivors of the 1942 round-up unable to bear the re-enactment and pleading to absent themselves. According to Reginald Beck, Losey gave up on the big crowd scenes in the stadium and left them to his first assistants. In Beck's opinion, these scenes required more organization and planned reactions.

The music was by Egisto Macchi (who had worked with Losey on *Trotsky*) and Pierre Porte. On 10 January, Losey reported a meeting with Maachi: 'He seems quite paranoid and an entirely different person from Trotsky [*sic!*]. Reginald [Beck] agrees. Franco [Solinas] loathed him and he Franco ... arrogant, rude and aggressive.' However, on 22 January Losey noted: 'Macchi is here again, and the trio/quartet is excellent, so I am reassured about him.'[6]

---

*The sequence at the Gare d'Austerlitz involved 345 males, ten Gestapo in leather coats, forty-five German officers and soldiers, four dogs, ten railway wagons ('*1 wagon en stationnement pour masquer les autres wagons modernes*'), plus '*walkies-talkies pour liaison*'. A sequence showing a train arriving at the station to disgorge 100 *gardes-mobiles* involved 200 extras, controlled by a *chef de manoeuvre*. The planned moves of the actors and of the camera, mounted on a crab dolly, were chalked on the ground.

Losey's eye remained fertile, his technique superb, but conceptually he relied on the moribund, 'real-unreal' dichotomies he had regularly applied to film after film, including his most recent, *The Romantic Englishwoman*. 'Stylistically,' he wrote in his production notes, 'the film breaks down into three visual categories which sometimes overlap':

'1. *Unreality*. These scenes are not, strictly speaking, unreal ... [but] they consist of a reality so far removed from the real world ... that they should have, almost, the quality of a fantasy reality ...' Examples: the early scenes in Robert's bedroom, the auction, the church, the château, La Coupole ...

'2. *Reality*. This is the stark world of the Occupation ... The reality sequences ... should have a documentary or newsreel quality. There are, however, scenes where the reality becomes Kafkaesque – grotesque, horrifying, unbelievable ...' Examples of 'reality' given are the Jewish Vendeur, the apartment of the Other Klein, the factory, the morgue.

'3. *Abstract*. These ... are the scenes that deal with the workings of the bureaucratic apparatus ... I propose always to start in close on some movement which leads the camera into a long view of an empty space ...'

Losey's three levels of reality-unreality-abstraction do not hold up as authentic categories but the main problem with *Mr. Klein* is the human beings. In this regard Losey's character notes are revealing. Robert Klein is 'a manipulator ... great charm and knows how to use it ... little, if any, compassion or capacity for love ... has grown rich out of other people's misfortunes ...' Nicole is 'very chic – very French. She observes everything and is involved in very little ... she would never risk her marriage and her comfort ... essentially an elegant, immobile creature – a Sphinx'. So far we have a manipulator and a sphinx. 'Pierre in a sense stands for all of the professional upper French bourgeoisie of this period ... he is basically fascist and anti-Semitic. He respects money, he accepts corruption. He takes no risks.' Manipulator, sphinx, corrupt bourgeois.

Losey stated the central theme of the film as one of 'indifference ... the inhumanity of the French towards sections of their own people ... Among the specific threads of indifference interwoven into the texture of the picture are: Klein's indifference to Jeanine's growing love and his total unawareness of her hurt; Klein's indifference to Nicole's jealousy ... Klein's easily dismissed pangs of conscience about [the Jewish Vendeur]; Klein's unawareness of the Concierge. His unawareness of the family in the château which, perhaps, may be Jewish; his blindness to the anguish of the Woman on the bus; his total unconcern about the Resistance ... In fact Klein really feels nothing deeply about anything except the *other* Klein who is entrapping him.' (But Delon's mannered glaciation, his handsome-enigmatic, *beau ténébreux*, personality undermines the social critique.) Losey then spreads the emotional constipation through the cast. 'There are many other instances of indifference in the scenario: the unconcern of Florence with Klein's fate, Pierre's indifferent toleration of Nicole's adultery, Nicole's cold observation of the inexorable fate which is overtaking Klein, Pierre's exploitation of Klein's situation . . . the indifference of the other Robert Klein to our Robert . . .' And more. Reading this, the central flaw of *Mr. Klein* becomes apparent: Solinas and Losey offer only a cursed fatality overhanging every member of the 'bourgeoisie'. The abidingly interesting question thus goes unanswered: why individuals capable of honour and decency in private life close their eyes and imaginations to genocide.

Tom Milne commented, generously: 'What we are watching is in fact the last days of the brittle, Fascist-oriented society portrayed beneath the surface of Resnais' *Stavisky* and now about to self-destruct through its own emptiness.'[7] That could also be taken as the theme of Renoir's *La Règle du jeu* (1939), but Renoir's country-house and shooting party is packed with quite affectionate, generous people who are losing their moral way because the 'game' is bigger than they are. By contrast, *Mr. Klein* is disturbingly exploitationist, capitalizing on what it purports to condemn: a luxurious bachelor establishment strewn with sex objects, a sumptuous château, a transvestite cabaret, coquets and coquettes – a Cook's tour of '*le tout Paris*'. Losey explained that he had set out to achieve 'the unrelenting fascination of a Borges labyrinth'.

Mr. Klein was one of four French entries at Cannes in 1976. Losey was photographed in a velvet suit with a triple-cravat, Patricia wearing an exotic silk-fringed crochet dress. Delon, however, did not turn up, complaining of 'a bad *grippe*' but – Patricia noted – 'none of us really knows'. The film did not receive even an 'honourable mention' from the Cannes jury, which heaped laurels on Scorsese's *Taxi Driver*. It seems that Delon did not consent to see the finished film until September: Losey held a screening for him. Patricia reported: 'Joe says Alain looked awful and was completely inaccessible. Said film was "just as he expected". No real comment, no real rapport.'[8]

Mr. Klein opened in October at the Biarritz, Publicis-Matignon and Paramount-Montparnasse cinemas. By 11 November, it was playing at seven Paris cinemas, including the Studio Jean-Cocteau and the Paramount-Galaxie. The November 1976 issue of the magazine *Cinéma* made *Mr. Klein** its main feature. The reviews blended admiration with misgivings. Writing in *Le Monde* (25 May), Jacques Siclier expressed disappointment. The film was

---

*The film is often called *M. Klein* in France. Anglo-Saxons are sometimes puzzled why a Frenchman should be styled 'Mr.', i.e. 'mister', but 'Mr.' also means '*monsieur*'.

aesthetically seductive but 'one retains an impression from it of malaise, of misunderstanding'. There was nothing Kafkaesque about the herding of Jews into the Vélodrome d'Hiver. On 1 November, *L'Express* carried a laudatory review by François Forestier: the film was admirably directed, supple in form, and Delon was perfect. If certain scenes carried a disagreeably mannered tone, the total effect conveyed the spider's web known as the Grande Rafle. Losey's style had achieved *'une perfection d'écriture'* – yet Forestier regretted that Losey's recurrent vice was to create 'perfectly insipid, marvellous arabesques woven on a void'.

The César awards were presented on 19 February 1977, in the Salle Pleyel, with some 1,500 guests. *Mr. Klein* was nominated for seven Césars, although Solinas's screenplay was not among them. Losey, then living at Fregene, was urged by the organizers to catch the 6.10 flight to Paris. Arriving at the airport, the Loseys found not the promised tickets, only reservations, promptly returned home, and picked up the ceremony on the French television channel. Trauner received the 'Best Art Direction' award from the sculptor César himself. When Losey was announced as 'Best Director', Romy Schneider collected the award: '*Merci pour Joe.*' Joe opened a bottle of San Giorgio many miles away. The phone began to ring. At about 10.30 came the summit moment – 'Best Film': the second accolade. Bernard Tavernier, who had to suppress disappointment about his own much nominated film, *Le Juge et l'assassin*, describes *Mr. Klein* as a masterpiece, 'the sense of fear, repression, alienation. You can't breathe – like Joe's asthma. Nothing is symmetrical, everything oblique, claustrophobic.' Michel Ciment calls it 'the best ever film about the Occupation'.[9]

Mr. Klein was the opening film of the London Film Festival on 15 November 1976. Introducing Losey as the guest of honour at the Savoy dinner, Sir Harold Wilson mentioned that 'Joe' had been hounded out of America but made no allusion to Losey's current flight into 'exile' from the Wilson government's tax provisions. Losey gave nine interviews, relentlessly grumbling about British tax laws but not reporting a nightmare he'd experienced while making *Mr. Klein*: he and Patricia were in his father's bed in La Crosse, but when he woke up (within the dream) he found Patricia gone to Brazil and Reginald Beck in the bed alongside him. Alerted by Beck, Losey looked out of the window in the still-dark dawn, and there was his wife 'in a lighted Mercedes with Harold Wilson . . .'[10]

In the *Observer* (14 November), Tom Milne hailed 'a film of astonishing virtuosity and control, superbly acted and saying much more about racism in its oblique way than a hundred liberal tracts'. *Mr. Klein* was finally released at EMI's Bloomsbury Cinema on 4 April 1977, presented by Gala Films. The fact that it was also shown within a week on BBC-2 TV was a clear signal of defeat. A friendly review by Julian Fox, in *Films and Filming* (February 1977) brought a bullying personal letter from the director: 'I rarely write to critics – and never to those I don't have some personal contact with.' Losey then fired a volley of challenges, among them: 'What gives you the right to presume that Delon's performance is more or less than precisely what I wished?' 'Why do you call the opening scene a cipher?' And so to the eleventh and last question: 'Have you, by any chance, seen the film more than once – and do you speak French?'

The film was distributed by Quartet Films of New York and released at the Eighth Street Playhouse in November 1977. The harshest of several hostile reviews was Pauline Kael's blitz in the *New Yorker* (2 February 1978): 'The title may sound like a Jewish washing detergent,' she began. 'Losey has only two modes of expression, the oblique and the obvious – and you can never be sure which is which . . . *Mr. Klein* is a classic example of his weighty emptiness.' The film did travel beyond New York, but as of 28 April 1978 the combined box office gross was only $108,021, from which advertising costs of $103,459 were deducted, leaving a net profit of only $4,561. The gross for the Eighth Street Playhouse was given as $66,298, compared with advertising costs of $75,236.[11]

Under the impact of commercial failure, Losey fell out with Raymond Danon, of Lira Films. In March 1977, clearly stung by his losses, Danon regretted that Losey's talent did not bring in the crowds. Losey replied hectoringly, demanding that Danon re-release *Mr. Klein* 'in at least seven *good* theatres in Paris, and not just in two bad ones . . . also that you have the dignity to take a double-page ad in *Variety*, announcing the six or seven Césars that you have had between me and Bertrand Tavernier. What other French distributor [*sic*] has had such an abundance . . .?' Danon remembers this 'disagreeable' letter. 'I was not the distributor. I had to take the *salles* given to me. The film was too symbolic and incomprehensible to the "*grand publique*". They are not initiates (*initiés*). I had raised a number of questions after the shooting, but he didn't or couldn't respond.'[12]

Eighteen months passed between the completion of *Mr. Klein* and the filming of *Les Routes du sud*. Renewed efforts to finance the Proust–Pinter project failed. During the hot, dry summer of 1976 the Loseys occupied a third-floor apartment in the rue Monsieur. Tyger had to be taken down to the little garden without benefit of a lift. A big, celebrity, visit to Mexico followed; in December 1976 the Loseys moved to Fregene, where they remained for the following six months; Losey flew to Australia for talks on *Voss*, fell seriously ill, and returned in a wheelchair.

After spending a week at Royal Avenue, familiarizing himself with the Losey papers, Michel Ciment embarked on the series of long taped conversations which became *Le Livre de Losey* – in effect, an informal autobiography.

Ciment's role in Losey's career as film critic, interviewer and, finally, advocate, is unique.* Initially, Losey's confidence in Ciment – in anyone perhaps – was qualified. On 9 July he wrote to Christian de Bartillat, president of Editions Stock, following a meeting with him and Eugene Braun-Munch. There was a hint of betrayal: 'Good as Michel [Ciment] is and much as I like him, I have a feeling he will produce something a little too standard and too dry.' By 17 August, however, Ciment had recorded twelve hours of tape covering Losey's career up to 1948. In November, de Bartillat suggested a contract with Ciment and Losey splitting the royalties. To this Losey agreed. In December 1976 the interviews moved from Paris to Fregene, where Losey was working on the script of *Ibn Saud* with Franco Solinas. Ciment recalled: 'To begin with, he wasn't at ease . . . But there was such freshness in his eye, so strong a charm in each of his gestures, that his frequent complaints, his bitterness, and certain errors of judgment were effaced by this unbreakable core of purity which seemed to have been his since childhood . . .'[13]

A year later, in December 1977, Ciment reported that he and his wife Jeannine were working flat out through the university vacation on translating the interviews into French. His defensive tone indicated pressure from Losey – whom he reminded that he wouldn't spend so much of his life on anyone else. Editions Stock published *Le Livre de Losey* in 1979, arousing widespread interest in France. Garzanti published an Italian version, but the slightly abbreviated Anglo-American edition did not appear until after Losey's death, in a film-book series for Methuen. In July 1980 Losey attended a retrospective of his work at the Prades Film Festival. Across the festival brochure appeared a large, orange blow-up: 'Yours sincerely, Joseph Losey.' But where – he complained to Ciment – was *Le Livre de Losey*? 'With an audience of 730 people at Prades, all avid for the book, there was not one single copy on sale. I think it's time you checked on the question of [the] royalty owed me. You will remember that I wanted you to have it, unless it was used for exploitation.' What this last phrase meant is obscure. Ciment, meanwhile, had risen to Losey's defence against Vincent Canby's two disparaging reviews of *Don Giovanni* in the *New York Times*.

Losey had returned to Paris from Fregene in June 1977. He and Patricia now reoccupied the ground-floor flat in the rue du Dragon, which Jack Loring had sold to Emile and Charlotte Aillaud. Close to the boulevard Saint-Germain, the rue du Dragon is packed with restaurants and shops; the double-door of No. 10 opens on to a courtyard and a severely classical façade

---

*Now Professor of American Studies at the University of Paris, Ciment is the author of books on Erich von Stroheim, Elia Kazan, Stanley Kubrick, Francesco Rosi, Jerry Schatzberg and John Boorman. A member of the editorial board of *Positif*, and a regular contributor to *L'Express*, he first met Losey in 1968 when he interviewed the director in London about his two recent films.

en face, with rambling improvisations to left and right, one of them the Losey apartment. The courtyard was good news for Tyger. But the flat was too small; Charlotte Aillaud regards it as more suitable for a bachelor, '*un garçon seul*'. Her husband, the architect Emile Aillaud, spoke no English, but he and Joe understood each other 'like animals'. Joe's frustration rubbed off, she noticed, on the films of the period in the form of a morbid pessimism, '*la délectation morose chez lui*'.[14]

In September Paris *Vogue* (No. 579) featured couture collections covered by Losey (previous incumbents of this fancy role had included Lord Snowdon, Roger Vadim and the artist Jacques Henri Lartique). Choosing Eve Arnold as his photographer, he visited eighteen couturiers; *Paris-Match* (August) showed him at a '*séance de coiffure*', working on a bare-breasted model. Dior he described as 'Feminine/very wearable/very beautiful – one of the best'. Saint-Laurent was 'Ideas/witty/brilliant/mostly unwearable'. Cardin was 'Great show/Bitter joke/(Pity, he used to make clothes.)' Put down were Balmain: 'Nothing' and Chanel: 'Madame Chanel is dead.' Eve Arnold recalls that the fashion people 'fawned' on him; creative and inventive, irritated by slow movers, he reminded them how his first wife had shown her collection in Paris in 1931. Forty-five years after Hawes published *Fashion Is Spinach*, he remained proud of her. Patricia noted that Joe's fee for the *Vogue* job was $5,000 worth of clothes for herself (with 30 per cent discount). Into fashion and into non-taxable 'fees', Losey promptly presented a new project to *Vogue* – having actors or socialites model Proust's characters in authentic settings. Negotiating in January 1978, his terms included 'complete coverage of expenses . . . a supplemental gift' plus a 30 per cent discount on *haute couture* clothes for Patricia to the value of $5,000, all adding up to a value of 7,525 francs – but this must not be 'an honorarium, or anything taxable, or anything which can be possibly construed as a payment to me'. He also preferred his 'living expenses' during photography to be non-declarable, but, as he noted, this had not been possible 'in the case of my European pictures'. Nothing more is heard of the Proust waxworks.[15]

Moving to Paris did not douse Losey's inclination to write off the life of a bon viveur against tax. The Georges Beaume agency accounts show the words 'Tax File' stamped on regular payments to florists such as Maryse Fleurs (1,590 francs in March 1976) and Fleuriste Lanéelle. Among restaurants, Cuisine Odette, Lipp and La Chamarre were regular haunts, with single meals costing, for example, 2,193, 2,017, and 1,434 francs. Between October 1975 and November 1977, ten 'Tax File' meals cost 12,384 francs (approximately £1,240, about £3,700 at 1990 prices).

In September 1977, preparing to film *Les Routes du sud*, Joe paid his regular health visit to Quiberon in Brittany. On the beach he threw Tyger's ball and two large dogs appeared out of nowhere to attack the dachshund. Joe went to the rescue, got his hands bitten, and was in a state of shock for several days.

Les Routes du sud

Jorge Semprun had written the screenplays for a succession of politically inspired feature films, including *La Guerre est finie*. Florence Malraux had urged him to work with Losey, but he felt reticent in view of Losey's criticisms of *L'Aveu*, directed by Costa-Gavras. In July 1976 Semprun sent Losey an idea for a film called *Les Souterrains ou The Underground* – an H.G. Wells-like dystopia. Losey was later offered Semprun's screenplay, *Les Routes du sud*, which dangerously resembled *La Guerre est finie*, directed by Alain Resnais in 1966. Yves Montand was once again to play an exiled Spanish Republican living in France. A worried Losey showed the script to Resnais who evidently reassured him, but Florence Malraux Resnais admits that the script was disappointing, *La Guerre* reheated. Michel Ciment comments that Losey's judgement was eroded by his attraction for 'names', Semprun's included.[16]

The contract, dated 25 May 1977, was between La Société Trinacra Films, the production company, and Société Citel Films (which disposed of Losey's services as *metteur-en-scène* and *co-adapteur*.) Trinacra's *président directeur-général*, Yves Rousset-Rouard, had launched the lucrative *Emmanuelle* films, the second of which was banned in France. Brought in as *coproducteur majoritaire* of *Les Routes du sud* by SFP, he wanted to prove himself: 'A real producer must do a film with Losey. I was proud to do it. I hired a château for Joe.' The final budget was 11.6 million francs. An agreement between Trinacra and Citel, dated 17 September, set Losey's fee at 525,000 francs, plus 5 per cent of net receipts. The film was officially credited as '*écrit*' by Jorge Semprun, with Joseph and Patricia Losey '*co-auteurs*'.[17]

From April to August a succession of Semprun synopses hauled themselves towards the story-line of the final screenplay. Losey showed it to Franco Solinas, who found it 'desperately bad politically'. Losey confided to Maurice Rapf, of Dartmouth College: 'I would *not* put [Semprun] in the category of writers I've worked with such as Pinter, Mercer, Tennessee Williams, Tom Stoppard and above all, Solinas.' Semprun and his wife visited Losey in Fregene early in June. Losey's lack of enthusiasm is reflected in his undated 'General Notes' on the first continuity script.[18]

1. 'What do we want the theme of the picture to be? Is it that betrayal of a philosophical/political nature (Stalin, etc.) is an excuse or rationale for personal betrayal . . . son by father, son by girl, husband by wife? Personally, I do not think this is our intention . . .'

2. 'Are we telling two stories – the personal one and the political one? There

is a danger of the two not meshing, and of the political story being cheapened by a too-trivial personal tale.'

3. 'Do we wish to tell a negative, nihilistic story? And if not – and I am sure the answer is "no!" – through whom is the positive statement to come? . . . But what statement?'

Despairingly, Losey called for more humour: 'all kinds of humour: fun, sophisticated, childish, bitter'. He wanted Julia 'to be more antic, more crazy, a character with whom I can improvise: suddenly, inexplicably, she does a dance . . . or stands on her head; or she paints a moustache with her lipstsick . . .' Losey's notes of 7 October concluded that what he had on his hands was 'a more or less classical problem between father and son . . . an Oedipus conflict at its most extreme . . .' But Semprun rejected this approach: 'It's not a matter of . . . a battle between males,' he replied. 'It's about lucidity between adults who have achieved a certain freedom of relationship.'

In reality it was about Yves Montand. Just as *Mr. Klein* had been a film for Delon, *Les Routes du sud* hinged on the participation of Montand. In an interview with *France-Soir* (28 April 1978), Losey commented: 'Montand . . . wants to direct everything himself. He tells other actors what to do. Miou-Miou and the young Laurent Malet did not know which saint to dedicate themselves to. I advised them always to say, "Yes, M. Montand", and to listen only to what I told them.' Rousset-Rouard recalls that Miou-Miou was at first amused by Montand's coaching, then vexed by its persistence. Losey – who claimed that Spain 'was closer to me than to Yves and I think he resents it' – told Michel Ciment that Montand was 'a very kind and very generous man. He was disciplined and professional, argumentative and tedious . . .' Gerry Fisher noticed that 'Montand lived with a separate person called Yves Montand. "I'd like to do such-and-such a film, but Yves Montand would never do it," he'd say.' There were also devious complications surrounding the casting of the young woman, '*la marginale*', Julia: Isabelle Adjani and Maria Schneider were the leading contenders, but there were negotiations which faltered and rendezvous which aborted; at the eleventh hour the part went to Miou-Miou.[19]

For the score Losey called in Michel Legrand, who wrote a jazzy trumpet solo main theme and a South American song lamenting the death of Allende, sung by Tania Maria. Alexandre Trauner refurnished his own house in Normandy to encompass the personality of the exiled Spanish writer. '[Joe] often came to see me and he liked the kind of rather bohemian wildness he found there.' Losey noted that 'the colors of Normandy are green, blue, basically cool. Monochrome. And as the autumn advances this will become more and more true . . .' Yves Rousset-Rouard noticed Losey's total confidence in Fisher – few instructions were necessary. For the second time Pierre-William Glenn served as cameraman (*cadreur*), while Carlo Lastricati returned as first assistant for the first time since *Modesty Blaise* (not credited) and *Boom!*[20]

Clearly Losey got on well with Yves Rousset-Rouard, but while shooting in Spain the producer vetoed for budgetary reasons an intended scene involving a big crowd gathered at the tomb of a Catalan Resistance hero. Rousset-Rouard telephoned Semprun and got him to confirm that no mass meeting had been allowed immediately after Franco's death. According to the producer, Losey defiantly took to the air in a helicopter to survey the cemetery, then passed a message that he'd dropped the scene for his own reasons. Oddly, this is news to Gerry Fisher.

Having viewed the film on 22 December, Rousset-Rouard informed Losey that the story was still weak, the beginning too long, the ending too literary. The Spanish doctor's hospital monologue on Francoism was 'completely insipid and very bad'. Montand's monologue in his office, prior to the sequence of the three girls and the guitar, was long and heavy. And so on: a litany of doom.

In April 1978, *Les Routes du sud* opened at the Publicis-Elysées, Le Paris, Publicis-Saint-Germain, Paramount-Opéra, Paramount-Montmartre, Paramount-Montparnasse and nine other Paris cinemas, plus cinemas of the *périphérie*. Writing in *Le Matin* (28 April), Michel Perez was wholly favourable about 'a poem of annihilation completely worthy of the creator of *The Servant*'. But the general critical reception was disastrous and devastating. Jean de Baroncelli, of *Le Monde* (29 April), noted rehashed themes and tones from Semprun's novel *La deuxième mort de Ramon Mercarder* (1969) and *L'Autobiographie de Federico Sanchez* (1977). Baroncelli was fatigued by the relentlessly harsh and cruel confrontations between father and son. Scenes were repetitive, personal dramas interacted poorly with history. 'And, more serious, a total absence of emotion.' Most brutal was David Overby, writing in *The Paris Metro* (25 May). Ideology was mixed up with undigested, dime-store, Oedipal stuff. The characters were merely 'cardboard-cutout clichés'; the film was visually elegant, but empty of force, ideas and imagination – 'artsy-crafty filigrees unsuccessfuly glued to the slow moving glacial block of the rest'.

On 14 February 1978 Losey had written optimistically to Pinter: 'There will be an English version and it will almost certainly go to Cannes in competition.' Patricia worked on the English subtitles, to the mild displeasure

---

*Thereafter production-quota factors had constantly intervened, although Losey's description of Lastricati to Nicole Stéphane – 'indispensable, there is no one in any language to touch him' – makes it clear he wanted him for Proust. *Les Routes* was to be their last collaboration: a distempered letter from Losey to Lastricati's wife suggests that he declined an offer to work on *Don Giovanni*: 'I have no quarrel with Carlo if he chooses not to work with me . . . I object to nothing excepting the manner in which this was done – with false reasons and false excuses.'

of Eric Kahane, but not even the American majors and distributors with a record of promoting French films would touch *Les Routes* – the only Losey film never distributed in either America or Britain. In November 1978 it was screened by BBC-2. Like Raymond Danon before him, Yves Rousset-Rouard faced financial disaster. The Trinacra balance sheet to 31 August gave total production costs including publicity expenses ('frais d'exploitation') at 11.6 million francs. Earnings were 5.9 million francs; the deficit was 5.7 million. Two years later revenue had barely moved. The accounts to 8 September 1980 showed sales at 6.11 million, the deficit at 4.7 million.[21]

On 14 November 1979, the Cinéma Spéctacles page of *Le Matin* published an interview with Laurent Malet who, while admiring of Losey, casually remarked that the director had not had his heart in the film. Malet received a letter: 'And if ever there were a role for you in a film of mine, you would have it . . . unless you alienate me by further silly remarks. I have had experience before of very new and inexperienced actors who get a great deal of attention and then feel deserted by their surrogate father when the film ends.' Malet responded by swearing his lasting affection and respect. Subsequently Losey wrote again: '. . . perhaps you're in too much of a hurry . . . a little too ambitious. Take a look at [*Le Livre de Losey*] and see how long it takes.'

42

Don Giovanni and *Boris Godunov*

Losey's *Don Giovanni*: a glorious feast of sound, sight and intellect, a film to be seen many times – or an insolent travesty of operatic convention? To make a filmed opera which belongs authentically to both media is a challenge untouched by the average televised recording of a staged performance. Ingmar Bergman's celebrated solution in *The Magic Flute* (1974) was to establish an eighteenth-century court theatre in parkland, playing up the enchanting make-believe of the stage, the delighted involvement of the 'audience' in the performance – while offering the real, cinema audience the additional privileged motions of the camera-eye.

Rolf Liebermann, General Administrator of the Théâtre Nationale Opéra de Paris since 1973, had an alternative vision – to make a film of *Don Giovanni* on location, using a pre-recorded score, a 'play-back'. Everything flowed from that concept and the technology which now made it possible: every motion of liberation, and every danger. For Liebermann the venture was attractive socially as well as artistically: 'In the movie theatre one has the possibility of first-rate sound and life-size images without any limits on the size of the audience.' Losey agreed: he, too, rejected the élitism of the opera house. It was Liebermann's inspiration to film *Don Giovanni* in and around Vicenza, where the Renaissance architect Andrea Palladio had strewn the town and the surrounding countryside with beautiful villas, palaces, churches and a theatre. The Palladian style appealed, and persisted, up to the early nineteenth century: order, harmony, equilibrium, a sense of nature, views multiplied by other views – a perfect counterfoil for the contest between anarchy and order, libertinage and convention, within the opera.[1]

Liebermann had been discussing orthodox opera productions with Losey since he moved to Paris in April 1975: Strauss, Debussy, Wagner. On 26 May Losey wrote to Maria Callas: 'In spite of your warnings, I find myself inclined toward opera. I wish it could be something I could do with you.' At the end of the year he was inspired to write two letters to the sculptor Henry Moore, the first tentatively proposing 'a theatrical collaboration between us which I hesitate to commit to paper', and the second, following Moore's friendly reply, mentioning a production of *Tristan und Isolde*, with Sir George Solti conducting, scheduled for the Paris Opéra late in 1978. But no Moore came of it. Early in January 1976 Losey experienced a dream in which he arrived for the first rehearsal of *Tristan* but Liebermann told him it needed neither a conductor (Solti) nor an orchestra. (This is presumably an anxiety-inversion of 'it doesn't need a director'.)[2]

Liebermann's first choice for director was Patrice Chéreau. Franco Zeffirelli, the obvious specialist in the field of filmed opera, was also considered. Visconti was dead. Losey had never attended a performance of *Don Giovanni*; indeed, by his own admission the only Mozart opera he had ever seen was *Così fan tutte*, at Salzburg. Mozart had called *Don Giovanni* a '*dramma giocoso*' – how '*giocoso*' (amusing) was Losey? But Frantz Salieri (Francis Savel), Losey's collaborator on *Mr. Klein*, was confident that he himself could provide the grim humour. In January 1978, Losey, Liebermann and Salieri travelled to Vicenza and Venice (staying at the Danieli). Palladio's Villa Rotonda, once a cardinal's belvedere, had been provisionally chosen as Don Giovanni's residence, but access was initially a problem because the owner, Count Valmarana, had installed burglar alarms and forbidden visits. A fee of $65,000 later resolved the difficulty. Otherwise the trip went well except that Losey lost his Gucci briefcase at the airport.[3]

Contracts were delayed largely because of arguments between Liebermann, who wanted a German investment, and Gaumont, who preferred a Franco-Italian co-production. There was also haggling over percentages, geographical domains, rights and guarantees. In the upshot Gaumont created its own Italian company, and the formal production credit of the film emerged as 'Opera Film Produzione (Roma)/Gaumont/Camera One/Antenne-2 (Paris)/Janus Films (Frankfurt)'. Losey fretted about the unsigned artists' contracts, not least his own. To Gaumont's Daniel Toscan du Plantier he regularly complained that preparatory work on the screenplay (*découpage*), the budget, the work plan, locations and personnel selection, had gone largely unremunerated, 'causing me constant embarrassment'. Losey's services were again vested in Citel Films, which signed a contract dated 5 July 1978 with Camera One and Gaumont. Citel was guaranteed a down payment of 300,000 francs against a royalty of 0.525 per cent of producer's net worldwide receipts. In addition, Camera One would pay Citel 150,000 francs 'in remuneration for JL's work as director . . .' Losey's total guaranteed fee was thus 500,000 francs. The budget for the film, meanwhile, had risen sharply from 13.2 million francs in January 1978 to 18.2 million in July. On 13 November, *Newsweek* reported that shooting had moved into a seventh week, sending the film one million dollars over its scheduled $4,000,000 budget. On 5 November 1979, a *Washington Post* report put the actual cost at $7 million (about 28 million francs).

The source of all Don Juan dramas was Tirso de Molina's *El Burlador de Sevilla y Convidado de Piedra* (1630), but the more immediate source for Mozart and his librettist, Lorenzo Da Ponte, had been a libretto by Giovanni Bertati. Following the success of *The Marriage of Figaro*, *Don Giovanni* had first been performed in Prague on 29 October 1787. The Vienna première followed on 7 May 1788, but came off after only fourteen performances.

Losey, Salieri and Patricia Losey set to work on the screenplay. It was Salieri
who designed the engraved credit titles, and conceived the sumptuous
costumes. Patricia Losey's diary complained that he didn't stick to the point
in discussion. Losey confided to Janine Reiss, head of musical studies at the
Paris Opéra, that his relationship with Salieri had deteriorated, and it's also
clear that Salieri's role as designer overlapped abrasively with that of the art
director, Alexandre Trauner.

Not a word of Da Ponte's libretto could be cut or altered. '*Découpage*' –
the shooting script – had to begin from there. For example:

45. EXT. LA ROTONDA. DAWN.	RECITATIF (cont.)
Reverse of shot No. 42. The camera is now at eye level, square on to La Rotonda, the three of them confronting each other, static.	DIRECT SOUND
D. Giovanni: *Stelle! che vedo!* Leporello: *O bella! Donna Elvira!*	
87. INT. DON GIOVANNI'S HOUSE (POIANA). DAY.	RECITATIF DIRECT SOUND
Don Giovanni's bathroom. Windows open. Don Giovanni is languid and detached.	

Leporello: *Io deggio ad ogni patto*
    per sempre abbandonar
    questo bel matto!
    (he is circling the bath tub)
        [. . . etc.]	

115. EXT. LANDING PLACE. CANAL. DUSK. High shot walking away from boat from approximately Leporello's pov.	FINALE (cont.) PLAYBACK TRIO (cont.)
3-shot of Black Valet leading them up.	Adagio

Conducted by Lorin Maazel, the music (arias and ensembles) was recorded
by the orchestra of the Paris Opéra and the Schola Cantorum. Financed by
CBS Records, the recording sessions took place in the nave of l'Eglise du
Liban in the rue d'Ulm. (Sixteen-track tape would allow CBS's sound
engineers to alter balances and achieve their own mix.) The writer Roland
Gelatt described the atmosphere: Maazel, racing back and forth between Paris
and his conducting duties at the Royal Opera in London, was 'drawn and
edgy'. Many of the singers disliked the sound of the playback system in the
control room. Losey, whom they hardly knew as yet, was 'a distracting and

disturbing presence'. Rolf Liebermann kept popping in to add his own comments and advice. On 9 September, Losey wrote to Maazel congratulating him on the mixed tape of the whole opera. It was marvellous, flawless, the right tempo, drive and energy, with superb performances. In retrospect, however, Losey was critical of Liebermann's set-up. The acoustics in the church were too 'reverberative' for scenes which were to be played in the open air.[4]

Twelve different settings were used from the Vicenza area and the two Venetian islands of Murano and Torcello – Losey being familiar with the dreamy winter lagoon and canals of Torcello from the time of *Eve*. (A barge served as a platform for the director and camera crew; frogmen shadowed the singers on the water, as required by insurance.) Losey's technique was to transplant and juxtapose facets of different Palladian buildings, creating a semi-fictitious architecture and a wholly fictitious topography. To give the impression that the Villa Rotonda could be approached by water, Trauner set a gondola on wheels – and reconstructed the great staircase leading up to the villa on the edge of a canal, with statues on the lower level. The rest was *trompe-l'œil* camerawork.[5]

Trauner laid out the main visual feast within La Rotonda, an astonishing mansion of perfect symmetry and commanding a low hill – each side of the central cube bearing the same classical portico and descending to the garden by the same staircase, the pillars flanked and winged by statues. The gently sloping tiled roof is crowned by a dome. Despite the parkland, today La Rotonda barely fends off the sprawl of suburban Vicenza and stands within eye-range of low-cost workers' housing. Trauner filled the villa with works of art representing both pagan and Christian mythology, setting the central moral confrontation against the ambiguous moral heritage – the conflict between Eros and Order – of the Renaissance and its increasingly hedonistic offspring: neoclassical, baroque and rococo. Don Giovanni, as possessor and possessed, is both validated and condemned by the the wealth of art he takes for granted.[6]

The Loseys installed themselves in the Villa Emo soon after recording ended in Paris, while the lead singers disappeared to fulfil other engagements. 'I got a group of young actors [from Vicenza, Venice, Padua and Milan] . . . I used the [1937] Busch recording and rehearsed with these kids . . . before the opera singers came to the locations.' He called this 'an act of desperation that turned out to be rather inspired' – he was absorbing the music in relation to the locations. Losey had long since complained to Liebermann about the schedule and the lack of rehearsal time. On the first day of shooting a tele-monitor checked the lip-sync of the singers against the playback, but Losey discarded it and relied on the aural eye of Janine Reiss. Pamela Davies, Losey's veteran continuity girl, bristled at first. (One French paper disparaged

Reiss's role as '*claveciniste . . . devenue script-girl musicale*'). Reiss describes Losey as 'a giant'; affectionate and caring, he used to put a shawl round her shoulders, or lead her to his warm caravan where she marvelled at the vodka bottles – a litre could vanish without visible effect. 'He envied me because I could read music. Such humility: "I shouldn't touch Mozart," he would say. Liebermann complained that I was enchanted by Joe.' (Don Joe.) Janine Reiss served as his ear and musical conscience. Only the recitatives were recorded on location, to the accompaniment of her harpsichord. Gaumont's Toscan du Plantier regarded this as an expensive luxury. 'At night it was too cold to play the harpsichord. We had to re-record and post-sync. This was my error.' It was difficult to keep the harpsichordist warm.[7]

Having pre-recorded their arias, the singers were free to move more expansively than when physically bonded to orchestra and stage. In certain scenes Losey capitalized on this freedom: gondolas, perambulations, but he also revealed a lifetime's knowledge of the cinema; the camera moves, the figure freezes. Toscan du Plantier recalls that Losey gave his singers few directives, the most common one being a growl, 'Don't move!' A creature of the stage, the singer's instinct is for supplementary gesture; Losey froze them like frescos in a Greek tragedy (Toscan's phrase): 'The cinema is introversion. The eye is everything.'

Losey laid down in advance several bold lines of interpretation for the benefit of the cast, as was his habit, but having done so he knew perfectly well that they were too abstract to be translated into performance and that such memoranda were essentially directed at himself. 'In confidence', he advised his assembled singers that Don Giovanni was a man *en fuite*, using sexual excess as an opiate – and profoundly troubled by social fermentations he was not consciously aware of. Da Ponte's tall story also needed a political gloss. 'So it's related to the beginning of an industrial society and the turmoil of that society; it's the end of an era.' According to the Gaumont publicity brochure Don Giovanni was also 'an anarchist rebel'. In a taped interview with Chris Nelson, of CBS, Losey explained: 'The fact that Mozart was hardly aware of the French Revolution, which was taking place at almost exactly the same time . . . has nothing to do with the case . . . at certain times in history things are in the air all over the world, and I believe *Don Giovanni* is a piece of rebellion, if not of revolution.' To ram home the dialectic, Losey pinned to the front of the film an epigraph from the prison notebooks of the Italian Communist, Antonio Gramsci: 'The crisis consists precisely in the fact that the old is dying and the new cannot be born; in this interregnum a great variety of morbid symptoms appears.' (The novelist John Fowles supplied Losey with the page reference from Gramsci.) Later, interviewed by French journalists, he retreated a step: the Gramsci epigram was merely to make the audience sit up and think – but it might seem 'pretentious'.

Shooting in Panavision, Losey was working with a first-class technical team overloaded by the demands of an international co-production. Thus Carlo Poletti is credited alongside Gerry Fisher as director of photography; Emma Menenti and Marie Castro-Vazquez alongside Reginald Beck, as editors. Losey and Fisher conversed intimately and with professional pessimism, creating tensions between Fisher and the Italians. (Supporting Fisher's application for an American work permit, Losey rated him 'among the top six lighting cameramen in the world'.) In October Beck fell ill with pneumonia and a bad lung infection, the beginning of the end of his working life. The sound editor was Jean-Louis Ducarme, who imported sound equipment for the evening rushes at the Cinema Palladio, Vicenza. Jean-Michel Lacor, the first assistant, was obliged to speak in many languages: '*Silencio! Giriamo!* Camera! Action! *Moteur! On tourne!*'

The production was immediately beset by two problems: poor organization and heavy rain. 'It was such a mess,' Losey told Gelatt. 'I realized there was nothing to do except stop . . .' The star singers were less than contented. Cold nights and caravans were not what they were used to – Ruggero Raimondi confided to Reiss his permanent fear of catching a cold. In a letter written from a hotel in Treviso on 6 October, Teresa Berganza complained (in Spanish) to Gaumont about the 'lamentable circumstances' – disorder, indiscipline – which had obliged Losey to break filming. If things were not put right by 12 October, she reserved the right to rescind her contract.

The opening scene, the pursuit of Giovanni and his servant Leporello by the ravished Donna Anna, shot at night under heavy artificial rain, makes brilliant use of architectural make-believe, setting the Commandantore's house in a great public building, Palladio's arcaded Basilica in Vicenza, then moving the resultant duel to the arches under the Loggia del Capitanatio on the other side of the Piazza dei Signori. Real rain and thunder impeded the shooting. Drenched to the skin by the artificial rain, the singers repeated this scene on three consecutive evenings. *Newsweek* quoted Raimondi: 'Every time I opened my mouth it filled with water. I kept thinking, thank God I don't have to sing too.' The Basilica scene is a masterly portrayal of the opening catastrophe on which the whole opera hinges – desperate figures plunging along stone colonnades at night, propelled then halted by the music, until the Commandantore falls dead on his own doorstep on the sword of his daughter's ravisher: 'I felt I'd never again be able to watch it on the stage,' Janine Reiss comments.

Gelatt describes the shooting of the grand finale in La Rotonda. The sound engineer, in a truck parked outside, gets the signal to start the tape. A surge of festive music fills the room. Raimondi begins to perform, singing in half-voice, followed by the camera. Trailing immediately behind, their heads ducking to avoid throwing shadows, are Losey, Liebermann, Fisher, the script girl, and Janine Reiss, whose responsibility is to check the lip sync. The hardest challenge for Fisher was to echo the opening glass-foundry scene (which was shot in an abandoned foundry on Murano, using butane gas) in Don Giovanni's final descent into hell (shot in the Rotonda, with only light effects to convey the engulfing flames).[8]

The dramas off-stage required no playback. When Gaumont and Camera One decided to process the film in a Rome laboratory, pleading co-production rules, Losey started drinking (Patricia recalls) and 'flipped'. On her advice he wrote a stiff letter to the young executive producer (*producteur délégué*) of Camera One, Robert Nador: 'I have never agreed . . . to the processing of negative or prints with any other lab than LTC. You *are* therefore . . . in violation of my contract . . . I am quite capable of stopping the shooting . . . if you do not give me the [following] assurances . . .' Losey prevailed: LTC laboratories were used, as originally agreed. Later, at dinner with Nicolas Seydoux, president of Gaumont, Nador and others, he stoned himself and insulted everyone. Nador and the chauffeur, Boda, had to support him to the car. Losey needed a rest: on 25 November he wrote to President Valéry Giscard d'Estaing: 'Would you be able to spare a little time to talk to me about the nationality problem of my film of Mozart's opera – *Don Giovanni?*'

The film was edited at the Studio de Billancourt, with further work in London on the original recording, involving Dolby, Maazel and some of the singers. Gerry Fisher was less than happy about all of the laboratory work, particularly the insertion of 16mm fillers, which – he warned Losey – were inevitably inferior in quality. He was especially puzzled by one 16mm insertion of Elvira just prior to Leporello's '*La Lista*', as he remembered having shot a close-up of it. On 5 June Losey wrote to Nicolas Seydoux, recommending Salieri's posters for the film, and reiterating his opposition to a Salvador Dalí poster which was 'cheap, vulgar and . . . obscene'. This poster hangs in a place of honour on the wall of Seydoux's seventh-floor office in the Avenue Charles de Gaulle, Neuilly.

In July Losey and Liebermann exchanged sharp memos. It was a crisis: the whole issue of pre-recorded, filmed opera was in question. Losey's memo of 12 July (which is not on file) evidently complained that the projection of the film at the Gaumont-Champs-Elysées cinema earlier that day had been a catastrophe. This was the first screening of a combined Dolby stereo print. 'The sound was a disaster,' Patricia noted. 'Rolf was hysterical.' Responding on 17 July, in the form of a general memo to all concerned, Liebermann agreed about the catastrophe but not the cause. The key issue was the marriage of reality and stylization, of cinema and Mozart – 'who did not compose film music'. The corrections that Liebermann had proposed when visiting the Billancourt studio had not, he said, been heeded – for example, naturalistic sound effects, such as the swish of oars through water, still intruded on the singing. He then went on to complain about the relationship of 'piano', 'forte' and 'crescendo' to the singers' distance from the camera. Although he had attempted to discuss these problems with Losey calmly and amicably '*à maintes reprises*' (on many occasions), it was 'a dialogue of the deaf'.

Losey's reply (18 July) began abrasively – would Liebermann like to take over the finishing of the picture? – but then calmed down into a discussion of the sound-track problems. In a letter to Toscan du Plantier, Losey blamed Gaumont: 'Our experiences with the Gaumont theatre projection, both mono-optical and magnetic-stereo, up to now have been appalling . . .' On 28 September, after the first showing in Paris, Losey was sufficiently nervous to suggest to Nador that a mono sound system was the solution: 'The stereo is too wide and disturbing . . . The sound is too harsh in Dolby . . . Dolby doubles sound-effects with the mistaken 2-track mixing . . . Try to be a little wiser in the future in your dealings and your actions.' Losey also accused Nador of causing 'frustration, worry and anger' with regard to the subtitles. Losey had insisted that 'where words are repeated in the libretto with no variation excepting the musical variations, the subtitles should be established in the beginning and NEVER repeated'. Yet Nador had engaged in 'back-room manoeuvres' to override him 'through fear, ignorance, panic or any other reason'.

Losey's *Don Giovanni* is a masterpiece ravishing to both ear and eye. Never was his extraordinary talent for marrying human emotion to the physical world more exuberantly displayed – but a Losey marriage is also a divorce. Beautiful settings become the silken web for callous seduction and cruelty; the festive spirit mocks a funereal gloom; a demon is at work.

On one level we don't believe a word of it; the story-line of Lorenzo Da Ponte's libretto is contrived and confining. Losey met the challenge by flooding the picture with sunlight and water, paintings and sculpture, his camera movements boldly responsive to Mozart's music, a dazzling fusion of the fine and performing arts. Confronted by the visual stasis of operatic convention, Losey eased apart the orchestral and the dramatic, reuniting them in the cutting-room on his own terms.

The theatrical conventions of eighteenth-century opera are not easily transcended. Da Ponte's characters tend to run round in circles and their grand passions (Elvira's spurned love, Anna's murdered father) get absurdly sidetracked by concern for Zerlina's virtue. Losey stated as his 'problem' that 'Mozart and nobody else either cared anything about the action. They cared about a set-up for an aria, or for a septet, or whatever it might be . . .' Paradoxically, his own instincts as a director were notoriously for a 'set-up' –

but one servicing the eye not the ear. The long takes and deep focus of his camera constantly offset operatic friezes – as when one of Donna Elvira's impassioned arias is filmed in a single, tracking shot, pursuing her through the corridors of the villa. Losey's chosen style ruptures neo-classical stasis by bursts of frenetic Romantic energy. In the climax to Act 1, bolts of lightning and the flickering lights of massed torches combine in a bravura display of chiaroscuro effects. In other passages he uses the camera to play up operatic stage traditions – as when Don Giovanni (disguised only by a hand-held mask) sends Masetto and his angry friends into choreographed retreat, each ebb and flow repeated five or six times with comic energy.

As ever, Losey is a cinematic admiral – the slow, hypnotic motions of the gondoliers, or Don Ottavio's voice preceding him as his boat gradually materializes out of the misty lagoon. (But the opening shot of sea and sky oddly lacks the enclosing rims of land characteristic of the Venetian lagoon – it was later shot from the Channel coast at Dieppe.) At the end of the story, each masked aristocrat is found in his or her separate boat, the water moving laterally, until we finally discover Leporello comically adrift in his own shabby craft. (Jose Van Dam constantly steals the show, the omniscient, unservile servant.)

The bass-baritone, Ruggero Raimondi, who had sung the part of Don Giovanni in at least four previous productions, displays a full histrionic range – now ravishing now sinister – his lyricism fractured by mordant outbursts. The tall Raimondi's half-human chalky make-up signals a carnivorous spider of sex who experiences triumph but never joy; his pursuit of women is saturnine, unsmiling, a driven addiction. Exercising droit de seigneur by threats when charm doesn't suffice, his 'I must have ten more women by tomorrow' owes more to neurosis than lust. Only once, dressed in white with a rakish hat for Zerlina's wedding feast, does he show signs of genuine exuberance, but the mood doesn't last. The good humour belongs to his servant, Leporello, who unfurls the long scroll of his master's conquests down the steps of La Rotonda. According to Toscan du Plantier, the first time Raimondi saw the film he wept: 'It's not me!' Patricia Losey recalls the same occasion (20 March 1979): Raimondi described his own performance as '*brutto*'. Losey asked Edda Moser (a coloratura) to play the part of Donna Anna as if she had invited Don Giovanni to her bedroom – then lied about it to her fiancé and to society, presenting the Don as her ravisher. To this end, Losey included her among the guests invited to inspect Don Giovanni's glassworks during the overture. She throws him a smouldering glance, from one viewing platform to the other, across the furnace. This leaves us with a lady who is secretly torn between passion for Don Giovanni and filial remorse – hence (perhaps) her brusque rejection of Don Ottavio after Don Giovanni has gone to hell. Give it, she tells her patient suitor, another year. Out of grief for her father? Or for her father's murderer? Losey's own heart clearly belonged to Kiri Te Kanawa (Elvira); his subsequent letters to her are touching. As for Teresa Berganza, she brings passion and joy, tenderness and girlish frivolity, to the part of Zerlina, but in her case the camera creates problems: Berganza in close-up is old enough to be Zerlina's mother; her vivacity cannot offset the matronly profile. 'She knew she was too old,' recalls Toscan du Plantier. 'When the camera came too close she felt got at.'

Don Ottavio (Kenneth Riegel) is a buffoon, pompous and vacillating, timid and bourgeois. 'I am your husband *and* your father,' he tells the bereaved Anna. Kenneth Riegel plays him like a parody of the puffing, dough-like *gentilhomme* – one sympathizes with Losey's decision to break the sense of Don Ottavio's aria *andante grazioso* by sending him out into the sunlit garden, to sing at full pitch to no one: 'Meanwhile go and console my dearest one . . . Tell her that I have gone to avenge her wrongs . . .' By having Ottavio hurl these brave promises into thin air, Losey emphasizes the man's impotence.

The Teatro Olimpico sequence, the make-believe Cardinal's court, is disappointing. Built by Palladio in 1580, and externally anonymous, the Olimpico boasts a semi-circular auditorium on the classical pattern, the fixed scenery a riot of arches, pillars, niches, capitals, statuary; the ceiling painted like the sky; and a three-dimensional backstage depicting the miniature narrow streets of a town which converge on the stage through seven symmetrically situated arches. Down these Don Giovanni (really Leporello in disguise) is pursued until 'cornered' on the open stage, the '*piazza*', where the court and citizens are waiting to judge him. But they have the wrong man. The subsequent sextet is static, dull. Losey seems bemused by the setting and generates no vitality.

Don Giovanni is not an erotic opera but Losey – like Fellini in *Casanova* – painted in arrestingly sensual moments framed in the image of (perhaps) Ingres, or Manet's *Dejeuner sur l'herbe*: a bare-breasted maid at an upper window serenaded by Don Giovanni; a slow-eyed girl washing herself at a fountain, arm raised and staring back pensively at Leporello; the naked girl asleep in Don Giovanni's bed, on whom he leans, like a bolster, at the start of Act 2. The problem with these indulgences is that they detract from Losey's claim that Giovanni's drives are less erotic than manic; to convey this dramatically, his targets and conquests should be gruesomely unattractive.

The silent Black Valet is an astute innovation (Salieri's in fact). While it is too much to call him (as does the script) 'the guardian – in metaphysical terms – of Don Giovanni's doomed soul', Eric Adjani's pale, silent, expressionless, yet all-wise, young valet wonderfully conveys fatality with a eunuch's serenity: the rich and powerful must be served; Adjani contrives to be both there and not there; inside the frame and outside; occasionally he yields an inch of ground out of social deference, but it is the grudging retreat of a cat. Losey had wanted his friend Rudolf Nureyev for the part.

The final banquet sequence is a masterly evocation of hubris and nemesis as the Don, calmly awaiting the statue/ghost of the Commandantore, unaware that this is his own last supper, moves contemptuously among sumptuous displays of wealth, bowls piled with food, marbled walls and floors surrounding the golden table, rich brocades and satins. A light 'alienation' touch is offered by Mozart and Da Ponte, when the musicians strike into *Figaro*. 'This one I know too well,' smiles Don Giovanni. The Black Valet alone is uncowed by the arrival of the stone Commandantore and by bolts of lightning which shatter furniture and overturn huge candlesticks; alert but fatalistic, he calmly watches his master topple into the bonfire of the vanities. Finally he closes the great doors of La Rotonda. Will he settle down to the remains of the feast; or lie down to sleep with open eyes while the house falls into ruin?

Losey checked in at the Pierre Hotel on East 61st Street on 27 October 1979. A week later he travelled by metroliner to Washington, where he and the Gaumont bosses attended a reception for 150 guests given by the French ambassador, followed by a gala fund raiser ($100 a ticket) sponsored by the Friends of the Kennedy Center. They flew back to New York for another gala opening at the Lincoln Center on 5 November. In Brazil, too, the film was honoured by the French embassy with a reception. Losey flew from Paris to Rio on 23 November for a gala opening, followed by a screening for university students organized by the Maison Française. The Brazilian press coverage was massive.

Reporting from Paris in the *Herald Tribune* (21 September), Thomas Quinn Curtiss had called the film 'a thing of beauty and a joy for three hours . . . the most original, intelligent and handsome motion picture . . . in a long, long while', but on 24 October *Variety* printed a review which, while lauding the singing ('uniformly superb') and Losey's 'masterly technical direction', nevertheless threw a wet blanket over the box-office prospects of a subtitled film 184 minutes long. The fatal blow was struck when Vincent Canby delivered a devastating attack in the *New York Times* (6 November). Canby began with a principle: that opera, like legitimate theatre, places the audience in a fixed position, and does not 'go wandering around on the stage, poking its nose into the faces of the performers . . .' Bergman's *Magic Flute* succeeded, according to Canby, because it never pretended to be more than an extension of a theatrical production. Losey's version would bring 'utter confusion' to anyone ignorant of the score or the story. On 11 November Canby attacked again in his Sunday column. 'Here is a movie to blacken the name of intellect.' The film was loaded with arbitrary movement and directorial flourishes. The portly Don Ottavio punts through the marshes 'destroying our appreciation of one of opera's loveliest arias, "Dalla sua pace", as we worry whether or not he's going to topple overboard'.

Losey wrote to the editor of the *Times* requesting publication of his Open Letter to Canby – and without deletions. To this there was no response – Losey had confused a public letter with a personal one: 'For about thirty years we have had an amicable acquaintance and I believe there has been a certain mutual respect . . .' In the Spring 1980 issue of the magazine *Cinematograph*, Canby referred to Losey's 'asinine letter' and 'that man' while pouring scorn on European film critics: 'I don't want my ass kissed and I don't want to kiss anybody else's ass.' Canby was by no means out on his own. Despite some comfort from Andrew Sarris, Losey called it 'the worst lot of notices that I've ever had and probably the worst I've ever seen'. The film opened in Los Angeles, at the Cinerama Dome Theatre, on 25 March. On 1 April, a buffet supper was followed by an Opera Guild gala with Losey in attendance. Charles Champlin damned it in the *Los Angeles Times* (23 March): 'The film underscores and confirms all the reasons people who don't like opera have for not liking opera. It is a posturing bore . . .'

A France-Inter interview with a group of French critics, about 21 November, revealed the extent to which Losey felt shattered by the American reviews. France alone understood an artist who had been hounded out of his own country. The political climate in America was now worse than under McCarthyism, he claimed. Conspiracy too – just as Canby had attacked him not once but twice in the *New York Times*, his predecessor Brooks Atkinson had done precisely the same when *Galileo* opened in New York in December 1947. 'It's a pattern. It has to be political.' He might not be able to make *Silence* in Chicago in such a climate. A woman critic responded that America would be the loser – and at that point Losey winced: 'Biting off one's nose to spite one's face,' he groaned. No one could translate into French.

The British première was delayed until September 1980, when it opened at Academy Cinema One. Despite Philip French's praise in the *Observer* (28 September) – 'The movie gripped me throughout, providing the kind of enjoyment I have rarely had before from filmed opera' – the balance of comment was adverse. Interviewed on BBC Radio's 'Kaleidoscope' (26 September), Jonathan Miller, himself celebrated for his innovative productions of classic works, argued that opera is written for an acoustic box and becomes absurd in realistic surroundings – for example recitatives during which people are walking through undergrowth pursued by a phantom harpsichord playing the *continuo*. (But is not every instrument in the orchestra pit of an opera house a 'phantom'?)

Gaumont screened the film for the BBC's TV arts supremo, Humphrey Burton, but he was not excited and generally preferred Unitel films shot in opera houses. The news from the box-office wasn't good either: the London cinema takings were about £7,000 per week for the first two weeks, then slightly over £6,000 per week for the next three weeks.[9]

In France salvation! The Gallic soul leapt to Losey's genius! Gaumont launched the film with huge pre-publicity. TF1 produced a programme on '*Le regard des femmes sur . . . Don Juan*'. Antenne 2 screened two programmes, including a thirty-minute documentary on the making of the film. A large part of the magazine *Musiques* (November) was devoted to it; *Positif* (October) published Losey's production notes. *L'Express* (10–16 November) accorded six pages to Losey and *Don Giovanni*. *France Soir* (11 November) collected notable opinions, including Bertrand Tavernier's: 'It's prodigious, reverberating with beauty and power . . . This film has truly brought me to an appreciation of opera, which I don't often watch.' Michel Tournier, who confessed himself normally averse to opera, called the film a masterpiece in *Le Figaro*. The Gaumont team conducted a royal progress across France. On 15 November Losey and Liebermann accompanied Nicolas Seydoux and Toscan du Plantier to '*l'avant-première*' at Lyon. Visiting Brussels on 10 December with Seydoux and Liebermann, Losey gave three TV interviews and a press conference. Seydoux often acted as interpreter. 'We talked a lot on planes and trains. We had "*grande complicité*". He had directed the film with Protestant rigour, with professionalism and passion – in quotes.' (The President of Gaumont is fond of this '*en guillemets*', as if afraid of cliché.)[10]

On 12 December *Le Matin* published an interview in which Losey was asked whether the original conception of the film had been Liebermann's. He replied, 'No, no, no. Rolf Liebermann had the idea of associating Mozart with Palladio. But it was I who did the synchronization (*les répérages*), wrote the screenplay, chose the costumes, supervised all aspects of the film.' Liebermann took deep offence, requested that his name be removed from all publicity and from the credits (*générique*), and cancelled a lunch with Losey, who wrote protesting that he was not a 'credit grabber'.

The second wave of provincial openings began in the New Year. A nation of film enthusiasts responded with joy. *Libération-Champagne* (10 January 1980) reported Losey's personal presentation of the film before an audience of 500 at Troyes: Palladio, Mozart, Losey, three geniuses joined in a single work of art. Red carpets everywhere: the front page of *L'Est-Éclair* (11 January) announced that Losey's train would arrive from Paris at 18.50. During the first week 1,615 tickets were sold. People emerged from the cinema declaring themselves 'dazzled' (*éblouis*). A school teacher who took twenty pupils from form '3e1' to see the film wrote two adulatory but extremely well-informed columns for the paper.

According to Toscan du Plantier, *Don Giovanni* had grossed over $6 million in France alone. Gaumont informed Losey that 700,000 tickets had been sold by the end of June, but his complaints about poor-quality prints and bad soundtracks persisted. Gaumont would later go on to make other filmed operas with superior soundtrack technology – but without Losey. As Seydoux puts it, if Joe had been difficult but the film had made a big financial return – fine. If Joe had been easy and the film had merely done as it did – fine. But . . .[11]

Boris Godunov

Losey's memories of Mussorgsky's lugubrious opera *Boris Godunov* reached back to his youth – in 1924 he'd heard Chaliapin at the Met. Facing a gap in his schedule, Rolf Liebermann, then approaching the end of his term as *administrateur général* of the Paris Opéra, attempted to bring La Scala's production of *Boris* to Paris, but this could not be done without the presence of its Soviet director, Boris Lyubimov, a dissident in bad odour with the Kremlin. On 8 January 1980, Liebermann inquired whether Losey would be interested in directing *Boris*, using Dmitri Shostakovich's 1939 orchestration. Losey accepted because the commission took effect immediately – he hated the horoscope of the normal operatic timetable.[12]

The contract between L'Opéra de Paris and Losey stipulated a fee of 250,000 francs, of which 50,000 was payable on 27 March, the rest in ten instalments. The Opéra put at his disposal two assistants and Janine Reiss, his musical counsellor for *Don Giovanni*. After several stage designers declined the commission for lack of time, Losey engaged his Paris landlord, the seventy-eight-year-old architect Emile Aillaud. The bold Losey–Aillaud innovation was to put the orchestra on the stage behind the singers and elevated above them within a giant crown, a metallic, trelliswork shell resembling the onion dome of a Russian Orthodox church and surmounted by the Cross of St Andrew. Filling the back of the stage was a larger dome whose colour changed from gold in the first act, the heyday of Boris's autocracy, to sunset red at his demise. The action – soloists and chorus – was thrust forward on to a platform covering the orchestra pit. Conductor and singers observed each other through television monitors. Fourteen marble-like columns could be shifted to suggest the scenes – a geometric design for the interior of the Kremlin, a nest to enclose the monk Pimen in his cell, a scattering of trees in Marina's garden.

Boris's police force terrorizes the peasantry in black cloaks and fezes. The Tsar bestrides the stage in a robe of Cubist design then confronts his lonely guilt in black, Olivier's Richard III without the hump, beleaguered in a plastic throne resembling a heavily cast, outsize garden chair. Aillaud (who drew inspiration from the paintings of Gustav Klimt) told *L'Express* that he adored Losey's films but Losey did not understand his French, while 'it's impossible for me to understand a word he says' – Losey's English being 'a sort of ejaculation'.

Losey's production notes include a quote from Stanislavski's comments on crowd scenes. The chorus should never behave 'in general' and must be individualized, each member finding a background for himself with which to identify – boyars, police, guards, Marina's entourage, widows, mothers, orphaned children, serfs, lumpen proletarians – and 'what it means to live in Moscow or Russia under a variety of tsars and doumas in the 16th century'. (A tall order; the fact that a musical chorus is unified by word and music hinders differentiation.) Losey's notes concluded: 'Power, self delusion, emotion of great egocentricity, conscience of religious nature, all enter into this story which has in it elements of "Hamlet", "Macbeth", the Medici, "Richard III" . . . the elements are classical and therefore readily accessible to a large audience.'

This was the final production of the Liebermann era, with a chorus of 130, fifty extras, and an orchestra of 104. Ruggero Raimondi played Boris, although this role is commonly assigned to a bass. Ten days before the opening the conductor, Seiji Ozawa, returned to Tokyo with a slipped disc, though Janine Reiss recalls that Liebermann suspected frustration about the indirect contact with the singers through the monitors. Ozawa was replaced by Ruslan Raitscheff, chief conductor of the Sofia Opera, who seriously damaged his hand on one of the TV monitors during rehearsals.[13]

The first performance was on 9 June, a gala benefit for the Fondation Claude-Pompidou. Madame Pompidou, widow of the late President, occupied the presidential box with Princess Grace of Monaco, Premier Raymond Barre, Jacques Chaban-Delmas and Madame Simone Veil – a Gaullist celebration (although Barre had refused the Opéra the additional funds it requested). Losey reported: 'The opera has opened. I was slightly booed. I went to Guy de Rothschild's sit-down supper for two hundred people. I was appalled by the wealth and power and the obnoxiousness of the élite. Much more terrifying than the White House.' By Janine Reiss's account this supper was given by Charlotte Aillaud: 'People were talking about anything except *Boris*.' Losey reported to Ozawa that the audiences at the first two performances had been 'stuffy, conventional, cold', but when the popular audience arrived for the third performance there was an ovation.[14]

The production was given eight performances from 9 June to 5 July. In *Le Monde* (11 June), Jacques Longchamp complained about 'reducing the music to a sound-track, as in the cinema'. The orchestra should be the essential personage in Mussorgsky's case, but here it was faded under a chorus of 200. And yet, the choral groups were 'marvellously arranged, each individual having an individual destiny and a collective destiny . . .' Arguing precisely the opposite in *Le Quotidien* (11 June), Marcel Clavence complained that the stage design provided no contrast between mass choral scenes and intimate ones. As for the orchestra, 'In Mao tunics, the musicians are placed inside as on a music kiosk' – a rather artificial (*forcé*) experiment, not to be repeated. The critics were divided – the Losey–Aillaud production also won enthusiastic support.

Losey's ambition was to direct another filmed opera – perhaps *Tosca* with Kiri Te Kanawa and Raimondi, both of whom expressed enthusiasm. But Gaumont was wary and he pursued stage-opera, *faute de mieux*. After visiting Herbert von Karajan at his home, Losey sought the Maestro's blessing for a film, 'a tragic love story', using the *Vier Letzte Lieder* by Strauss. In September 1980, Lorin Maazel, now Director Designate of the Wiener Staatsoper, notified Losey of performance dates for *Lulu* in Vienna, scheduled for October 1983. Although Losey was not enthusiastic about an 'art nouveau' approach to *Lulu*, Maazel nevertheless recommended it for Vienna and promised him a 'great cast' and the best terms, 'which means no stage director will receive a higher fee or expenses'. In May 1981 Losey reported to Kiri Te Kanawa: 'Maazel's Vienna offer was so "mingey" [*sic*] I may not be able to accept it.' But he did, then, travel to Vienna to discuss *Lulu* with Maazel. The conductor politely lost patience: 'We simply cannot wait any more, and I therefore cannot . . . consider offering you the stage direction of the opera for the 83/84 season . . .' Losey put up further vague proposals: 'almost any Verdi . . . almost any Mozart . . . Puccini . . . some Strauss'. He wrote again in May 1982, suggesting *Leonora*, *Girl of the Golden West*, and *Lady Macbeth*. There is no evidence of a reply. In February 1983 Losey expressed concern to Massimo Bogianckino, the director-general designate of the Paris Opéra, about a report of plans to stage a new production of *Boris*. Losey requested the opportunity to change and improve his own one. Bogianckino declined the offer – a reprise of an old production was usually 'less interesting and worse than the original'. To Kiri Te Kanawa Losey continued to unburden his operatic woes – by post.

Drink lay behind almost every social disaster. At the Venice Festival of September 1979, where the subject of discussion was 'Cinema of the 1980s' – the future – Losey had rambled on about how little money he'd made, starting with RKO in 1947: 'And I got paid nothing, I got paid $15,000 a year by the studio.' Writing to an Italian editor, he later described 'the whole Venice experience' as 'such a débâcle . . . that I would much prefer you don't publish *anything* of what I said'. In a letter to his music agent, Frédéric Sartor, Losey described another drunken débâcle (which occurred in Paris, according to Patricia): '. . . that disastrous social evening when I had to excuse myself from the table to "pee" and remember nothing until 4.00 a.m. the following day. It all looked like a bad 30's set – the table uncleared and not a person in the house to be roused. I took my departure at about 7.00 in the

morning. I don't even know the name of the people . . . I strongly suspect
that I was doped.' Now he broke with Sartor who (he informed Kiri Te
Kanawa), 'did nothing except get me an interview via Concorde with Jim
Levine and the lovely woman at the Met., Joan Ingpen . . . I don't know what
I have done wrong . . .'[15]

PART XII

The Bitter End

43

Taxes, Quarrels and Trout

In France he enjoyed the adulation. Taxi drivers and bakers recognized him – seven years after his death the night desk-clerk at the Hotel Saint-André-des-Arts can remember a dozen of his films. He remained an American-in-Paris: *Le Matin*'s reporter, interviewing Losey at the Café Flore, commented that only an American could still think of giving his interviews there. But Losey's humour and warmth communicated themselves to other patrons: 'Soon people who were listening became interested, drew closer, and enjoyed themselves.' Losey himself enjoyed and didn't enjoy the great occasions: in April 1978 a *fête masquée* was given by friends of the Aillauds to which Patricia went dressed as a sea creature and Joe in his Dior evening clothes, made up to look a bit devilish. Patricia danced while Joe drank champagne 'and then we left. It was all too much for Joe.'[1]

Their relationship was sustained by activity, by tomorrow's diary, the working energies of an autumnal marriage. Recalling the small flat in the rue du Dragon, Richard Roud, Peter Smith and Gavrik Losey stress a confinement at odds with Losey's stature of spirit and body. Monique Lange remembers him sitting 'like a schoolboy', reading books and scenarios, at a small desk situated in an extension of the bedroom. Sometimes he took her to eat in Il Teatro, the Italian restaurant in the rue du Dragon. To Michel Ciment, who ate with him *chez* Lipp, at the Muniche, or in one of the small restaurants of the rue du Dragon, it was obvious that Losey felt isolated, exiled; in *Mr. Klein* he had imposed a stylized, imaginary, phantom Paris, 'which expresses all the anguish of a perscuted man'.[2]

Shortly before shooting *Don Giovanni*, Losey engaged a new secretary older than himself and unlike any of her predecessors. Ann Selepegno, a tough, gutsy American, had worked for John Huston and Anatole Litvak, then fallen on hard times. 'She was a little Caesar, a ballsy, tough Hollywood broad, very bossy, and she wanted an exclusive relationship with the man she worked for,' recalls her friend, the artist Richard Overstreet. 'She and Joe would have long fights. Her weakness was gambling. She'd card-gamble her money away. Joe would pistol-whip her with his tongue, though she answered back.' By most accounts Ann Selepegno was hostile to Patricia. 'Ann felt that Patricia had hitched her wagon to a star, liked to be the glamour puss in Paris, *and* to write scripts,' explains Overstreet. In Selepegno's view church and state should remain separate. 'Ann didn't like wives,' comments Barbara Bray. The editor Marie Castro, a friend of Selepegno's, describes it as a case of '*une sécrétaire montée en grade*' putting the actual secretary in her place; this echoes the case of Philippa Drummond a decade earlier. Normally affable and affectionate, Patricia also had a regal streak which probably hardened as the years passed, until it was directed at Joe himself. Her diaries lack any trace of self-scrutiny.[3]

Patricia's diary references to Ann Selepegno (now dead) are uniformly friendly: 'She really is a "doll", as she would say.' The doll was then helping to entertain Patricia's mother in Vicenza; one may guess this may not have been exactly the desired role. In September 1978, Patricia sympathetically noted Ann Selepegno's distress at the explosive way Joe treated her when drunk. Four years later, when he declined to take his secretary to Japan for *La Truite*, Selepegno was deeply hurt. That Patricia was the victim of envy among other women one cannot doubt – she was more than once congratulated on being Madame Joe Losey (though close and observant friends regarded her as indispensable to him and extraordinarily long-suffering).

Patricia's increasing involvement in Losey's work, and therefore with his colleagues' work, clearly fostered resentments. Although her aspiration to write screenplays was not achieved until *Steaming*, the contracts with Joe, but with no other director, proliferated during their years in France, partly a reflection of her language skills, partly of financial expediency, partly of Joe's anxieties on her behalf beyond his own lifetime. Arriving in France, she translated Solinas's screenplay of *Mr. Klein* into English for a modest remuneration. Although (as she points out) she did not work on Semprun's script for *Les Routes du sud*, official credits name Joseph and Patricia Losey as 'co-authors'. Her contract with Citel was now separate from Losey's, but he arranged payment patterns to suit his tax purposes: 'It would be best,' he instructed his agent Georges Beaume, 'for everybody if Patricia received all of her money during the period of shooting and if I received equivalently less, because ... will already have an income of around $75,000 this year.' This was done: during November–December 1977 Joe's income from the film was 90,000 francs, whereas Patricia's was 125,000. With *Don Giovanni*, the 'adaptation' was credited to Losey, Patricia and Frantz Salieri (Francis Savel). 'And I actually did the writing,' she recalls. Her diary entry for 27 May 1978 mentions 'a constant underground stream of tension between Joe and Francis and me'.[4] For a fee of 75,000 francs she translated Da Ponte's libretto into

English and worked on all literary aspects of the production. CBS Editorial Services in New York used her translation for the booklet accompanying the CBS Masterworks record. With *La Truite*, she was credited as the 'Associate Producer'.

Patricia and Joe were no longer happy in each other, though joined by partnership, loyalty and bonds of dependence. How to cope with Joe's anger, melancholia and often brutal dismissiveness was her permanent problem. Some believe no one else could have lasted the course. But the price was there: she attended a psychoanalyst in Paris, and wrote in her diary for 1 June 1978 of 'Joe's unutterably awful humour which no compassion or sympathy could relieve or excuse'. According to Marie Castro, Patricia did not conceal her exasperation with Joe and conveyed the impression that her life was close to insupportable. Increasingly Joe struck friends in France and England as 'helpless'. He took refuge in long lunches with cultivated young sparks such as Overstreet, who describes any such session as a *'bateau ivre'*, and with old flames. Losey told Overstreet that he had a woman friend whom he 'loved', and who lived two hours from Paris – almost certainly an American, and of Joe's generation, Overstreet recalls, though her name escapes him: 'Joe confided the woman's address to Ann Selepegno in case of emergency.' A fair guess may be made as to her identity. On 3 January 1979, Losey returned to Paris after a family reunion in London; two days later he telephoned Patricia complaining of feeling abandoned and un-nursed (she was preoccupied by her children's difficulties and did not follow him until the 23rd). Clearly it was during this brief period of separation that (as he reported to Françoise Sagan) he visited his former lover, the artist Hanna Axmann, whom he hadn't seen for twenty-five years.[5]

These years were soured by an ugly quarrel with his old friend Henri Langlois, director of the Cinémathèque Française. Losey had loaned many film prints and other materials to the Cinémathèque for exhibition; his film contracts stipulated that one perfect 35mm print be deposited as a personal loan with the CF. In November 1970, Langlois's colleague, Lotte Eisner, a lady of seventy-four and *conservateur en chef* of the CF, visited Royal Avenue and took away a large number of items for a special exhibition, including drawings and set designs. In September 1972 the CF staged a *'hommage'* to Losey. The Loseys spent the weekend in Paris, visited the museum escorted by Langlois, took him to eat oysters and had lunch at La Grande Cascade with Paloma Picasso. The bill came to £80 (£425 in 1990 terms). 'Joe mad,' Patricia wrote.

But Losey now began to complain: the CF had not acknowledged receipt of recent film prints, nor answered his letters of inquiry. Correspondence from June 1973 shows the National Film Archive in London competing for his prints. Clyde Jeavons, head of the NFA's Film and Television Acquisitions, assured him that the NFA guaranteed preservation of prints – unlike the CF. Freddy Buache, director of the Cinémathèque Suisse, also entered the competition; CS stored prints under scientifically monitored good conditions, issued prompt receipts, and allowed withdrawal at any time. As Langlois's energy declined, his refusal to delegate responsibility became more serious; in June 1976 Losey informed him that he did not intend to deposit a print of *Mr. Klein* until he had the CF's statement of receipt for *A Doll's House*, *The Romantic Englishwoman* and *Galileo*. Belatedly Langlois confirmed receipt of these films but it was now '*Cher Monsieur Losey*' – and he confused *Maison de poupée* with a film unconnected to Losey, *La Mouette*. On 26 November Buache wrote to Langlois by registered letter requesting that seven named prints be temporarily transferred to Lausanne for a Losey '*hommage*'. On 6 January 1977 Losey complained: 'We used to be friends – even admirers . . . yet when Buache and I both asked for prints for a retrospective, neither of us gets the courtesy of an answer . . .' Langlois died six days later. Losey sent a cable of commiseration to Mary Meerson, but his efforts to obtain the prints met with no more success than those of Buache, who advised him that the affair had thrown the Council of the CF into disarray: '*celà créa un choc à Chaillot*'. In 1978 Losey began to consider legal action. The CF's prevailing chaos was now apparent – a stream of conflicting messages indicated possession of only five Losey films plus a black-and-white version of another. Losey had meanwhile altered the contract for *Mr. Klein* to stipulate deposition of the complementary print in the Cinémathèque Suisse. Subsequently *Les Routes du sud* and *Don Giovanni* were also deposited with the CS.

On 19 April, Anatole Dauman, '*Le Président en exercice*' of the CF, wrote an instruction to the *directeur délégué*, Hubert Astier: 'Anxious to dissipate as quickly as possible the bad feeling between Joseph Losey and the Cinémathèque Française, I request you to kindly deliver to Freddy Buache the eight films by this director that we seem to hold.' Dauman insisted that this be done 'next week' and underlined these words. Nothing happened. Losey resumed the situation for Dauman's benefit: 'There are no records of anything . . . vandalism in the guise of egoism.'[6] By June 1979 he had launched a legal action. Maître Angelo Boccara requested payment of 5,000 francs in advance. On 6 July Maître Jacques Morin, '*Huissier de Justice, audiencier près le Tribunal de Grande Instance à Paris*', ruled that the CF must return all the ('many many') films and documents which Losey had deposited between 1965 and 1976.

On 6 August Hubert Astier instructed the CF's lawyer, Maître G.P. Langlois (brother of the late Henri), to inform Losey that these films had entered France without Customs duties, because deposited in the CF. Losey would now have to 'regularize' the Customs situation. This threw Losey into a fury. The CF agreed only that it held five prints: *Galileo, A Doll's House, Secret Ceremony, Figures in a Landscape* and *The Assassination of Trotsky*. Transfer of the five to the Centre National de la Cinématographie was effected on 9 October. In February 1980 they reached the Cinémathèque Suisse. Finding his drawings and documents in good condition, Losey decided to leave them for the time being with the CF's Musée du Cinéma at the Trocadéro. It emerged that the CF had some 30,000 film prints in storage, of which only 13,000 had been catalogued.[7]

The tensions and complexities of Losey's financial manoeuvres are vividly reflected in his relationship with his Paris agent. The Société Georges Beaume represented Losey in France throughout the 1970s. In an interview published in December 1974, Beaume described himself as '*un agent artistique*', representing ten actors and actresses, five directors and four authors. His clients included Simone Signoret, Annie Girardot, Nathalie Delon, Jacqueline Bissett, Monica Vitti, Alain Delon, Romain Gary, Jacques Lanzmann – and Joseph Losey.[8] Clearly an élite agency. Beaume worked enthusiastically on a number of Losey's projects which came to nothing, including Proust, Simone de Beauvoir's *Les Mandarins*, and *Ibn Saud*. He was intimately involved in Losey's decision to vest his services in Société Citel Films, of Geneva, which accepted a commission of only 5 per cent as a personal favour to Beaume. Citel's initial contract for Losey's worldwide services (other than the UK and Eire) ran for three years from 1 December 1974. Beaume negotiated his contracts to direct *Mr. Klein, Les Routes du sud* and *Don Giovanni*.

Beaume's problem was that Losey, now living in France, was using him not only as an artists' agency but also as an accountant – the function performed by Norden & Company in England. Indeed, Beaume was paying '*virements*' (20,052 francs in December 1977) to Norden on Losey's behalf. Losey declined to pay his own bills or write cheques. Beaume's statements show the Bureau regularly paying the four-star Hotel Diététique at Quiberon in Brittany, which Losey visited periodically, engaging the services of Dr Lavenant (at a cost over 10,000 francs in 1976). Beaume also paid for domestic repairs, the telephone bill at 10 rue du Dragon, travel agents, Losey's chauffeur at the Cannes Festival, florists, restaurants.

In November 1977 Losey instructed Beaume to verify all commissions required for his American tax returns, including Ros Chatto's. This provoked an exasperated protest: Beaume found himself increasingly embroiled in the complexities of Losey's financial life: 'transfer of funds to other countries, the control and receipt of same after their transfer, the mysteries of their movements'. The result was bad feelings 'at every level'. On 15 December he wrote again: 'I cannot allow problems which in no way concern me to poison our professional relationship.' He did not want to be 'a certified accountant, a tax specialist or a business manager . . . at any price'.

Beaume faced yet another snarl: while binding his wife into his film work, for tax reasons Losey wanted Patricia's earnings to be treated separately: 'Yves Rousset-Rouard [executive producer of *Les Routes du sud*] you and the TRINACRA accountant have all been informed repeatedly . . . that any contracts for "Madame Losey" (who should be called PATRICIA, not Madame) should never be combined in contract or letter with mine . . . She has an entirely separate contract with CITEL for entirely separate functions . . .' And more: 'It was established months ago that the [*Don Giovanni*] contract must be amended in such a way as to provide for money to be transmitted to CITEL – *not* to you . . . you have absolutely no authority, legally or otherwise, to accept "traites" on my behalf . . . I do *not* want any more money to be handled by your office, with the exception, possibly, of expenses . . . I am utterly shocked that you should let matters come to this state.'

Beaume responded: '. . . these problems which exceed the normal competence of our office have reached the limit of the insupportable'. Yet, despite it all, Losey's diary for 1979 records frequent meetings with Beaume: at the Losey's apartment; lunch at Prunier, dinner at the apartment, a rendez-vous at Beaume's office in December to meet an actress. The break came – or began – in 1981. In November Beaume expressed indignation – '*profondement outré*' – that Losey had discussed the renewal of his Gaumont contract for *La Truite* with a lawyer (Pierre Bréchignac) without consulting him. (According to Bréchignac's wife, Marie Castro, it was Patricia who appoached him, but according to Patricia it was Ann Selepegno who took Losey to the lawyer.) The fact that Beaume on this single occasion expressed himself in French was clearly an indication of his anger: 'I expect the worst from anyone. But I owe it to myself to say to you today, most sadly: not this! not you!'[9]

If British taxation was insupportable, French taxation rates did not appeal to Losey either. In America supertax was only 51 per cent and a wide variety of expenses (such as medical payments) were tax-allowable. His US tax returns for the years 1977 through 1984 were handled by Heather Timms, of Arthur Andersen International, Paris, who explains: 'His tax liability for the UK, France or any other country, except the US, was outside the scope of our client arrangement.'

Heather Timms also explains the French tax regulations then prevailing: 'Prior to January 1, 1979 in accordance with Article 164–1 of the domestic French income tax code, US citizens residing in France were able to exempt from French income tax any income from sources outside France, as they were subject to US tax on their worldwide income.'[10]

Losey's problem was that income from *Les Routes du sud*, *Don Giovanni* and *La Truite* was derived from French companies registered in Paris. His solution was not to admit his own residence in France. Indeed for eight years the Loseys did not declare themselves resident anywhere. (It is not known who advised them on this, but Laurence Harbottle has affirmed to this writer that universal non-residence was a legal possibility.) In 1980 Losey explained to his California lawyer, Leon Kaplan: 'We have a pied-à-terre available to us when we are in Paris . . .' In completing form RF3EU he confirmed his American liability to the French authorities; as a non-resident he was subject to a 25 per cent withholding tax in France, which was deducted from his American tax.[11]

As regards the 'pied-à-terre in Paris', the Loseys rented 10 rue du Dragon from the Aillauds – but his Paris bank was instructed to write to him 'c/o Aillaud'. His account at Barclays Bank, SA, 157 Boulevard Saint-Germain, remained classified 'non-resident' for the eight years. Annually the bank sent him a form: *'Vous êtes titulaire d'un compte "Non-Résident" . . .'* Annually Losey declared *'ma résidence principale n'est pas en France'*. Although this bank inquiry may strictly have related to exchange control rules, the declaration was consistent with avoidance of a French tax liability – which from 1 January 1979 became mandatory on US citizens resident in France on their worldwide income.

The pretence of non-residence was maintained on other fronts. Losey's contract with Gaumont to direct *La Truite* stated his address as care of Leon Kaplan's law firm in Beverly Hills. *'Bulletins de déclaration annuelle, au titre de l'année 1982'*, in respect of income from this film, give his address as Beverly Hills, California 90210. Gaumont even addressed letters to him care of Kaplan while he was living in Paris.

On American Tax Form 2555, he had stated his country of residence as 'outside US' – although a specific country of residence was required. It's a fair guess that 'outside US' was part of the general strategy of masking residence in France. His long-serving California business manager, Henry J. Bamberger, writing to Heather Timms of Arthur Andersen, confessed himself 'personally not pleased with the designation'. 'I wasn't happy,' he recalls. 'But I don't think he was prescient or clever. He sought professional advice.'[12]

There was also the question of Exchange Control, the movement of funds. The Réglementation des Changes (1973) stipulated that foreigners became subject to French Exchange Control regulations governing the movement of income and capital after two years' residence in France – *'lorsqu'elles sont établies en France depuis 2 ans . . .'* ['*elles*' referring to '*personnes*']. Prior approval from the Bank of France was thereafter required. But Losey never admitted he was 'established' in France. His payments in respect of *La Truite* were sent by Gaumont direct to Société Bancaire Barclays (Suisse), 6 Place de la Synagogue, Geneva, where his French francs account number was 333 750/ 2–3500–07.[13]

What was the purpose of the Loseys' Swiss bank accounts and of the arrangement with Citel, Geneva, in which '*société*' Losey vested his directorial services for his first three films with French companies? Capital lodged in Switzerland was free of exchange control regulations and the currency was strong. In February 1978 Joe and Patricia flew to Geneva and back the same day, to see Roland J. Jouby about money and contracts. In November Patricia travelled to Geneva and was met at the airport by Jouby with a new car. But Citel did not escape Joe's recriminations; in January 1979 he sent a four-page litany of complaints to Jouby: 'I am paying a total of approximately 40 per cent to be sure that such net income as I have is properly handled and accounted for ... I never know how much money I have or where it is ... You opened an account in Patricia's name in a new bank without her consent, signature or knowledge ... You deposited both her money and *mine* in this account ... I told you I wanted the use of a car and that I could get a 40 per cent discount from Leyland ... This car was finally delivered *in late November* with no discount and only a one-year guarantee ... My wife and I ... and my accountant, Heather Timms, have met with you in Geneva ... all this at my expense.' By the time Losey made *La Truite* his services as director were no longer vested in Citel.

La Truite

Frédérique grows up on a trout farm in the Jura, the daughter of a peasant. From an early age she learns the predatory nature of men and how to' capitalize on her own charms while giving nothing away. Later she carries this tactic into the world of international business (or businessmen) with chilling success. Her blood is as cold as a trout's.

La Truite was Roger Vailland's last novel. Losey had discussed a film adaptation with the author before his death in May 1965, and consulted Frederic Raphael about writing a screenplay – but got a negative response which was also prophetic: 'What is basically tedious is the worldweary attitude of Vailland himself, the seen-it-all man.' The characters did not develop and the origins of Frédérique's sexual 'closed door policy' would have to be radically re-examined. Dirk Bogarde's opinion of the novel resembled Raphael's – 'heartless, no love, too intellectual'.[14]

But Losey's eyes were glazed in stardust. In June 1966 he met Brigitte Bardot (whose agent was asking for $400,000 plus a percentage, or $500,000 without one) at her home in Saint-Tropez and discovered her 'fabulous untouched reserves ... as an actress and as a person'. The story of *La Truite*, he explained to her, was about Force, Power, Glamour . . . he sent a telegram to Bardot at La Madrague, Saint-Tropez: 'Love happiness and congratulations to you and Gunther Sachs . . .' On 10 January 1967 he reported to her the failure of his efforts with Paramount, Universal, MGM, Harry Saltzman, Sam Spiegel and Robert Aldrich. Losey did not abandon hope. In April 1968 he sent bunches of double cream freesias to Bardot at the Mayfair Hotel, gave her lunch and showed her *Secret Ceremony*. A few months later he wrote to the actress: 'An exquisite nude picture of you in the *News of the World* reminded me I haven't seen you even clothed for some time . . . Let's meet — in the nude.' A decade passed.

Losey first met the actress Isabelle Huppert at the Bayreuth Wagner Festival in 1977, in the company of Daniel Toscan du Plantier, and notionally cast her for the role. Bertrand Tavernier recalls: 'I told Joe not to do *La Truite*. Toscan du Plantier asked me to produce it. I turned it down.' Patrick Modiano and another potential scriptwriter also declined the assignment. Florence Malraux then introduced Losey to Monique Lange (author of *Les Cabines de bains* and of screenplays for Robert Rossellini, H.-G. Clouzot, Roger Vadim and Claude Lelouche). 'I knew the book was empty and artificial,' she admits, but she nevertheless accompanied Losey and Alexandre Trauner on a location trip to Berlin — Toscan du Plantier thought he'd found a German producer. The first draft of her screenplay is dated 18 January 1981, revised in February, again in June, finalized and translated into English for Losey's benefit in November. The Lange–Losey script transposes the story from the 1960s to the present day, substituting Japan for America, but faithfully conveys the cold, cynical aridity of Vailland's novel. 'Joe had rigour, a genius for construction and ellipsis,' Lange has written. 'He taught me to drop everything useless or explanatory.'[15]

Despite a draft contract dated 9 April 1981 between Losey, Partners Production (Paris) and Gaumont, the deal was far from firm and Losey sent complaining letters to Nicolas Seydoux, President of Gaumont. On 30 June, disaster: Partners Production (Ariel Zeitoun) let its option lapse. Patricia Losey wrote to her London accountant on 14 July: 'The French film was completely off last week until Joe had lunch with Mitterrand and I now think it will be on again.' In June Losey had congratulated Jack Lang on his appointment as Minister of Culture; in December he thanked Lang for his help in getting *La Truite* going — but did not indicate how. Seydoux is agnostic on this point: 'Certainly there was a state subsidy.'[16]

'I was long hesitant,' Toscan du Plantier recalls, 'and still am! The plot of the novel was weak. The most beautiful girl in the world could not have behaved like Frédérique. But Isabelle Huppert was my companion at that time and I was devoted to both her and Joe. Japan was expensive. I failed to find a Japanese financial partner. I turned to television.' Monique Lange describes Losey waiting for telephone calls which never came, working in conditions he found humiliating and in a language which he didn't really like. Toscan du Plantier phoned him one morning: 'If one arranges a co-production with television, *La Truite* will be made, I've had enough of you telling everyone I'm a shit.' Losey replied: 'If you answered my phone calls, I'd tell you myself.'[17] In September Losey flew to Japan for '*un voyage de répérages*' with his art director, Alexandre Trauner. On 8 January 1982, Gaumont finally contracted Losey's services as director for 225,000 francs, plus an escalating percentage, with a minimum guaranteed sum of 450,000 francs. A further 50,000 dollars was assigned to him as co-author. The film emerged as a Gaumont–TF1–SFPC co-production. The budget was $3 million, of which $450,000 was to be spent on the Japanese sequences.*

In his production notes Losey hoped the film would be 'funny, sad, disturbing, sensuous and sensual'. Frédérique's mocking laughter, he predicted, 'will be one of the hardest things to achieve, for actress and director'. (He was right.) He entertained the idea that Frédérique could figure alongside the heroines of Tolstoy, George Eliot and Henry James; but none of the main characters in *La Truite* has more life than a Champs-Elysées poster. For Losey the deserted wife Lou was to be 'the great tragic figure of the film . . .' Lou and Rampert, indeed, were 'not unlike Macbeth and Lady Macbeth . . .' He warned the cast that notes such as these 'always fall into the hands of journalists or critics who want labels'. However, as usual, he deposited the same notes with the magazine *Positif* (October 1982) for publication.

After a gap of thirteen years, Henri Alekan again worked with Losey. Alekan was keen to explain the virtues of the new Louma system; electronic transmission allowed the frames seen by the camera to be viewed on a monitor. After Alekan laid on a demonstration at Billancourt, Losey decided to shoot three sequences with the Louma, starting with the bowling alley – in a single day the whole sequence was achieved with the utmost fluidity. But Losey still objected that the system involved loss of contact between the director and the actors. They discussed colour. 'Enough of naturalism,' Alekan quotes himself as saying, 'let's go resolutely for hard transitions.' Losey agreed but, 'alas', would not jettison what had already been planned with Trauner. 'Once more,' lamented Alekan, 'I found myself behind the times on colour and not in the avant garde.' Nevertheless, he won a César for his photography.† Richard Hartley wrote a musical score which Losey described as: 'First some popular music, rock, reggae, and so on. Then the

---

*At Losey's insistence Gaumont appointed Yves Rousset-Rouard as *producteur délégué*, but his role was nominal (as compared with *Les Routes du sud*) and he says he never expected success. Christian Ferry was the *producteur exécutif*; Pierre Saint-Blancat *directeur de production*.

music for Frédérique, which is largely electronic with an elaborate use of twenty-four tracks, linked to the water of the Jura, the water of Japan . . . fountain streams, trout breeding. I would place it among the best scores – if not *the* best – I had for any of my films . . .'[18]

In December 1981, Isabelle Huppert had reported to Losey that she was viewing his films at the Cinémathèque Suisse and reading *Le Livre de Losey*. But an undated letter from Nyon indicates her doubts concerning the revised script, particularly the post-Japan scenes – she found difficulty in locating the points at which to release her imagination. Lange recalls conflicts in the Jura between Losey and Huppert over wigs. Toscan du Plantier comments: 'They were false conflicts like dogs barking over territory. He was macho and wanted to impose his authority. She is very tough, too. She was a spoiled girl (*fille gâtée*) because I was the leading French producer.' Later she wrote to Losey from the Holiday Inn, Kyoto: '. . . what happened *had* to happen . . . For the moment I still feel very much hurt but I remember what you told me. No more tears . . . anger, fury, violence but no more tears. Every moment spent with you on the set and out of the set is mythical, that is whatever you do, whatever happens, it is big, it is important, it is cinematographic.' Lange tells how the actor Jacques Speisser arrived in Japan bringing a letter from Toscan du Plantier to Huppert: he was getting married.[19]

For Losey the constant travelling from one location to the next became an ordeal – although the locations meant more to him than the characters. At one juncture he told his first assistant, Jean Achache, that he felt so ill that he must rearrange the schedule to shoot an important scene first: 'If I'm dead, you can do the others.' After studio work at Joinville, they flew to Tokyo on 17 March. On 1 April they travelled by rail to Toyoto – employing an all-Japanese crew assembled by Yukio Fujii. While filming in Tokyo on 25 March Losey suffered a double fracture of the collar-bone – he received 308,000 francs from medical insurance.

At the end of *La Truite* the actors' credits come up against their freeze-faces – as in a soap opera. This seems as eloquent a comment on the film as Monique Lange's: 'Joe wanted to show the cruelty of human relations in capitalist society.' *La Truite* was shown as the official French entry at the Venice Festival on 5 September, and was badly received; in Paris the film opened at the Gaumont-les-Halles, the Gaumont-Richelieu, the Gaumont-Colisée, and six other cinemas. Losey embarked on the usual round of interviews, including the trivialities required by *Elle*, whose Anne Chabrol

†The first day of shooting was 26 January 1982, at the Truiterie d'Arcier in the Jura, which Losey had first visited in 1966. On 2 February the crew moved to Besançon for the bowling scenes. On 17 February it was Charles de Gaulle airport, two days later the Cimitière de Haute Isle, for the burial of Lou.

asked him, 'Are you a driven man? You like bars . . . have you any regrets?'
Martin Henno, of *Le Figaro's* rubrique, 'La Vie au Masculin', sent him ten
questions about the male characters in the film. Would he define masculine
elegance? Why choose Dior, not Lanvin? Who was the best-dressed man in
the world? What was the ideal *garde-robe* for an elegant man?

The critics were generally hostile. In *Le Matin* (7 September), Michel
Perez complained of 'this cosmopolitan refinement dear to Losey' and
dismissed the characters as fleshless bores. In Perez's view the association of
sexual freedom with socially privileged libertinage was an outdated theme of
the sixties (Fellini). In *Nouvel Observateur* Michel Mardore complained: 'The
stereotypes become exasperating. Events take place arbitrarily, without progres-
sion, like selected morsels. Nothing is credible and style disappears. Only a
constant coldness remains on the screen . . .' A few voices were raised in the
film's defence: in *Les Nouvelles Littéraires* (30 September), Georges Charensol
called it a great work of art; he was 'stupefied' by the hostile reaction of other
critics. On 10 November, six weeks after opening, Losey reported to Mark
Urman: '*Truite* was taken out of all Paris theatres as of today, even though it
was doing quite well in small theatres. Gaumont obviously dumped it.' Two
days later he told Elizabeth Vailland: '. . . they will lose a lot of money as a
result of their stupidity'.

La Truite was shown at the 1982 New York Film Festival. Triumph Films
– a new Columbia-Gaumont company – accommodated Losey at the Sheraton
Netherland Hotel in a suite once occupied by Fellini. In the *Village Voice*
(18th October), Andrew Sarris commented: 'Ellipsis and distortion figure in
[Losey's] work as methods of dismantling his dramatic structure to the point
that the outbursts of emotion and violence are transmitted into an unmotivated
hysteria.' The film opened at the Baronet in May 1983 – normally the theatre
for films like *The Exorcist*. Writing in the *New York Times* (5 June), Vincent
Canby was scornful: 'I'm not kidding. Never has a film by a major director
been so abjectly "serious", but what it's serious about, I've no idea.' It was
'relentlessly chic'. After three weeks only 30,000 tickets had been sold; the
film vanished. Despite a screening at the London Film Festival (which Losey
attended) there was no commercial release in the UK – the fate of two out of
three of his French-language films. In December 1982 he travelled to Oslo
for the Norwegian première, the dinner menu at the Grand Hotel consisting
of *La Truite fumé du chef; La selle du cerf King and Country; Céléri au
fromage The Go-Between; Parfait glacé Don Giovanni.*

As his career flagged, there was still the small comfort of retrospectives. In
1980 he was lionized at Prades, where eight of his films were shown. In
January 1981 he attended a '*mois Britannique*' at Saint-Etienne. In September
Les Criminels was reshown at four cinemas in Paris. In October *Gypsy* was

screened at La Royale cinema, despite Losey's last-minute attempt to ban a film he disowned. He berated Pierre Desgraupes, of Antenne-2, for unauthorized cuts in *L'Anglaise romantique*, screened on Christmas morning 1981 at 6.15 a.m. *La Bête s'éveille* (*The Sleeping Tiger*) was re-released in France in 1982.

Monique Lange recalled him constantly using the word 'agony' to describe the months of waiting and indecision before he was able to shoot *La Truite*. 'My God when I am going to begin work again!' Or: 'I await the next catastrophe.' She adds that he had a genius for creating gloom (*atmosphères de plomb*); friends were powerless before this permanent anger – but he was also capable of infinite tenderness. He complained that six of his films were on sale as cassettes in Paris but, 'needless to say, I've received no money from all this'. Relations with his major French patron, the Société Gaumont, were in ruins. Visiting New York, Toscan du Plantier told junior colleagues that he was fed up with the demands, complaints, frustrations of the 'old drunk'. 'Of the great directors I've known,' wrote Toscan du Plantier after Losey's death, 'Joe is the one who asked most of me.'[20]

In November 1980 Joe had spent ten days of enforced rest in hospital. A year later he attended a London hospital for X-rays, opinions and minor surgery. His condition was diagnosed as cervical arthritis, bronchitis, infected sebaceous cyst. In January 1982 he was in the Hospital of St John and St Elizabeth, London, for tachycardia and menorrhagia. While shooting *La Truite* in Japan he fell down some stairs and broke his collar-bone. Peter Smith found him overweight, out of breath, anxious to link arms for support in the street. But the spirit was indefatigable. In mid-winter, December 1982, Losey and Henri Alekan arrived in Vicenza, the location of *Don Giovanni*, to make a thirty-second publicity film for a hair conditioner. Alekan noticed that he was tired and relied on Patricia, '*sa femme dévouée*'. The Palladio villa was sweating 'like a Turkish prison'. Losey decided to turn it into a studio, covering all windows and openings – night for day. Alekan's last memory of Losey was of him standing close beside the camera, wearing a heavy coat, attentively observing the smallest gestures of the young actress shaking blonde curls stirred by a ventilator.[21]

It was time to return to London. Losey had left Britain with the arrival of a Labour government which raised his tax liability, and now he took himself back to Mrs Thatcher's Britain, where income tax was in sharp decline and exchange controls abolished. 'Under the present government,' Patricia Losey recalled in 1988, 'it was possible to return.'[22]

44

Further Disasters

Throughout his years in France, Losey had continued to dream in his native language; and to reach across the sea in pursuit of the great English-language film projects which drifted, like becalmed galleons, several million dollars beyond his trembling, outstretched hand. Gallantly, indefatigably, he swam towards them – until a light, cold breeze or a sudden tempest carried them out of sight.

Take the spirit-breaking saga of *Ibn Saud*. He had first been approached in 1975, by the Egyptian film director Youssef (Joseph) Shahine, on behalf of a Saudi syndicate wishing to finance a feature film about King Ibn Sa'ud. Although the financial backing probably came from Sheikh Kamal Adam, the 'principals' chose to remain anonymous and to negotiate through a London solicitor, Peter Stone, of Wright and Webb, Syrett and Sons. Vesting his services in Citel, Losey signed a $100,000 contract in March 1975; on the following day King Feisal of Saudi Arabia was assassinated. The iron gate of the desert kingdom descended.

Losey talked to a number of writers including Ring Lardner, Jr and Robert Bolt, to whom he wrote: 'I am not going ahead with you. It appears that you are too indoctrinated with Lawrence and too little familiar with Philby . . .' He enlisted Barbara Bray for research and visited Algeria to talk to the writer Kateb Yacine, whom he'd met in 1967. But Yacine's script was a disappointment. 'The story was shapeless,' Bray recalls. Patricia read Yacine's French script to Losey in April and concluded: 'It's agit-prop, and theatre, not cinema, has no structure . . . Incredibly disappointing.' (And particularly for her own hopes of translating it. Evidently there was an element of rivalry; her diary for 20 May 1975 records Losey telling Barbara Bray that her work on *Ibn Saud* was 'totally unacceptable'.) After further negotiations with Ring Lardner, Jr, who according to Losey was 'asking [for] $200,000 and 5 per cent of the profits', he turned to Franco Solinas, the screenwriter of *Mr. Klein*, to achieve a Shakespearean rendering of a single critical year, 1929, when a major battle occurred between 'progressive' and 'fundamentalist' forces in Saudi Arabia. Solinas encountered a mental block. In April 1978 Losey wrote with tact and tenderness, begging the Italian writer to 'WRITE. Sit down and put it down.' Otherwise it would be a case of self-indulgence, masochism . . . Solinas's reply (June 1978) was dignified, proud and slightly twisted: 'For the moment I force myself to ignore your displeasures and delusions . . .'[1]

Now a fatal row blew up. Patricia Losey had been translating Solinas's pages from the Italian as they arrived. Barbara Bray, who didn't appreciate Loscy's habit of 'smuggling' his wife in to the work she herself was engaged on, suggested that the script should be translated into English by the poet and radio playwright (Prix Italia, 1951) Henry Reed, who was also the translator of Leopardi and Ugo Betti. The solicitor Peter Stone duly informed Losey that his clients wished to engage 'a first-class literary translator . . .' Losey was furious: 'I was amazed and appalled at the brutality of your libel by implication of Patricia's work . . .' To make matters worse Losey persistently referred to Henry Reed as 'Herbert Reed', i.e. the art critic, Sir Herbert Read. Having read 'Herbert Reed's' translation, he declared it inferior to Patricia's: 'I cannot begin to describe to you my rage and perplexity . . . [your] insensitivity and discourtesy . . . after *Don Giovanni* I am quite sure that Patricia's name will be far more known to the mass audiences than Herbert Reed's [*sic*].' Worse was to come: the principals put the entire project, plus finance, in the hands of the London-based Palestinian entrepreneur Naim Atallah, who promptly dropped Losey. Atallah explains that he was advised by distributors he consulted that Losey was a brilliant but ageing art-house director, inflexible, intractable, irritable. A furious Losey claimed to be contractually co-author of the script with Solinas and demanded damages of $100,000 (this claim was later withdrawn). Solinas pleaded with Atallah on Losey's behalf but in vain. Bray decided to collaborate with Atallah as associate producer. 'Why should I cut my own throat?' she comments. The rest is epilogue. The film was never made. Franco Solinas died in September 1982, at the age of fifty-five. Three years later his children published his script.[2]

Losey's first discovered dealings with Graham Greene occurred in 1965, when he proposed that Greene should write a screenplay of *The Wings of the Dove*. Greene politely declined but the following year Losey was engaged by Daniel Angel's Keyboard Productions to direct a stage musical of Greene's *Brighton Rock*. Losey flew to meet Greene in Nice in March 1966 but nothing came of it. Six years Losey approached both Greene and Lillian Hellman to write a screenplay of Joseph Conrad's *The Secret Agent*. Hellman bowed out after asking for $75,000 which, she explained, was less than half her last fee. Greene's asking price was a modest £15,000 but the producer, Jerry Bick, offered him an insulting £7,000 and Greene cut Bick dead in Paris, outside Rampaneau's. Six months later Losey inquired whether Greene would be interested in writing a screenplay of Patrick White's *Voss*. Greene declined. In March 1978, Losey opened a correspondence with Greene about his new novel, *The Human Factor*. 'It is the first time in five years I've heard Harold [Pinter] so enthusiastic on any project I proposed to him. I expect I am once again too late. This is apt always to be the case with you – or me.' He was right; Otto Preminger already held an option and shot the film with Nicol Williamson and Richard Attenborough. Losey told the *Los Angeles Times* (19 March 1980) that 'all the grand old English actors – "the knights and would-be knights" – feared to play one key role after the exposure of the Queen's art expert Anthony Blunt'.[3]

In 1980 Greene published his novella, *Dr Fischer of Geneva and the Bomb Party*, having sent Losey the page proofs 'on the chance that it might interest you ... I am not sending a copy to Preminger!' Although Losey responded rather guardedly, asking for a few weeks to think about it, Greene passed on his own ideas for casting and proposed a screentest for a young actress, Martine Cloetta, whose photo he enclosed. 'She is not my mistress!' he added. Losey failed to raise the money. Meanwhile the writer John Mortimer approached John Pringle, of Consolidated Productions, who acquired the rights and, with them, Greene's commitment to Losey as director. Pringle offered him an advance of $50,000 against an eventual fee of $225,000 plus $7\frac{1}{2}$ per cent of the profit. 'I know, Joe, that this is less than the amounts you mentioned in Paris ...' Losey found himself working with a screenwriter, Mortimer, whom he had not himself chosen; Mortimer recalls that Losey's own ideas about *Dr Fischer*, 'in particular his opening, might have gone too far and were not really compatible with Graham's book'. On 3 November, Pringle and two colleagues sent Mortimer a memo on his first draft, mailing a copy to Losey, whose temper was now on the boil. Patricia Losey's diary predicted a 'last-ditch' meeting between Losey, Greene, Pringle and Mortimer, in Cannes on or about 29 December. 'I tremble for it.' The journey began badly for Losey with a five-hour flight delay at Orly airport. According to Pringle, Losey came off the plane drunk and conducted himself like one of the megalomaniac Hollywood impresarios he despised. Losey behaved so badly at dinner, rude to everyone but directing his abuse mainly at Mortimer's script, that Greene afterwards advised Pringle to call it a day with Losey. (Mortimer was not present at this dinner.) The next morning Pringle found Losey in his big hotel suite disconsolate, guilty and confused, his stomach bulging from his dressing-gown – and broke the news.[4]

On the way home to Paris Losey fell on three granite steps and came off the plane in a wheelchair with severe swellings of the knee, elbow and eye. He had blown it. To Greene he wrote: 'Everything seems to have gone sour on *Dr Fischer*. To the extent that any of it was my fault, I am very sad.' Greene did not reply. The project went ahead, although slowly, without Losey – and then without Mortimer, turning into a BBC-backed television film which Pringle hated. Two further years had elapsed when, in March 1983, Losey wrote to Greene: 'Our last telephone conversation was so curt ... I hesitated to pursue anything with you ... since the Cannes disaster I have not heard one word . . .' He then changed gear into distemper. His contract contained a $100,000 indemnity clause but he had so far been offered a mere $20,000 in settlement. In reply, Greene professed ignorance, but with a sting in the tail: 'I may add that you are lucky to be getting so much money out of *Dr Fischer* . . . I would be very glad to receive even $20,000.' (Pringle believes that Losey's final compensation was $50,000.) Shortly before his death Greene would merely say: 'I liked Losey and I don't remember the disagreement. But at eighty-five one's memory is very bad.'[5]

James Kennaway's novel *Silence* was incomplete when the author died in 1966. A decade later Losey persuaded John Heyman to finance a script by David Rayfiel, whose work with Tavernier on *La Mort en direct* (*Death Watch*) had impressed him. Heyman tried hard but Losey was restless, aggressive, bringing a protest from the producer: 'I would not have spent $130,000 of my own money with another $90,000 due next month on the option unless I really believed in the project and in you . . . [but] every time I made suggestions [about the script] in New York it seemed to cause offence . . .' For the year ending December 1979, Heyman's Film Writers Company paid Losey £35,000 for work on unrealized projects, principally *Silence*, for which his full fee was to be $225,000.[6]

Set in Chicago, *Silence* portrays the love of a young black girl and an older white doctor, both of whom are escaping a race riot in mid-winter. While in Los Angeles in March 1980 Losey told the press: 'Essentially it comes down to a belief I have that the major world problems are finally left to individuals to solve among themselves.' In November he told *American Film*: 'I'm doing a picture – I hope . . . in Chicago, where the music is Chicago blues . . . a film in four colors. White, black, red and yellow.' But Heyman could not find a backer. Alain Bernheim of MGM and Elmo Williams of Gaylord Productions declined. With Heyman's concurrence Losey took the script to Alan Carr, producer of *Grease*, who bought out Heyman's financial stake. In October 1981 Losey again met Carr in Hollywood and discovered that his vague plans for producing *Silence* had been postponed. In his bitterness Losey seems to have misled French colleagues such as Bertrand Tavernier into believing that Heyman had callously decided to sell Rayfiel's screenplay to 'the horrible producer Alan Carr' – who then told Losey that he no longer belonged to the American film community! This was not the case; however, Rayfiel adds that Carr only began to push the project after Losey's death.[7]

The Loseys moved back to London in January 1983. Monique Lange accompanied Joe while Patricia stayed in Paris to finish packing. Tyger's enforced absence in quarantine depressed him. 'Why is it I suffer so much for this dog?' he asked Lange. On 17 March the Loseys gave a party to celebrate

Tyger's release from quarantine. Losey grumbled to Bogarde that he was no one in England although a god in Paris. He needed work. There had been several personal meetings with Jeremy Isaacs, head of Channel 4 TV – but Losey had 'never been able to come to terms with their bureaucracy'. When Bogarde unburdened his own grief – his companion Tony Forward was suffering from both cancer and Parkinson's disease – Losey showed no interest: 'He told me he'd been watching *The Servant* on television with a stop watch.'[8]

In May Losey spent a week in New York with Patricia, promoting *La Truite*. 'He can move only with difficulty,' Louis Archibald reported, 'his face shows some of the folds of pain. He lets his eyes dart round the room, somewhat like a very bright, very canny toad scanning the territory.' Losey took himself to the University of Wisconsin for an honorary doctorate of Humane Letters. Commencement was at 9.30 a.m. on 22 May, at Camp Randall Stadium. The Citation briefly described Losey's career, then came the Conferral: 'Joseph Losey: Your devoted artistry has yielded works of intelligence and beauty. Accordingly, the University of Wisconsin confers on you . . .' (Dirk Bogarde's commendation had been sought and warmly given: 'he is one of the greatest living cinema Directors'.) Losey wrote his speech on six cards, beginning with his maternal-side connections with the University – his grandfather, Judge Higbee, was a political progressive and a personal friend of Senator LaFollett. But he also reminded his audience that he himself went to Dartmouth, his father to Princeton, his paternal grandfather to Amherst. When Russell Merritt, of the University's Center for Film and Theater Research, sent him an article from the Madison *Journal*, by Steve Paulson, Losey irritably took it apart: 'I wasn't "partly responsible" for bringing *Galileo* to the American stage – I was wholly responsible.' He also objected to the phrase, 'his bitterness is still evident' – he'd never been 'bitter'. Merritt himself was quoted in Paulson's article: '[Losey] is one of the great champions of wealth for what it brings, for making it sexy.' Losey was incensed.[9]

The British Film Institute arrived to collect his papers. Janet Moat, a member of the BFI Library staff who subsequently spent three years sorting and cataloguing his archive, remembers him speaking bitterly of a disappointing life and of unfulfilled projects.[10]

Dennis Potter's *Track 29* (originally *Tears Before Bedtime*, then *Track 39*) was to be Losey's first film set and shot in America for thirty years, a BBC co-production with Potter's Pennies From Heaven Inc. In April 1983 Losey received a miniscule advance payment of £1,500 from the BBC. The setting was Texas; his production file includes a pile of model railway magazines and a book about psychedelic vans. In June Losey sent Potter notes on the script covering eleven typed pages and disputatious in tone. 'This is an impossibility from a technical point of view ... I do not believe in dissolves here ... No Texan would refer to food as "halfway edible" ... there are so many things which I take to be intended as Americanisms which are not ... in spite of your brilliant and incontestable accuracy of ear ...' The film was to be shot in Dallas, from 5 August. On 23 July Losey told John Heyman: 'I have a very good cast, Vanessa [Redgrave], Lee Marvin, Anthony Higgins, Louise Fletcher ...' He was apprehensive about the BBC co-production, 'which is messy, and the consequent absence of any personnel that I have previously worked with ...' The project collapsed in the first week of August when Losey was in Mexico on his way to Dallas. Trade union problems figured but, according to his secretary, Victoria Bacon, the last straw was the BBC's insistence on 16mm film, whereas Losey, Potter and Rick McCallum, the American producer, wanted 35mm. Another blow.[11]

In September Losey accepted an invitation to the San Sebastian Film Festival and also accompanied his host, Guillermo Cabrera Infante, to his cinema course at the Menendez Pelayo University, Santander. 'I have recently spent a few weeks at Spanish Film Festivals,' he reported to an American friend. 'It appears that I am a "legend", a "mystic figure", an "international veteran" – none of which adds up to anything useful.' To Peter Smith, and almost identically to Hardy Kruger, he wrote: 'I find the present world of films and theatre intolerable and I don't know how much more I can take.'[12]

Steaming

Losey ended with a film whose affectionate and relaxed view of the human condition could not have been anticipated from the glacial cynicism of *La Truite*, the mocking generational war of *Les Routes du sud*, the chilling *sauve-qui-peut* of *Mr. Klein*, or the sexual masochism of *The Romantic Englishwoman*. Not since *King and Country* had Losey so warmly embraced the human condition. In that film, of course, no woman is seen; in *Steaming*, no man.

The setting of Nell Dunn's play, *Steaming*, is a London baths built in fine style at the turn of the century for the enjoyment and good health of the People, but now a bit tatty and ripe for demolition. Women of various social backgrounds who regularly converge to cleanse and soothe themselves, and to talk about their lives, now face a threat to close their beloved baths. The play had opened at the Theatre Royal Stratford on 1 July 1981, transferred to the Comedy Theatre, and run for over 700 performances. The Loseys were taken to see *Steaming* in March 1983 by the producer Paul Mills. Patricia Losey recalls: 'I was so enthusiastic and certain about it that I asked Joe that night to let me do the adaptation.' After Mills had commissioned a script which no one liked, Patricia again put her own name forward. Richard Dalton, whom John Heyman later appointed as executive producer, comments: 'Joe told me

he wanted to launch her on a writing career. Nobody else wanted her to write
it. She had no track record at all.' Jack Loring, too, remembers Joe explaining
his anxiety to establish Patricia as a screenwriter before he died.[13]

The screenplay replaces the single stage-set with a variety of settings
within the public baths: the Main Entrance and Rest Room, with adjoining
cubicles; the Steam Room; the Cooling-off Room; the Hot Rooms; the
Staircase; the Plunge Pool. Of the leading characters, Sarah (Jane in the play)
is the most 'adapted'; no longer a mum working at home on a university
degree and living off a grant, she re-emerges as a successful solicitor who
explains her late arrival: 'My two assistants went out for long business
lunches ... No wonder young men get heart attacks.' This reincarnation
more sharply separates Sarah (Sarah Miles) from her unhappily house-bound
and recently deserted old school chum Nancy (Vanessa Redgrave), who is
miserably groping to discover what a woman's choices really are.

In the play the women's petition to the Council, protesting the closure of
the baths, fails. In the film, the women return from the Council meeting with
a reprieve: 'They voted five thousand pounds for urgent repairs and they've
given us six months to get the attendance up ...' Victory incurs the penalty
of *Girls' Own* dialogue:

> SARAH: And how do we persuade more people to come here?
> NANCY: We tell all our friends. Well ... What about all those people in
> your office, Sarah?

Patricia Losey has explained the happy ending as a proposal from John
Heyman. However, the preservation of 'uneconomic' public amenities, despite
property developers hungry to replace the baths with a modern leisure centre,
was hardly the characteristic dénouement of the Thatcher Decade.

Steaming was produced by John Heyman's World Film Services, the first
realized collaboration between Losey and Heyman since *A Doll's House*. The
total budget was set at £1,417,013. Losey's contract (Launcelot Productions)
with World Film Services was for £82,000; Patricia Losey's fee as screenwriter
was £10,000. She delivered her first draft on 3 January 1984. Heyman hated
it and wanted to bring in a new writer. Losey argued that the film script 'had
always been circumscribed' by protective clauses in Nell Dunn's contract –
but this was a red herring. Richard Dalton is clear in his mind that Losey
threatened to withdraw if Heyman called in a new writer, but there is no
evidence in the correspondence and Patricia Losey denies it. Relations became
fraught: a large composite set was already under construction and pre-production costs were running high. Heyman loathed the visualized women's fantasies
in the screenplay; Losey agreed that some could go, but not all – he wanted a
top artist to design them. David Hockney was approached but declined.

Losey wrote to Alexander Trauner on 11 January, explaining that the fantasies should be 'in quite a different style from the rest of the film', but nothing came of this – farewell to the fantasies.

Farewell, also, to Heyman's investment. Paramount pulled out. 'We decided to use our own money. Paramount lent me $2.8 million. I paid it back and lost $1.2 million, plus interest, out of my own pocket.' In America, New World paid only $350,000 for distribution rights.[14] Frantic rewriting took place throughout January and February, extending into the period of shooting, which began at Pinewood on 27 February. The script entered what Dalton calls the 'pink page syndrome' – or the committee stage. Friction was evident between the Loseys. On 21 February. Patricia wrote to Joe: 'The dreaded scene 12 [which depicts Sarah's first entrance] was [re]written by a committee.' On 13 March she wrote again: 'Please do not cut anything else or add anything. Dreams cannot be bowdlerized.'

The film itself is clearly undermined by too many hands on the script. Heyman and Losey wanted the threat to close the public baths to be made clear at the outset. Patricia Losey therefore reintroduced Nell Dunn's single male character, the caretaker Bill, who – heard but unseen – tells the supervisor, Violet: 'You know the Council's going to close us down.' But at the editing stage Losey eliminated Bill, who'd been filmed in longshot from a crane; instead we have Violet (Diana Dors) carrying a symbolic toolbox and muttering enigmatically to herself, 'They can't pull this place down. They can't.' But she says nothing about it to the women and no more is heard of the threatened closure for half an hour of film, until Violet announces the news in shocked tones. In short, a hash. The editing in general is poor: a five-day gap between two early scenes looks like 'one minute later'.

The film works well when the women are talking about life 'with David' or whoever; the virtue of the play lies in the relaxed airing of personal dreams and woes by informal group therapy. Like Chekhov and Gorky, Dunn can command an audience's attention without resort to plot suspense. The film begins to creak with the contrivance of an amateur theatrical when the 'action' dimension – the threatened closure of the baths – is introduced. Sarah, the efficient lawyer, arrives at the baths: 'I've got news! I'll tell you all together at our meeting.' New, 'filler' lines are not redeemed by being delivered from rowing machines or exercise bicycles.

NANCY: Actually I feel terribly keyed up about the meeting. Do you think Josie'll be all right?

Although the film preserves much of Nell Dunn's wit, skilfully capitalizing on sudden confessions, reversals, fantasies and Mrs Meadows's (Brenda Bruce) know-nothing cockney wisdom, it becomes clear that the humour which provokes laughter in the theatre merely raises a smile in the cinema. It

is the 'gap' between stage and auditorium which provides the magical
intimacy of theatre: the audience laughs not only at the characters but at the
actors' refusal to acknowledge the audience's presence or laughter: the camera
destroys this innocence.

The total casting budget was £220,000, of which £100,000 was heading for
Vanessa Redgrave. (Other lead parts were allocated £34,000 each, with
£24,000 for Diana Dors.) Losey's first discovered film proposal to Redgrave
was in December 1967. Seven years later Losey and Patrick White, mentally
casting a film of *Voss*, discussed Redgrave for the part of Laura. Losey's
chauvinistic comments were parhaps influenced by his recent row with Jane
Fonda and were symptomatic of his reactions to actresses of independent,
radical outlook. White found 'Vanessa Redgrave as Revolutionary very hard
to take.' Losey agreed: 'I find her stance hard to take also and I am not at all
sure I am prepared to cope with her . . . she's rapidly growing more eccentric
and improbable . . . and is persona non grata with a lot of film people, I
gather. This may present obstacles but doesn't influence my opinion of her.'
(It should be added that Vanessa Redgrave continued to command star roles
for major studios throughout the 1970s.) By the early 1980s Losey was clearly
keen to work with her; in 1983 she was to have played in a Losey film which
fell through at the eleventh hour, Dennis Potter's *Track 29*.[15]

For *Steaming*, Redgrave – like two other actresses whom Losey approached,
Julie Christie and Jacqueline Bissett – adamantly refused nudity or implied
nudity. Nor would she contemplate a 'body-double'. On 7 January, Losey
reported to Heyman that he had spent 'hours' with Redgrave 'off and on over
various days and weeks'. His worries focused on the climax of the film. 'If she
insisted on covering herself up at the end, when the whole point of the
character [Nancy] was that she [no longer] felt inhibitions about herself, then
she would look an idiot and it [would] rebound on the film.' (John Heyman
comments, in retrospect: 'Since Mr Losey and I had already accepted her
refusal for both frontal nudity and a body double, it hardly came as a surprise
when she refused to appear nude in the film.') On 4 January, Losey cabled
Heyman proposing Jenny Agutter instead of Redgrave. Heyman replied: 'We
sold the picture on the basis of Vanessa Redgrave who is a prestige star for
American purposes.'

Losey's general fears were not dispelled during shooting (although Vanessa
Redgrave remembers that she admired the script). Christopher Challis, the
director of photography, recalls: '. . . there was considerable discussion prior'
to shooting when it was I believe agreed that Vanessa would not appear in the
nude, apart from one scene in the swimming pool. It was at the start of
shooting this scene that Vanessa came on the set and handed Joe a note. He
exploded with rage, tore it up, and threw it on the floor. What the contents
were I do not know. We subsequently completed the shooting of the scene.'

Vanessa Redgrave herself recalls: 'I had my periods the day of the pool scene and I asked if I needed to go into the pool, I was performing every evening at the theatre, and I made a suggestion as to maybe what I could do instead if Joe permitted.' After Losey's death Guillermo Cabrera Infante published in France derogatory comments which he claimed Losey had made to him: 'She's a bitch. If I give her an instruction, or better if I ask her, to go from A to B, a logical move, she says yes. She does it in the rehearsal. At the first take she goes from A to C or D or wherever she pleases. The bitch!' We can accept Redgrave's own verdict on these comments as untrue as well as offensive, but they reflect Losey's morose state of mind. Certainly her performance as Nancy was highly and characteristically professional in the manner of her work both before and since. She herself is generous about Losey: 'I admire Joe Losey's films very much indeed. It was our common misfortune that the film we finally met on was *Steaming*. My father [Michael Redgrave] always spoke with admiration of Joe.'[16]

Unhappiness also attended the casting of the sparkling cockney character Josie, played by Georgina Hale in the London stage version. Hale sent in a list of queries and complaints about the screenplay. Why, for example, in shot 55, did Josie enter the baths 'like an American presidential candidate' and declare, 'Hi everybody! The posters look great! I like those balloons. Save me a T-shirt'? Why so American? Hale asked. Losey passed her objections to his screenwriter, whose responses were generally accommodating, but Patricia showed impatience about shot 55: 'Would Georgina prefer to make an entrance like Mrs Thatcher arriving at the committee rooms of her party in her constituency at Finchley?' Georgina Hale came to a read-through at Pinewood – the only actress present who'd appeared in the original stage play. According to Dalton, her attitude to the script was such that Losey decided to take her off the film. Brenda Bruce, who sat next to Hale, felt that she went over the top during the read-through, and Losey, obviously irritated, complained that he couldn't understand the accent she was using. 'Georgina behaved appallingly at the reading,' recalls Patricia, 'objected to everything not in the play, read her lines in an odd voice, succeeded in infecting Sarah and Vanessa – and did herself in.' Patricia insists, however, that the decision to fire Hale was Joe's and Paul Mills's. In the event, Patti Love was brought in to play Josie and produced a sparkling performance.[17]

Losey had not worked with Gerry Fisher, his most prized and long-serving director of photography, since *Don Giovanni*. Fisher was ready to film *Steaming* but John Heyman refused him his customary credit on cinema posters and advertisements: advertising credits would be confined to director, script and music. As Fisher recalls it, when he asked sardonically, 'Who's writing the music, Beethoven or Mozart?' Heyman put down the phone. According to his own notes, Losey told Heyman on the telephone that he was 'pissed off' with Gerry Fisher and his 'preposterous' demands regarding billing . . . it might be time for a break with Gerry . . .' Douglas Slocombe, meanwhile, politely excused himself on the ground that he was a chronic asthmatic who could not face the fog of steam required in the baths. In reality he disliked both the play and the script: 'I thought it was a nasty, cheap thing for Joe to do, and I thought doing this will kill him. It did kill him. Sad to end on that note.' Christopher Challis, the excellent photographer of *Blind Date*, was brought in at the eleventh hour. In his view, the composite set turned out to be a cinematic disaster. 'We repainted it, but nothing would float, you couldn't move anything. It was wedged in with backing. The script described atmospheric weather outside, but it was impossible to get it from inside.' Losey was both ill and ill-tempered throughout.[18]

On the eve of shooting Losey told his cast and crew that 'In my experience cinema can be used in many ways: one of them is to increase enclosure rather than the Hollywood cliché of "opening up". This film, like *King and Country*, attempts that.' Losey and Challis achieved a delicate blend of water, skin, towels, tiled walls: greys, greens, pale oatmeal, the sharp turquoise of the plunge pool; snow and rain filtered through skylights and windows. Losey's notes refer to the painters Renoir and Modigliani, but the muted colours of the film are not the smouldering tones of Renoir. Perhaps the most memorable shot is of Josie (Patti Love), dressed in severe white and black for her coming ordeal as petitioner to the Council, rehearsing her lines in seclusion before a mirror in a shabby cubicle of sombre-textured walls. Her red lips burn with life and renewal – Renoir indeed.

Shooting completed, Diana Dors sent Losey a touching note of thanks with her gift. She died soon afterwards. The obituary which appeared in *The Times* was long, detailed and fair, praising her bravery while dying from cancer, but on 10 May Losey wrote to protest against 'such a put-down and casual dismissal of her work . . . a completely negative obituary'. Clearly unaware of Joe's own grave illness, Sarah Miles sent him a postcard from Nepal to report that she had fallen in love with a huge bull elephant called Shamshar. Three days after the card was posted, *The Times* printed an obituary of Joseph Losey.

The film opened on 7 June 1985 at the Classic cinemas in Haymarket, Tottenham Court Road and Chelsea, and at the Odeon, Kensington. Censorship (absurdly) banned children under eighteen; the language of women, not their nudity, was held unacceptable. David Robinson's review in *The Times* (7 June) typified the general critical reaction; he regretted that Losey's 'troubled and uneven career . . . ended on a dying fall . . .' The effect of the 'near view of the camera' was to make the original play seem more contrived and less sophisticated. *Steaming* was first shown in America at the New York Film Festival on 9 September 1985. Writing in the *New York Times* (9 November),

Vincent Canby called it 'a very small, unexceptional movie to end the exceptional career of Joseph Losey . . . not even a cast of first-rate actresses can give these dreary characters a dimension of interest that hasn't been supplied by the screenplay . . .' Losey's last film was shown out of competition at the Cannes Film Festival on 11 May 1985. The word '*hommage*' was used but the real homage that year was for the late François Truffaut. *Steaming* was released in Paris only in VO (Eric Kahane's subtitles). Every review was touched by sadness at Losey's departure; many recognized that he had been gravely ill when shooting the film. Michel Mardore commented in *Nouvel Observateur* (17 May) that Losey had treated the feminine condition with tenderness and humour, while ignoring the 'MLF' (or feminist movement) – but one wouldn't want to take this '*toilette finale*' for the testament of a great director. Last word to Michel Ciment: Losey, an artist 'reputed sometimes stupidly to have painted a universe of men', had produced 'a tender, sensual, warm portrait of seven women' in a classic, natural style.

The final reel had been shot and we may pause to take stock of the fluctuations in Losey's critical reputation. The 1950s had brought him little comfort in the press. The prevailing British critical fashion of the era demanded the toning down of emotion, the ballasting of the story line with realistic detail and location atmosphere. Successive BFI *Monthly Film Bulletins* treated Losey's films as potboilers – often 'ludicrously pretentious' or 'exaggeratedly hysterical'. His champion against all detractors was Raymond Durgnat, who attacked 'the sensitive plants of the *Sight and Sound* group', a 'clique' which had written Losey off but was now [after *The Servant*] forced 'to take seriously yet another director whom it scarcely understands'. Durgnat treated Losey as a philosopher-architect carving his own chosen tablets rather than as master-builder often unhappy about his commission and the quality of his stone. 'Paradoxically (given its "Neitschean" [Nietzschean] energy), there is no indulgence of amorality of any sort in Losey's work . . . There is a streak . . . of a (highly sublimated) moral sadism . . . Losey's achievement is in bringing to modern, to British, society the puritanism underlying the Western.' But whereas the 'fundamentalist puritanism' of rural America was fashioned out of sharp moral dichotomies, good and evil, lust and innocence, by contrast the dynamic or 'cubist' puritanism of a Losey made the soul 'a battleground' where victory was lucidity.[19]

The French critic Gilles Jacob also took up the theme of 'lucidity', discerning in Losey's work a desire to 'reduce everything to essentials . . . this champion of lucidity suppresses everything which might not seem to serve his meaning . . . Losey's world is not so much a world as a setting for an allegory . . . The question which still remains is whether the world should be answerable only to a metaphysic of intelligence; whether the sole aim of a work of art should be significance.' Losey himself was particularly alive to his own 'puritan' credentials when talking to the French – but rarely, if ever, to British or American critics. 'I come from a Puritan milieu and Puritanism never abandons you. With a Puritan formation everything becomes tragic.' Explaining *Eve* to a French audience, he referred to 'This Biblical influence of the "Bible Belt", this terrible Puritanism . . .'[20]

In the mid sixties there were two Losey seasons at the NFT within three years. In August 1969 Thames Television invited him to be one of 'six of the world's foremost movie directors' for a forthcoming series – though all of them English-speaking. The film was shown at 11.30 p.m. in April 1970. *Current Biography* (December 1969) named him England's 'foremost film director'. In 1974 Jonathan Baumbach described him as 'the most cinematic, and I suspect most important, of contemporary English directors, his films increasingly assured and personal despite the distancing device of their cool formalism'. An issue of *Cinéaste* in 1981 declared: 'Losey has to figure in any continuation of Leavis's Great Tradition, or the English Moralists. Hence Losey, unlike Hawks, Hitchcock or Ford, is a crucial figure, not only of film culture, but of high culture overall.' Edgardo Cozinsky regards Losey as one of the few stage directors to have evolved 'an original method in the new medium'; like Bergman, he 'enriched film language' and was 'one of the most important directors of all time'. Philip French's obituary referred to 'one of the great *œuvres* of our time'. In Michel Ciment's view, he was without doubt 'one of the most gifted directors of all time'.[21]

But Losey's increasingly self-indulgent symbolism provoked misgivings even among his admirers. Gilles Jacob lamented *Eve* and the last section of *The Servant*: 'But what is one to make of the clumsy plastering of symbolism in *Eve*, where the couple can't find themselves on a bed without eliciting the crudely obvious sexual symbolism of a fountain?' Losey was 'submitting to the mannerisms which nag away inside him . . . The concise, shock-cutting style of his American period has given way to a convulsive baroque . . . dry violence alternating with flowery extravagance . . .' As Cozinsky points out, Losey's symbols at their most effective are 'objective correlatives': the Elisabeth Frink sculptures in *The Damned*; the Royal Artillery War Memorial behind the credits in *King and Country*; winter Venice in *Eve*; the op art and comics in *Modesty Blaise*. Karel Reisz comments: 'What he did with *The Prowler* was what the script demanded – and I really liked *Accident*. Losey's peculiar sourness towards the English worked very well. But then he became a dandy.' Jean Pierre Coursodon noted that as his films became more decorative and self-consciously mystifying (as in *Secret Ceremony*), his alienation was no longer Brechtian but a distancing of himself and his characters from the world, reconstructing reality in terms of obsessive fantasies, particularly the masochism of master and slave. In David Thompson's view the later work suffered because Losey was taken, and took himself, too seriously. Applause 'lured into the open Losey's uncertain yearning for signifi-cance'.[22]

The opposite point of view was most eloquently expressed in France, notably by Michel Ciment. Interviewed in *Télémoustique* (15 February 1980), and asked why Losey's films had not penetrated '*le grand public*', Ciment responded that they were not 'sufficiently demagogic. The public (and the critics as well) are most easily seduced by what's facile or by what flatters them. To tell the truth is always uncomfortable, above all to say it with profundity, honesty and in all its complexity . . . technique, research, thought, knowledge, all this vexes people. They have the impression of not belonging to a club of initiates.' But was Losey's later work not guilty of coldness? 'No,' responded Ciment, 'let's say simply that emotion is held at a distance. Losey's cinema constitutes a sort of equilibrium between sensibility and reflection.' For Ciment the supreme example was *Mr. Klein*: 'One doesn't know at any moment whether Klein is a Jew or not. That's typically Losey . . . If he'd been content to show the fate of a Jew under the Occupation, the film would have had 200 or 300,000 more spectators. But no, he preferred to be a dialectician . . . The majority of people are absolutely resistant to the dialectic, they're religious, they have need of a faith, no matter what, and of certainties.'

Thomas Elsaesser offered a final summation: Losey 'belonged to the very few great directors of world cinema' yet he remained at the margin of both the commercial film industry and the art cinema: 'Hence the anger, frustration and occasional self-pity of a man who had to spend much of his time in the ante-rooms of people he despised.' Elsaesser then offered a somewhat daunting critical vocabulary to convey the best possible case for Losey's later work: Losey's claim to be an 'auteur' was a matter of 'hermetic closure, self-reference, or the accumulation of the signs of a private ideolect . . .' Here, then, was a 'serious attempt to stay on the surface, constructing contexts and subtexts so intricate and dense that the "auteur" would disappear, or at any rate hide, in order for the work to be everywhere in its echoes but nowhere in it essence'. The critic Richard Combs may have been saying the same thing, but less charitably, when he wrote: 'It is hard to resist the impression that, rococoist to the end, Losey spent his last ten years adding curlicues, filigree work, thematic and stylistic whorls to what he had done before . . . a sudden sense of introversion . . .'[23]

Yet Losey's film work cannot be assessed simply in terms of chronological periods. *The Go-Between* came after *Secret Ceremony*; *A Doll's House* and *Galileo* came after *Boom!* and *Trotsky*. It was always a question of the particular script: offered a screenplay of genuine quality, he cut down on the flash and filigree. One cannot doubt that his *Voss* would have been a masterpiece had the old man been permitted to make it. The Proust project is another question . . .

45

He's Just Passed Us

Losey had first been approached about the disposal of his archives by Gene M. Gressley, the director of the University of Wyoming's Division of Rare Books and Special Collections. Gressley made two inquiries, in July 1970 and again in September 1972. Losey responded negatively in December of that year.

The University of Wisconsin's Center for Theater Research had also expressed interest in March 1969. In January 1974, Tino Balio, director of the Center, confirmed his interest but regretted that no funds were available.

Losey immediately reopened the correspondence with Wyoming. Gressley explained that the university would like to make a bid but doubted whether it would be competitive: no sum was stated. In April 1980 he invited Losey to put a value on the collection. Losey told Gressley that ten years earlier it had been valued at $50,000 but it was now more extensive and valuable. In September Gressley offered $45,000. A month later Losey proposed that such a price would exclude all of his film prints and any future material post-*Don Giovanni*. On 14 November, Gressley politely withdrew the offer in view of the difficulties.

In May 1982, at Cannes, the director of the British Film Institute, Anthony Smith, proposed to Losey that he should deposit both film prints and papers with the BFI in London. What transpired was a gift not a sale, the BFI contributing £5,000 towards the expense of collating and transporting the papers and stills. Any material that Losey specified as confidential 'and any personal correspondence to or from you [Losey]' was not to be available without the authorization of Losey or his heirs. In practice, some papers were placed in separate folders as 'Private and Confidential', but the distinction between business and personal correspondence proved to be unsustainable.[1]

According to the terms of the donation, Losey transferred to the National Film Archive of the BFI the eight prints he owned, but which were now deposited with the Cinémathèque Suisse in Lausanne. But he almost certainly felt embarrassed about asking Freddy Buache, Director of the CS, to hand over prints which he had only recently acquired from the Cinémathèque Française. Anthony Smith, too, did not wish to offend Buache: '. . . we therefore feel that any instructions for the transfer of the prints to us would be better issued by yourself,' he wrote to Losey. After Losey's death Smith appealed to Losey's solicitor, Laurence Harbottle: 'It is very embarrassing for us to consult another archive over a matter of this kind . . . It may be necessary for you . . . to be quite insistent if these materials are ever to be brought to London . . .' No transfer has taken place and the CS currently holds six prints not in the possession of the NFA.*[2]

By the last year of his life, Joe's marriage to Patricia was worn to the bone. For years the victim of his rages and insults, she had become increasingly impatient with this old and very sick man; visitors to Royal Avenue found her frosty and removed from him. She now had her own quarters in No. 27. Friends likened him to a bewildered child. Patricia was herself unwell: low-level hepatitis and hypertension. Between April and November 1983, she had twenty-three consultations with Dr Barrie Cooper in connection with what his account described as 'DaCosta's syndrome', at a cost of £1,400, including drugs. Writing a new screenplay for *Steaming* put her under exceptional stress. Joe told Harold Pinter how one day he'd been moving about the house, maybe in the kitchen, when Patricia appeared and asked him what he was doing.

> [STEPHEN: What are you doing?
> ROSALIND: I'm making the tea.]

Patricia informed him she couldn't write while he was shuffling around, so he took himself out and sat with the Chelsea Pensioners (Royal Hospital being the opening shot of *The Servant*). Patricia comments: 'It irritated Joe that I was busy writing *Steaming*.' As latterly in Paris, he made frequent calls to friends and sought comfort with old flames, notably his first British girlfriend, Jill Bennett. Patricia comments: 'He visited Jill only during the day. She rang up one night threatening to kill herself. I told her to go ahead. She and I laughed about it the next day.'

Between August and November 1983 Joe received eighteen visits from a State Registered Physiotherapist; standing and walking on his damaged knee was an agony – a special lift for him was constructed on the set of *Steaming*. Loring watched a day's shooting: 'It was unbearable, the conditions of that sound stage, the smell of rancid, moldering mildew.' On 7 April the Princess Grace Hospital conducted an X-ray diagnosis. A chartered physiotherapist saw Joe eight times between 11 and 27 April. Having finished the shooting, post-syncing, mixing and editing, he dragged himself to New York to show the final cut to John Heyman in Paramount's screening-room. 'Joe insisted on showing me the film before it was dubbed,' recalls Heyman. 'He was coughing and wheezing. There was no need for him to come over.' From the

---

Mr. Klein, Les Routes du sud, A Doll's House, The Assassination of Trotsky, Figures in a Landscape and *Don Giovanni*. In addition the CS's print of *Secret Ceremony* is probably longer than the NFA's, listed as 'short version 91969 feet'.

Regency Hotel Joe called Jack Loring: he wasn't capable of getting dressed or leaving the room – would Loring call room service and order a lot of things he might like? 'He couldn't swallow. I got masses of oysters and vodka. He tried to eat. He said, "I'm sicker than I've ever been. This time I'm really sick."' The horror of the journey back to London, alone, can be imagined. He visited the London Clinic, Devonshire Place, for an endoscopy, and had an ultrascan of the liver at the Imaging Centre, Harley Street. For six weeks all tests were negative. Then TB of the stomach was diagnosed – curable. The British United Provident Association met the bills, but the frontiers of insurance were reached. On the day of Joe's death, BUPA International set limits to the cost of treatment: in Joe's case this related to 'muscular degeneration, osteo arthritis and carpal tunnel syndrome'; in Patricia's case to 'psychiatric disorders'.

Nell Dunn, who loved Joe's informality, his intimacy, his courage, visited him in the Princess Grace Hospital when drugs were producing frightening hallucinations. She read Yeats to him and was struck by his rage at dying. Moira Williams, often at his bedside, read A.E. Housman: 'He was deeply moved. He didn't know much poetry'. (Lines from Housman are heard early in *King and Country*, voice-over.) In general, she recalls, he was 'furious about the mis-diagnosis. They'd thought it was hepatitis and he'd even given up booze. He said he felt "So fucking undignified".' He also expressed deep regret that he was leaving nothing for Gavrik and Joshua. But when Richard Dalton found Gavrik at his father's bedside, and excused himself, Joe waved Gravrik out and Dalton in: 'I've got nothing more to say to him'. Joe hadn't entirely lost his touch.[3]

On Monday 18 June, they brought him home to Royal Avenue. It was cancer and no more could be done. He needed help in getting to the toilet and Dalton could not support his weight. Whenever he went up to the top floor bedroom, Joe would clutch his hand, not wanting him to leave. 'Something's wrong,' he gasped.[4]

The morning of Joe's death, Friday the 22nd, Dirk Bogarde telephoned to inquire; Joe died while he was on the line. Dalton joined Gavrik, Joshua and Joe's secretary, Victoria Bacon, in the ground-floor dining-room. Gavrik inquired about the will and the funeral arrangements. Patricia – 'I was still in charge' – asked them all to follow her out to the tiny, paved garden while Joe's body was carried out to the hospital van (he had bequeathed his eyes to science – fine, but painful to those left behind, since the surgery had to be rapid). No one followed her. Later they did gather in the garden. According to Loring, Gavrik had removed Joe's watch from his wrist and now offered it to Loring (who says that Joe had more than once offered it to him, but he'd demurred). Loring urged Gavrik to keep it. Patricia's account is different in emphasis: with a gesture that she found presumptuous, Loring offered Joe's

Cartier watch (a gift from Elizabeth Taylor) to Gavrik. Moira Williams recalls that later that day Patricia telephoned her and unburdened herself about the watch. Under pressure, Gavrik handed it back to her. That evening Jack Loring stood them all to dinner in the English Garden restaurant, 'to get us all as drunk as possible'.[5]

Losey's will, dated 9 April 1980, gave his whole estate to his wife and requested her (without imposing any legal obligation) to bequeath two-thirds of what she inherited from him to his sons and one-third to her own children. But of course the overwhelming balance of the Losey estate – the two houses in trust (worth £400,000 each), and the contents absolutely, had been hers since their marriage in 1970. Despite an £82,000 fee for *Steaming*, Joe's estate was valued at £1,600 – bad news not only for the old enemy, Inland Revenue, but also for Gavrik and Joshua Losey.

John Heyman made a two-way flight from New York on the day of the funeral. He and Richard Dalton arrived at Putney Vale crematorium half an hour early, not a soul in sight. They studied the gravestones. A hearse came and parked itself discreetly. Dalton said, 'I think he's just passed us.' Joe's will was adamant: no religious dimension and therefore no priest. But who can replace the priest, the binding and presiding comfort? For Dalton the non-religious, non-service was unbearable, an aching succession of empty spaces. Joe's last secretary, Victoria Bacon, who had insisted the chapel Cross be taken down, played a tape of music from Joe's films compiled by Richard Harvey. Nell Dunn recited the Yeats poem which Joe had so much appreciated in hospital. But there was no eulogy. Heyman found the atmosphere 'very cold – yet Joe himself was a hairy, huggable bear. It was awful – no feeling of warmth or caring.' Patricia placed a wild rose on the coffin, which then departed on its conveyor belt. No one moved. The doors opened.[6]

Pamela Davies, Joe's faithful continuity lady, herself fatally ill, sent Bogarde a long monologue on a cassette, full of loathing and vituperation concerning the funeral arrangements. After a while Bogarde could listen no longer and destroyed the tape.[7]

Bogarde's inspiration was the much larger memorial gathering held at Shepperton Studios in September. Joe's world convened, romantics and sceptics, to eat, drink and talk. To Patricia, Bogarde later wrote: 'Whatever we all felt, thought, or sensed, happiness rather than gloom was the order of the evening, and I am certain that Joe would almost have approved . . . And everyone spoke of him with such terrifically warm pleasure that he HAD to be a decent bloke underneath all the humph and huff and blowing.' Bogarde was delighted that there had been 'No weary Knight bleating from the pulpit; no quotes from John Donne . . . Just an end-of-picture-party for a most particular man.'

Cut.

PART XIII

The Films: A Viewing

46

The Films
in Part III

THE BOY WITH GREEN HAIR, THE LAWLESS, THE PROWLER, M

The Boy with Green Hair

The Boy is structured on long narrative flashbacks. We start at the end of the story in a police station where two cops are having no luck getting sense out of young Peter (Dean Stockwell), who has no hair at all. The kindly Dr Evans (Robert Ryan) offers the boy his hamburger and milkshake, then invites him to begin at the beginning. Peter's memories focus on a candlelit cake, a Christmas tree, the present of a dog, and his father's hands carving the roast: a series of comforting Christmas-card images of a good American home.

Peter's voice-over narrative explains that his parents had gone away (in fact to help orphans in blitzed London). We see a telegram arrive; Peter is not informed of their deaths. A succession of houses indicates that Peter has been shunted from one aunt and cousin to the next. Finally he settles in with an old Irish vaudeville actor (Pat O'Brien) who likes to be called Gramp, and who now works as a waiter in a faded tuxedo. There follows an imagined scene, related by Gramp to amuse Peter, set in a fantastical theatre dressing-room, of Gramp being congratulated and decorated in distant days by a clownish Ruritanian King. The wall is purple; Gramp wears eighteenth-century breeches, the King sixteenth-century hose and spectacles. They perform a song-and-dance routine.

Gramp must go out to work at night, leaving Peter alone. Peter declares himself unafraid of the dark – then lays his precious baseball on his bed. He's a plucky little boy.

Morning. Gramp introduces Peter as his grandson to neighbours and tradesmen. Like the aunts' houses, this is fairytale 'typification': the milkman, barber, doctor, grocer. Several pat the boy on his nice mop of hair. Group psychology is indicated by repetition. After some hesitation, Peter is persuaded by Gramp (a soggy-faced bore) to enter his new school, where he meets the kind, beautiful and progressive teacher, Miss Brand (played by Barbara Hale in what turns out to be an almost wordless part). The other kids stare at the newcomer – in succession.

The school stages a war-orphan drive. The gymnasium walls are covered with posters and pictures of children hurt by war: 'Remember Greece', 'United Jewish Welfare Fund', 'United China Relief'. A classmate innocently tells Peter that he, Peter, is also an orphan, citing Miss Brand. Unaware of his parents' deaths in the London Blitz, Peter angrily attacks the boy. Gramp and Miss Brand are compelled to inform him of the truth, though Gramp as always does the talking. 'You should be proud of them. They stayed to save little boys and girls.'

Peter, who is earning money as an errand boy for a grocery store, overhears ladies discussing war and peace, preparedness or international collaboration, mainly in voice-over: this was the contentious scene at RKO. Back home, Peter asks Gramp if everyone will be killed in a new war – then picks up a green potted plant and places it on the dining-table. This is obscure.

Flashforward to Dr Evans and the police station. 'You won't believe the next bit,' Peter says. 'I was taking my bath . . .' – we see him in the bath lathering his hair in soap. Then he dries his hair and looks in the mirror: his mop of hair has turned green. Chirpy string music. He makes faces at himself in the mirror. There follows routine comedy as Gramp makes breakfast and doesn't turn round to see the green hair, then the milkman doesn't notice the hair, then he does and drops all his bottles (off screen). There follows another 'set-up' sequence in which everyone in the small town gapes at Peter's green hair. In the school playground the normal noise-level subsides as the kids crowd round Peter, laughing, jeering – the stock confrontations. Miss Brand is angelic: don't some of us have black hair, some blond, some brown, some red – and some green? Each category is invited to raise its hands (more of the one-after-another routine). Ostracized, Peter goes home and tears up a sealed letter from his father 'To be read on his sixteenth birthday'.

With tears in his eyes Peter runs out of the town into lonely countryside. To heavy strings, he crosses a glade then lies down and cries. He hears a voice: 'Peter, Peter.' Looking up he sees the same ragged – but not very ragged – war orphans who appeared in the posters on the gymnasium walls. The tall one, Michael, says: 'It's the boy with green hair . . . We were waiting for you. Your green hair is beautiful. It means Hope. People will take notice of you. There is a reason you-did-not-know' (the dead speak slowly in flat tones). The orphans are posed round an old brick chimney with a lighted fire – another studio set. Losey later recalled: 'What I wanted to get . . . was absolute horror, real terror, the kind of thing Joris Ivens and John Ferno did in their film of the Chinese–Japanese conflict, *The Four Hundred Million* . . . the dreadful starvation, the concentration camp feeling . . .' But this was a family film.[1]

The orphans instruct Peter to spread the message: 'War is very bad for children . . .' Peter rebels: 'I don't want to be different. I want to be like everyone else.' In a rapid Brechtian touch, Losey has Peter turn straight into the camera.

Back in town, more stock set-ups as Peter delivers the orphans' message to one citizen and friend after another. The music swells. But then Gramp receives protests and finally Peter submits to the barber's chair while the townspeople gather to watch. Close up of electric shears. Green hair falling. Now everyone looks guilty. At this juncture a didactic intervention, typical of the script, is cut from the film. Enter Miss Brand: 'They didn't know. They didn't understand. The truth has never had an easy time – never – not even two thousand years ago . . .' (As in Losey's radio days at NBC, Marxist scriptwriters continued to pitch for Jesus, but the passage was cut.)

Back home Gramp offers Peter warm milk but the boy feels betrayed and leaves by night – which brings us back to where we started, in the police station. Seated in the next room are Gramp, Miss Brand and the town doctor. Gramp produces the torn letter from Peter's father and reads it aloud. The message is sugared, portentious, and everyone is moved. 'Remember us as having died for something fine and worthwhile – if those who did not die will not forget. Remind them, Peter.' The last shot shows Peter arriving home arm in arm with Gramp to the sound of 'Nature Boy'. The running time is only 82 minutes.

The Lawless (*The Dividing Line, UK*)

'This,' the opening caption announces, 'is the story of a town . . . in the grip of blind anger and [which forgets its] American heritage of tolerance and decency.'

Plantation immigrant workers – disparaged locally as 'fruit tramps' – are queueing for wages from their employer, Jim Wilson (Walter Reed). Wilson wants our two young heroes to put in some weekend work on time-and-a-half. They refuse. 'Lazy good-for-nothings.' A convoy of old cars leaves the plantation. We join the topless, battered roadster of Paul Rodriguez (Lalo Rios) and Lopo Chavez (Mauricio Jara). 'I wish I had forty acres.' Lopo has served in the army: 'I forgot what I was coming back to.' They pass a sign: Welcome to Santa Maria the Friendly City.

At an intersection they collide with a convertible driven by two young Americans, Harry Pawling and Joe Ferguson (John Sands), who immediately become abusive: 'Can't you read English? All you Cholos are the same – no licence, no brakes, no insurance.' The four boys fight. A decent motorcycle cop arrives and has to book Lopo, who was at fault for the accident.

Lopo visits the modest press-room of the local Spanish-language newspaper, *La Luz*. He reminds our heroine, the reporter Sunny Garcia (Gail Russell), to make sure she comes to the dance that evening.

Paul arrives home in the district known as Sleepy Hollow. His parents are of Aztec ancestry. His father warns him to steer clear of Americans and to stay out of fights. Paul ('Pablito') protests that he too is an American; in his bedroom we see a picture and medal ribbons of his paratrooper brother, who died a hero in Normandy. Paul goes outside to the back yard, bounded by a rail track, and washes himself in a primitive, lean-to shower. Losey then shows his style by cutting from one shower to another – Joe Ferguson is seen as a silhouette in an ultra-modern glass cubicle (perhaps a pre-figuration of Tony's distorted shower-silhouette in *The Servant*).

Joe's father (John Hoyt) walks up the drive of his large garden – the social status of the two families is neatly contrasted. Mr Ferguson, clearly a liberal, is sceptical about Joe's account of the fight. The boy expresses scorn for 'fruit tramps'.

In the spacious, well-equipped office of the local newspaper, *The Union*, the new editor-proprietor, Larry Wilder (MacDonald Carey), is bantering with a prejudiced reporter, Jonas Creel (Guy Anderson), whom he has assigned to cover the 'Good Fellowship' dance which Creel predicts will be 'a fruit tramp brawl in Sleepy Hollow'. Larry decides to cover it himself. Leaving the building, he switches on the *Union* sign outside.

Larry joins Sunny, whom he doesn't know, in the line outside the dance hall, offering her his spare ticket: 'No strings attached.' They dance to sambas and rumbas. Sunny expects trouble and hints that Larry is here simply to witness it. Harry, Joe and the white boys arrive. Soon there is a massive racial brawl. Losey's first spectacular crowd shot comes up as everyone spills out of the hall into the night. Paul Rodriguez hits a cop by mistake. Panicking, he hijacks an open ice-cream wagon then steals a car. Ham music – the score was by Mahlon Merrick – merges with the wail of police sirens as the desperate Paul drives through the night.

Larry enters City Hall and finds the worried parents of the arrested boys of Sleepy Hollow huddled within. A lawyer is urging them to plead 'Guilty'. The journalist Sunny rebels on their behalf; she wants justice, not racial justice. Larry takes the worldly view – no point in going to trial. (When the hero finally sees the light, it will be High Noon.) Larry follows the simmering Sunny outside. She accuses him of easing his conscience. 'If you'll take the chip off your shoulder,' he replies, 'I'll drive you home.' Now Sunny conveniently outlines Larry's distinguished journalistic career, which she apparently knows by heart, country to country. Has he now opted out? Larry, tired of fighting for causes, explains that he's looking for the peace of a small town, for the smell of burning leaves in October and fresh ploughed fields in March. More immediately, he wants to kiss her.

The police car carrying Paul skids off the road and catches fire. Paul runs off into the night. Cut to police radios and reports in Santa Maria. 'FRUIT TRAMPS RIOT', announces a newspaper headline (not *The Union*'s of course).

Out-of-town reporters are using *The Union*'s telephones. One of them, Jan Dawson (Lee Patrick), a formidably cynical woman, is pouring prejudice down the line. Later she will reveal a nice turn of humour.

The chase continues: fire engines, ambulances, sirens, Paul running through the night. Reaching a farm, he hides in a barn. Sixteen-year-old Mildred, doing her homework, hears her dog barking, goes outside, takes fright when she sees Paul, strikes her head on a door jamb, faints. Cut to Mildred, head bandaged, surrounded by reporters eager to hear that the fugitive Paul has attacked her. Such a story suits Jensen, her bigoted farmer-father; he enters the ranks of villains and vigilantes alongside farmer Wilson. Cut to a crowd in Santa Maria watching Mildred being interviewed on television. The TV reporter is a prejudiced demagogue, his commentary inflammatory. The TV station puts up a $500 award for the capture of the fugitive Paul Rodriguez. (Television companies refused to lend their equipment and Losey had to fake the TV cameras.)

In Larry's smart proprietor's office the solid-citizen fathers are lobbying the editor not to mention that their sons were involved in the dance brawl.

Dredged-over bottom land: the manhunt. Windrows of cobblestones long ago spewed out by dredger. Clumps of willows and cottonwoods. Police, farmers with shotguns. Rotting, rusting dredging equipment in stagnant pools. The hunters close in on the desperate fugitive to the sound of feet tramping on shale – brilliantly done. Larry turns up, pursues Paul, who is stumbling through a river, past a crane and a bridge – Losey's eye at full stretch – and gets shot at by mistake while climbing on to a barge. He finds Paul cowering and weeping.

The camera moves in on the TV camera finder, then shows Paul's capture by sheriffs as viewed through the finder, an image which becomes less sharp as we see it broadcast on television in the main street of Santa Maria before a large crowd.

A cortège of police cars brings the captive Paul to the DA's office. The mob jostles to get at him. The motions of the camera are clean and urgent. Losey photographs Paul's shadow on the police-cell wall as an identification tag is put round his neck.

At Sunny's request, Larry uses his influence with the police chief to allow Mrs Rodriguez to see her son. Here Losey achieves the subtlest motion of the film, first pulling back to show shelves of law books outside the DA's office, then Paul's parents passing between them, then the door of the DA's distant but brightly illuminated office opening to admit them – then closing. Sunny kisses Larry in gratitude.

Time for some writing. Larry spins his editorial chair then starts dictating while Jonas Creel, wearing his perennial hat and chewing gum, types out Larry's sentimental story about Paul, a good kid whose brother died 'on the Normandy beach head', who works in the fields and takes his pay home to his good mother. The presses run, newsboys race towards the crowds still gathered outside City Hall. Larry is found in his empty offices, taking the phone on desk after desk as the abusive calls rain in. Led by Farmers Jensen and Wilson, the mob closes in on *The Union* – spectacular street shots, cars overturned, an urgent choreography of bigotry: the crowd coming from behind the camera and surging past it; the mob chasing Lopo and a friend through a park; the approaching mob observed through the reversed lettering of the *Union* sign. Losey's direction is spectacular, although the long sequence in which the newspaper's offices are ransacked – tables overturned, typewriters hurled, windows smashed, linotype and press destroyed – feels contrived.

Returning to total devastation, Larry comes upon Wilson still smashing windows and slugs him unconscious. He finds Sunny lying in the debris and lifts her into his arms. She's fine!

Addressing the now-ashamed townspeople from the high steps of *The Union*, Larry tells them he's leaving town. 'Try to find some shame.' They drift away rather theatrically, according to their given cues. Larry wanders through the wreckage, glass crunching under his feet. Enter Ferguson, anxious to dissuade him: 'We need someone to yell at us when we forget our decency and look the other way. Don't run out on us, Larry.' (Of the Ferguson character, Losey commented: 'He was my two grandfathers. He was the kind of person you find in every small town – absolutely incorruptible and capable of sacrificing what he has for what he believes.')

Larry arrives at the *La Luz* print shop and announces that he wants them to print *The Union*. As Sunny moves into his arms, her crusading character is reduced to the final destiny of Romance.

The Prowler

A police motor patrol consisting of two officers, Webb Garwood (Van Heflin) and the honest Bud Crocker (John Maxwell), receives a call one night from a pretty housewife frightened by a prowler in the garden of her handsome suburban residence. Susan Gilvray (Evelyn Keyes) is married but childless and idle. As soon as he enters her home Garwood signals his cynicism by his loose gait, his knowing grin and by the way he picks up a framed photograph. Later that evening he returns to the house, without his colleague, on the pretext of checking up. He discovers that Susan's wealthy husband is a late-night radio commentator whose voice he knows well. Every sequence involving Heflin and Keyes until the murder of the husband is shot at night, under bright artificial light, an extended metaphor for Susan's vulnerability and Garwood's predatory power as he eases her towards infidelity.

We hear Gilvray's high-pitched voice riding the night-time radio waves, selling, cajoling, comforting. Its seductive patina, on behalf of this product and that, forms an ironic counterpoint to the seduction of his wife: 'As for seasoning – well – tell you, my wife likes nothing better than fresh, garden-grown parsley – and for just that right touch of contrast – a few of our ice-crisp hearts of celery . . . The cost of living is going down. I'll be seeing you, Susan.'

On that dependable and homely note Gilvray always signs off. Officer Garwood absorbs the situation and makes his calculations. Discovering that both he and she are from the same school in Terre Haute, Indiana (he was a football player of failed promise), he says bitterly: 'You had lawns and sidewalks out in front. I lived down on Covington. My old man's idea of success was a buck twenty an hour – union scale. He was a maintenance worker – in the coalfields. He had a dozen chances to cut loose and make himself some *real dough* – wildcatting. He was too yellow to risk his buck twenty an hour. He never made it.'

When Garwood runs out of cigarettes, Susan mentions that her husband keeps his in a locked drawer. Plucking a pin from Susan's hair, Garwood smilingly opens the locked drawer, discovering not only cigarettes but also Gilvray's will, leaving to his wife $60,000 in cash and securities, the house, and a $20,000 investment in his brother's drug store. (But why does the wealthy Gilvray lock up his cigarettes in the same drawer as his will?)

Garwood and Susan become lovers. Although the production code did not allow an explicit depiction of adultery, by convention a camera dropping down from a standing embrace signalled that sex was imminent.

Premeditated murder follows. We suspend disbelief. Garwood arrives stealthily by night at the Gilvray home, makes prowler-like noises, and shoots dead Gilvray when he appears on the front porch nervously clutching a tiny gun. Garwood pleads a tragic accident – he had mistaken the owner of the house for 'the prowler'. Susan is torn between scepticism and love; at the inquest she hesitantly subscribes to Garwood's story that they had never met before the night of the killing. Garwood's co-patrolman Bud Crocker listens incredulously, and indeed the coroner has the first visit of Garwood and Crocker to the Gilvray house on record – yet no further investigation is set in motion.

Trapped between love and grief, Susan withdraws into a widow's isolation, but Garwood's slick and brash charm prevails. Susan confesses to him that she's four months pregnant, her intention being to pass the baby off as her late husband's – but Garwood, having spoken to John Gilvray's brother (Emerson Treacy) in his pharmacy, has learned that Gilvray was not able to have children. Garwood presses Susan to marry him without delay and to join him in concealing the timing of the baby's birth. Losey particularly remembered 'the beginning of what critics like to call my "baroque" style' – a wedding scene with a funeral taking place across the street in another church. 'That's all shot in one take, on a crane . . . and it's a 180 degree shot.'

Sheer vitality enables Losey to transcend the awkward improbabilities of the plot. The urgency of the direction and the acting in Garwood's newly acquired motel, where the couple spend their wedding night in single beds, is compelling. Car lights flicker through the motel curtains, an ambulance wails, and cheap modern comfort mocks the couple's guilt. To escape scrutiny, they leave for a remote ghost town in the Mojave Desert. For a while they are happy honeymooners and Garwood's affection for Susan seems genuine, but a mistake with a record player, and the sudden sound of her dead husband's voice, throws Susan into premature labour, gales lash the shack, dust swirls across the desert, and Losey's cinematic language – a symbolism both of style and detail – displays its power. Now the ugly gun-carrying side of Garwood resurfaces – he is even prepared to kill the doctor whom he is obliged to summon after failing to cope with the delivery himself! (But how could one murder be concealed by yet another?) Prostrate and fearing a miscarriage, Susan comes to realize that Garwood murdered her husband, Gilvray.

> GARWOOD: So what? So I'm no good. But I'm no different from anybody else. You work in a store – you knock down on the cash register. You're a big boss – the income tax. Wardheeler – you sell votes. A lawyer – bribes. But whatever I've done I've done for you . . . How am I different from those other guys? Some do it for a million, some for ten . . . I did it for sixty-two thousand.

Losey regarded this 'moment of truth' as what the film was all about.[2] For Garwood the game is up. Losey has Van Heflin desperately rummaging for the spare key to his car while layers of ambition, greed and despair are verbally stripped away. Escaping by car, Garwood finds the narrow road blocked by a beaming, slow-witted Bud Crocker who has driven out with his wife in search of algae. The two cars are locked like butting stags as tyres spin in the dust. As the cops appear on the horizon, Garwood turns and scrambles up an almost vertical pyramidical mine-tip down whose slippery surface he repeatedly slides. The shots ring out.

M

Huge, harsh music strikes like a fist. The credits come up against a succession of subtly composed images. The lonely M (David Wayne) is seen always from behind, his face concealed beneath a twisted trilby. He shadows a child into a washroom; he stands over a child drawing with chalk on the sidewalk; he prowls a seaside fairground, he rides in a funicular. Newspaper headlines whirl – CHILD KILLER SOUGHT. Oblivious to every reality except his own obsession, M climbs into the funicular over a bundle of newspapers.

M's image is caught in an amusement arcade mirror; he looks out of the window of a shoe-shine parlour; in these opening sequences he occupies only a peripheral position within the frames, a man evading scrutiny, caught at oblique, furtive angles.

Little Elsie Coster emerges from school a fatal fraction after her usual companions. She plays with a ball as she walks. M picks it up, she smiles. When they reach a street carnival, Elsie rushes to a blind balloon-seller. M buys her a balloon, his smiling face visible for the first time. The blind balloon-seller takes note of the sound of M's enticing flute – the pied piper – as he leads Elsie to her fate.

We see children's feet on a vacant lot. The children are singing: 'The man grabs the child . . . the man stabs the child . . .' Overlooking the lot is an apartment block. Elsie's mother, Mrs Coster (Karen Morley), shouts down angrily at the frivolously singing children; moments later Losey provides spectacularly expressionist shots of the apartment block as Mrs Coster descends the staircase in a panic.

At this juncture the film changes gear. The intensity of Losey's direction yields to the documentary obligation – stock-typical scenes of suspected child-murderers being waylaid and mobbed in the streets. Despite Losey's fast eye, we are back with the sequential sociology of *The Boy with Green Hair*.

Alone again with M, Losey's intimate genius revives. In a remarkable sequence shot in half-light, M lies on his bed in a vest, a Beethoven piano concerto on the radio, his charged fingers intently fastening a child's shoelaces to the string of the bedside lamp, his face in shadow. Symbolism intensifies as he carries a plasticine model of a child to the dressing-table, on which stands a photo of Mom, her solid features set in an inescapable reproach. He gazes at himself in a mirror as if unable to identify or understand the reflection. Pathetic and fevered, M winds the string round the neck of the doll against the relentless ticking of a clock – until the plasticine head is severed. (In 1976 Losey referred to 'the scene that has been chopped up in most prints where he masturbates using a shoe string . . .')[3]

Back to the stereotypes. The populist mayor (Jim Backus) wants a quick arrest and hectors the police chief (Howard da Silva) to get results. The police raid various bars and sleazy joints looking for the child-killer: gangland's boss Marshall (Martin Gabel) becomes alarmed. In a sleek, modern boardroom (a contrast to the seedy, smoke-filled mob dens filmed by Lang), Marshall briefs the cinema audience by the tired expository expedient of telling his colleagues who and what they are: 'We're in a business with a take of millions (*points to Riggert*) from your race wires and book making (*points to Sutro*), from your gambling houses (*points to McMahan*), from your baseball and football pools (*points to Pottsy*), from your slot machines.' The mob enlists its satellite Ajax taxi service to find the killer.

M is found playing his flute on a park bench against a panoramic view of the city – an extraordinary shot which contrasts with the ensuing return to the documentary mode as a doctor-psychiatrist applies word-association tests to a variety of suspects recently released from psychiatric hospitals. M is among them. 'Committed State Hospital. Released 11/7/49.' The police raid his apartment but he is out. A detective finds a child's shoelace tied to the lamp string. Rows of children's shoes are discovered in the wardrobe.

M possesses a knife with which he carves figures of children out of soap, before squeezing them to pulp. Approaching a small girl he produces the same knife – then uses it to skin an orange for her, in a continuous 'snake'. The summer sky is relentless. In an empty café garden we find M moved by the plight of an injured bird. He cradles it then spills his whisky and begins to sob, his head on the table. M's fatal mistake occurs when he approaches yet another little girl (Janine Perreau) buying a balloon from the blind balloon-seller who, recognizing M's flute, sets in motion a chase involving relays of street boys. Taking the little girl with him, M descends a spectacular flight of steps then runs through a tunnel and into a vast warehouse built round an open courtyard at ground level, each floor flanked by elegant iron railings and joined by an old iron elevator. The gang and its minions come in pursuit – as in Lang's version, the chase is interminable. M is reduced to a sweating, hunted refuge-seeker locked in a storeroom packed with hosiery, dolls, naked clothes dummies, and an innocent child. Losey's night photography is remarkable – the sharp shadows, the silhouettes of trilby hats, panoramic shots of mobsters thundering up the great staircases, then dragging M out on to the roof.

They take him to a large underground garage. Painted on the wall above the central exit ramp is a prominent sign, KEEP TO THE RIGHT. Lang's M remained silent throughout the Kangaroo court scene. Losey wanted a long speech of self-explanation from the killer. Sitting on the concrete ramp, his face bruised, M explains, pleads, shrieks, weeps – with the camera positioned slightly below him, so that his legs and feet remain visible. This is a continuous take lasting more than four minutes. The final plea was the work of Waldo Salt, 'who did most of the writing, whatever the credits say,' Losey added. 'And when David Wayne finished his speech, the whole cast – all extras, all the crew – applauded him.'

The Big Night, Stranger on the Prowl

See Chapter 11.

47

The Films
in Part IV

THE SLEEPING TIGER, A MAN ON THE BEACH, THE INTIMATE STRANGER, TIME WITHOUT PITY, THE GYPSY AND THE GENTLEMAN, BLIND DATE, THE CRIMINAL, THE DAMNED, KING AND COUNTRY

The Sleeping Tiger

A quick shot of a dark alley. A dimly glimpsed mugger with a gun has it twisted from his hand by the intended victim.

A taxi pulls up to a large, ugly, mock-Tudor residence. Mrs Glenda Esmond (Alexis Smith) arrives home; she's 'Madam' to the maid, Sally. Frank Clements (Dirk Bogarde) is glimpsed in a mirror – it appears that Glenda's husband, Dr Clive Esmond (Alexander Knox) has not only overpowered Frank (the young mugger) but managed to frog-march him home. The cool, tall Glenda listens sceptically as her husband explains his motives for this improbable house-arrest: 'It's a complete waste of a human being. If he goes to prison again, he's finished.' Frank's prison record (mysteriously acquired 'by research') reveals a long history of robbery with violence despite a respectable family background (the army) and a good school.

Glenda, who owns a Jaguar, informs the insolent Frank that she'll teach him to ride horses or to sing hymns but she won't tolerate his rudeness. She takes him riding but Frank is too stung by her contempt to bother about escaping. 'I know your sort so well,' he tells her. 'You're a tight wire and it wouldn't take very much to break you.'

Filling his pockets with the Esmonds' cigarettes, he goes out at night – wearing gloves – and arrives at a seedy nightclub whose name, the Metro, is daubed on the brick wall. Jazz, girls, dancing. Frank smokes, bebops.

Now comes the first visit from the suspicious police inspector, Simmons (Hugh Griffith). The inspector and the doctor continually address one another as 'Inspector' and 'Doctor'. Frank begins to confide in Dr Esmond – his father was a cheat, a mean, petty, cruel bully, a hypocrite. Worse, Frank has suffered a stepmother who was 'tall, blonde, very smart. I hated her. She hated me, too. But I got even with them.' Sounds like Glenda.

At one in the morning a smartly dressed Frank leaves behind a brimming ashtray to perform a jewel robbery with an accomplice. They break the jeweller's window watched by a wealthy, elderly man in evening dress – a nice Losey touch. Next day the inspector arrives. Frank denies having done it.

Clad in a close-fitting shimmering artificial silk dress, Glenda asks Frank to take her to the Metro Club. 'Go upstairs,' he replies, 'and put on something a little cheaper.' On arrival at the Metro they dance, Frank smoking a cigarette which is burning short in the hand resting on Glenda's back. Later they go riding again and roll about in the leaves – then more of the nightclub jazz musicians. Afterwards Glenda drives very fast, the speedometer climbs, a police car pursues, she evades it.

FRANK: It's a police car. You've got to stop.
GLENDA: Ha! You bore me.

In a contrivance reminiscent of *The Prowler*, Frank opens a desk drawer which is not only conveniently unlocked but contains a revolver and a cheque book. With jazz off, he buys a new pair of dark glasses in Piccadilly, then stages a robbery, using the doctor's gun. Enter the inspector. Dr Esmond provides Frank with a false alibi, observed by some polished carvings.

Suddenly moved by the doctor's devotion, Frank begins to babble obligingly about his tormented childhood. He had stolen a bicycle and when the police arrived his father had handed him over. 'When I came out of prison I was going to kill him but he was too strong, he twisted my arm behind my back and made me kneel down in front of her. Then he beat me. I'll always remember the pattern on that carpet. Blue and yellow.'

Lugubriously, Dr Esmond then offers a protracted analysis while Frank cowers in his arms. Then they go fishing. Glenda drives off to see Frank who is staying at the fishing inn. They embrace. 'I don't love him, I don't think I ever did.'

Terrific music as a distraught Glenda paces her bedroom and then, having cut her own face, descends a splendid Losey staircase to blame Frank for the assault. The doctor takes his gun (now returned). Music! We hear a shot. The doctor descends.

THE WIFE: You killed him! He's dead!
THE HUSBAND: That's what you wanted, isn't it?

Aha. But Frank had already left the house. Glenda – in whom the real sleeping tiger resides – drives off in her Jaguar, pursued by Dr Esmond. She picks up Frank, walking virtuously through the night in his duffle coat. Swerving to avoid a lorry, she crashes through a huge hoarding of the Esso tiger and overturns. Frank alone survives this studio set-up, with car wheels spinning, smoke, strangers peering at the wreckage . . .

A Man on the Beach

To have viewed Losey's *A Man on the Beach* is a misfortune – its twenty-nine minutes weigh like sixty. Produced in Eastman Color by Hammer at Bray Studios from a screenplay by Jimmy Sangster, it was noticed, very briefly, in the BFI's *Monthly Film Bulletin* in April 1956.

The film opens with cheerful accordion music by John Hotchkis, as a grand old Rolls Royce cruises along a harbour causeway during the holiday season. Cut to the interior of a casino where smart clients are gambling. The very tall, mannish dame (Michael Medwin in drag) who has arrived in the Rolls is addressed by the manager as 'Your Ladyship'. 'She' (Max) knocks him out, creams the money on his office desk into a briefcase, returns to her open Rolls, instructs her chauffeur to drive to a disused gravel pit in the hills, then kills him after a terrific fight – one pair of hands on the loot being tidier than two. Although shot in the arm, Max pitches both the unconscious chauffeur and the car over the steepest cliff in the world. Spectacular shot of car falling and breaking up.

The injured Max staggers along the beach with his loot and reaches a well-furnished villa, where he finds no one at home. The search for the incumbent drags on tediously until he returns home resembling a blind Donald Wolfit. Max threatens him with the gun, is gratuitously abusive, and sets fire to one of the Wolfit character's poems. Dialogue and action are equally amateurish, inconsistent, awful. Everything is spelled out, usually several times. Losey called it 'a half-hour divertissement' and described Donald Wolfit as 'a bigger pain in the ass than Paul Muni, if possible . . .'[1]

The Intimate Stranger (*Finger of Guilt*, USA)

The story is related in the same flashback formula as *The Boy with Green Hair*, but with less voice-over: the narrative is the hero's recapitulation of events while trying to sort himself out on a doctor's couch.

Reggie Wilson (Richard Basehart), once a cutter/editor in Hollywood, with a reputation as a womanizer, is married to Lesley (Faith Brook), the English daughter of the studio boss for whom Reggie now works as a producer in Britain. Lesley plays tennis and is bored; she's also jealous in general, and in particular of the American actress Kay Wallace (Constance Cummings), an old flame starring in Reggie's current film.

Unknown to his wife, Reggie Wilson has been receiving a series of compromising letters from a woman who claims to have been intimate with him in the past. So depressed does he become that he fears he may be suffering from dual personality (hence the consultations with a doctor). Biting the bullet, Reggie goes home to his expensively ugly, 1950s sitting-room, with its post-thirties brick fireplace and its wooden-armed furniture – and tells his wife about the letters. Tensions mount: Reggie's suspicious studio-boss father-in-law, Ben Case (Roger Livesey), comes to dinner (grand silver urns and candlesticks) and screens a film he's just completed. This allows for double-level conversation about whether endings can be changed and whether characters should be granted a second chance in life. Having been fired from Hollywood for an affair with the boss's wife, Reggie's current job is his second and last chance.

Now his wife Lesley begins to receive the woman's letters. In desperation, Reggie sets out for Newcastle, to seek out his tormentor, taking Lesley with him to prove his good faith. They find Evelyn Stewart (Mary Murphy), a young American TV actress, who seems totally sincere about her past relationship with Reggie, which she describes with convincing circumstantial detail, coolly insisting that Reggie married Lesley only in order to further his career.

Reggie passionately denies everything but Lesley, distraught and convinced, runs out to her car. Reggie now goes to the police, chain-smoking his denials. Later that night, Evelyn meets him 'by chance' in the street and leads him to the local pub, a meeting observed by Lesley, who happens still to be in Newcastle. That's it! By the time he arrives home, Lesley has gone to live with her father. Worse, Ben Case has promptly cancelled his son-in-law's next picture.

About to go on leave of absence or jump from a window, Reggie catches sight of Evelyn Stuart walking through the studio lot. He shadows her progress through props and sets until she, ambitious and unscrupulous, tracks down the man she has come to find: Ben Case, the studio boss!

Eavesdropping on their conversation, Reggie discovers that Evelyn has been instructed to pursue her imposture if she wants to work in films. Now she has come for the pay-off: a part. Reggie leaps from the shadows to confront them both – but it soon emerges that the real villain is Ben Case's embittered old assistant, Ernest Chapel (Mervyn Johns), who has never liked Reggie and resents his sudden influence over the boss. It was Chapel who put Evelyn up to the whole stratagem, pretending to act on Ben Case's orders. So it's showdown time amid the splendid paraphernalia of studio lots – control rooms, loudspeaker system, blinding lights, extras in costume – with much running and slugging until all the pieces fall into place, villainy gets its just deserts, and virtue is rewarded to loud and happy jazz trumpets.

Both camerawork and choreography are conventional: characters light cigarettes, pour drinks, hunt for their slippers and – in Richard Basehart's case – frequently dress and undress. Losey's rampant eye reveals itself sparingly, as if snapping back its shields, but he gives it free rein at the climax. Trapped by Reggie's rapid closure of all studio-lot exits, the blackmail-

ing Evelyn runs through the lots in black dress and white raincoat, and is
frightened by sound-track gunshots as she gets entangled with a group of
extras dressed as British Tommies. The camera-angles are superb and the
climactic fight between hero and villain on a studio stage is photographed
with the protagonists spectacularly blown-up by studio lights.

Time Without Pity

A brief prologue hurls us into a smart flat sporting a violently symbolic
painting – dogs savaging a bull. We see Jenny fall dead on the sofa and
Robert Stanford (Leo McKern) staring down at her, pop-eyed with alarm.
Cut.

A long shot of a BEA Viscount landing. A haggard David Graham
(Michael Redgrave) alights from the plane and is met by his son's lawyer
(Peter Cushing). Graham finds himself separated by an iron grille from his
son Alec (Alec McCowen), who has been set up (by Stanford) and condemned
to death for Jenny's murder. Their dialogue is highly instrumental; for
example, we learn that Graham has been a patient in a sanatorium for
alcoholics. Alec also discharges other, lifelong grievances: a sad history of
paternal neglect. Due to be hanged the following day, Alec wants no help
from his feckless father.

Convinced of his son's innocence, and deeply ashamed of his own conduct,
David Graham embarks on a desperate search for the real murderer. His first
stop is the Revuedeville follies, where the murdered girl's sister (Joan
Plowright) conveniently works as a chorus girl, flanked by other pairs of
shapely legs. Losey makes the most of the dancers, the décor, the backstage
atmosphere, the predictable hostility of the girls to Alec's father. Graham
next visits the tycoon Robert Stanford (whose guilt was established in the
prologue) at his glossy motor showroom. A brash, self-made mogul, Stanford
is absorbed by the impending road-test of his new prototype. We also meet
his adopted son Brian (Paul Daneman), a friend of Alec's, who is both afraid
of his adoptive father and torn by conflicting loyalties. He drives Graham in
an open sports car down Piccadilly to Stanford's luxury flat, where the
murder took place.

Graham, a writer by profession, now emerges as the most genre-cursed of
characters, the amateur detective. We suffer a good deal of awkward dialogue
about a letter on blue notepaper, including such lines as 'What are you
hiding?' Searching the flat for clues, Graham finds a loaded pistol belonging
to Stanford (see *The Prowler*, *The Big Night*, *The Sleeping Tiger*). Graham is
making a phone call to Montreal about the mysterious letter on blue notepaper,
when Stanford bursts in with his north-country accent and inferiority
complexes – the stigma of commerce, the fear of losing power, virility and

beautiful wife. In a display of frenzied bullying, fist-waving and striding about, he seems to be parodying monomania to measure up to the bull on the wall.

Graham turns up at a committee room in the Houses of Parliament where the anti-capital punishment lobby is conveniently giving a press conference. By coincidence Stanford's wife, Honor (Ann Todd), is also in the audience, determined to save Alec by listening to a speech from a bow-tied gentleman reeking of Hampstead pinkery. Honor follows Graham to a pub where he finally drinks himself into oblivion, falls unconscious – and wakes up in Stanford's flat, where the tycoon is shouting and bellowing as usual.

Emerging from his drunken stupor, Graham takes Honor to the prison. Alec agrees to see her but not his father. She tells Alec she loves him, and they kiss. Now father and son embrace. Two fine actors struggle with melodramatic dialogue about what happened last Christmas Eve and whether there was 'ever something' between Alec and Honor. Alec then becomes hysterical and again turns on his father. Redgrave is at his brilliant best as he emerges, tearful, from the prison, momentarily free of Barzman's off-the-page lines.

But what has it all got to do with 'Who killed the girl?' In an awkward sequence, Stanford starts following Graham and giving him lifts in his car. They visit the Home Office where a senior civil servant refuses a delay of execution, then call on an embittered newspaperman who throws darts at a board and won't help. The stocky figure of Stanford now stands permanently behind Graham's shoulder, or lurks in the rear of the frame – evidently the monomaniac captain of industry is uneasy. He follows Graham to a pub. Losey observes them in close-up; Redgrave superbly blots out his misery and guilt in drink. Reassured that he will not be exposed, Stanford becomes introspective and reports his own dreams, evidently deriving his pleasure from Graham's helpless knowledge that he is the murderer. The scene was shot in Soho under pressure 'and Redgrave's breakdown was so terrifying that it completely shook me', Losey recalled. It has been the longest day, ever.

At dawn (one supposes) Graham reaches Stanford's test track. Here the tycoon gives free rein to his commercial ambition and craving for dominance, zooming manically round the circuit as the sky lightens. Obviously at some juncture Graham will stand in the middle of the track as the car approaches, and will then have to jump for his life. So he does. (Apparently Redgrave refused a stand-in for the moment of danger, though terrified. Losey later remarked, 'I wish I'd had my present skills when I did that sequence.')

The exhausted, desperate Graham follows Stanford to his office. Wearing mechanic's overalls, Stanford laughs and snarls truculently about lack of evidence, while admitting he done the who-dun-it and offering Graham shares in the firm if he'll keep quiet. Finally Graham stages a fight with the intention of forcing Stanford to kill him. The tycoon is kneeling on the ground trying to place the gun in the dead man's hand when in rush his wife and adopted son, Honor and Brian, just in time to witness 'everything' and to make the vital phone calls to the police and the Home Office. David Graham has sacrificed his life for his son's.

The Gypsy and the Gentleman

Sir Paul Deverill (Keith Michell) is your typical Regency wastrel baronet, keen on beer, wenches, gambling, horses, and with his boots up on every table. We find him in a drinking den among the lower orders, waging 200 guineas that he can subdue a slippery, squirming piglet. Subduing the gypsy woman, Belle (Melina Mercouri), will be something else.

Now he's in a carriage with his future father-in-law, Ruddock (Newton Blick), promising to treat his fiancée, the prim Vanessa (Clare Austin), as a husband should. Although he doesn't love her, she will bring a handsome dowry; the profligate owner of Deverill Court, with its grand façade, pillared portico and splendid staircase, is strapped for cash. And no wonder! While laying bets at a bare-fisted boxing match, he's pickpursed by the brazen gypsy beauty, Belle. Chased and caught by a crowd of locals, Belle clings to Sir Paul, who gallantly announces that she stole nothing.

Night. Torrential rain. Conspiring with her gypsy lover, Jess (Patrick McGoohan), Belle stops Sir Paul's carriage and proudly informs him that her father was a gentleman, her mother a gypsy who left her in a ditch. She laughs constantly, cynically and whorishly. That night she turns up in Sir Paul's room – 'I'm not used to sleeping alone' – and next morning steals his purse yet again. Tossing every inch of her body, she challenges him to 'Make me the lady of the house, make me my father's daughter.' So they go riding together, she in a fine hat and his mother's riding clothes. Belle throws a big tantrum about not being a liar: Mercouri's steam rises higher than that of the horses.

Now Sir Paul and Belle are found at the dinner table, she sporting his mother's pendant and sitting in the chair normally occupied by his sister, Sarah. Belle is now Lady Deverill. When Sarah bitterly reveals her brother's debts, Belle stops laughing and immediately smashes everything in sight – 'You liar!' – despite the huge house, jewels and servants she already enjoys. For good measure she slaps a serving girl.

Enter a crooked lawyer (Mervyn Johns, the villain of *The Intimate Stranger*). It emerges that rich Aunt Caroline has died and left her entire fortune to Sarah, provided she marries before she's twenty-one. Otherwise it all goes to the dissolute, sexually besotted and habitually drunken Sir Paul. A dastardly plot is hatched.

Belle, meanwhile, gallops madly, side-saddle, here and there, Mercouri permanently over the top, whip raised, expelling her fellow-gypsies from 'her' land. Revenge follows: gypsies break into the house and vandalize it – a greedy, ignorant, vicious bunch; the few decent characters in the story are strictly genteel. One of them, Sir Paul's sister Sarah, suffers a succession of kidnaps and rescues while her brother moans, 'Now I'm Satan's man.' When Sarah is committed to an asylum by a corrupt doctor, her wise aunt (Flora Robson) comes to the rescue. Belle and her lover set off in pursuit of Sarah's carriage but their own plunges into a river – and not before time. Sir Paul, also in pursuit, leaps from his horse to save Belle, but she only cries out for her drowning lover, whereupon Sir Paul decides that one drowning is worth three. The final image is of Belle's cloak borne away on the water.

Blind Date (*Chance Meeting*, USA)

The young Dutch artist Jan (Hardy Kruger) descends from a London double-decker bus on the Embankment, accompanied by a dissonant blast of jazz followed by chirpy springtime music (by Richard Rodney Bennett.) Beds of daffodils symbolize Jan's innocent happiness. Losing a shoe, he hops around cheerfully then skips like a sprite of spring to his mistress's mews flat, oblivious to the disapproving stares of London's buttoned-up pedestrians. Finding the door open but no one at home, he tosses his raincoat down without noticing the corpse under the coverlet. He studies the plush décor with wonder – paintings, a huge, ornate bed, a splendid bathroom with lacquered walls, silhouettes, ceramics, a Chinese dragon head over the basin, carvings, mirrors. Not a word spoken so far.

Enter Inspector Morgan (Stanley Baker), with the lazy, insolent alertness of a cinematic detective. Jan becomes indignant: 'You amaze me, Inspector. Are these the questions one gentleman asks another?' Morgan, the son of a Welsh chauffeur – 'I know how to recognize those who receive and those who give' – now shows him the corpse. Jan averts his face and collapses, convinced that the dead woman is his beloved Jacqueline. Police experts explain that she died so recently that Jan himself is the prime suspect.

Jan now tells his own version of the story to the sceptical Inspector Morgan in a series of flashbacks. We see him picked up in a Bond Street gallery by a woman of the world (Micheline Presle) calling herself Jacqueline Cousteau and smoking a cigarette in a sultry manner. 'I always wondered what Holland exports apart from tulips,' she purrs. In reality, as we shall later discover, she's Lady Fenton, the wife of an influential civil servant, Sir Howard Fenton.

Back to the present: Inspector Morgan is examining a small Van Dyck, which Jan explains is a study for a larger painting. Morgan and Jan see

different apartments, Morgan's eye translating the sunken bathtub, the mirrored boudoir, the huge bed, the frilly black lace underwear as the décor of a
classy prostitute. Jan, meanwhile, continues to utter lines like 'Do you get a
bonus for quick solutions?' and 'I tell you, she wasn't the kind you think.'

After a brief flashback of Jan loudly berating uncritical spectators of
modern art in the Tate Gallery we move to his studio, the artist hard at work,
irritated by the knock on his door. Jacqueline glides in: 'Doesn't a gentleman
take a lady's coat?' The camera examines Jan's own paintings and sculptures,
including a Joszef Herman-like picture of coal miners. Macdonald recalled: 'I
did the paintings . . . and I also tried to show Hardy Kruger how to look as
though he was painting them . . . The abstract in the gallery which [Jan] was
so rude about was also by me; if you get someone else to do them, or borrow
things, then you have difficulties about being rude.'² But Jan has no such
difficulty when Jacqueline arrives at his studio:

JAN: You talk too much. Stop jabbering and get to work.

Jacqueline is outraged: she doesn't expect to be spoken to 'like a barmaid'.

JAN: Barmaids work and I respect them for it. You can get out.

When she offers him money he angrily throws it on the ground and makes
her pick it up. (Three years later Losey was to have a new shot at this with
Stanley Baker and Jeanne Moreau, in *Eve*.) Losey shot these studio scenes,
the lovemaking, in a strong white light, lending a striking openness to Jan's
passion and forming a good contrast to the ambivalent lushness of
'Jacqueline's' ornate apartment. But enter now the censor, John Trevelyan.
'He had a bowler hat and a carefully wrapped umbrella, and a very long,
marked face, and he chain-smoked, dropping ash all over his waistcoat.' Then
in his first year as Secretary of the Film Board, Trevelyan would not permit
Kruger to be seen putting on his jeans and buckling his belt – 'It looks like
his prick' – and also cut a long, almost static shot from outside the studio
window of Kruger and Presle making love, modelled on Rodin's *The Kiss*.
Trevelyan 'never after that cut anything out of any of my pictures, and he
had the grace, some years later, to say he'd made a mistake'.³

End of flashback. The suave Assistant Commissioner, Sir Brian Lewis
(Robert Flemyng), arrives at the dead woman's flat. Alone with Morgan, Sir
Brian discloses that the bank account of the murdered woman, Jacqueline
Cousteau, shows frequent deposits from Sir Howard Fenton.

MORGAN: Friend of yours.
SIR BRIAN: Oh, we know each other.

Sir Brian insists that Fenton could not possibly have murdered the woman:

'It's a question of background.' (No Hercule Poirot, he.) Sir Brian proposes a blatant cover-up to protect Fenton, a top establishment figure currently engaged in delicate negotiations in Bonn on behalf of HMG. 'You're good,' Sir Brian compliments Morgan. 'Cigarette? Yes, you're the kind who either goes to the top – or doesn't go at all.' Morgan's Welsh blood boils.

At Heathrow Airport, Sir Howard Fenton steps out of his plane and waves his VIP hat like Neville Chamberlain (not a customary gesture for a diplomat, unless he's really a politician: it isn't clear). His elegant wife, Lady Fenton (Micheline Presle again), follows him down the steps. On a Welsh hunch, Morgan has brought Jan along to witness the scene. 'That's Jacqueline!' Jan exclaims, astonished that his 'dead' mistress is really alive.

'Do you know this man?' Morgan asks Lady Fenton, indicating Jan. 'No,' she replies coldly. Jan pleads with her: he is suspected of murder, she must help him. (Everything is spelled out, reiterated.) But she turns away, haughty, icy. We next find her at the police station (Morgan is putting his career on the line). She denies that she has ever called herself Jacqueline Cousteau. When Morgan decides to leave her alone with Jan, the better to eavesdrop on them, Jan comes out with the correct solution to the whole mystery: 'You knew your husband was going to leave you – for her. There was only one way out – kill her.' But Jan still loves *his* Jacqueline (a long speech) even though she has set him up as the murderer of the real Jacqueline. At this juncture the ruthless murderess softens, emitting the giveaway word, 'Jan!' Re-enter a grinning Inspector Morgan. (This scene was so improbably scripted that the three principal actors had to be recalled, after completion, to shoot it again without remuneration.)[4] Jan walks out of the police station a free man to swelling music, leaving us to ponder what a Losey film with a good script might be like.

The Criminal (*The Concrete Jungle*, USA)

A group of prisoners is found playing cards – Johnny Bannion (Stanley Baker) and his chums waging dream-money on their chips. The three floors of the prison, linked by iron staircases, are filmed in fluid movements through a succession of arresting compositions. Close-ups and dialogues are achieved with real style – as when a black ('coloured') prisoner plays calypso music on a guitar. Losey described Robert Krasker's lighting as 'extraordinary all the way through the film'.

Kelly, a former inmate who has 'shopped' a number of prisoners, arrives back inside. Bannion whistles: Kelly stands terrified. Chief Warder Barrows (Patrick Magee), an enigmatic but sadistic Irishman with bushy eyebrows and a mobile mouth, turns a deaf ear to the beating up of Kelly – Losey kept it off-screen – which is overlaid by the banging of pots and pans. Barrows rattles his massive ring of keys on the metal staircase. The sound-effects match Alun Owen's dialogue; for the first time in a British film Losey transcends studio-stereotypical speech patterns.

At liberty, Bannion is met by a smooth American villain, Carter (Sam Wanamaker). Clearly Bannion's world 'outside' is a thick cut above the one awaiting his prison pals, working-class Irishmen like the cheerfully gregarious Snipe (John Molloy) and the mystical Pauly Larkin (Brian Phelan). It is on Snipe's advice that Bannion now plans to rob the Tote at the racecourse. A party is held in his hard-edged flat, the men in smart suits and the girls to match. Here Losey's appetite for visual extravagance surfaces – Carter is shown examining a girl's arms and legs through a kaleidoscope, and when Bannion's discarded girlfriend Maggie (Jill Bennett) arrives, uninvited and doped, she too is put through the kaleidoscope. 'Get rid of her,' snaps Bannion. Losey later apologized for this episode, blaming its arbitrary quality on cuts; originally Maggie had met Bannion at the prison gate but was brusquely rejected because of her reported infidelities.

The guests have left. Here Stanley Baker's fine character acting inside the prison yields to the box-office lady-killer as he embarks on a tour of his heavily designed apartment, which (Losey explained) had been 'done over and over again' to achieve what he wanted, but without success: '. . . it was not intended ever to convey luxury . . . it's rather tatty'. Indeed, Foster Hirsch noted that the rooms are small and 'the ceilings seem almost to be closing in on [Bannion] . . . the recurrent low angle shot of the large-framed actor in his apartment suggests his entrapment'.

Carrying a sunray lamp and goggles to bed, Bannion discovers the beautiful Suzanne (Margit Saad) already in occupation.

> SUZANNE: I think I must have passed out.
> BANNION: Before or after you got undressed?

In a scene with humorous touches, Bannion, Carter and the gang rendezvous under a park pagoda. Carter acts on behalf of the Big Man from Highgate who can wash stolen money but requires an exorbitant percentage. Bannion arrives at the racecourse for the great robbery disguised as a bookie from Tattersall's, carrying binoculars and a suitable bag. This scene was done at Hurst Park on the grand scale – a big American car with tail fins, the horses thundering down the track, the grandstand, the tipsters – and with Albert Dimes on hand as criminal consultant. Mysteriously, Carter's gang allow Bannion, carrying the loot, to give them the slip by discarding his appointed getaway car. Losey later commented: 'A lot of people have said, "Well, what did he do with the other man? How did he get rid of him? . . ." I think this is just sheer nonsense; nobody really cares.' The comment points up a recurrent problem with the thriller dimension of Losey's mixed-genre films: his own impatience about the detail which provides sequential logic.

The camera pans across snow-covered winter landscape. Pacing out his chosen spot in a spectacular long shot, Bannion buries the loot in the middle of a field. Losey drew a distinction between the anarchic 'loner' (Dimes–Bannion) and the 'corporation' crooks like Carter who inevitably triumph. Bannion is duly arrested. Chief Warder Barrows now separates Bannion from his Irish pals, consigning him to inmates associated with Carter, led by the bald, sad-eyed Frank Saffron (Grégoire Aslan). The codes of prison intercourse are revealed – codes of honour, enmity and obligation. Losey conveys the implacability of prison routine by rigorously closed framing and the deliberate repetition of camera movement, as in the crane shot that is repeated whenever a former inmate returns to prison. The texture of brick, paint, ironwork excites Losey's talent for deep focus long shots and the wide-angled lens: a high shot of prisoners in the exercise yard; prisoners packing coir for stuffing mattresses; Catholic Mass in the hall. Here a priest intones in Latin while the gang leaders talk business on their knees; Chief Warder Barrows is also observed at prayer, yet is it prayer?

Scheming to provoke Bannion to violence – 'You're a liar, a thief, a waste of time . . .' – Barrows plants a knife on the vulnerable Pauly Larkin, leading to a punch-up. Collapsing into a trance, Pauly falls – or throws himself – from the third to the ground floor: suicide is his only alternative to murder. When the body hits the ground, everything freezes, warders and prisoners alike. During those ten seconds there are three shots, including Pauly falling – the film's supreme moment.[5]

A huge riot follows Pauly's death, chairs are burned, cell furniture destroyed, warders are overpowered. Dankworth's music becomes modernist, atonal, strident. Held captive by the rioting prisoners, Barrows is seen being led away to a cell, then is glimpsed smoking serenely – cinema language for a job well done. He has engineered the riot in order to persuade the prison governor that Bannion must be moved to another prison – in effect a plot designed to effect Bannion's escape into the waiting arms of Carter. The message handed to Bannion as he leaves prison is: 'One riot, one transfer, one fast black car. Total £40,000.' But can the spectator understand the complexities of the conspiracy? Cut from the film is shot 129, when Bannion remarks to Barrows, 'You've been in on mutinies before,' and also cut is a script amendment in which Saffron is explicit about arranging Bannion's escape. The price of these cuts is confusion.

A gangster in a long-nosed car and carrying a gun shadows the prison van transporting Bannion and overpowers the guards. Driving away through countryside frosted in thin snow, the gangster brings Bannion to a houseboat moored on a canal. However, in a disturbing failure of continuity, there is no sign of snow around the barge, although plenty of Dankworth's haunting jazz. Inside the barge the suave Carter smokes a cigarette in an ivory holder; the beautifully attired Suzanne is his hostage. 'Your sort doesn't fit into an organization,' Carter tells Bannion. 'Saffron and I are only cogs in the machine.' Carter produces a false passport plus an air ticket to 'nowhere' for Bannion if he will reveal where the money is hidden.

The villains drop their guard (as villains must) and Bannion makes his escape with Suzanne, Carter's gang in hot pursuit on a very cold day. Shots are fired, Bannion is cut by flying glass but drives on to the field where his loot is hidden – the last place he'd make for with hounds on his tail! (In the script this is laboriously explained.) Leaping from his car, Bannion shoots dead one gangster but is then hit himself and staggers across the vast, snow-covered field. Observed in long shot, he crawls like an insect, black on white, followed by Suzanne in black stockings and high heels, her white dress matching the snow. Cut are lines given to the dying Bannion as he tears off his Catholic medallion, whispering the Act of Contrition in three bursts while Carter urges him to disclose where the money is. Gone, also, are Bannion's last words, '. . . and because they offend thee, my God, who art all good . . .' As Bannion dies we hear the voice of Cleo Laine: 'All my sadness, all my joy, come from loving, come from loving a thieving boy . . .' Carter and an accomplice start digging desperately in the frozen, furrowed field. The credits come up against a high shot of the prison.

The Damned (*These Are the Damned*, USA)

The credits come up against the sea, sky and enormous cliffs on the south-east side of Portland Bill. This is the view from the window of the sculptress Freya, the one truly free spirit in the story. James Bernard's score shifts from the eerie to the ethereal as we view Weymouth in long shot from out at sea. The camera slowly tracks down the Queen Victoria clock tower until it finds the gangleader, King (Oliver Reed), and his minions lounging on the plinth, wearing black jackboots, jeans and nylon zip-jackets – but King himself sporting a smart tweed jacket, like an officer in mufti. The only girl to hand is King's sister, the slender Joan (Shirley Ann Field). The song, heard over, is 'Black Leather'. An American tourist, Simon (Macdonald Carey), is found on Weymouth promenade. He is duly picked up by Joan and mugged by the gang. The mugging is preceded by King's ritual routine of neo-military marching, whistling and panache, King himself fired up by incestuous rage against the innocent tourist.

Cut to the verandah of a sea-front hotel, where the sculptress Freya (Viveca Lindfors), dedicated to peace on this earth, is in conversation with Bernard (Alexander Knox), a distinguished but dour civil servant totally dedicated to a secret Government project which anticipates an inevitable nuclear war. ('He once had a love affair with Freya, a passion which still lingers in him but which she has long ago outgrown . . .' explains the script.) She despises Bernard's 'mysterious project', which (he warns) she must never know about on pain of death. The sequence ends with a shot of a sculpted bird on a table alongside glasses and teapots, with the sea through the window.

Rock music as we discover King and his gang in a shooting arcade with one-armed bandits and fruit machines. Joan leads the gang on a daredevil motorcycle dash along the Weymouth causeway until she finds Simon (a former insurance company executive) tinkering with the engine of his boat. Despite yesterday's mugging, she lingers; King thrusts his dagger between his sister's eyes: 'Do you think I'd let a man like that put his dirty hands on you?' Joan jumps from the quayside into Simon's boat; a helicopter shot of the bay shows them heading for the shelter of a cove.

At this juncture the film begins to alter the sequence of scripted scenes, and to eliminate some entirely. Cut, now, to Bernard's secret 'Project'. A sign reads 'Edgecliffe Research Station – No Unauthorized Personnel'. On television monitors in Bernard's office we see the nine captive children in their sealed bunker-classroom. Their high voices are without exception upper-class and pious in the prep-school tradition. When Bernard addresses them on closed-circuit television, the children immediately protest about their circumstances. The girl Victoria says: 'But you always talk about "when the time comes". What we want to know is . . . when *does* the time come?'

Bernard replies: 'You will be told everything in time . . . You will have to trust me, children, and allow me to be the judge.'

After some corny dialogue at sea, Joan leads her American up to the sculptress's Birdhouse Terrace; he dutifully examines Freya's bird-men before climbing into her bed with Joan.

SIMON: . . . I have never found . . . this . . . kind of quietness before.
It's as if I were no longer afraid of dying.

Freya arrives back at night, laments the loss of a bottle of '57 wine, hopes the people who occupied her bed were happy, then finds herself confronted by King, also staring at the bed. Freya's tolerance angers him:

KING: I know your kind. Smart talking and bad living. People with no morals.

To make his point he smashes one of her statues with a hammer. This is an excellent sequence, tensely acted by Lindfors and Reed, King perplexed by the sculptures, Freya a convincing blend of fear and confidence. As Philip French has pointed out, this scene anticipates the homicidal encounter between Alex and the Cat Lady in Kubrick's *A Clockwork Orange*.[6]

Pursued by King and his gang, Joan and Simon blunder into the electric fence of the Project, setting the alarms and dogs going. What now transpires is both fantasy in the creative sense and in the negative aspect of film cliché. Soaking wet, Simon and Joan have been rescued at the base of the cliff by the captive children, whose radioactive properties enable them – unknown to the authorities – to open and close the cave door merely by approaching it. Each child touches Simon and Joan to feel and marvel at the warmth of their skins. Joan recoils – the children's hands are ice-cold. One of the girls explains that they are all eleven, have their birthdays in the same week, don't mind being wet and don't know what a cold is. Stuck to their walls are pictures of couples labelled in childish handwriting – 'Victoria's parents', 'Elizabeth's parents' – but the 'fathers' and 'mothers' are pathetically drawn from different periods of history.

We discover the main floor of the Project, a huge room with inverted triangular pillars. The machinery level is a vast rock cave with heavy buttresses, the floor filled by a mass of machinery, generators, ventilating pipes. Losey and his uncredited production designer, Richard Macdonald, make the most of the slow journey through the underground passages and into the classroom, laboratory and dormitories – an open-planned, futuristic set-up with high-quality furniture, modulated lighting, hi-tech ceiling lamps. Such design values, of course, were inconceivable in any British 'school' of 1961.

Meanwhile, up in the real world, Freya is working on her sculpture while the impassive Bernard exposes his own dire obsession: he regards nuclear war as inevitable. 'We *must* be ready when the time comes.' Just how a dozen radioactive (and therefore immune) children will save the nation is not explained – but the sculptures against the sky above the cliffs are stunning. Losey does not linger on them, an aspect of film craft later abandoned in his 'baroque' period.

Exposed to the radioactive children, King, Simon and Joan begin to suffer nausea. Major Holland (Walter Gotell) has now spotted the adult intruders on the TV monitors. Aghast at the breach of national security, Bernard addresses the children on the TV system: 'Big people are dangerous to you . . . They are warm and nothing that's warm can live with you.' Now the children riot. Simon and Joan lead the children out of captivity, pursued by guards in radiation suits, with much shouting and swift camera-work. Bernard in his tweed cap monitors it all from the top of the cliff as the compassionate Freya clasps a child's ice-cold face. The children scream continuously as they are recaptured and carried back to the Cave. The now radiation-sick Simon confronts Bernard: 'You're the man who knows about violence, aren't you? Why are you doing this?' He and Joan head out to sea again, doomed to a hideous death, shadowed by a helicopter. Another chopper pursues the delirious King towards the causeway where he plunges over a bridge into the sea. (The stuntman almost died; the car turned over in the water, 'so that he was sinking head down into the mud . . . He had only a matter of seconds to get out and his seat belt didn't release.' Losey was shaken, vowing never to use a stuntman again.)[7]

The final accounting between Bernard and Freya is framed by her sculptures outlined against the sky, as angry and beautiful as the woman who made them. Bernard explains the mystery of the children's provenance: 'Their mothers were exposed to an unusual level of radiation by an accident. To survive the destruction that is coming, we need a new kind of man . . .'

FREYA: You are so in love with death you are dead yourself . . .

Bernard fires the fatal shot as the camera pans up from the sea across the magnificent, rugged rocks. The startled gulls fly upwards from the cliff. Freya falls beside her giant statue. The shrieks of the recaptured children – 'Help us, help us!' – haunt the ear to the finish, as the camera lazily pans along the sea-front, among the holiday-makers and deckchairs, past the clock tower where we began.

Eve

See Chapter 18.

The Servant

See Chapter 2.

King and Country

Losey begins with an extended prologue using a wide range of cinematic techniques, including stills, freezes and jump-cuts. From a bronze figure of Peace crowning the Wellington Arch (solo harmonica) he darts in on details from the Royal Fellowship monument to the Royal Artillery – we hear the murmuring forgetfulness of London's traffic. Beneath a massive cannon a frieze of helmeted British soldiers lie frozen in stone, in time, and some in death. Cut in are enormous explosions which freeze and dissolve into the ravaged landscape of the trenches. The rain pours down. A horse lies dead in its harness. Someone is playing a harmonica (music composed and played by Larry Adler). A skeleton in uniform slowly dissolves into the reclining figure of Private Hamp (Tom Courtenay): this image being a premonition of the end, but in reverse. The harmonica is Hamp's.[8]

Passchendaele, 1917: Hamp is awaiting trial by court-martial for desertion. Detailed to defend him is Captain Hargreaves (Dirk Bogarde), an officer of

conventional, even socially supercilious, outlook, a rather vain moustache
perched above his lip, and with no enthusiasm for his distasteful (because
futile) brief – 'If a dog breaks its leg, you shoot it' – but endowed with a
strong sense of duty and a capacity for grasping a common humanity across
the barriers of class and rank.

Invited to outline his life, Hamp responds hesitantly. A bootmaker who
had left school at the age of twelve, and now a father, he had signed up
'because the wife and mother dared me'. Losey intervenes with tinted sepia
stills: Hamp's small child, dressed up for the photograph; a humble corner
shop in Islington; the 'friend' who wrote to report the infidelity of Hamp's
wife. The friend is seen in bed, his spiv eye alight with triumph – no doubt
he is the lover; yet this 'still' surreptitiously displays movement, the smug
passage of a tea-cup to his lips. With the corner shop in Islington comes a
flash of the slogan, 'Women of Britain Say Go'. Here again the impression is
of a still photograph, yet the woman's broom moves across the front step.
Diffidently – searching for words, thoughts, in the forbidding yet encouraging
presence of the officer – Hamp suggests that for 'our sort, not your sort, sir',
the prospect of war had not seemed much worse than life at home. 'But we
were wrong.'

The 'stills' are both evocative and problematic; do they belong to Hamp's
consciousness or to the narration? The same doubt applies to a flash-still of
the King and the Kaiser riding together, pre-war, and a double exposure, as
Hamp recalls past battles to Captain Hargreaves, showing mud and corpses.

The long interrogatory between Courtenay and Bogarde is acting at its
best; a painful human dilemma emerges from the cramped world of the dug-
outs – mud, pools of rainwater, soldiers being dispatched on fatigues by
bellowing NCOs. Throughout the film Losey wonderfully evokes the filthy,
wet discomfort of a battalion which is not even in the line, but resting. Heavy
gunfire rumbles continuously, the endless, irresolvable human storm.

Like Captain Hargreaves, Hamp has been at the front for three years,
longer than most. And he was a volunteer, too: does this entitle him to special
consideration? Tom Courtenay achieves pathos through a flat, colourless and
essentially baffled recitation of who he is and what, finally and fatally, he has
done: turned and walked away from the front, past guards and look-out posts,
heading blindly for England, until they picked him up near the coast. Had he
planned it? Not exactly – he'd simply had enough. He is painfully honest;
more than once he'd thought of shooting himself in the leg – 'We talked
about it' – but he never did. A loyal soldier who had been splattered with the
remains of the friend advancing beside him towards enemy lines, Hamp does
not know why he finally walked away. But, surely, 'They can't shoot me.' At
one point this young man, whose expressiveness is normally pared down to
the bone, offers an uncharacteristic metaphor, likening a soldier struggling

and sinking in mud to a boiling egg. This was clearly Bogarde's contribution to the script, reflecting his own father's post-war horror of boiling eggs.

The field court-martial takes place in a dug-out. Presiding is the Colonel (Peter Copley), who maintains a cold neutrality towards the prisoner. Only one officer present is a brute: called as a witness, the Medical Officer (Leo McKern) dismisses Hamp's action as 'cold feet' – laxative pills are his prescription for all nervous disorders. McKern's portrayal of the Medical Officer is almost as exaggerated as his depiction of the entrepreneur in *Time Without Pity*. He bullies, blusters, shouts, while Hargreaves presses him to agree that Hamp was suffering from shellshock, a recognized medical condition, at the time of his desertion. The MO has seen it all before: 'Cold feet, funk! He was fit for duty!' The camera observes the exchanges unobtrusively while Hamp listens, bemused but always confident of 'justice'.

Meanwhile, in the Tommies' quarters, a sleeping soldier's ear is bitten by a rat. Stones rain down on the cornered rat as it clings to safety above a pool of water. Later the soldiers beat a dead mule to force rats out of its belly.

Back to the court-martial. Hamp, a kind of Billy Budd, cannot lie to the court about his motives – except, ironically, when clutching at Hargreaves's leading questions. There are other witnesses, including Hamp's platoon commander, the cool, self-regarding Lieutenant Webb (Barry Foster), whose main thought is not to have his own platoon turned into a firing squad, but Webb's defence of Hamp as 'a good soldier' cuts no ice. As the prosecuting officer, the faintly supercilious Captain Midgley (James Villiers – his third film for Losey), points out, Hamp's record over three years is neither good nor bad: a blank. As the contest continues, Hargreaves becomes increasingly isolated from his fellow officers; Bogarde carries Hargreaves into enigmatic alienation; utterly conventional in his values, highly decorated, he begins to assume the personality of a loner, with a hint of the artist whose khaki tunic and cap don't, in the end, quite fit. He is, as they all acknowledge, doing his duty; but he is also possessed – moral shellshock – to the point where his CO has to rebuke him more than once for insolence. The court-martial finds Hamp guilty of desertion but adds a plea for mercy.

Hamp thanks Hargreaves for 'having spoken the truth so well'. The rejoinder is devastating: 'I only did my duty – a pity you didn't do yours.' Hamp's shattered expression may be the high point of the film – acting beyond the call of duty. The death sentence is confirmed by General Headquarters – the brass-hats 'pinned down in Paris', as one of the soldiers wittily puts it – to 'encourage' the others. Finding the Colonel reading poetry on his bunk while sipping whisky from a tin mug, Hargreaves, who has stumbled in the mud, displays his emotion to the point of over-stepping the mark. He gulps the whisky offered to him:

COLONEL: Rather short on ceremony, aren't we?

Does shooting a soldier encourage the others? The CO's answer is 'Yes – but I'm not sure.' Hargreaves drops a cigarette into thick mud; no need to lay his boot on it; the mud is monarch here. (According to Bogarde, the script originally called for him to threaten the Colonel with a rusty knife: 'I had to explain to Joe that one doesn't do that.')

The soldiers, meanwhile, display their ironic contempt for military justice by staging their own 'court-martial'. In pouring rain they have put on trial a rat trapped in a wire-topped bucket. Imitating the prosecuting and defending officers, they also ventriloquize the rat's 'answers'. As Foster Hirsch later commented, Losey's deep focus compositions and his preservation of natural time by means of long takes create a palpably realistic environment strengthened by the sound track, which is wonderfully alive with the continuous noise of rain, dripping water, distant voices.[9]

The other ranks bring purloined booze to Hamp's detention cell to comfort him and to forget their own fears through his last night and maybe theirs. They get roaring drunk; a soldier cradles Hamp and then, in a macabre rehearsal of the execution, he is blindfolded and made to play blind-man's-buff, groping in search of his comrades – until they vanish at the approach of the platoon commander, Lieutenant Webb, and the stand-in Padre (Vivian Matalon). The blindfolded Hamp grabs the Padre by mistake. From the Padre he hears pieties about sinners repenting and a scourging God chastizing those he loves. He prays with closed eyes but Hamp's own eyes remain open and bewildered until the final Communion of bread and wine; sceptically he swallows the ridiculous sliver of bread, the 'body of Christ', before gulping the 'blood', which he then vomits up. The officers administer a merciful morphine jab in the arse:

WEBB: Where's the soul, Padre?
PADRE: It's here.

Dawn arrives with a typical Losey shot: a soldier carrying a tea-mug peers out through a jagged hole in a mud wall and sees the officers assembling outside, in the rain. Tied to a chair and blindfolded, Hamp is carried through squelching mud; the chair slips, dumping him in the slime. His own platoon must serve as the firing squad – a grey dawn of desolation and ruin. Lieutenant Webb pins a strip of white cloth over Hamp's heart then closes his eyes on his own command – 'Fire!' Hamp and his sandbagged chair keel over into a puddle. The MO pulls off the blindfold and finds Hamp's eyes still flickering with life (how many of the firing squad have averted their aim?). Webb must deliver the *coup de grâce* but can't; Hargreaves steps forward, takes Webb's revolver, kneels over Hamp and lifts his head.

HARGREAVES: Isn't it finished yet?

HAMP: No, sir. I'm sorry.

Hargreaves places his revolver in Hamp's mouth. Losey moves in very close; we see only Hargreaves's face as his finger squeezes the trigger through two empty chambers before the final report. Still strapped to the chair, Hamp slowly sinks below the surface of the mud. We are spared his shattered skull.

A voice over conveys the official letter of commiseration: Private Hamp has been killed 'on active service'. Losey bangs home the point with a tinted still of an old warhorse general in his staff car – then leaves us to desolate memories of writhing flesh mimicked in dead matter; men stiffened by coatings of mud like clay sculptures nearly dry.

48

The Films
in Parts V-X

GALILEO, BOOM!, SECRET CEREMONY, THE ASSASSINATION OF TROTSKY, A DOLL'S
HOUSE, THE ROMANTIC ENGLISHWOMAN

Galileo

Starting with a high shot of the studio stage-set, the camera conducts us into
Galileo's room in Signora Sarti's house. Her son, the boy Andrea (Iain
Travers), brings milk and bread to the great scientist who has set out to prove
the new Copernican system. Master and pupil are simply moulded in brown
and beige (camera movement minimal), joined by their mutual appetite for
knowledge and by Andrea's rapt attention to the Ptolemaic model. Galileo,
while washing and having his breakfast, is gauging the boy's capacities –
Losey was fascinated by the forthcoming span of father–son 'conflicts,
competitiveness, and the excesses of adulation and condemnation'.

With the arrival of the wealthy Ludovico (Tim Woodward), 'we see the
drudgery . . . for Galileo even to contemplate teaching an uninterested or bad
pupil – no matter how much he needs the money'. But his daughter Virginia
(Mary Larkin) will fall in love with Ludovico.

Losey offers his first 'alienation' shot, Galileo talking VCU into camera: 'I
predict . . .' He seethes with energy; he wants more time for research, less
private teaching, a better stipend from the University of Padua. The bourgeois
Curator of the University, Priuli (Colin Blakely), rejects this request:
'Mathematics is a profitless art, so to speak. Not that the Republic [of Venice]
does not esteem it most highly.' The Curator's attitude to scientific research
prefigures classic Thatcherism: 'For the knowledge which you sell, you can
demand only as much as it profits whoever buys it from you.'

Scene 2: We move 'outside' (but still within a studio set) as Galileo
presents his telescope to the Republic of Venice at the Great Arsenal.
Nothing much is made of this dramatically, but we notice a glimpse, in the
rear of the frame, of real ships on real water. As Hirsch puts it, 'Losey thus
slyly counterpoints the real and the artificial.' Losey noted that Galileo's
attitude towards the power structures of Venice, Florence and the Church 'is
not so different from that of any artist who is dependent on patronage and
commerce . . . Galileo must get what he needs by craft and cunning and by

stealing and plagiarism, if you like, even as Brecht himself frequently did.'[1]

In Scene 3 we are introduced to Sagredo (a moving performance by Michael Gough), who understands the implications of the telescope for research but foresees the consequences – charges of heresy against Galileo. Sagredo warns Galileo not to move to Florence, 'Because the monks are in control there.' Galileo disregards his advice; Signora Sarti, Andrea and Galileo's daughter, Virginia, will accompany him to Florence. In Scene 4, the absurd obscurantism of the old scholarship is represented, at Florence, by the Mathematician and the Philosopher, who contest Galileo's theories before the Grand Duke and will not tolerate Aristotle being 'dragged through the mud'. Galileo responds that 'Truth is the child of time, not of authority.'

In a nice shot of the singing chorus (one tall boy and two small), which introduces each scene, two helmeted centurions open a fine pair of doors to a marble hall – framing the boys in perfect symmetry. Losey presents the chorus from a variety of arresting angles, distancing them from the action.

Scene 6 takes place in 1616 at the Vatican's Collegium Romanum. In a marvellous vignette the angry Old Cardinal (John Gielgud) tells Galileo, 'You wish to degrade the earth . . .' In Scene 7, Galileo is confronted by Cardinals Barberini and Bellarmin (Michel Lonsdale and Patrick Magee) effetely holding masks on sticks, a lamb and a dove respectively. Here the colours are richer, less austere, reds and blacks moving in slow, lazy (but alert) arrogance, marble pillar to pillar. Again Losey carefully sets a 'quiet' camera at the service of the exchanges, avoiding in this case his own production notes: 'Brecht once said that Cardinal Bellarmin ought to look like Harry Truman. Bellarmin is a businessman.' As for Barberini, the future Pope, he is 'an enlightened intellectual, but the enlightened intellectuals, like so many [liberals] in the 40s . . . did not stand up – they ran.'

Afraid of the Inquisition, Galileo resigns himself to a form of liberal house arrest: he is permitted to pursue his research, but not to publish or teach it. Edward Fox plays the bearded Cardinal Inquisitor with smooth precision, notably when talking paternally but menacingly to Galileo's deeply religious and devoted daughter, Virginia. According to Losey, Galileo treated his daughter 'as an object of no importance . . . He is a male chauvinist: an attitude which Brecht to some degree was guilty of himself . . .' In Scene 9 Ludovico breaks off his engagement to Virginia, alarmed by the potential impact of Galileo's teaching on the peasantry, while Galileo turns his mind back to science. But this is less 'male chauvinism' than its close cousin, paternal self-absorption. (Topol has likened Losey in certain moods to Galileo's lack of fatherly feeling.) However, as Losey noted, 'The shock makes [Virginia] an enclosed and inaccessible person . . . frightened and concerned for her father's safety ... fanatical guardian against his "heresies" ...'

Returning to Scene 8: here Brecht introduces the Little Monk (Richard O'Callaghan), himself a scientist of peasant stock, thrown into turmoil by Galileo's discoveries. Discussing the Little Monk, Losey suggested a comparison with the worker priests 'who have opposed fascism in Germany and Portugal and so courageously recently in Africa and South America'. But Brecht's text offers almost the opposite evidence: the Little Monk sees in the 'Holy Congregation' the comforter of the poor, 'true maternal compassion, great goodness of soul'; yet, fascinated by physics, he convinces himself that God and the Church will permit the truth.

Galileo is encouraged by the enthronement of a new Pope, Barberini, himself a mathematician, to resume his research into the forbidden subject of sun-spots. Young Andrea, meanwhile, has grown to manhood, although Judy Parfitt, as his mother Signora Sarti, does not visibly age despite her multiplying worries. Indeed no one seems to age while Andrea progresses from boy to Tom Conti. Topol's hair is bleached a goldish hue, a rather glamorous ageing; set against the English cast, he tends to sound faintly 'foreign', as if he were something other than an Italian among Italians.

The marketplace carnival follows (Scene 10), with modernist, Eisler-like satirical ballads, involving choreographed episodes of Rabelaisian mock-copulation. The common revellers hold aloft Galileo's spheres and giant puppets of a sad, silenced scientist. Losey decided to shoot the carnival at night, for reasons of 'simplification and stylization', with bonfires and torches distorting the masks.

Some scenes begin with captions: 'Galileo Refuses Support of Rising Bourgeoisie and Relies on Grand Duke and Pope' (Scene 11); 'Pope Urban VIII Yields to the Clerical Establishment' (Scene 12). The film's executive producer, Otto Plaschkes, later complained of inconsistency – why were there 'no titles over Bellarmin's Ball, the Little Monk Scene, the Experiment Scene, and the Final Scene'?

The Florentine Court can no longer resist the Inquisition's request to examine Galileo in Rome. Beautifully staged and designed is Scene 12, a long and statuesque sequence in which the liberal Cardinal Barberini, now Pope Urban VIII, stands on a marble plinth while being robed and listening to the seductive arguments of the Inquisitor. Behind him, above sumptuous marble walls, banners can be seen moving beyond a balustrade. As Losey noted, 'the "shuffling feet" should convey the weight of the millions subject to Roman Catholic dogma'.

Scene 13: will Galileo recant under duress? In the palace of the Florentine ambassador to Rome, Losey becomes more cinematically abstract and theatrical. 'Chess pieces are magnified as shadows on a luminous backcloth as

Galileo's disciples await the outcome – the next turn of history's screw. The devoted Virginia continues to pray, on a raised plinth, for her father's prudent repentance, her Ave Marias alternately exultant, despairing, abject, joyous in the exact reverse of the mood of Galileo's students who 'pray' that he will not recant. This scene is shot on an open stage against a bare cyclorama, the shadows of Galileo's disciples and of the pious Virginia looming in contention, wrestling for the great man's soul – until the tolling cathedral bell signals victory for the Church.

As Galileo approaches, his shadow is huge and heavy, replete with grief, shame and relief. Andrea berates him: 'Winebag! Snail-eater! Have you saved your precious skin?' It is here that Galileo offers his famous response to Andrea's 'Unhappy the land that has no heroes':

GALILEO: No. Unhappy the land that needs heroes.

Losey noted that this should be said aside, or vainly to thinning air, or, he suggested, somewhat in the spirit of Hamlet:

> The time is out of joint: O cursed spite
> That ever I was born to set it right.

Accident

See Chapter 22.

Modesty Blaise

See Chapter 23.

Boom!

Waves smash and boom on rocks behind the credits. We see the magnificent Goforth island from the sea. Chris Flanders (Richard Burton) jumps aboard a boat hired by journalists hoping to get on to the island. Discovering that he is an impostor, with no introduction to Mrs Goforth (Elizabeth Taylor), the journalists throw his bag into the sea; he dives after it. Bullets hit the water round the boat.

Dictating her memoirs to her secretary, Blackie (Joanna Shimkus), Flora ('Sissy') Goforth employs an affected English accent, then lapses into raw American when yelling.

Chris Flanders, dripping wet, is scrambling up the vertical rock face when attacked by murderous dogs controlled by Goforth's sadistic neo-Nazi dwarf, Rudi (Michael Dunn). Bleeding profusely, he is rescued by Blackie and

housed in a wing of the villa out of Mrs Goforth's sight. Periodic coughing
indicates Goforth's terminal illness, which miraculously leaves the rest of the
investment in good shape. Festooned in jewels worth millions, she lacks
noblesse oblige towards her employees (who wear a uniform bearing griffin
insignia): 'What are ya grinning about, huh?' The cool interior designs and
Arabic woodwork are better value.

Receiving the Witch of Capri (Noël Coward in a dinner-jacket), Goforth
performs a short Kabuki dance she once did at a charity show – but her
features retain the immobility of a still photograph out of *Picturegoer*. Their
dîner-à-deux in gorgeous, carved chairs overlooking the sea, is much spun
out with Oriental musicians, gull's eggs, and a sea monster served on a dish.
She recalls her last dead husband, the poet Harlan Goforth: 'And our days
would begin at sundown when Alex got out of bed and put on the robe of a
Samurai warrior . . . and I would get out of bed as naked as Eve and pick up a
little pearl-handled revolver.' A servant picks a large candelabra off the table
and carries it straight into the camera.

Chris Flanders, the Angel of Death, is discovered in the guest room, a
burning-red, Chagall-like mural above his bed (the central figure may be
Aphrodite, goddess of love). Enter the Witch in his dinner-jacket, bird-
howling; Chris Flanders receives him in the late Mr Goforth's black kimono
and samurai sword. The Witch falls asleep on the bed under the 'Chagall'.
Chris proceeds to seduce Blackie.

When Goforth dictates her memoirs at night, her voice screeches through
all the loudspeakers. Chris surfaces on the terrace in his kimono, gallant and
obsequious, and administers her pills. The Witch reappears to berate Chris
for all the rich women he's murdered – but invites him to visit Capri. Chris
keeps leaping about on the balustrade overlooking the massive rock precipice,
scaring Sissy Goforth, waving the sword, shouting things like 'Bonzai!', luring
her into a small funicular then pulling the lever to leave them swaying
hundreds of feet above the wave-lashed rocks. Taylor and Burton are constantly posing in arresting settings. The colours are ravishing – black, white,
yellow, ochre – gorgeous carvings, pillars, capitals, Easter Island heads
looking down on waves foaming over grinning rocks, and a fabulous living-
room with long, swaying curtains, glass-shaded candles. Burton reciting from
Coleridge does not atone for his own poetic wisdom: 'Death is one moment –
and life is so many of them.' Goforth promptly sprays herself with perfume,
then instructs Blackie to post armed guards since 'Mr Flanders keeps looking
at my rings as if calculating their value.' When Chris receives a phone call
about the death of another old lady, Goforth becomes hysterical, reads his
black address book, tells him to leave at once, then asks for the kiss which
sells a million tickets.

In her darkened bedroom, Douglas Slocombe's camera silhouettes the

Taylor body stretched invitingly against the rectangular shapes along the inner wall, punctuated by modernist window apertures to the poem-sea, poem-sky (as Chris poem-calls them). Golden light floods the long slow scene in which Chris refuses her invitation to bed – if it's to be the price of a square meal. (He's hungry.) 'No man,' she informs him, 'ever refused me before.'

Now she lies dying. Slowly he strips her of her jewels as the camera circles them in a continuous take and he tells her stories from his past, including the tale of the blind Hindu teacher who wanted to be led into the sea. She dies. Chris tosses one of her priceless rings (jewellery by Bulgari) over the cliff in a glass of wine. The wind is forever in Burton's hair.

Secret Ceremony

Miss Leonora (Elizabeth Taylor) is found examining herself in a mirror while removing her prostitute's blonde wig. Dressed in black (by Marc Bohan, of Dior), she takes the bus to a large Catholic church set in surroundings of bleak dereliction. On the bus she becomes the object of obsessive attention from a pale waif of a skinny girl (Mia Farrow) who finally clasps her with a cry of 'Mummy'. As framed family photographs will soon reveal, Leonora closely resembles Cenci's dead mother.

Scurrying and hiding on spindly legs, Cenci follows Leonora into the church where a baptism is taking place in a hue of coppery light. Leonora kneels to pray and the officiating priest is heard to sneeze. Outside the church Cenci intently observes Leonora laying flowers on her own (drowned) daughter's grave: a flashback to the burial of the daughter follows later. The parameters are established: the woman lacks a child and the child/girl/woman lacks a mother. But whereas Leonora is solidly locked into grief, loneliness and poverty, the wealthy Cenci is both deranged and – as close-ups of Farrow's flickering eyes will suggest – playing a diabolical game.

Cenci leads Miss Leonora away from the cemetery to the vast Engelhard family mansion which she inhabits alone. The camera embarks on an extravagant visual orgy which will not cease until the end of the film: tiled walls, stained glass, art deco conceits, a cupola. Led to 'her' bedroom, Leonora opens a vast, multi-mirrored wardrobe packed with fur coats and expensive gowns Taylor-made for an unending fashion parade in shocking colours. While the astute Cenci fixes her new mother a huge breakfast, complete with sausages, on a silver tray, Leonora test-bites medallions worthy of theft. Then she eats, decorously wiping her mouth. Abruptly Leonora seizes the proferred role of mother, but in a common, screaming idiom – she even slaps the girl with enough force to send her to her knees. Contrition follows and – for us – utter disbelief. Whereas Farrow acts with the whole of her face, Taylor's expression varies between freeze and vituperation. Drawing on her talent for body-language, Losey lays her across the vast art-deco bed then plunges her into a bathtub large enough to accommodate Cenci too, plus some floating toy ducks. Mischievously Cenci 'drowns' them, evoking hysteria from Leonora. Gallantly Taylor attempts to convey a common, decent woman's difficulty in playing the part of a wealthy, well-bred look-alike whom she has never met, while periodically lapsing back into Leonora and constantly fending off the girl's desire to hug and kiss. The fey Cenci's mode is close to baby-talk and the succession of short-skirted numbers she wears lie half-way between Swinging Chelsea and a James Barrie nursery. Threatened with another cuddle, Leonora whispers a prayer for Cenci to repeat.

Soon we hear threatening male banging on the front door – this is the vulgar and lecherously slouching stepfather, Albert (Robert Mitchum), who has returned from America unaware that his wife, Cenci's mother, is dead. Fearful and protective, Leonora observes him from a window – he tosses away his flowers, but he keeps calling and sending Cenci cards carrying a heart pierced by an arrow. When he eventually forces an entrance, he loses no time in telling Cenci how her mother disliked sex but was proud of her breasts. Cenci titters, we gag. Increasingly assertive, Albert boasts to Cenci about his lifestyle on an American campus: '. . . and all the little sophomores think I'm just a baffled – benign – old puff from England, dabbling in cybernetics. Until we get to the parking lot. Then I grab them . . .' He warns Cenci, 'If you don't watch out, I might turn into a fag.' His phone calls to the house are beaten off by Leonora in her most foul-tongued idiom. Demon-eyed, Cenci continues her predatory adoration of 'Mummy' by brushing Leonora's hair with two spectacular brushes.

How innocent is Cenci? In a long sequence she talks to herself in the huge kitchen, voices the thought that her virginity is the only thing she still possesses, then stretches herself in her nightdress across the vast kitchen table and acts out resistance to rape: 'Take your hands off me.' Leonora, of course, observes this display through a window, as no doubt Cenci intended. When bored with the human drama, the camera lovingly studies the architecture.

Enter the two wicked and faintly comic aunts (Pamela Brown and Peggy Ashcroft), sisters of Cenci's long-dead father, hectoring and bullying the girl, systematically popping precious objects into their bags, raiding the bedroom wardrobe for furs. Leonora goes to confront them in their fabulous antique shop, her genteel English accent lapsing into American idioms.

Back in the church, she is seen through the confessional grill (a cinematic cliché) as she bitterly laments her life as a prostitute and prays that she may make a new life with Cenci. But Leonora's religious dimension is as unconvincing as everything else – sacramental décor. Meanwhile Cenci, maddened by Leonora's absence, is found petulantly breaking up the furniture, observed by Gerry Fisher's camera in a protracted single shot. When Leonora returns she may conclude that Cenci has been raped by Albert.

Albert and the two aunts lay flowers at the grave of Cenci's mother. This is filmed from a high camera angle between winged angels in white marble. Albert kicks the grave – why, he demands, isn't it pure marble? (But should he care?) The aunts appear bemused, as well they might, at finding themselves stranded with the twisted mouth of Robert Mitchum in a catch-as-catch-can screenplay.

The scene changes to a luxurious seaside hotel. Cenci and Leonora dress for dinner. Cenci is observed by Leonora in her slip, as skeletal as ever, but when she appears in the dining-room she is hugely pregnant and grinning from ear to ear. Under her dress she has stuffed not a pillow but a folded Muppet-type 'frog'.

Out on the sands the now persistently 'pregnant' Cenci is observed at a distance riding a horse at a gallop (which may seem odd). Albert, meanwhile, has tracked them to the seaside and slouches sneeringly in on Leonora who has laid out a picnic in the dunes (for which she's wearing a low-cut mini-dress). During the confrontational dialogue that follows, both Taylor and Mitchum ride new depths of vicarious vulgarity and obscenity, including two memorable lines from Albert, 'You look more like a cow than my wife' and 'I couldn't even rape a randy elephant.' (Who could?)

Now more of Cenci and Leonora lying about in bed: after a massage and kisses, Leonora turns vulgar-rough again and pulls the Muppet-frog from the girl's nightdress, the result being a mess of stuffing on the carpet and a reversion of Cenci to 'reality': 'Why,' she demands, 'are you wearing my mother's dress?' Returning home, Cenci lays out an overdose of pills with linear precision on a polished desk top and a moment later they are gone (perhaps the best sequence in the film). By the time Leonora returns to plead for readmission Cenci is struggling for life in a chair while clinging to her new arrogance: 'Are you applying for a position?' Leonora departs, unaware that Cenci will die moments later, up in the Venetian balcony under the gaze of a copper crucified Christ. (Losey discarded several flashbacks after shooting them, including Leonora's daughter drowning while her mother and her father, a Marine, make love on the beach behind a windbreaker. We are also spared Leonora confessing to a sniffling priest that she killed his beloved cat because it stole her butter.)

Laid out in an open casket, Cenci is inspected by the aunts. Leonora arrives and places herself in the shadows; when Albert enters the funeral parlour she stabs him through the heart. He drops dead beside the coffin. Finally we are with Leonora lying on her prostitute's bed, swinging the lampshade above her head, and relating a Tabori fable of two mice in a pail of milk.

Figures in a Landscape

See Chapter 24.

The Go-Between

See Chapter 27.

The Assassination of Trotsky

The film begins with stills of Trotsky at various phases of his career, then the credits come up against a blank screen. We hear Trotsky (Richard Burton in effect), dictating an epigram as the Mexico City May Day Parade, 1940, appears on the screen: men in ponchos, men in uniform, women in uniform, chanting, red flags, old buses, suburbs. We see a huge painting of slaves toiling in chains, while a priest grasps a crucifix and rebels swing from a tree.

Cut to Jacson (Alain Delon) shaving in his hotel bedroom. He glances out of the window at the parade, then engages in some friendly wrestling with Gita (Romy Schneider), who's demanding attention. The parade goes on; Gita appears in the crowd and shouts, 'Trotsky!' Jacson, wearing dark glasses, pulls her away. There follows a violent clash between Communist demonstrators and Trotskyists: an effigy of Trotsky is trampled underfoot. Gita, an ardent Trotskyite, is seen furiously fighting in the mêlée.

Now our first glimpse of Trotsky pacing along a balcony behind a protective wire screen; his posture suggests a donnish impatience with the masses. As he enters his office, where male secretaries are at work, we hear his dictation on play-back, about Stalin issuing orders to finish him off as soon as possible. (Most of Trotsky's dictated passages, throughout the film, concern himself, more egoist than Marxist – however, he goes on to denounce 'imperialists, Fascists, Wall Street plutocrats' – the last word coming out of Burton as 'plurocrats'.)

The camera pulls back from close-up to show the full extent of a vast mural around a courtyard. Our Communist artist-commando, Ruiz (Luigi Vannucchi), is at work in hat, apron and floppy moustache, a handsome fellow, as artists should be. He looks rather bored. The great murals are the work of Jose Clemente Orozco.

A nightclub: the sinister Jacson is seen through a grille. Cut again to Jacson, in a smart suit, pacing up and down outside the Trotsky residence; Trotsky with his rabbits; a shot from above of the garden; Gita comes out of the gate and into Jacson's waiting car (he's posing as a wealthy 'import-export' man). What Gita does *chez* Trotsky we don't gather.

Trotsky tells his grandson, Seva (Marco Lucantoni), that he should have been alive in 1917: 'I could have done with a young man like you.' For our convenience, Trotsky then summons Sheldon Harte (Carlos Miranda) down from guarding the roof and gives him a history lecture culminating in the promise that the international workers' state will eventually triumph. Harte, who has come from New York to guard Trotsky, looks as if this is a revelation.

The painter Ruiz is discovered walking purposefully in the city outskirts beside walls covered in cartoon paintings; in an urban wasteland the conspirators unload and transfer a huge crate; against the background of yet more murals, they put on their false police uniforms, laughing, and arm themselves. Losey works in some nice choreography here, bringing together the enigmatic Jacson and the Communist agent Felipe (Guilio Del Prete) in a succession of carefully contrived settings: decaying buildings, balconies, staircases, more murals, and the top of the cathedral overlooking the city (whose ominous great bell is a straight steal from Buñuel's *Tristana* – the very sound of it is liable to send Jacson/Delon rushing up and down a fancy staircase, all by himself). Felipe lifts Jacson's dark glasses to see whether the eyes behind them belong to a true professional: they belong to a favourite Delon role, pure psychopathology, catatonic inversion.

Headlamps in the night: Ruiz's attack on Trotsky's house is done in the thriller genre, culminating in gunfire, shattering glass and an anti-climactic shot of Trotsky and Natalya (Valentina Cortese) dumbly holding each other in their nightwear when it's all over. The police, led by Colonel Salazar (Giorgio Albertazzi), search Trotsky's house and grounds the following morning. Salazar is sceptical: how did Trotsky and Natalya survive despite walls blasted with huge bullet-holes? Why did none of Trotsky's guards fire back? Burton talks his way out of it with a few actor's tricks.

The bullfight scene is so protracted that it seems to belong to a drama documentary or travelogue. It begins with shots of carved bulls on the stadium roof, silhouetted against the sky, a motif to which Losey returns. A poker-faced Jacson and an increasingly distressed Gita are found in the huge audience. The bullfight moves slowly through all its phases. When Gita complains about the blood and brutality, Jacson shouts at her hysterically; they chase each other out of the stadium through empty gangways and concrete architecture. Losey then shows us the horses dragging the dead bull through the sand and, later, the carcass being carved by butchers.

Now a holiday scene in a pleasure park, water, an organ grinding in the background: the camera pans slowly, Jacson and Gita relaxing in a pleasure boat. He turns over to gaze in the still water and we see the head and shoulders of Stalin; judging by his expression, Jacson also sees him, but Gita notices nothing. A rare moment of humour arrives when Jacson, hanging around in the street outside Trotsky's house, waiting for Gita and bantering

with one of the Trotsky guards above, explains his profession: exploiting and
stamping on the masses. This disreputable activity Delon mimes brilliantly.
Gita finally introduces Jacson to the Trotsky household.

Jacson is drifting through a junk market; his neutral gaze falls on an ice-
axe. Soon Delon and another weapon, his knife, are doubled in a bedroom
mirror. Cut to Trotsky's hand fingering the revolver that lies on his desk
beside the alarm button; he then resumes his interminable monologue about
himself while the camera examines the back of his head. Soon Jacson–Delon
will start appearing at suspicious angles at open windows.

Jacson has written an article and now sits on the corner of Trotsky's desk
as the old man begins to read it – a posture suggesting insolence and menace.
(Significantly, Gita's critical position towards Trotsky on the issue of defending the USSR against fascism is spoken on her behalf by Jacson. Her real
identity as a political intellectual is subsumed in the film by sexual hysteria.)
The ice-axe is hidden in the sleeve of Jacson's coat, which he hasn't taken off.
As Trotsky corrects Jacson's essay like a don, the assassin moves behind the
great man's chair, and awkwardly removes his coat. Back of Trotsky's head.
Jacson's nerve fails him, or is it scruple? and he leaves. This is brilliantly
directed; Losey progressively moves Jacson from the side or rear of the frame,
underscoring his outsider status, until he joins Trotsky in pictorial centrality,
poised to kill. Jacson returns for another tutorial and bangs the axe into
Trotsky's skull. Jacson–Delon's face in the final frame dissolves into a still
photograph of Jacson Mornard Mercader.

A Doll's House

From the opening shot across a skating pond, Losey captures not only the
intense cold of a northern winter but the ability of a prosperous community
to keep itself warm with furs, well-constructed houses and great, tiled stoves.
Nora (Jane Fonda) and her friend Kristine (Delphine Seyrig) run out of the
snow into a café, Nora scoffing cakes and terribly happy about her forthcoming
marriage to the wonderful Torvald. She hurries away into her busy life,
leaving Krogstad (Edward Fox) to steal in out of the cold and confront
Kristine's rejection of him. To these human exchanges Losey adds another
action of his own, a succession of double images in the café window reflecting
both the 'Brueghel' scene outside and the cosy eating and gossiping within. A
wonderful start. As Hirsch comments, 'the deep focus which presents two
fields of action within the same frame, immediately announces the kind of
visual generosity and complexity that Losey will use throughout. Spatial
depth is reinforced by Losey's use of natural sounds: the jangling bells of
horses' harnesses and of sleighs traveling through the snow, tolling church
bells, the music of Christmas revelers . . .'[2] Ticking clocks augment the

underlying tension in the domestic scenes; the sound engineer, Peter Handford, produced an unnerving tinkling that cuts through the air like a blade through ice at crucial moments.

The action moves forward to the point where Ibsen begins. Seven years have passed – but no one looks a day older. Returning from an unhappy marriage and years of grinding employment, Kristine is uncreased by time and toil despite Nora's 'You've changed so, Kristine!' Delphine Seyrig captures Mrs Linde as the embodiment of womanly common-sense and self-sacrifice, as well as the energetic do-gooder who decides that everyone will benefit if the truth is told. Yet she gives the faint impression of an outsider – a foreign governess, perhaps – rather than a native Norwegian. Edward Fox plays Krogstad as a basically decent character with a weakness for blackmail and debasement whenever fate deals him yet another cruel blow. The screenplay helps to humanize him, notably the scene showing him playing with his children in the snow; Fox subtly captures the alternating 'Jekyll and Hyde' in Krogstad, the erosion of his confidence to the point where only Kristine's love can restore him to the man he might have been.

The décor in the solid, bourgeois home of Torvald and Nora Helmer indicates the imposition of his personality, not hers. The rooms are sedate, heavy, ordered, richly brown, with no place for Nora's spontaneity. Indeed Nora normally appears in black dresses, an ironic counterpoint to her lively personality, as if the condition of wife is a version of widowhood. Losey's camera tends to keep its distance from Fonda's pretty but rather set features, in which the lustrous eyes do most of the work. The middle-distance camera allows her to capitalize on her greatest strength, body-language, to convey the life-embracing energy of a young wife determined, if possible, to 'love' her husband against all the odds. At the height of Nora's 'Krogstad crisis' she straightens her hair in a mirror before grasping the nettle of Torvald. Tracking in on the mirror, Losey establishes this as the moment when she is most deeply aware of her divided self and the choice confronting her. But one never quite shakes off the impression that the lugubrious Torvald has married an American.

David Warner's portrayal of the vain, insipid Torvald, who regards himself as the natural centre of Nora's universe, and who refers to her as his little doll, his little pet, his little darling, is chilling. Torvald's lank, smarmed hair and monstrous sideburns suggest a repulsive vanity, not least when he simpers lasciviously or wets his lips at the prospect of managing the bank. Mean-hearted and bourgeois, puffed up with white-collar male pride, he is unable to distinguish between a woman and a child. Moments before his downfall, the spectacle of Nora performing her tarantella at a Christmas party reduces him to smirking lust: this seductive doll is my doll. Losey films the final confrontation in the simplest two-shot way, the camera unobtrusively at the service of the acting and the dialogue. Both Fonda and Warner rise to the occasion.

The Romantic Englishwoman

Titles and credits come up against a view from inside a train travelling through snow-covered hills lined with fir-trees. Images are reflected in the train's windows – variations on a woman and the landscape. As the train pulls into a station, Elizabeth Fielding (Glenda Jackson) asks, 'Where are we?' while a customs officer examines the bags of a handsome passenger – this is Thomas (Helmut Berger) – who has taken the precaution of hiding one of them in the toilet: drugs.

We arrive in Baden-Baden, coachmen in top hats, on a sombre morning. Elizabeth travels in an open fiacre under a grey sky, with Thomas twice glimpsed in his white raincoat. When she enters her hotel he is already there, watching. Thomas climbs out of a dormer window on to the hotel roof to hide a small packet of (presumably) heroin in a drainpipe.

We move to the interior of a well-appointed house in England, where Lewis Fielding (Michael Caine) is listening to Herman, a film producer (Rene Kolldehof), expounding his idea for a film script about a New Woman who has gone on a voyage of self-discovery. Lewis calls the storyline 'boring', adding that he doesn't know why his wife has gone to Baden-Baden 'where there are a lot of Huns'.

Back to the glittering chandeliers of the hotel in Baden-Baden: carvings, satin curtains, fine ceilings, pilasters, black-tie gambling, elegant croupiers, gilded guests, tapestries, the camera panning slowly to take it all in. Outside we see majestic steps, high columns, Elizabeth in a black cloak moving among electric lights shaped like old gas lamps. She takes a telephone call from her husband in the lobby. Lewis has a vision of his wife being fondled in a hotel lift with her skirt rising up her legs. Moments later she actually finds herself in the lift with Thomas. The photograpy of the lift and lift shaft is brilliantly and enigmatically done.

Elizabeth arrives home in good sensible clothes, moving among modernist paintings *après* Magritte and *objets d'art*. Tense with suspicion, she examines bed sheets and wash bags for signs of infidelity. When Lewis returns she flirts with him in the garden by night, throws off her dressing-gown, and has begun to make love on the lawn when interrupted by an oblivious neighbour (Tom Chatto) who talks about the hard winter and his fruit trees.

Lewis, seated in front of his typewriter, is harassed by domestic intrusion: vacuum cleaners, dishwashers, cleaning ladies. A visit from Isabel (Kate Nelligan), a newspaper reporter inquisitive about Elizabeth's trip to Baden-Baden, provokes him to shout and rant. 'Piss off,' he says. 'Lewis, you fucking pig,' Isabel says – adding that 'women are an occupied country'.

In bed Lewis informs Elizabeth that he's writing a film thriller about a woman who goes to Baden-Baden: he wants to know what really happened between her and the poet Thomas: 'I want to write him in.'

In flight from drug gangsters, Thomas arrives in England and steals some Lewis Fielding paperbacks from a station bookstall, flipping through them on the train, then tossing them out of the window. When he telephones the Fielding home, Lewis invites him to tea. Elizabeth is indignant.

Lewis is writing in his study, with gold watch and huge cigar, earphones relaying a classical cassette. His fictional heroine (called 'Caroline' in Tom Stoppard's script but identical to Elizabeth on the screen), is explaining that she came abroad to discover herself – women are an occupied country. As she turns to the poet a great lorry comes roaring towards the plate-glass window of the hotel where they are sitting. Lewis rejects the scene – 'Bollocks' – and tosses the page into the basket. But why is 'Caroline' physically identical to Elizabeth? It's also unclear how the poet in Lewis's 'Baden-Baden' imagination exactly resembles Thomas/Helmut Berger before Lewis has set eyes on him.

The real Thomas arrives: 'Nice house.' Lewis leads him into an urbane conversation about gigolos and what they do. Thomas remarks that Lewis's fiction demonstrates that coincidences don't occur only in bad art. The two men banter about the burden imposed on possessors to hang on to what they've got – jobs, wives, cigars. Lewis invites him to stay for dinner – then for the night. A curved mirror observes the dinner table. Elizabeth is outraged. What is Lewis's motive? To test her fidelity? To provoke her infidelity? Simply to observe Thomas for his screenplay? The flashbacks to the Baden lift become so baffling that even scholars studying each frame in tranquillity are puzzled.[3]

In a bedroom scene Lewis asks Elizabeth if she's discontented. She replies, 'I would be, but I don't feel I have the right.' Later she reads the identical words from the mouth of the fictional wife on the page of his typewriter.

ELIZABETH: Bloody hell! Do I get a percentage?'
LEWIS: You already get half. What more do you want?

The parasitic Thomas is meanwhile making himself at home.

LEWIS: As a matter of interest, do you ever say thank you?
THOMAS: I must be grateful that you have all this? You are the one who should be grateful.

The au pair, Katrina (Beatrice Romand), takes a fancy to Thomas, further angering Elizabeth – in this respect Thomas is a Gevacolor reproduction of the insolent young male intruder played by Bogarde in *The Sleeping Tiger* twenty years earlier.

Elizabeth takes Thomas out for an evening and pays the bill. Arriving home, they kiss for the first time, then step into the garden-level gazebo, which is visible from Lewis's study window. Lewis either sees his wife, or imagines his fictional 'Caroline', being screwed by Thomas, or 'Thomas', in the gazebo: either way, it happens, as in the Baden lift, in the vertical. Close up of Elizabeth (or 'Caroline') in ecstasy. Then another shot of the Baden lift and a reverse zoom – all very deep on the surface.

Elizabeth and Thomas escape to France in her Austin Mini. (When Glenda Jackson read the script she was unconvinced that Elizabeth would leave with Thomas.) At this juncture the film breaks its own back, flopping into the mode of a half-parodic thriller-travelogue with helicopter shots of Lewis's Rolls Royce Corniche driving through mountain mists at dawn – much lovely photography of villages and hills. Rolling on a Riviera bed with Thomas, Elizabeth laughs and lets money fall from her hands (echoes of *Eve*). Insultingly Thomas recounts how he deceives old ladies by describing himself as a poet (*Son of Boom!*). The camera, meanwhile, is soaking up ornate gardens, statues of the Virgin, cypress trees, with glorious sweeps of azure sea in deep focus under slanting light.

Clearly the script was frequently abandoned: for example, new mauve pages for shot 261 indicate Thomas vaulting over the balcony and falling to the rocks below. His attaché case bursts open to reveal his poetic notebooks.

ELIZABETH: Yes . . . he was a poet.

But this is not in the film. In the film the punch-line is delivered to Elizabeth well before the final shot:

LEWIS: He phoned me. He told me where to come.

49

The Films
in Parts XI and XII

MR. KLEIN, LES ROUTES DU SUD, LA TRUITE

Mr. Klein

Against the credits, a large tapestry of a vulture pierced by an arrow.

Occupied Paris, 1942. A middle-aged woman, probably Jewish, is being medically examined by a dispassionate doctor for 'Semitic' characteristics. The woman's face is pulled and measured with ruler and calipers – like a horse. She is made to walk naked across the consulting room in a cold light composed of faint grey, green and brown – Losey and Trauner subtly assemble the colours of a corpse. The woman listens to the doctor's verdict, her face visible above the dressing screen. 'Hips wide and flabby could be Jewish could be Armenian could be Arab.' The final medical report will reach her through the Prefecture of Police.

(Michel Ciment commented that this first scene, shot at the Hôpital Cousin de Mericourt, Cachan, was essential. 'Without it the film breaks down.' However, the weak aspect is that the ordeal is inflicted on a woman we will never meet again; Klein himself will not be subjected to the indignity of this medical examination before his deportation as a 'Jew' – an inexplicably lost opportunity.)

The second sequence establishes the good fortune (racial, financial and sexual) of the wealthy art dealer Robert Klein. We see his young mistress, Jeanine (Juliet Berto), in a silk slip, wriggling impatiently on a bed while the voices of two men bargaining filter up the stairs. Cut to Klein (Alain Delon) in a silk dressing-gown, conducting his business in a leisurely, rather contemptuous fashion, buying a painting from a fleeing Jew (Jean Bouise) – *Portrait of a Gentleman* by Van Ostade. Klein is bidding the Jew down from 600 louis to 300. The gold coins are hidden in an ornament in the bathroom; Klein shoos Jeanine out of sight before extracting the money – a telling touch – before contemptuously scattering the money on the table, forcing the Jew to scoop it up. But, as the Vendeur departs, Klein's downfall begins: a Jewish newspaper, *Information Juive*, has been misdelivered to his address and lies on the floor inside the front door. The Jew hands it to him with a significant glance: you, too?

Klein's profession carries us to an elegant sale-room, where the auctioneer is describing the same tapestry shown with the credits, at the centre of which is a vulture whose heart has been pierced by an arrow: the arrow represents a flicker of Remorse but the vulture (Evil) continues to fly. The auctioneer analyses the symbols – indifference, cruelty, arrogance, avidity – for the benefit of the smart audience. This is symbolism with a vengeance.

Klein goes to a bar: on a window is affixed an official notice, Accès Interdit aux Juifs (Entry Forbidden to Jews). Against evocative street sounds (the growl of 1940s bus engines) he makes a phone call, then visits the Jewish newspaper to complain to the manager, who informs him that the subscription list is in the hands of the police. Klein takes himself to the Commission for Jewish Affairs, where he is not invited to sit down. The police have on file another Robert Klein at a different address (which the official does not divulge). By making his inquiry, Klein has merely put himself into the police dossier. Nervously he touches his hat as he leaves; the seal of his immunity is broken.

There follows another 'establishing' scene in which Klein himself is not involved: officials are seated around a big table in the police conference room with a huge map of Paris on the wall. Despite sinister lighting and chords on the sound track, this scene violates the hermetic convention of 'Kafka Rule One' – we should never know more than K. knows.

With air-raid sirens wailing, we return to Klein's stylishly furnished house where the bored Jeanine is still pining for attention. She reads bits of *Moby Dick* aloud while wriggling on the bed; Klein, meanwhile, works at his desk with his back to her. Jeanine reads him an erotic passage from another work: '. . . imposing and incredible . . . that enormous, tumescent, superb, living, penis'. (Sex and belles-lettres – chic.)

Having discovered the Other Klein's address (22 rue des Abbesses) on the wrapper of the Jewish newspaper, Klein goes in search of his namesake. When he arrives, the concierge (Suzanne Flon) is talking to a pair of plain-clothes men pursuing inquiries about the Other Klein. She claims that he has vacated the apartment, which is now to let. Robert Klein inspects the shabby interior with its evidence of rats, its leaking lavatory and peeling wallpaper. In a pretentious coincidence, he discovers a copy of *Moby Dick* on a shelf, and within it an empty envelope from a photographic studio. When developed the print will show a man and a girl astride a motorcyle, accompanied by his Alsatian in a sidecar. Klein finds a number of other clues: a bullet in a drawer, a dog muzzle.

Returning home, he is confronted by a smart, fashion-conscious woman simmering with anger about finding Jeanine installed. This is Nicole (Francine Berge), mistress to Klein and wife of his dough-faced, cigar-smoking lawyer, Pierre. With women Klein/Delon is invariably cool, debonair: 'he caresses her

[Nicole] absent-mindedly', says the script. In his flat the beau monde is dancing and drinking to jazz on the gramophone, but Pierre (Michel Lonsdale) is disturbed by the arrival of the two detectives we have already encountered outside the Other Klein's apartment. Later:

JEANINE: Why should anyone have anything against you?
(*Robert rouses himself and smiles at her provokingly.*)
ROBERT: Why not?

Another historical-establishing shot not involving Klein follows as groups of police officers and cars rehearse *the Grande Rafle*. Losey uses these rehearsals to spectacular visual effect; his shots of pre-dawn Paris are chilling, beautiful, repellent. We see an empty covered Vélodrome, neatly divided into alphabetically ordered sheep pens, ready to receive the Jews.

A postman delivers the mail in pouring rain. The camera devotedly observes the debonair Delon wandering to his front door and leafing through the day's mail, jazz on the gramophone. (A star is an actor whose most minor motions are accorded the status of a scene.) Klein receives a letter from a woman, intended for the Other Klein. He opens it: 'My darling, what interminable months!' it begins. For our benefit Klein reads it aloud. The writer, who signs herself 'Florence', urges her beloved Robert to catch the 6.15 train for Ivry-la-Bataille 'at all costs'. Klein duly catches it. A chauffeur is waiting with an exotic British car. The ground is ice-bound with patches of snow as the car approaches a vast château whose lights are blazing. (This is the Château d'Esclimont, at Saint-Symphorien-le-Château, near Chartres.)

An old flunkey opens the door, greeting him as 'Monsieur Klein' and offering us a long shot down a magnificent hallway to a striking painting. A wobbling camera tracks along the corridor, as if fixed in Klein's eye-sockets, as he follows the servant and glances sideways at fleur-de-lis wallpaper and fine paintings – and ominous spaces where paintings have until recently been hung. The camera turns a corner, then another, without a cut or break; as it finally enters the great drawing-room, its motions wonderfully convey Klein's hesitation. Wearing only a business suit, he is confronted by *tenue de soirée*. The camera trawls slowly through the dreamy stasis of the very rich. This is blatantly the Effect Syndrome, the aristocracy in its finery listening to music – 'a wave of pure, delicate notes' emanating from a harpsichord, violin and flute, with the fourth instrument, a cello, unused in its case. Was the Other Klein scheduled to complete the quartet?

The lady of the house, Florence (Jeanne Moreau), carries a tray with two champagne glasses to her unexpected guest, who is engaged in polite exchanges with her smoothly courteous husband. The three discuss the mistaken identity, the wrong address, the two Kleins. Clearly the Other Klein is a close friend of the family as well as Florence's lover. Our Klein apologizes and requests to be put on the next train (there is none) or in a hotel (all requisitioned). So he must stay the night: 'You were also invited, after a fashion,' he is told. This leads to the next set-up: a dining-room reminiscent of Visconti's *The Damned*; Losey gives it a single take then cuts.

(To Michel Ciment he explained: 'I wanted that scene . . . to show that there were also Jews who had enough money to simply buy their way out of France without changing their life-style much . . . when I saw the corridor with some of the pictures lying on the floor (the La Rochefoucauld family, who lived there, were getting ready to leave for Argentina), the scene . . . came to me straightaway . . .'[1] When setting up the film Losey had written to la Duchesse Pozzo di Borgo: 'Madame la Duchesse, Perhaps you will remember my visit with a large entourage to your magnificent hôtel particulier when I was looking for locations for "Proust". You received us most graciously and offered us some exquisite marc from your Corsican vineyards.' Now, in *Mr. Klein*, he wished to portray 'a great family, perhaps Rothschild . . . I would very much like you to be the centrepiece of one of the scenes, the matriarch . . .' He apologized for his 'presumption and impertinence'.

Next we find Klein in bed, naked. A fire burns in the grate. Enter Florence, in quest of her own incriminating letter to the Other Klein. Robert refuses to return the letter until she divulges where the Other Klein is to be found. Holding a candle to scrutinize his features, Florence cites the Other Klein's categorization of human beings, comparing her guest to an insect then to a vulture. But the script is lifeless:

FLORENCE: A vulture.
ROBERT: Why? Do I eat dead bodies?
FLORENCE: You don't give back letters.

Robert now understands that the Other Klein's tactic is to conceal himself by disappearing behind someone else. The sound of a motorbike is heard approaching the château. Florence hurries from the room, descends the stairs, and is seen in long shot running across the freezing front drive towards the distant motorbike (spooky, night-owl music). Klein watches from his window, faintly mirrored in the glass. Abruptly the zoom lens pulls in to a medium shot as Florence and her lover briefly embrace before the latter roars away. (But this camera-jump can be available only to us, not to Klein.)

In the morning a group of riders on horseback are assembling for the hunt; dogs bark; despite the war, life as usual. The camera pans over the walls, pictures and *objets d'art* in the bedroom until it finds Klein dressed in a grey trilby hat standing (or posing) next to an oval, eighteenth-century portrait. A gloomy Florence is with him. Pouring coffee, he threatens to go the police. Florence yields and writes down the Other Klein's address. It is Klein's own: 136 rue du Bac. Another historical-establishing shot follows: sorting the files of Jews at the police prefecture. Clack of typewriters. Superfluous.

Inside a baroque church a bishop is administering a confirmation mass –
the consecrated host – to children dressed in white. The Latin sacraments,
the angelic faces of the boys' choir and the rows of proud parents indicate
superficial bourgeois piety. We find Pierre and Nicole in the congregation –
their son is among the children. The lawyer murmurs to Klein: 'Are you
French? French? Or are there by any chance . . . any Jews in your family?'

This question takes Robert Klein to see his father in Strasbourg. The old
man (Louis Seigner), wearing a homburg and swaddled against the winter in
his bath-chair, is being wheeled beside a canal by an elderly retainer. Robert
Klein takes over the chair while his father, oblivious to his arrival, continues
his loud monologue on the theme of Indifference, the ultimate evil. A
greenish light reflects off the canal on to Robert's camel-hair coat. They cross
a weir. At the family home we have a view of Strasbourg cathedral spire
(bells, off), medals in a cabinet, a chiming clock. The old man insists, 'I have
told you: we have been French and Catholic since Louis XIV!' But clearly he
is suppressing his awareness of alien elements in the family tree.

Klein is next discovered with Jeanine in a Paris nightclub watching an all-
male cabaret (Frantz Salieri's La Grande Eugène), surrounded by bloated
bourgeois and German officers with cropped hair. A transvestite figure with a
face painted grey-green is singing the part of a widow, falsetto, outside a
closed food shop. The music is by Mahler. Enter an oily Jew (a poster of Jud
Süss in the background). The audience laughs (perhaps an echo of the anti-
Semitic cabaret performed by country house guests in Jean Renoir's *La Règle
du jeu*). The oily Jew steals a necklace from the lamenting widow. When
Klein joins in the laughter Jeanine rebukes him. As she drags him out of the
nightclub the lovely legs of a chorus line come prancing on.

We are in La Coupole. No wartime rationing here. Ladies stuff themselves
and their Tygers. A page boy moves between the crowded tables with the
name 'Klein' written on a board. The camera tracks with him, picking up the
greedy, absorbed, made-up faces of the ravenous bourgeoisie. Pierre urges
Klein to stop waiting for certificates of origin through the post, and to stop
playing the detective. Klein replies with one of the improbable lines that litter
the script: 'I'm a good Frenchman and I believe in our institutions.' Nor will
he submit himself to medical examination: 'I'm not a horse, I won't have my
face measured.'

Klein retraces his steps to the rue des Abbesses. Agitated, he forces his way past
the protesting concierge, offering her a deposit on the Other Klein's flat, which
she refuses, then pacing the shabby apartment with a new urgency. He examines a
woman's high-heeled, sequinned boot discovered in a cupboard. Finding a fuse-
cord, he puts a match to it then stamps it out. He tears off a page of music covering
a broken window pane. Three clues. The telephone rings. After a long hesitation
he answers it; an unknown woman wants the Other Klein. She rings off.

Klein is found at the stage door of a theatre, looking for the Other Klein's girlfriend, 'Isabelle'. From an exchange with a flirtatious chorus girl he discovers that 'Isabelle' is now working in a factory at Balard. He takes the early train; the Balard Métro sign comes up against bleak winter trees – the camera catching the chill light filtering through high windows as the workers arrive for the morning shift. A crowd of cheerful factory girls, excited by the arrival of 'Alain Delon', gossip about the photograph of 'Isabelle'. A young woman exclaims: 'It's the new girl . . . Françoise! She's in the detonator workshop . . .' But another woman, obviously dedicated to the Resistance, shreds the photograph: 'No. It's not Françoise.' A brilliantly realized scene.

In a succession of brief takes the Paris police are discovered rehearsing the coming pick-up of the Jews before dawn – an iron iconography in funereal grey: a clock tower registers 4.30. We return to the conference room of the police prefecture where the wall map of Paris is scored with arrows. Voices specify the time required to bring Jews from each quarter of the city (another violation of the hermetic K-Effect).

Returning home, Klein loses his temper on discovering that the police have confiscated his passport and are impounding his paintings. His lawyer, Pierre, tries to calm him. Losey uses the staircase as a focal point to convey the invasion and rape of a man's life. The bitterly jilted Nicole finds a sheet of music lying on the piano – the one he'd stripped from the window of the Other Klein's apartment – and plays it loudly, clumsily. It turns out to be the Internationale (though Nicole is oblivious). The police inspector is furious. 'Stop!' But Robert in his rage tells her to go on playing.

Later that night, the camera roams the interior of the house, its walls denuded, until Klein is discovered in a chair, alone with Pierre, who advises him to sell up and head for Mexico. Pierre knows where to obtain a false passport and visa; he will also find a buyer for Robert's business and art collection.

Exterior, day, a news kiosk. Klein, a coat slung over his shoulders, is buying a paper when he is picked up by a hungry stray dog, an Alsatian, resembling the one that the Other Klein is known to have owned. It devotedly follows him home, despite his kicks (does Robert smell the same as his persecuting *doppelgänger*, identical toilette and cologne?) Once home, he fondles and feeds the dog affectionately.

A tearful Jeanine is packing to leave, wilting from his indifference. While she makes a last effort to capture his attention, he remains absorbed in his newspaper, studying the anti-Semitic reports with detachment. Abruptly his mood changes to animation. We don't know what he has read, but we later learn that a group of terrorists have blown themselves up while planning to dynamite Gestapo headquarters. Jeanine departs, suitcase in hand, defeated. Klein hurries to the mortuary in search of the corpse of the Other Klein. A sympathetic attendant pulls out a body; Klein, handkerchief over his nose, peers at the pulp. What does the pulp prove? Why does it confirm his belief that the Other Klein is now dead? Why are the bodies of Resistance terrorists left in the public mortuary, without police surveillance?

Robert Klein has now decided to accept Pierre's advice and leave France. The Gare d'Austerlitz is dark, sombre, full of soldiers, police and thundering boots. Klein brings the dog. Pierre arrives with a false passport, a boat ticket, and what is left of Klein's money after selling off his assets and – Robert himself is in no doubt – taking a big cut for himself (but how has he sold off a collection impounded by the police?). The lawyer does not seem such a bad chap, quite devoted to Klein – who hands him the dog.[2] Behind them we see a young woman (Francine Racette) leaning out of the train saying goodbye to a man whose back is turned to us. Ah!

Inside the train, Klein takes a seat opposite the young woman. The camera chops evocatively, lurching and alive with the gathering pace of the train. Klein now recognizes the young woman as the one in the photograph, the Other Klein's Isabelle/Françoise. She being a female, he becomes super-cool, although there is a vengeful look in his eye. 'Bonjour Isabelle. Bonjour Françoise.' She stares back bleakly, then shows him papers describing her as Nathalie. She admits to knowing the Other Klein but then protests, 'No, you're not a friend of Robert's. You saw each other at the station. He was there . . . And you didn't speak to each other!' (Yet the Other Klein's Alsatian would surely have reacted to his real master's presence a few feet away.) The exchange between Klein and the woman ends when she slaps him in the face. Smoothly he lifts his hat: '*Excusez-moi.*' Convinced that the other Klein is alive, rather than dead in the mortuary, he decides to leave the train and return to Paris.

Rousing Pierre from his bed (with a bitter Nicole lingering in lingerie in the background), Klein at last speaks to the Other Klein on the telephone and sets off through empty streets for the rue des Abbesses, darting into doorways, flattening himself against walls – only to see his quarry being bundled into a police car. The concierge is in tears. Pierre arrives – it is he who has tipped off the police. 'It was necessary. He's a criminal. I was worried for you.' With a cry of rage Robert makes as if to strangle him.

At dawn the police come for Klein. It is the day of the Rafle. Jews are being packed into fleets of municipal buses and herded to the stadium. A woman on the bus expresses her anxieties to Klein. Losing his temper, he snaps at her: 'Nothing to do with me.' As the buses thread towards the stadium we see queues forming at the vegetable market – the rest of Paris is going about its business. A fine shot of Klein's stationary bus, seen through a market stall, its engine growling – the camera work is powerful, evocative. Klein slips a note out through the bus window; a market porter stares at it.

The buses snarl into the Vélodrome, packed with Jews wearing yellow stars. There are painful scenes as the police separate families, pulling children from their mothers, herding each species into its pre-allotted pen. Drifting with the crowds as megaphones call out individual names, Klein hears, 'Robert Klein!' repeated, then catches sight of another man's arm raised in response. Klein pushes through the crowd in pursuit of his Other. (But would the police or the Gestapo deposit a just-captured Resistance fighter, only hours after his arrest, with the mass of Jews heading for the extermination camps? And would such a hardened and resourceful character meekly obey the megaphone?)

Accompanied by the police inspector in charge of Klein's case, Pierre has appeared in the stadium, probably bearing papers of non-Jewish certification and still trying to save his client. But, mesmerized by his pursuit of the Other Klein, Robert ignores Pierre's shouts – Losey uses jerky hand-held cameras to convey a sense of desperation as the frames become increasingly packed with doomed people. 'I was thinking as well of Gustave Doré's etchings for Dante's *Inferno*, which I grew up with.'

Klein is jammed into a cattle wagon with the Jewish Vendeur who sold him the *Portrait of a Gentleman*. The faces of the Jews are pressed to the bars of the cattle wagons passing through a tunnel of darkness and steam towards death. What we hear, voice-over, is Klein haggling with the Vendeur: 'Six hundred Louis . . . No, three hundred . . .' – this reprise is as heavy-handed as the final crash in *Accident*. The cattle wagons fade into a darkened screen – and extinction.

Les Routes du sud

The credits come up against a bleak sea shore, wet sands. Out of nowhere emerge three mounted Red Army soldiers, galloping towards the camera – but the viewer cannot identify them, except as soldiers. Cut to a close-up of the screenwriter Jean Larrea (Yves Montand), on the same Atlantic beach, eyes closed, his imagination at work, against the thundering waves. The Red cavalry of his imagination fade as they reach the camera.

Jean Larrea drives away in his Range Rover. Now we see the German soldier Korpick scrambling across the dunes to the sea, surrendering to the Red cavalry. (But why does this imagined action, set in 1941, not take place in the plains or forests of Poland – why across the Atlantic beach on which the screenwriter Larrea stands? The producer, Yves Rousset-Rouard, comments: 'Perhaps it was the Baltic shore.') Larrea drives back to his comfortable and photogenic country house, where his wife has received a phone call, a summons from the Spanish Resistance. Eve (France Lambiotte) tells Larrea that their son Laurent – making a rare visit to the family home in Normandy

– has need of his father: she will make the journey to Spain. Anxious to continue work on his screenplay, Larrea acquiesces but Laurent (Laurent Malet) sulks when his mother tenderly informs him of her journey. Spain again? 'You're in 1975,' he grumbles. He's still sulking when they bid Eve farewell at Cherbourg station.

Laurent, a rebel without a cause, bicycles off to practise shooting in a field. Larrea follows him by car, bringing his revolver. After some target practice (at which the father prevails), Laurent asks why his parents are always running messages in Spain: nostalgia? habit? Why not assassinate Franco instead? Larrea responds with what Laurent doesn't want – another personal story about the Spanish Civil War.

Larrea's screenplay again: the Russians receive the order to shoot Private Korpick – a good Communist who has come to warn the Red Army of Hitler's imminent attack – as a spy or *provocateur*. Korpick is executed. Laurent, however, is weary of his father's obsession with Stalin's crimes: 'And are there producers who pay you to write *that*? Well, those guys must be masochists.'

But those guys are not. Enter (as in *The Romantic Englishwoman*) a film producer, a Hungarian Jew, to read Larrea's new script. He expresses polite admiration but won't buy it. They drink champagne.

News of Franco's execution of five young anti-fascists in September 1975. French television shows counter-demonstrations in Paris; suddenly Laurent comes on-screen, being beaten by the CRS riot police. On the steps of the Palais de Justice in Paris, Larrea meets Laurent and his lawyer but inexplicably elects to mock his son: 'I thought Spain was folklore, old-fashioned nostalgia . . .' Laurent is equally scathing: 'Perhaps I should thank you for being a famous father. It impresses the magistrates . . . Thank you, Jesus Christ.'

On the street the pretty Julia (Miou-Miou) picks up Larrea and presents herself as Laurent's friend. Larrea is curt with her, but she pursues him back to his apartment (to lively jazz, off) in order to explain Laurent's rebellion: 'There is still only one woman who counts for him, and *you* sleep with her . . .'

On his telephone answering machine, Larrea hears news of a car accident in Spain. His wife Eve is dead.

Arriving at the Barcelona hospital, Larrea and Laurent are handed Eve's travelling bag and personal effects by a nun (the jazz continues, despite the bereavement). While the TV screen shows a Falangist rally in Madrid, an anti-fascist Catalan doctor explicates: 'Today is the first of October. Thirty-nine years ago today, General Franco was appointed Chief of State by the Grace of God . . . To prove that he's still alive he had five young anti-fascists shot last Saturday . . .'

Larrea and Laurent are instructed by the Resistance to report to the Miró Foundation. Here, from a high terrace, they observe the city at their feet, hazed in industrial smoke. Enter Miguel (Jose Luis Gomez). Passing through the beautiful white galleries among the Miró paintings, Miguel informs Larrea that Eve's fatal assignment was to help an underground leader escape from the police. Performing her task perfectly, only subsequently did she crash her car. Laurent, who has been drifting along, heartbroken, displays his anger. Fine acting here from Yves Malet. More Miró paintings.

Back in France, Larrea sits in his Range Rover on his Atlantic beach at dusk and lets the waters rise around it, gazing at a half-sun sliding into the sea, with the vibrant Tania Maria's voice off. There follows a rather baffling flashback of Larrea murdering two French collaborators during the war, a farmer and his wife. It takes place at night, in shadow; only at the last moment do we see the assassin's face – Yves Montand's, and not a day younger, despite the passage of thirty-three years. Laurent and Julia turn up, soaked to the skin. Larrea is as nasty as possible (but why?): 'What the fuck are you doing here?' Laurent loses no time in reverting to his habitual scorn: 'Twenty years devoted to the cult of Stalin and then twenty years devoted to wondering how he could have fallen into that trap . . .' Angry and distraught, Larrea spins the barrel of his revolver, presses it to his temple and pulls the trigger. He then tosses it to his son, who is about to take up the challenge when Julia grabs it. The gun goes off and shatters an object. Larrea and Laurent hug one another in horror, Laurent weeping in his father's arms. Laurent departs, Julia stays. She tells Larrea that he has chosen to be what he is. 'You want to be loved,' she says, 'but your money is you.' Upstairs they make Laurent's bed together, then she lies on it and unzips her jeans.

Julia brings more '*marginales*' into the house (though happily some pseudo-lesbianism and a chic reference to Walter Benjamin are cut from the film). Julia and her friends are watching an old video of Larrea and Eve being interviewed on TV. Larrea observes their boredom; enraged, he sends the two visitors packing.

Larrea is gazing out of the window of his Paris flat when Laurent appears with his mother's travelling bag and her diary, discovered in a hidden compartment. Larrea must now confront the truth about his wife's affair in Spain with Miguel. Later Larrea finds Julia hunched up inside his car in an underground car park.

JULIA: I know you're going back to Barcelona . . . Take me . . .

Here the storyline fatally substitutes Julia for Laurent; an explosive chemistry yields to a pallid geometry. Arriving at a Mediterranean hotel in Catalonia, they hear martial music: Franco is dead. Julia curls herself on the bed and asks Larrea what he wants: to suffer? vengeance?

In Miguel's spectacularly luxurious flat Larrea and Miguel drink champagne and chain-smoke their way through the immediate political prospects, Parkinson's Law and the revolution of the masses. Julia sits apart, aloof. She tells Larrea: 'Every time they talk about politics, it's like a prayer wheel ... They're petty-bourgeois intellectuals ...' – then she quotes, improbably, from Malraux's novel, *L'Espoir*.

Back in Paris, Laurent and Larrea hold a firelight conversation about nothing much.

Don Giovanni

See Chapter 42.

La Truite

The titles come up against the rushing water of a trout farm. A slender young woman with long, carrot-coloured hair and a blank, sulky expression is squeezing the roe from the trout. This is Frédérique (Isabelle Huppert). With faint disgust she watches from a window as an old lecher lifts the skirt of a squealing girl in the yard.

Cut to a ball hurtling down a gleaming, ultra-modern, upmarket bowling alley and crashing into white skittles. The camera tracks to show Frédérique playing with her effete young homosexual husband, Galuchat (Jacques Speisser). Frédérique is wearing a white T-shirt with '*Peut-être*' printed on the breast and '*Jamais*' on the back. Enter a smart party of four: Rampert (Jean-Pierre Cassel), his wife Lou (Jeanne Moreau), his younger business colleague Saint-Genis (Daniel Olbrychski) and Mariline (Lisette Malidor), a tall, fashionable black beauty with a shaven head. (Losey had interviewed this *Folies Bergères* star for a role in his unrealized film, *Silence*.)

The suave businessman Rampert is instantly fascinated by Frédérique, to the weary distress of Lou. Saint-Genis also registers a calculating interest. Mariline rolls a bowl listlessly, then stretches her long arms across a large fish tank. Rampert challenges Frédérique to play for money. 'I'd like to play with you and lose,' he tells her. 'I want you to beat me.' The background rock from the discothèque is continuous. Jeanne Moreau goes home in a taxi, weeping.

Flashback to the Jura. The local count (Roland Bertin) is smoking a cigar while trying to seduce Galuchat under the observant eye of Frédérique. Galuchat evades a kiss but lays his head limply on the count's shoulder. Two lecherous old men are fishing – Frédérique's father and Père Verjon. It's a wicked world, all round.

In the bowling alley Rampert is talking big business deals. Frédérique bops on the dance floor with the young Saint-Genis. 'I won't sleep with you,' she tells him. Promptly he invites her to accompany him to Japan on a business trip. The offended Mariline descends a staircase elegantly: 'I can't bear women who ignore other women,' she drawls, passing between statuettes.

Losey repeats the motif: cut to a grand reception where guests are descending a splendid staircase-plus-statue. Financiers and diplomats sit down to a sumptuous dinner in honour of the opera singer Ruggero Raimondi.

YOUNG LADY: There's a rumour you're leaving opera for cinema.
RAIMONDI: I'm like Don Giovanni. I like taking risks.

A galloping Frédérique and her battered suitcase turn up at Charles de Gaulle airport to take the plane to Japan with Saint-Genis. In the almost empty plane she wears a seductively guarded expression – her first flight. So to the streets of Tokyo, neon lights, athletes marching and chanting in unison, sampans, and general travelogue. A TV soap opera is playing inside the limousine. Saint-Genis enters a bank while Frédérique is driven away to their hotel, observed through a plate-glass window. In a handsome, marbled interior with leather chairs, Saint-Genis embraces his elderly Japanese boss, president of a multinational, Daigo Hamada (Isao Yamagata). Losey works his way through the statues and *objets d'art* of Saint-Genis's office; a carved white cat sits on his desk while his two Japanese female secretaries bow and twitter.

At the hotel Frédérique observes a panorama of the city through glass walls, then embarks on a guided tour of the luxurious suite, open floor spaces, sliding doors, low furniture.

Back in Paris, in a hairdressing salon, Saint-Genis's black mistress Mariline tells Lou, 'I've just sold a dubious Dali to a Brazilian.' Multiple mirrors monitor their exchange.

Frédérique returns to the Tokyo hotel from a shopping spree and meets a jolly American millionairess (Alexis Smith) who instructs her in the ways of love in a French accent as thick as a tatami mattress. 'I've made love 33,000 times in thirteen capitals. I never do it in cities of less than 100,000 inhabitants. I've touched the sky. Give up your Western sense of sin.' In case this isn't enough, the giggling Frédérique drops her armful of parcels 33,000 times.

She is discovered watching Japanese wrestling on television while having her hair cut very short and observing herself in three mirrors. She wears a succession of new outfits. In a smart Tokyo restaurant she makes eyes at the handsome cook across spits of roasting meat. Soon they are together in the red-light quarter, where they meet the cook's mother. Much hugging and giggling. Inside a bar we catch a glimpse of an old soak (Losey himself) who

raises his glass to Frédérique as she enters (sad to watch). Later she and her
cook are driven by chauffeured car to meet up with Saint-Genis and his sleek
Japanese girlfriend in another ultra-smart set-up – more glittering leather
sofas. The camera roams over a vulgar silver statuette of a nude female as
Frédérique goes into her bopping routine.

Saint-Genis takes Frédérique on a boat trip up a turbulent, rock-strewn
river. She won't let him kiss her. The oarsmen groan rhythmically. We
sympathize.

Back in the Jura – flashback – old Verjon is chopping wood while
Frédérique parades herself round him, seductive and cold. Inside a barn
stacked with wood she shows him her small breasts. His eyes popping out of
his head, he reaches into his pocket for money, his shaking fingers separating
the notes, which she tucks inside her *poitrine*. Fine acting from Jean-Paul
Roussillon, the star of the film.

In Japanese woodlands Saint-Genis brings Frédérique to the country
mansion of the elderly entrepreneur, Daigo Hamada. Servants bow, beautiful
vases glimmer – another guided tour. In the garden Frédérique examines
some large carved dogs.

Childhood: giggling girls run into shot. In a loft Frédérique and two
girlfriends bare their breasts to an altar and take a libation – this is the secret
club of those who have vowed to exploit men (Société des Vraies Luronnes).
They then stick needles in photographs of the various local lechers. Père
Verjon gets one in the penis. More giggling.

In their Tokyo bedroom Frédérique takes off her dress to view her naked
self in a multiplicity of mirrors, with Saint-Genis undressing in the
background. Side by side in bed, she frigidly keeps her back to him, fights
him off, jumps out of bed and curls up like a skinny cat in a chair.

An urgent cable from Rampert summons her home from Tokyo; left
behind, Galuchat has attempted suicide. On the flight back to Paris she
travels with the tycoon Hamada, who tells her what a haiku is. 'I'd really have
liked a father like you,' she says – then another flashback to the Jura where
she is found selling a fake story of her pregnancy by someone unnamed to
Père Verjon, from whom she wants 800 francs to fix it. Visiting Père Verjon's
chaotic and grubby room overlooking the millrace, she flirts with the drunken
old man, hops around, fingers his unmade bed in a nasty way, takes his
money, giggles and starts throwing his stuffed pikes out of the window,
including his prize mounted one – it hurtles over the millrace while Verjon
weeps on the floor.

Galuchat is convalescing at the Ramperts' fabulous and ugly-modern
country residence while begging for drink and pills. Frédérique arrives and
keeps changing her clothes. The maddened Rampert appears in a double-
breasted suit and grabs her: '*Laisse moi!*' she commands. 'I'm doing it all for

you,' he pleads. 'I broke a bank for you.' 'I hope it's a big bank,' she says. This is baffling – Rampert hasn't had an opportunity to spend a cent on her since the bowling encounter.

Lou and Frédérique do a lot of walking about arm-in-arm. Lou remarks that if Rampert had 'gone with men' rather than women he would have had fewer problems. Moments later, in apparent contradiction, she tells Frédérique, as they walk across the lawn at night, that there are no longer homosexuals and heterosexuals: 'You're sexual or you're not.'

A massive conjugal row carries Lou and Rampert through virtually every room in their house and past some obvious furniture until Rampert strikes her on the back of the head. They both lie on the floor, groaning. Lou dies. Jean-Pierre Cassel staggers about in a display of embarrassing ham acting. Later we find him in a psychiatric ward awaiting trial. 'I want to see Lou,' babbles the broken bourgeois.

Cut to a panoramic helicopter shot of a cortège of black limousines arriving at Daigo's Japanese trout farm. Frédérique and Galuchat now manage it for an international company whose American and Norwegian executives are on a visit – but Galuchat is a pathetic drunk and Frédérique is in charge. 'Let's go,' she says in American, taking the company President's arm. View of snow-capped mountain.

Steaming

See Chapter 44.

Joseph Losey's Professional Directing Credits

Little Ol' Boy, by Albert Bein (1933, Playhouse Theatre, New York). ITEM 31.

A Bride for the Unicorn, by Denis Johnstone (1934, Brattle Theatre, Cambridge, Mass.). ITEM 31.

Gods of the Lightning, by Maxwell Anderson (1934, Peabody Theatre, Boston).

Jayhawker, by Sinclair Lewis and Lloyd Lewis (1934, National Theatre, Washington; Garrick Theatre, Philadelphia; Cort Theatre, New York). ITEM 30.

Hymn to the Rising Sun, by Paul Green (1936, Civic Repertory Theatre, New York). ITEM 33.

Conjure Man Dies, by Rudolph Fischer (1936, Lafayette Theatre, Harlem). ITEM 46.

Triple-A Plowed Under (1936, Living Newspaper, Federal Theatre Project, Biltmore Theatre, New York): ITEMS 33–4.

Injunction Granted (1936, Living Newspaper, FTP, Biltmore Theatre, New York). ITEMS 33–4.

Who Fights This Battle?, by Kenneth White (1936, Theatre Arts Committee, Delanee Hotel, New York). ITEM 33.

Sunup to Sundown, by Francis Faragoh (1938, Hudson Theatre, New York). ITEM 35.

Pete Roleum and His Cousins, 20-minute animated puppet cartoon film (1939, Petroleum Industries Exhibition, New York World's Fair). ITEM 27.

A Child Went Forth, 18-minute film (1940, National Association of Nursery Educators).

Youth Gets a Break, 20-minute film (1940, National Youth Administration). ITEM 28.

Day of Reckoning, radio drama series (1943, NBC Radio). ITEMS 29, 33.

A Gun in his Hand, 19-minute film (1945, MGM).

The Great Campaign, by Arnold Sundgaard (1947, Princess Theatre, New York). ITEM 32.

The Life of Galileo, by Bertolt Brecht (1947, Coronet Theatre, Los Angeles; Maxine Elliott Theatre, New York). ITEM 37.

The Boy with Green Hair, 82-minute film (1948, US; 1950, UK. RKO-Radio. A Dore Schary Presentation). *** ITEM 1.

The Lawless (UK: *The Dividing Line*), 83-minute film (1950, Paramount. A Pine-Thomas Production). *** ITEM 2.

The Prowler, 92-minute film (1951, United Artists. Production Company: Horizon Pictures). ITEM 3.

M, 88-minute film (1951, Columbia. Producer: Seymour Nebenzal). ITEM 4.

The Big Night, 75-minute film (1951, US; 1953, UK. United Artists. Producer: Philip Waxman). ITEM 5.

Stranger on the Prowl (*Imbarco a mezzanotte*), 80-minute film, Italian version 100 minutes. Released under a directorial pseudonym, Andrea Forzano. (1952, 1953, United Artists. Production Company: Produttori Cinematografaci Tirrenia).***

The Sleeping Tiger, 89-minute film directed by Losey under the pseudonym Vincent Hanbury (1954, Anglo-Amalgamated. Production Company: Insignia). * ***

The Wooden Dish, by Edmund Morris (1954, Phoenix Theatre, London).

The Night of the Ball, by Michael Burn (1955, New Theatre, London).

A Man on the Beach, 29-minute film (1955, Hammer Films). *

The Intimate Stranger (US: *Finger of Guilt*), 95-minute film directed by Losey under the pseudonym Joseph Walton in the UK, direction attributed to Alec Snowden in the US. (1956, Anglo-Amalgamated (UK), RKO-Radio (US). Production Company: Anglo-Guild). * ***

Time Without Pity, 88-minute film (1957, Eros. Production Company: Harlequin). *

The Gypsy and the Gentleman, 107-minute film (1958, Rank. Producer: Maurice Cowan). * ***

Blind Date (US: *Chance Meeting*), 95-minute film (1959, UK; 1960, US. Rank (UK), Paramount (US). Production Company: Independent Artists). * *** ITEM 6.

The Criminal (US: *The Concrete Jungle*), 97-minute film (1960, UK; 1962, US. Anglo-Amalgamated (UK), Fanfare (US). Production Company: Merton Park Studios). * ITEM 7.

The Damned (US: *These Are the Damned*), 87-minute film (1963, UK; 1965, US. Columbia. Production Company: Hammer/Swallow). * ITEM 8.

Eve (*Eva* in Europe), a film cut to varying lengths (1962 France, 1963 UK; 1965, US. Gala (UK), Times Film Corporation (US). Production Company: Paris Films/ Interopa). * *** ITEM 9.

The Servant, 115-minute film (1963, UK; 1964, US. Warner-Pathé. Production Company: Springbok/Elstree). * *** ITEM 10.

King and Country, 86-minute film (1964, UK; 1965, US. Warner-Pathé (UK), Allied Artists (US). Production Company: BHE Productions). * ITEM 11.

Modesty Blaise, 119-minute film (1966, 20th Century-Fox. Producer: Joseph Janni). * *** ITEM 12.

Accident, 105-minute film (1967, London Independent Producers (UK); Cinema V (US). Production Company: Royal Avenue, Chelsea). * *** ITEM 13.

Boom!, 113-minute film (1968, Universal. Production Company: World Film Services–Moonlake Productions). * *** ITEM 14.

Secret Ceremony, 109-minute film (1968, Universal. Production Company: World Film Services). * *** ITEM 15.

Figures in a Landscape, 110-minute film (1970, UK; 1971, US. Cinema Center Films (CBS). Production Company: Cinecrest Films) ** *** ITEM 16.

The Go-Between, 116-minute film (1971, MGM-EMI (UK); Columbia (US). Production Company: World Film Services). * *** ITEM 18.

The Assassination of Trotsky, 103-minute film (1972, Anglo-EMI (UK); Cinerama Releasing Corporation (US). Production Company: Josef Shaftel/Cinetel/Dino de Laurentiis). ** *** ITEM 19.

A Doll's House, 106-minute film (1973, British Lion (UK); ABC-TV (US). Production Company: World Film Services). ** ITEM 20.

Galileo, 145-minute film (1975, US; 1976, UK. Cinevision/American Film Theatre. Production Company: Ely Landau Organization). * *** ITEM 21.

The Romantic Englishwoman, 116-minute film (1975, Fox-Rank (UK); New World Pictures (US). Production Company: Dial/Meric Matalon). ITEM 22.

Waiting for Lefty, by Clifford Odets (1975, Dartmouth Players, Warner Bentley Theatre, Hanover, New Jersey).

Mr. Klein, 123-minute film (1976, Quartet Films (US); Gala (UK). Production Company: Lira Films, Paris). ** ITEM 23.

Les Routes du sud, 97-minute film (1978, Trinacra/FR 3, Paris/Profilmes, Barcelona). ** ITEM 24.

Don Giovanni, 176-minute film (1979, France and US; 1980, UK. Gaumont. Production Company: Gaumont/Camera One/Antenne 2/Janus Films). ** ITEM 25.

Boris Godunov, opera by Mussorgsky (1980, Paris Opéra). * (video) ITEM 43.

La Truite, 105-minute film (1982, France; 1983 US and UK. Gaumont (France), Triumph Films (US). Production Company: Gaumont/TF 1/SFPC.) ITEM 26.

Steaming, 95-minute film (1985, Columbia. Production Company: World Film Services). ITEM 26(a).

References

ABBREVIATIONS

JL   Joseph Losey
PL   Patricia Losey
Int.   interview
Letter   letter to the author
Item   item number in the BFI's Losey Collection
BFI   British Film Institute
BMFB   BFI *Monthly Film Bulletin*
Cahiers *Cahiers du Cinéma*
Ciment   Michel Ciment, *Conversations with Losey*, London, 1985 (*Le Livre de Losey*, Paris, 1979)
F&F   *Films and Filming*
Hirsch   Foster Hirsch, *Joseph Losey*, Boston, 1980
Milne   Tom Milne, *Losey on Losey*, London, 1967
NS   *New Statesman*
NYT   *New York Times*
S&S   *Sight and Sound*

Note: A reference number refers to all the cited references within a paragraph of text.

INTRODUCTION

1 Tom Milne, *Losey on Losey*, London, 1967; Michel Ciment, *Le Livre de Losey*, Paris, 1979, updated and translated as *Conversations with Losey*, London, 1985. See also James Leahy, *The Cinema of Joseph Losey*, London, 1967; Pierre Rissient, *Losey*, Paris, 1966.
2 Beverle Houston and Marsha Kinder, *Film Quarterly*, Fall 1978.

Part I Royal Avenue

I   *THE SERVANT*: I

1 Ciment, pp. 241–2.
2 Dirk Bogarde, *Snakes and Ladders*, London, pp. 225, 229, 235.
3 JL quoted by Jacques Brumus, 'Joseph Losey and *The Servant*', *Film* (No. 38), p. 28; Harold Pinter, int., *Isis*, 1 February 1964; Ciment, pp. 224–6.

4 JL, int., *Isis*, 15 January 1964; Ciment, p. 232.

5 Ciment, pp. 227–8; H. Pinter to JL, 19 March 1963.

6 D. Bogarde, int., *Isis*, 15 January 1964; comparative film budgets: *S&S*, Spring 1964.

7 D. Bogarde, int.; N. Priggen, int.

8 H. Pinter, int.; D. Bogarde, int.

9 D. Bogarde, *Snakes*, pp. 299–300, 231.

10 D. Bogarde, int.

11 P. Mayersberg, int.; D. Bogarde, int.

12 Ciment, pp. 233–5; Sarah Miles, int. by Sue Lawley, BBC Radio, 15 August 1991.

13 S. Miles to JL, undated; JL to S. Miles, 30 July 1963.

14 Milne, p. 141; Donald Wiedenman, *Daily Telegraph Magazine*, 18 April 1969; W. Craig to JL, 2 July 1969; JL to W. Craig, 8 July 1969.

15 D. Bogarde, int., *Isis*, 1 February 1964; Brian Phelan, int.

16 N. Priggen, int.; D. Bogarde, int.; Ciment, p. 168.

17 Ciment, pp. 155–6.

18 D. Slocombe, int., *Isis*, 1 February 1964; D. Slocombe, int.; Chic Waterson, int.

19 JL, int., *Isis*, 25 January 1964.

20 JL to J. Proctor, 24 January 1964.

21 Misha Donat, *Isis*, 1 February 1964.

22 J. Dankworth, *Isis*, 1 February 1964.

23 JL notes, about 21 March 1963; J. Trevelyan to JL, 22 March 1963.

2 *THE SERVANT*: II

1 Gilles Jacob, 'Joseph Losey or the camera calls', *S&S*, Spring 1966, p. 65.

2 Ciment, p. 235.

3 D. Bogarde, int.

4 JL, int., *Isis*, 1 February 1964; G.Jacob, 'Joseph Losey or the camera calls', p. 65.

5 Evan Jones, letter, 7 January 1991; D. Bogarde, int.; JL, int., *Isis*, 1 February 1964.

6 Harold Pinter, *Five Screenplays*, London, 1971, pp. 57–9; Francis Wyndham, *Sunday Times Magazine*, 29 November 1964.

7 H. Pinter, int., *Isis*, 1 February 1964.

8 JL note, 17 May 1963; JL to H. Pinter, 2 July 1963, 25 August 1963; H. Pinter to JL, 14 August 1963; 25 August 1963; JL to R. Roud, 17 August 1963.

9 JL, int., *S&S*, Spring 1964, p. 67; JL, Radio 3 int., March 1975; PL. int.; D. Bogarde, *Snakes*, p. 240.

10 Richard Roud, 'Remembering Losey', *S&S*, Vol. 54, No. 1, 1984/5.

11 Eugene Archer to JL, 13 May 1964, 29 May 1964.

12 Ciment, p. 229.

13 Ely Landau to JL, 7 July 1967, 15 January 1968; JL to E. Landau, 13 November 1967; Albert Heit to JL, 19 December 1967; JL to Robin Fox, 13 February 1968, 22 July 1968; JL to Bernard Delfont, 24 October 1969.

Part II The American Years

3 FATHER TO THE MAN

1 Ciment, pp. 2–3, 5; Mary Losey Field, notes re Ciment book, 18 May 1985.
2 Ciment, pp. 3, 6, 12; M. Losey Field, notes re Ciment.
3 Ciment, p. 9; JL, int. by Helen Dudar, *New York Post*, 21 October 1972; *Moviemen*, Thames Television, 24 March 1970.
4 Ciment, pp. 6, 19; M. Losey Field, letter, 4 March 1991.
5 Ciment, pp. 9–13.
6 M. Losey Field, notes re Ciment.
7 JL to Sandra Martin, 24 April 1979; JL to Gerry Burke, 19 January 1978; M. Losey Field, letter, 4 March 1991.
8 For offprints from the 1925 *Booster* I am grateful to Loretta Nockels, of Central High School.
9 Ciment, pp. 13, 14, 16, 22.
10 M. Losey Field, letter, 4 March 1991; JL, int. by Joan Barthel, *NYT*, 26 March 1967; M. Losey Field, int.; Ciment, pp. 2, 5–9, 17–18, 23, 96.
11 Monique Lange, 'En attendant Losey', *Positif*, No. 293/4, July–August 1985, p. 6; Ciment, p. 9; JL, int., *L'Express*, April 1972, typed insertion.
12 M. Losey Field, int.

4 DARTMOUTH

1 Ciment, pp. 13–14, 16–19; Jean H. O'Neill, letter, 26 September 1989.
2 JL to William R. Evans, 16 January 1978.
3 Ciment, pp. 22–4.
4 Ciment, p. 30; Monique Lange, 'En attendant Losey', p. 7; Ned Calmer to JL, 28 September 1971; JL to PL, 11 January 1976.
5 Ciment, pp. 21–2, 24–5; JL, *Daily Mail* int., 11 September 1971, typescript, Item 18; M. Losey Field, letter, 4 March 1991.
6 Ciment, p. 24; May Sarton, *I Knew a Phoenix*, London, 1963.
7 Ciment, p. 25.
8 M. Losey Field, int. and letter, 4 March 1991; Ciment, p. 19; Françoise Sagan writing in *L'Égoïste*, date unknown.
9 M. Losey Field, letter, 4 March 1991; M. Lange, 'En attendant Losey', p. 6; F. Sagan, *L'Égoïste*; M. Losey Field, letter, 4 March 1991.

5 NEW YORK

1 Ciment, pp. 21, 26, 27–9.
2 Ciment, pp. 28–9; John Hammond, *On Record. An Autobiography*, London, 1981; JL to Peter Smith, 24 August 1977.
3 Simon Callow, *Charles Laughton. A Difficult Actor*, London, 1987, pp. 42, 43.
4 Ciment, p. 31; Elsa Lanchester, *Elsa Lanchester Herself*, London, 1983, pp. 102–3;

J. Loring, letter, 21 September 1991; E. Lanchester to JL, 2 July 1968, 21 August 1968; JL to E. Lanchester, 9 July 1968.

5 JL to Maurice Rapf, 10 May 1976; Ciment, p. 32.

6 M. Losey Field, int. and letter, 4 March 1991.

7 Ciment, pp. 33–4.

8 JL to George Freedley, 31 March 1937; Ciment, pp. 34, 35; Norman Lloyd, *Stages*, Metchuen, NJ, 1990, p. 21; Virgil Thomson, int.

9 Ciment, p. 35; J. Hammond, *On Record*, p. 145; JL, int., *Helsingin Sanomat*, 3 March 1935.

10 Ciment, p. 36.

11 May Sarton, *I Knew a Phoenix*, p. 192.

12 Norman Lloyd, 'Remembrance of Theatre Past', *Gambit* (Los Angeles), October 1975, pp. 17–18; Ciment, pp. 34–5.

6 TO RUSSIA AND BACK

1 Ciment, pp. 36.

2 Ciment, pp. 36, 40, 191.

3 I am indebted to Mauri Pasanen for his translation from *Helsingin Sanomat*.

4 Ciment, pp. 38–9.

5 JL, 'L'Oeil du Maître', *Cahiers*, No. 114, December 1960.

6 JL to Hella Wuolijoki, 14 July 1935.

7 JL to A. Goldman, 20 July 1972; A. Goldman, *Minnesota Review*, New Series I, Fall 1973, pp. 70, 72; Ciment, p. 351; Thomas Quinn Curtiss, letter, 15 January 1992.

8 Ciment, p. 41. I thank Karen Etcoff for her searches of the Chicago *Daily News* and *Variety*.

9 Ciment, p. 41.

10 Ciment, p. 40; Vappu Tuomioja, letter, 2 January 1990; Erkki Tuomioja, letter, 2 January 1990.

11 Ciment, p. 204; R. Durgnat, 'A Mirror to Life', *F&F*, June 1959; *Télémoustique*, 15 February 1980; *Nouvel Observateur*, 19–25 November 1979.

12 Ciment, pp. 42–3.

13 JL to Sandra Martin, 24 April 1979; Virgil Thomson, int.

14 M. Losey Field, int., 27 April 1990; Joris Ivens et Robert Destanque, *Joris Ivens ou la mémoire d'un regard*, Paris, 1982, p. 155; Ciment, p. 38; JL, Preface to 'Injunction Granted', *Minnesota Review*, New Series I, Fall 1973, p. 52.

15 Lewis Fraad, int.; Irma Fraad, letter, 24 October 1991; Rosalyn Fraad Baxandall, letter, 16 October 1991.

16 N. Mosley, int.

17 JL to F.V. Field, 29 December 1972; F.V. Field, letter, 14 February 1992.

18 Ciment, p. 39.

7 THE LIVING NEWSPAPER

1 Ciment, p. 43.

2 Rudolph Fisher, *Conjure Man Dies* (script: Item 46).

3 John Houseman, *Run-Through*, New York, 1972, p. 188; JL to J. Houseman, 19 February 1973; Barbara Leaming, *Orson Welles*, New York, 1983, pp. 98–114; Ciment, pp. 44–5.

4 John Willett, *Brecht on Theatre*, London, 1964, p. 84; JL, memo, Item 21.

5 John Houseman, *Unfinished Business*, New York, 1986, pp. 97, 288; JL to J. Willett, 24 October 1974; Ciment, p. 51; *Théâtre Populaire* (Paris), No. 53, 1964, p. 6, quoted in John Willett, *The Theatre of Erwin Piscator*, London, 1978, p. 153.

6 JL, 'The Individual Eye', *Encore*, March–April 1961, p. 11, quoted by Arnold Goldman, 'Life and Death of the Living Newspaper Unit', *Theatre Quarterly*, Vol. III, No. 9, January–March 1973.

7 Ciment, pp. 47, 52–4.

8 A. Goldman, *Minnesota Review*, pp. 73–4; see also the text of *Triple-A Plowed Under* in Pierre Rohan (ed.), *The Federal Theatre Plays*, 2 vols., New York, 1938.

9 JL to A. Goldman, 17 January 1973; Ciment, pp. 52–4; JL, Preface to 'Injunction Granted', *Minnesota Review*, New Series I, Fall 1973, pp. 51–2.

10 A. Goldman, *Minnesota Review*, pp. 74–5.

11 Ciment, p. 54; Virgil Thomson, int.

12 Nearly forty years passed before the text of *Injunction Granted* was published in its entirety by the *Minnesota Review* (New Series I, Fall 1973), with a Preface by JL, who passed his mimeographed production script to Professor Goldman. See pp. 51–2, 76, 48–50. Goldman cites Hallie Flanagan, *Arena: the History of the Federal Theatre*, New York, 1940. See also Jane De Hart Mathews, *The Federal Theatre 1935–1939*, Princeton, 1967, from p. 111. JL to John Houseman, 19 February 1973.

8 LOVE AND OIL

1 Ciment, pp. 46–7.

2 Ciment, pp. 48, 45.

3 Ciment, p. 59.

4 JL to George Freedley, 31 March 1937.

5 Bettina Berch, *Radical by Design. The Life and Style of Elizabeth Hawes*, New York, 1988, pp. 48, 52.

6 Ciment, p. 39; JL to P. Mahoney, 17 March 1978.

7 Newspaper fragment, New York Public Library collection; Ciment, p. 191.

8 Richmond, Va., *News Leader*, 16 August 1939; F.V. Field, letter, 14 February 1992.

9 Ciment, p. 61.

10 JL to S. Lumet, 24 March 1966; Ciment, p. 51.

11 B. Berch, *Radical by Design*, pp. 70–71; B. Berch, int.; Carlos Baker, *Ernest Hemingway*, London, 1969, pp. 397–400.

12 Ciment, p. 55.

13 JL to Raymond Borde, 3 April 1981; Ciment, p. 57.

14 *The Tale of Reynard*, Museum of Modern Art Files: Losey.

15 J. Hammond, *On Record*, p. 232.

16 Ciment, p. 29; J. Hammond, *On Record*, pp. 93, 145–6, 232; Jemison Hammond, conversation.

9 WHOSE WAR?

1 *A Child* is available from New York University Film Library.

2 *World Telegram*, 31 October 1940; Newark, NJ, *Call*, 1 December 1940.

3 JL to S. Martin, 24 April 1979; Ciment, p. 47.

4 On Edward C. Carter and Serge Semenenko, Item 92; Ciment, pp. 59–62; Yola Miller Sigerson, letter, 23 March 1990; JL to S. Martin, 24 April 1979.

5 Ciment, p. 61; application to join Army Air Corps, Item 92; Lewis Fraad, int.

6 12 radio scripts, Item 29.

7 Ciment, p. 62.

8 Ciment, pp. 61–3.

9 Ciment, pp. 63–4.

10 On JL's military service, French TV transcript, Item 230.

11 J. Hammond, *On Record*, passim.

12 JL to P. Mahoney, 1 March 1978, 20 March 1978; P. Mahoney to JL, 11 March 1978; B. Berch, *Radical by Design*, pp. 128–34.

13 B. Berch, *Radical by Design*, pp. 98, 99, 71–2; Gavrik Losey, int.; Irma Fraad, int.

14 B. Berch, *Radical by Design*, pp. 104, 139; Ciment, p.63.

15 JL to P. Mahoney, 17 March 1983.

Part III The Film Director

10 HOLLYWOOD

1 Luisa Stuart Hyun, letter, 13 February 1991, and int.

2 Ciment, p. 64; Luisa Stuart Hyun, letter, 13 February 1991, and int.

3 JL, memo on *A Gun in his Hand*, Item 100; Ciment, pp. 66, 102.

4 Gavrik Losey, int.; Luisa Stuart Hyun, letter, 7 September 1989; PL, int.

5 Ciment, pp. 65–6; Dore Schary to JL, 23 April 1947.

6 Ciment, p. 82.

7 Ciment, pp. 65, 90, 395, 401, 403; Andrew Sinclair, *Spiegel*, London, 1987, p. 52.

11 SENDING A MESSAGE, SIGNS OF A SIGNATURE

1 *Ebony*, December 1948; Ciment, p. 84.

2 JL to Marjorie Worcester, 7 October 1976; publicity article in typescript, Item 1.

3 Ciment, pp. 78, 82; Milne, pp. 71–2; Ben Barzman, 'Pour Joe', *Positif*, No. 293/4, July–August 1985, p. 11.

4 P. Rathvon's notes, Item 1.

5 Ciment, p. 88.

6 Ciment, p. 92.

7 Ciment, pp. 93–4.

8 Hirsch, pp. 36–9.

9 Ezra Goodman, Los Angeles *Daily News*, 13 July 1950.

10 Ciment, p. 10; D. Trumbo, int., *NYT* 15 December 1968.

11 NFT programme note, 1988; JL, int. by Paul Mayersberg and Mark Shivas, *Cahiers*, No. 153, March 1963, pp. 5–6.

12 See Hirsch, pp. 47–51, and Raymond Durgnat, *A Mirror for England. British Movies from Austerity to Affluence*, London, 1970, pp. 247–59.

13 *Cahiers*, No. 111, September 1960, p. 12; Thomas Elsaesser, 'Joseph Losey: Time Lost and Found', *BFMB*, June 1985, p. 173; JL to S. Cohn, 8 March 1954; S. Cohn to JL, 22 April 1954.

14 Losey's letter about Lorre is cited by James K. Lyon, *Bertolt Brecht in America*, Princeton, 1980. See also James K. Lyon, *Brecht's American Cicerone. Bertolt Brecht and Ferdinand Reyher*, Bonn, 1978, p. 197.

15 Ciment, pp. 114, 110.

16 Losey, int., *S&S*, Autumn 1961; Ciment, p. 112, 110; David Thompson, *A Biographical Dictionary of the Cinema*, London, 1975, p. 333; Hirsch, pp. 39–46.

17 Ciment, pp. 114–17.

18 Ciment, pp. 117, 120.

19 Gilles Jacob, 'Joseph Losey or the camera calls', p. 65; Hirsch, pp. 52–3; Ciment, p. 120.

20 Ciment, p. 126; Leon Clore, int.; P. Rissient, *Losey*, p. 88; B. Vorhaus, int.; David Meeker, int.

21 Ciment, pp. 89, 126–8, 130.

22 Henri Alekan, 'Miettes de souvenirs', *Positif*, No. 293/4, July–August 1985, p. 16.

23 See Lorenzo Codelli, 'Un film rétrouvé', *Positif*, No. 293–4, July–August 1985, pp. 58–9, and Raymond Durgnat, 'Losey', *F&F*, April 1966, p. 26.

12 NIGHTMARE: SURVEILLANCE, SUBPOENA, DIVORCE

1 JL to Albrecht Betz, 26 September 1974.

2 Milne, p. 42; JL to S. Martin, 24 April 1979.

3 Ciment, p. 65, 108; JL to Albrecht Betz, 26 September 1974; Luisa Stuart Hyun, int., Boston *Sunday Globe*, 22 August 1971.

4 Milne, pp. 42, 90–91.

5 Ciment, p. 71.

6 Ciment, pp. 71–2.

7 Ciment, p. 69.

8 Ciment, pp. 39, 320, 323.

9 JL, int., *NYT*, 26 March 1967; page proof of article by Keith Howes for *She*

magazine, undated (July 1973); Ciment, p. 107; *Los Angeles Times* and *Hollywood Reporter*, 19 March 1951.

10 Jean Butler, letter, 21 March 1991.

11 *Citizen News*, 23 April 1951; Ciment, pp. 106–9, 122.

12 Ciment, p. 126.

13 Victor Navasky, *Naming Names*, New York, 1980, p. 337; JL to Sidney Cohn, undated (January 1955).

14 Luisa Stuart Hyun, letter, 13 February 1991.

15 Henry J. Bamberger to JL, 20 November 1952; FBI Washington, DC, report, 24 July 1953.

16 Milne, p. 129; FBI Losey File; JL to Sidney Cohn, undated, (January 1955).

17 V. Navasky, *Naming Names*, p. 108; Sidney Cohn testimony, 1977, in Carl Foreman Collection, BFI.

18 Ciment, p. 118.

19 Henry J. Bamberger to JL, 30 September 1952, 14 October 1952, 6 November 1952.

20 JL, int. by David Sterritt, *Christian Science Monitor*, 18 July 1973; Ciment, p. 133; Keith Howes, article for *She*; H. Bamberger to JL, 20 November 1952.

21 Paul Radin's testimony: HUAC, *Hearings*, Los Angeles, Part 5, 12 March 1953.

22 V. Navasky, *Naming Names*, p. 338; H. Bamberger to JL, 9 September 1952, 22 September 1953; JL to S. Cohn, 11 January 1955; FBI Los Angeles memo, 31 August 1953; Luisa Stuart Hyun, letters, 7 September 1989, 13 February 1991; V. Navasky, letter, 24 April 1991.

23 Paul Radin's 'Briar Patch' proposal, Item 105; Carl Foreman, 'Memoirs', in Carl Foreman Collection, BFI; V. Navasky, *Naming Names*, pp. 338, 339, 357.

24 *The Disturber*, directed by Barrie Gavin.

25 Jean Butler, letter, 9 April 1991; Luisa Stuart Hyun, int.; Mary Losey Field, letter, 15 May 1991; Ciment, pp. 116–19.

26 Luisa Stuart Hyun, int. and letter, 13 February 1991.

Part IV The Blacklist Years

13   AN EXILE WITHOUT A NAME

1 Ciment, p. 134; C. Foreman, 'Memoirs', in C. Foreman Collection, BFI.

2 JL to George Elvin, 11 March 1968.

3 Sam White, *Evening Standard*, 14 May 1976.

4 Ciment, pp. 131–2.

5 *Positif*, No. 183/4, 1976, p. 20; Ciment, p. 133; JL to H. Bamberger, 19 March 1955.

6 *S&S*, Spring 1990, p. 117; Hannah Weinstein to JL, 21 January? 1954.

7 Sidney Cole, int.; JL to J. Proctor, 24 January 1975; Ruth Lipton, letter, 4 April 1991; JL to Yola Sigerson, 10 May 1984.

8 Leon Clore papers, 'Time Without Pity'.

9 JL to Hugo Butler, 4 September 1956.

10 JL to S. Cohn, 22 June 1958; FBI Losey File; S. Cohn to JL, 25 June 1968.

14   BRITISH STUDIOS AND STAGES

1 JL to H. Butler, 23 October 1953; Ciment, p. 135.

2 Ciment, pp. 135–6; D. Bogarde, int.

3 Ciment, pp. 135, 138; D. Bogarde to PL, 22 June 1984.

4 Ciment, p. 164; D. Bogarde, *Snakes*, p. 223–4; R. Durgnat, 'Losey', *F&F*, May 1966, p. 30.

5 R. Durgnat, *Losey*, p. 28; Hirsch, p. 64; Ciment, pp. 138–9.

6 Ciment, p. 187; JL to Peter Smith, 17 January 1973; T.C. Worsley, *NS*, 7 August 1954; Peter Cotes, int.; Dorothy Bromiley Phelan, int.

7 JL to H. Butler, 24 February 1955.

8 Ciment, pp. 140–41.

9 L. Clore, int.

10 Ciment, p. 146.

11 Harlequin production budget, Leon Clore papers; Alec McCowen, int.

12 Paul Daneman, int.

13 Freddie F. Francis, int.; L. Clore, int.

14 R. Durgnat, 'Losey', p. 28; R. Durgnat, *A Mirror*, p. 189; Hirsch, p. 67; Ciment, pp. 145–6.

15 *Cahiers*, No. 111, September 1960, p. 9; Ciment, pp. 151–2, 157, 160; David Deutsch, int.; Ciment, p. 151; Pieter Rogers, int.

16 JL, int. by Gordon Gow, *F&F*, October 1971; Ciment, p. 153.

17 R. Durgnat, 'Losey', p. 30; Hirsch, pp. 72–4.

18 Ciment, p. 154; JL to John Wexley, 2 January 1959.

15   BLIND DATE: RETURN OF THE BLACKLIST

1 D. Deutsch, memo on *Blind Date*; Ciment, pp. 174, 182; C. Challis, int.; D. Deutsch, int.

2 Ciment, p. 177; D. Deutsch, int.

3 Ciment, pp. 170, 177; D. Deutsch, int.

4 J.B. Matthews, 'Did the Movies Really Clean House?', *Legion Magazine*, December 1951.

5 *NYT*, 25 August 1952; FBI Losey File.

6 Ben Barzman on Lela Rogers, cited in *S&S*, Spring 1990, p. 117.

7 Ciment, p. 169.

8 *Film Daily*, 11 February 1960; *NYT*, 9 February 1960.

9 R. Lipton, int.; Daniel Mainwaring to JL, 11 October 1960; H. Butler to JL, 28 October 1960.

10 Cleo Trumbo, letter, 22 May 1991; P. Rissient, int.

11 Arthur Miller, *Timebends*, London, 1987.

12 Charles Duke to JL, 24 May 1961; JL to C. Duke, 21 and 31 May 1961.
13 Ciment, pp. 131–2; PL, int.

16   CAPTIVE STATES

1 *Cahiers*, No. 111, September 1960, pp. 6–7.
2 Ciment, p. 237.
3 Ciment, p. 200.
4 JL, int. by Gordon Gow, *F&F*, October 1971.
5 *F&F*, Spring 1964, p. 54; Ciment, pp. 201–6.
6 JL to G.J. Burke, 24 April 1967.
7 D. Bogarde, int.
8 E. Jones, int.; D. Bogarde, int.
9 T. Courtenay, int.
10 D. Bogarde, int.; JL, int., *Cinéma* (Paris) 65, No. 93, February 1965, p. 43.
11 JL to Luigi Chiarini, 20 October 1964.
12 N. Priggen, int.; JL to R. Fox, January 1968; D. Bogarde, *Snakes*, p. 242.

17   JOE AND THE WOMEN

1 R. Lipton, letter, 4 April 1991.
2 Norden & Co. statements; D. Bromiley Phelan, int.
3 *La Crosse Tribune*, 22 June 1984; *NYT*, 23 June 1984.
4 Stanley Foreman, int.
5 Gavrik Losey, int.; Charlotte Aillaud, int.
6 R. Lipton to JL, 23 September 1963; R. Lipton, letter, 4 April 1991.
7 Erkki Tuomioja, letter, 2 January 1990.
8 JL to Vappu Tuomioja, 23 December 1972.

18   EVE

1 Ciment, pp. 207–13; Jean Butler, letter, 21 March 1991; Milne, pp. 152–3; Evan Jones, int.
2 Ciment, pp. 156, 167–8, 220–21.
3 J. Dankworth, int., *Isis*, 1 February 1964; M. Legrand, int.
4 JL to David Rayfiel, 24 August 1979; Ciment, pp. 221–2; M. Legrand, int.
5 P. Rissient, int.
6 JL, memo on *Eve*, 25 April 1963; Correspondence: files of Harbottle & Lewis; Ciment, pp. 22, 214–15, 219.
7 See Hirsch, pp. 60–64.
8 Kenneth Tynan, *Observer*, 10 April 1966.
9 Ciment, pp. 212, 218.

Part V The Sublime

19   *GALILEO* WITH BRECHT

1 See Anthony Pagden, 'Trial and Error', *TLS*, 23–9 September 1988.
2 James K. Lyon, *Bertolt Brecht in America*, p. 156; JL, memo, undated, Item 21.
3 Barbara Leaming, *Orson Welles*, pp. 325–8; Simon Callow, *Charles Laughton*, pp. 178–80; JL to James K. Lyon, 13 January 1971; FBI Los Angeles Report, 12 December 1946; JL, memo, undated, Item 21.
4 JL, 'L'Oeil du Maître'; Guy Flatley, 'Remembrances of Joseph Losey's Past', *Los Angeles Times*, 9 March 1975, cited in J.K. Lyon, *Brecht in America*, p. 221.
5 James K. Lyon, *Brecht's American Cicerone*, pp. 85, 112–17; S. Callow, *Charles Laughton*, pp. 180–82.
6 Ciment, p. 68; J.K. Lyon, *Brecht's American Cicerone*, p. 117.
7 J.K. Lyon, *Brecht in America*, p. 184; Milne, p. 166; Norman Lloyd, *Stages*, p. 121.
8 N. Lloyd, *Stages*, p. 112.
9 J.K. Lyon, *Brecht in America*, pp. 186–8; A. Knox, letter, 15 March 1991; John Houseman, quoted by S. Callow, *Charles Laughton*, p. 186.
10 N. Lloyd, *Stages*, p. 123.
11 JL, 'L'Oeil du Maître'.
12 J.K. Lyon, *Brecht in America*, p. 190.
13 ibid., p. 176.
14 ibid., pp. 329, 334; Ciment, p. 70; Martin Esslin, *Brecht. The Man and His Work*, New York, 1961, p. 80.
15 Ciment, pp. 70–71.
16 James K. Lyon to JL, 23 April 1971.
17 Ciment, p. 75; JL to Robin Fox, 22 December 1957.
18 J.K. Lyon, *Brecht's American Cicerone*, p. 128.

20   *GALILEO*: THE BARREN YEARS

1 Paul Gregory to JL, 29 October 1954.
2 Ciment, p. 76; S. Callow, *Charles Laughton*, p. 196.
3 P. Rogers, int.; JL to Bertha Case, 7 April 1961.
4 *Los Angeles Times*, 3 and 9 September 1975; JL to Guy Flatley, 10 April 1975; JL to E. Lanchester, 21 September 1983.
5 JL to Lotte Lenya, 20 March 1961; L. Lenya to JL, 22 March 1961; JL to Helene Weigel, 21 March 1961.
6 Milne, p. 167.
7 Milne, pp. 167–70; JL to H. Weigel, 19 February 1963.
8 JL to K. Tynan, 4 September 1962; Stefan Brecht to JL, 31 October 1962; JL to Alfredo Bini, 4 November 1962; Jean Butler, letter, 21 March 1991.
9 JL to John Heyman, 21 May 1974; Barbara Bray, int.
10 O. Plaschkes to JL, 26 April 1974; O. Plaschkes, int.

11 Ciment, p. 335; JL to J. Heyman, 16 May 1974; *Galileo* Script, Item 21.
12 R. Dalton, int.; O. Plaschkes, int.; R. Hartley, int.
13 JL to Ely Landau, 15 November 1974; JL to Chaim Topol, 24 January 1975.

21   PINTER AGAIN: *ACCIDENT*

1 Ciment, pp. 99–103, 259–60, 261; H. Pinter, int.
2 *Guardian*, 6 August 1966.
3 N. Mosley, int.
4 N. Mosley, int. and letter, 16 November 1990.
5 D. Bogarde, int.
6 J. Terry to JL, 18 March 1966, 10 March 1967; J. Terry to R. Fox, 10 June 1966; J. Terry to W.V.A. Gell of LIP, 24 March 1966; R. Fox to J. Terry, 25 March 1966; L. Harbottle to JL, 12 April 1967; Horizon contracts in files of Harbottle & Lewis.
7 JL to John Northam, 17 February 1966.
8 PL, int.; JL to Maurice Bowra, 29 April 1966; JL to Dean of St Edmund Hall, 26 July 1966.
9 R. Carr, letter, 12 December 1990; R. Carr to JL, 7 June 1966; R. Carr, letter, 12 December 1990.
10 H. Pinter, *Five Screenplays*, p. 230.
11 R. Carr to JL, 7 June 1966; N. Mosley, int.; R. Carr, letter, 12 January 1990.
12 John Russell Taylor, 'Accident', *S&S*, Autumn 1966, pp. 179–84; Ciment, p. 272; D. Bogarde, int.
13 Margot S. Kernan, 'Accident', *Film Quarterly*, Summer 1967, p. 61; Ellen Baker to JL, 14 July 1975.
14 JL to Moura Budberg, 3 June 1966, 13 July 1966; Ann Firbank, int.; D. Bogarde, int.; JL to Clare Davidson, 9 September 1966; JL to R. Beck, 16 September 1966.
15 JL to Duke of Northumberland, 13 July 1966, 28 July 1966.
16 J.R. Taylor, 'Accident'; JL to C. Dillon, 9 September 1966.
17 Unsigned undated carbon typescript, 'Joseph Losey and *Accident*'; JL, int. on location, Item 13; J.R. Taylor, 'Accident'; V. Starr to JL, 15 May 1973.
18 G. Fisher, int.; R. Beck, int.; R. Dalton, int.; Ciment, p. 266.
19 Ciment, pp. 270, 272; J.R. Taylor, 'Accident'.

22   *ACCIDENT*: THE FILM

1 Ciment, p. 271.
2 N. Mosley, int.; R. Carr, letter, 12 December 1990.
3 Ciment, pp. 270–71.
4 Ciment, p. 268.
5 H. Pinter to JL, 20 February 1967.
6 JL to N. Zervudachi, 28 February 1967, 3 March 1967; *Scottish Express*, 15 March 1967.

7 JL to J. Proctor, 10 March 1967; J. Proctor to JL, 13 March 1967; H. Pinter to JL, 16 March 1967.

8 R. Fox to JL, 22 May 1970; JL to R. Fox, 21 May 1970; N. Priggen, int.

Part VI Boom and Baroque: The Burton Years

23 OP AND POP

1 J. Janni, int.

2 Peter O'Donnell's publicity brochure for the film; E. Jones, int.; Evan Jones projects, Item 106.

3 *NYT*, 1 January 1967; D. Bogarde, int.

4 J. Trevelyan to JL, 4 February 1966.

5 Ronald Bergen, 'Jack Hildyard', *Guardian*, 14 September 1990; Ciment, p. 254.

6 JL, memo to J. Janni and others, 10 March 1966, Item 12.

7 N. Priggen, int.; J. Janni, int.

8 J. Janni, int.; D. Bogarde, int.; PL, notes, 22 February 1992.

9 'Joseph Losey and *Accident*', typescript, no byline, Item 13; *Révolution africaine*, 3–9 April 1967; Ciment, p. 253.

10 Milne, p. 145.

11 JL to A. Brien, 30 November 1971; Ciment, p. 143; JL commercials, Item 88; Ciment, pp. 422–3.

12 G. Beaume to Transcontinental Production, 5 May 1980.

13 Ciment, p. 397.

14 Ciment, pp. 90–94; JL to Hardy Kruger 29 March 1967; *The Man from Nowhere* contract, Harbottle & Lewis files.

24 PUBLIC AND PRIVATE CEREMONIES

1 J. Heyman to Audrey Wood, 2 June 1967.

2 Ciment, pp. 274–5.

3 Milne, p. 130; M. Bragg, *Rich. The Life of Richard Burton*, London, 1988, p. 239.

4 M. Legrand, int.

5 Ciment, p. 275; D. Bogarde, int.; M. Bragg, *Rich*, p. 244; D. Slocombe, int.

6 *Cinéma*, June 1969; JL, NFT lecture, July 1973; JL, int., BBC Radio 3, March 1975, Item 79.

7 *Moviemen*, Thames TV, 24 March 1970; Ciment, p. 294; J. Heyman, letter, 22 October 1990; N. Priggen, int.

8 Ciment, pp. 294–5.

9 Hirsch, pp. 167–73; Marco Denevi, *Secret Ceremony*, trans. Harriet de Onis, in *Prize Stories from Latin America*, New York, 1964.

10 JL, 'Cérémonie sécrète,' *l'Avant-Scène*, June 1969, pp. 8–9; Ciment, p. 269; J. Heyman, int.; Boston *Herald Traveler*, 8 February 1970.

11 Pauline Kael, *Going Steady*, London, 1970.

12 Gordon Gow, *F&F*, October 1971; *Film*, April 1973.

13 *Today's Cinema*, October 1970; J.C. Youngerman to JL, 6 February 1970, 17 December 1970.

14 Ciment, p. 275; M. Bragg, *Rich*, pp. 232, 243; Donald Wiedenman, 'Close-up of a Legend', *Evening Standard*, 17 June 1969; D. Bogarde to PL, 22 June 1984.

15 D. Wiedenman, 'Close-up of a Legend'.

16 JL to E. Taylor, 21 December 1970.

17 D. Bogarde to PL, 22 June 1984; *Isis*, 1 February 1964; D. Bogarde, *Snakes*, pp. 299–300.

18 Ciment, p. 276.

19 D. Bogarde, int.; D. Bogarde to JL, 10 September 1966.

20 D. Bogarde, *Snakes*, p. 28.

21 Ciment, pp. 324–5; D. Bogarde, *An Orderly Man*, London, 1983, p. 129; D. Bogarde, int.

22 D. Bogarde, int.; D. Bogarde to PL, 22 June 1984; PL, int.

23 D. Bogarde, int.; PL, int.

24 D. Bogarde to JL, 31 March 1974, 14 April 1974, 24 April 1974; PL, diary, 18 May 1974.

25 D. Bogarde to PL, 22 June 1984.

26 Ciment, p. 297.

27 Ciment, pp. 296–7, 300; Henri Alekan, 'Miettes de souvenirs', p. 18.

28 JL to R. Fox, 28 August 1969; J. Kohn to R. Shaw, 7 May 1969; J. Kohn to JL, 16 July 1969, 22 August 1969.

29 JL to J. Kohn, 12 March 1970; J. Kohn to JL, 17 March 1970; Ciment, p. 297.

30 JL, 'Dialogue on Film', *American Film*, November 1980, p. 54.

31 PL, diary.

25 THE BARON'S COURT

1 Ints. with: M. Legrand, L. Langley, V. Lindfors, R. Roud, PL, F. Malraux, A. Firbank, R. Handelman.

2 Ints. with: S. Wanamaker, D. Deutsch, P. Cotes, B. Blair.

3 JL to B. Davidson, 22 December 1972; F. Raphael to JL, 7 July ?year; F. Raphael, letter, 8 January 1992.

4 F. Raphael, 'The Hauteur of the Auteur', *TLS*, 18 October 1991.

5 Ints. with: J. Loring, P. Smith, B. Cooper.

6 Ciment, p. 180; P. Mayersberg, int.; B. Bray, int.; A. Knox, letters, 15 March 1991, 21 March 1991; PL, int.; M. Lange, int.; N. Mosley, int.

7 E. Jones, letter, 29 December 1990; D. Bogarde, int.; F. Malraux, int.; J. Loring, letter, 21 September 1991; C. Aillaud, int.; N. Stéphane, int.; James Archibald, *The Aquarian*, 22 June 1983.

8 P. Rogers, int.; P. Drummond, int.; PL, notes, 22 February 1992.

9 PL, int.

10 B. Tolusso, letters, 24 March 1992, 15 April 1992.

11 Ibid.
12 PL, int.
13 B. Tolusso, letters.
14 Ibid.; PL, int.
15 PL to Mr Barr, 7 August 1963; J. Butler, letters, 24 August 1991, 6 April 1992; B. Tolusso, letters.
16 PL, int.; J. Butler, letters; B. Tolusso, letters.
17 B. Cooper, int.
18 PL, int.
19 PL, int.; B. Tolusso, letters.
20 JL, Memoir, September 1976, Item 18; PL, diary.
21 *NYT*, 17 November 1968.
22 JL to Marchioness of Bute, 21 April 1969.
23 JL to G. Burke, 1 June 1966; M. Losey Field, letter, 15 May 1991.
24 JL to T. Cowan, 27 November 1980; PL, int.
25 JL to Frank Chapman, March 1973; G. Burke, int.

Part VII The Palme d'Or

26   *THE GO-BETWEEN*: A NORFOLK SUMMER

1 L.P. Hartley to JL, 15 August 1968.
2 JL to H. Pinter, 22 October 1963; Pinter to JL, 22 October 1963; John Russell Taylor, 'The Go-Between', *S&S*, Autumn 1970.
3 Obituary for L.P. Hartley, *Sunday Times*, 17 December 1972.
4 J.R. Taylor, 'The Go-Between'.
5 JL, Memoir, September 1976, Item 18; R. Dalton, int.; J. Heyman, int.
6 Pinter contract, Harbottle & Lewis files; J. Heyman, int.
7 L.P. Hartley to JL, 27 March 1970; Ciment, p. 240; H. Pinter, int.
8 JL, int. by Gordon Gow, *F&F*, October 1971; J. Loring, int.; Julie Christie, int.
9 D. Kerr to JL, 2 June 1970; JL to D. Kerr, 9 June 1970; E. Fox, int.; J. Heyman to JL, 11 June 6, 1974.
10 Ciment, pp. 296, 310–11; JL to G. Fisher, 24 September 1970.
11 Harold Pinter, *Five Screenplays*, pp. 331, 337, 348.
12 J. Russell Taylor, 'The Go-Between'; Mel Gussow, 'An Interview with Harold Pinter', *Performing Arts*, June 1972, pp. 25–6.
13 R. Beck to JL, 28 October 1970.
14 M. Legrand, int.; JL to R. Rodney Bennett, 11 February 1971; R. Beck, int.; M. Castro, int.; Noel Sinyard, *Filming Literature. The Art of Screen Adaptation*, London, 1986, p. 78.
15 J. Heyman, letter, 22 October 1990; J. Heyman, int.; H. Pinter to J. Heyman, 21 July 1971; J. Heyman, letter, 22 October 1990.

27 *THE GO-BETWEEN*: DEADLY NIGHTSHADE

1 H. Pinter, int.
2 M. Riley and J. Palmer, *The Films of Joseph Losey*, typescript, to be published by Cambridge University Press, New York; H. Pinter, int.
3 For an interesting but unsustainable interpretation of Mrs Maudsley, see N. Sinyard, *Filming Literature*, pp. 78–80.
4 Ciment, p. 131; D. Bogarde, int.; D. Bogarde, *An Orderly Man*, p. 136.
5 B. Tavernier, int.; W. Alford to JL, 4 June 1971.
6 J. Heyman, letters, 22 October 1990, 20 December 1990; D. Bogarde, *An Orderly Man*, p. 88.
7 Laurence Schifano, *Visconti*, London, 1990, p. 401.
8 J. Christie, int. in *Women's Wear Daily*, 30 July 1971.
9 JL to T. Cowan, 14 December 1971, 6 January 1972; JL to J.C. Youngerman, 9 November 1972.
10 JL to Lady Harrod, 30 November 1971.
11 V. Starr quoted by JL to J. Heyman, 6 June 1974; Columbia Pictures TV inter-office memo, from Veronica Gardos to Joe Abruscato 26 July 1974; V. Starr to JL, 31 December 1974.

Part VIII Assassins

28 POLITICS AND FILM

1 JL to Charles Schneer, 8 February 1960.
2 A. Sillitoe to JL, 28 September 1960.
3 JL, 'Mirror to Life', *F&F*, June 1959; *Cahiers*, No. 111, September 1960, p. 12; Milne, p. 48; Ciment, pp. 81, 97.
4 Michel Ciment, int.
5 Milne, p. 92.
6 Jean Pierre Coursodon, 'Losey', *American Directors*, Vol. 2, New York, 1983, p. 206.
7 JL to Paul Kohner, 12 January 1978.
8 Richard Combs, *S&S*, Summer 1975, p. 141; Ciment, pp. 181, 238.
9 John L. Davis to JL, 14 May 1969; JL to J. Ivens, 21 May 1969; J. Ivens to JL, 1 June 1969.
10 M. Ciment, int.
11 JL to F.V. Field, 29 December 1972.
12 Int. by Judy Stone, *Saturday Review*, October 1972; JL to D. Mainwaring, 23 February 1973.
13 M.P. Barrett to JL, 15 February 1974.
14 R. Bolt to JL, 3 December 1974; N. Gordimer to JL, 22 January 1984; JL to N. Gordimer, 7 March 1984; PL, diary, April 1978.

15 JL to Yannik Piel, 11 and 27 January 1983.

29 *THE ASSASSINATION OF TROTSKY*

1 N. Mosley, letter, 16 November 1990, and int.
2 F. Raphael, 'The Hauteur of the Auteur'; JL, int., *Saturday Review*, October 1972.
3 Alex Buchman's film was shown on Channel 4 TV, 30 August 1990.
4 JL, memo, 27 June 1971; JL to J. Shaftel, 16 November 1971; JL to L. Harbottle, 21 April 1971; N. Mosley, int.
5 See also JL, int. by Leslie Rayner, *Film* (London), April 1973.
6 JL to J. Archibald, 24 December 1960.
7 JL to Alberto Pedret, 17 November 1971; N. Priggen, int.; R. Beck, int.
8 M. Bragg, *Rich*, pp. 358, 378–9; Elizabeth Taylor note undated, Item 67; R. Burton to JL, 10 December 1971.
9 Ciment, pp. 263–4, 320.
10 JL to Naomi Laish, 5 October 1983.

30 *A DOLL'S HOUSE*: FILMING WOMEN

1 Ciment, p. 329; JL to N. Stéphane, 19 February 1973.
2 Ciment, p. 329.
3 JL to Anthony Havelock-Allen, 1 August 1972; J. Heyman, int.
4 Ciment, p. 330; R. Dalton, int.
5 G. Fisher, int.; R. Dalton, int.; D. Seyrig to JL, 19 December 1972; *Guardian* obituary of D. Seyrig, 17 October 1990.
6 *She*, page proof, undated, Item 79.
7 J. Heyman, letter, 22 October 1990.
8 Ciment, pp. 198, 264; JL to R. Fox, 11 October 1963.
9 Jean Butler, letter, 16 June 1991; F. Malraux, int.; Ciment, p. 176; D. Bogarde, int.
10 F. Truffaut to A. Hitchcock, 31 August 1967, in François Truffaut, *Correspondance*, 1988, Paris, p. 344; Ciment, pp. 215, 220.
11 JL to Jeanne Moreau, 22 December 1969; PL, diary, 20 May 1975.
12 M. Castro, int.; M. Urman, int.
13 Ciment, p. 203.
14 Ciment, pp. 153, 249–53; Eve Arnold, int.; J. Janni, int.
15 D. Bogarde, int.
16 JL, int., *NYT*, 26 March 1967.
17 JL, int. by Noëlle Beichard for *Elle*, typescript.
18 Ciment, p. 123.
19 JL, int., cited in *Image et son. Révue de Cinéma*, No. 202, February 1967, p. 78 (retranslated from the French); Ciment, p. 212.
20 Ciment, pp. 173, 188.
21 *Image et son*, February 1967; see also *22e Rencontre Cinématographique de Prades*, July 1981.

Part IX Losey's Art and Industry

31   A DIRECTORY OF DIRECTORS

1 Michel Ciment, *Kazan on Kazan*, London, 1973, pp. 83–5; Arthur Miller, *Timebends*, p. 529.
2 D. Bogarde, int.
3 JL, int., *Screen International*, 7 August 1982.
4 *Cahiers*, No. 111, September 1960, p. 3; Ciment, pp. 40, 131; P. Mayersberg, int.; C. Challis, int.
5 P. Houston and J. Gillett, *S&S*, Autumn 1961, p. 187.
6 Francis Wyndham, *Sunday Times Magazine*, 29 November 1964; Milne, pp. 58–9, 119; R. Roud, int.
7 *Sortir*, No. 2, 20 October 1976.
8 JL on Antonioni, *NYT*, 1 January 1967; M. Antonioni, int., *NYT*, 17 November 1967; JL, int., *Variety*, 27 January 1971.
9 JL, int., *Le Figaro littéraire*, 13 October 1962; see Edgardo Cozinsky, 'Joseph Losey', in R. Roud (ed.), *Cinema. A Critical Dictionary. The Major Film Makers*, Vol. 2, New York, 1980, p. 634; JL to Terence Greer, 30 November 1973.
10 JL on Resnais, *S&S*, Autumn 1961; *Guardian*, 6 February 1967; PL, diary, October 1978.
11 JL to Dwight Macdonald, 27 November 1960; Ciment, p. 200; JL, int. by John Heilpern, *Observer*, 23 September 1973.
12 JL, letter in *Sunday Telegraph*, 29 December 1963; JL, 'An American Abroad', *The Times*, 21 January 1964; JL to R. Fox, 31 August 1967.
13 JL, int., *Observer*, 23 September 1973; JL, int., *Daily Mail*, 11 September 1971; JL to R. Wise, 30 November 1971; JL to L. Malle, 19 June 1972; JL to P. Kael, September 1974.
14 Milne, p. 165; *NYT*, 17 November 1968; JL, int. by Jeremy Ross, typescript, 1970, Item 79.

32   BY DESIGN

1 Ciment, pp. 160–63; R. Macdonald, int.
2 R. Macdonald, 'Le Pre-Designing', *Cahiers*, No. 111, September 1960, pp. 13–15.
3 Ciment, pp. 141, 152, 155, 160; R. Macdonald, int.
4 D. Deutsch, int.; Ciment, pp. 144, 169.
5 Ciment, pp. 93, 155; R. Macdonald, int.
6 P. Leech to JL, 24 January 1963; JL to G. Elvin, 24 June 1965, 26 October 1965; G. Elvin to JL, 19 October 1965; R. Macdonald to G. Elvin, 11 July 1963, 29 October 1965; W. Alexander to ??, 25 November 1963.
7 Ciment, pp. 163, 166; R. Macdonald, int.; N. Priggen, int.; R. Roud, 'Remembering Losey'.
8 Ciment, p. 279; R. Roud, int.; R. Macdonald, int.

9 R. Macdonald, int.; JL to J. Pepper, 15 June 1971.
10 R. Dalton, int.; Ciment, pp. 162, 161; R. Macdonald, int.

33   METTEUR OUI, AUTEUR NON

1 D. Wiedenman, 'Close-up of a Legend'; G. Moorhouse, *Guardian*, 13 September 1969; Jonathan Sa'adah, 'Losey on Location', *Dartmouth Magazine*, 1976.
2 Ben Barzman, 'Pour Joe', pp. 10–11.
3 JL, int., *Cahiers*, No. 111, September 1960, p. 6; JL, int., *American Film*, November 1980, p. 57; see also *22e Recontre Cinématographique de Prades*, July 1981.
4 G. Jacob, 'Joseph Losey or the camera falls', p. 64; Freddy Buache, *Cinema anglais autour de Kutrick et Losey*, Paris, 1978, pp. 208–9.
5 R. Dalton, int.; J. Heyman, int.
6 Ciment, p. 104; N. Priggen, int.
7 Ciment, pp. 270, 280.
8 R. Mills to JL, 11 May 1966.
9 JL to E. Landau, 14 June 1974.
10 Ben Barzman, 'Pour Joe', pp. 10–11.
11 Ciment, p. 112; JL to B. Huie, 7 March 1975.
12 André Bazin, 'De la politique des auteurs', *Cahiers*, No. 70, April 1957.
13 D. Trumbo to JL, 20 January 1969; JL to D. Trumbo, 29 January 1969; Thames TV, *Moviemen*, broadcast 24 March 1970.
14 Michel Ciment, 'Mémoire de Losey', *Positif*, No. 293/4, July–August 1985, p. 32.
15 JL, int., *S & S*, Autumn 1961; JL, int., *F & F*, October 1963, p. 54; JL, int., *American Film*, November 1980.
16 JL to D. Angel, 1 October 1958, 10 December 1966; JL to J. Lemmon, 20 October 1973; PL, int.
17 JL, int., *F & F*, October, 1963, p. 54; Penelope Houston, *S & S*, Summer 1966, p. 143.
18 Ciment, p. 195.
19 M. Lange, int.

Part X The Years of Frustration

34   PROUST AND PINTER FOUND AND LOST

1 M. Ciment, int.; D. Toscan du Plantier, int.
2 JL to P. White, 9 April 1974; B. Bray, int.; N. Stéphane, int.
3 R. Lantz to JL, 23 May 1973; JL to N. Rockefeller, 22 March 1973; N. Rockefeller to JL, 9 April 1973.
4 N. Stéphane to JL, 5 April 1973; JL to N. Stéphane, 9 April 1973; N. Stéphane, int.

5 JL to G. Greene, 3 December 1973; G. Greene to JL, 13 December 1973; L. Olivier to H. Pinter, 22 November (probably 1972); D. Bogarde to JL, 3 May 1974.
6 PL, diary, 20 October 1975; N. Seydoux, int.; N. Stéphane, int.
7 JL to R. Dalton, 28 October 1975.
8 H. Pinter to JL, 17 March 1977; JL to Francine Crescent of *Vogue*, 19 January 1978.
9 L. Langley, int.
10 JL to M.-H. and A. Guimares, 8 August 1973.
11 D. Mercer to JL, 17 July 1975, 21 September 1975.
12 PL, diary, 20 October 1975.
13 J. Heyman, int. and letter, 22 October 1990; Richard Gumpert of WFS, Inc. to Judy Daish, 31 May 1984.
14 H. Pinter, int.

35   SAVE ME THE WALTZ

1 Jean-Daniel Roob, *Alain Resnais*, Paris, 1986, pp. 126, 128; F. Malraux, int.
2 Guillermo Cabrera Infante, 'Joe Losey américain', *Positif*, No. 293/4, July–August 1985, p. 12; JL to G. Cabrera Infante, 24 May 1972.
3 JL to C. Ponti, 27 November 1973; R. Burton to JL, 14 July 1973.
4 Luis Buñuel, *My Last Breath*, London, 1985, p. 194.
5 JL to Ruth Misrachi de Davidoff, 17 June 1974; JL to R. Burton, 24 June 1974; J. Heyman, int. and letter, 22 October 1990.
6 JL to S. Kamen, 11 May 1982.
7 M. Linnit to H. Miller, 29 April 1976.
8 JL to K. Hellwig, 12 February 1977.
9 PL, diary; D. Warner, int.; F. Malraux, int.
10 D. Mercer to JL, 23 March 1976.

36   PARTING THE TRAFFIC

1 M. Wale, *Financial Times*, 18 August 1971; G. Gow, *Scotsman*, ? August 1972.
2 JL to PL, December 1976 and January 1977.
3 JL to B. Davidson, December 1972.
4 C. Parker, int.
5 M. Lange, 'En attendant Losey', p. 7.
6 Royal Avenue Residents' Association: Items 81–3, 85.
7 JL to Contessa Ceschi, 10 April 1978.
8 PL, letter, 28 January 1990.
9 Marsha Hunt Presnell, letter, 3 September 1989.

37   FADING LOYALTIES

1 B. Cooper to L. Harbottle, 1 October 1964, 1 April 1965; L. Harbottle to B. Cooper, 14 April 1965; files of Harbottle & Lewis.

2 B. Cooper, letter, 17 December 1991.

3 Norden & Co. to JL, 24 May 1977.

4 R. Chatto, int.

5 JL to R. Fox, 31 August 1967.

6 JL to J. Heyman, June 1974; L. Harbottle to JL, 17 July 1974.

7 D. Bogarde, int.; G. Burke, int.; Angela Fox, *Slightly Foxed*, London, 1986, pp. 184–5, 196–7.

8 R. Chatto, int.; J. Loring, int.

9 C. Parker, int.; Moira Williams, int.

10 B. Tavernier, int.

11 JL to Irene Webb, 27 March 1978; L. Kaplan, letter, 14 February 1990; JL to S. Kamen, 2 October 1981; JL to Carol Levi, 10 November 1981.

12 F. Malraux, int.; M. Lange, int.; M. Castro, int.; B. Tavernier, int.

13 JL to Ruth Misrachi de Davidoff, 21 January 1975; JL, memo, 24 April 1974.

14 J. Christie to JL, undated, January 1974; J. Christie, int.

15 T. Wiseman, letter, 31 December 1991.

16 Ibid.; JL to C. Levi, 17 August 1974.

17 Richard Combs, *S&S*, Summer 1975; Ciment, pp. 341–2; R. Macdonald, int.; JL to D. Angel, 4 November 1974.

18 G. Jackson, letter, 3 January 1991.

19 PL, diary, 20 May 1975; B. Bray, int.; T. Wiseman, letter, 31 December 1991.

38   THE BARON'S BUSINESS

1 H. Bamberger to JL, 9 September 1952.

2 JL to L. Harbottle, 25 January 1965.

3 L. Harbottle to S. Baker, 22 October 1968.

4 JL to H. Bamberger, 14 February 1972.

5 JL to J. Heyman, 16 January 1973, 28 December 1973; JL, memo, 24 April 1974.

6 G. Burke, cited by L. Harbottle to Trust Company of Bahamas, 2 October 1968.

7 JL to T. Cowan, 2 June 1971; JL to J. Heyman, 3 October 1971.

8 Ciment, p. 134–5.

9 C. Foreman to JL, 27 October 1954.

10 JL to H. Bamberger, 17 December 1955; H. Bamberger to JL, 30 December 1955.

11 P. Rogers, int.; S. Gorrie to L. Harbottle, 9 July 1964.

12 S. Gorrie to L. Harbottle, 9 July 1964; H. Bamberger to JL, 29 October 1964; Gorrie & Whitson to JL, 5 January 1965; Gorrie to Harbottle, 15 March 1965.

13 Harbottle & Lewis to JL, 15 August 1967.

14 JL to G. Burke, 8 January 1968; L. Harbottle to J. Heyman, 9 and 10 January 1968; Norden & Co. to Trust Corporation of Bahamas, 28 July 1969, 23 December 1969, 23 January 1970; TCB to Norden, 6 January 1970.

15 M. Weekes of Bookers to L. Harbottle, 20 October 1972.

16 Norden & Co. to Controller of Surtax, 6 June 1974.

17 G. Burke, int.

18 C. Parker to JL, 2 July 1975.

19 R. Sheldon, MP, to Alan Sapper, 16 July 1976; JL, int. by Quentin Falk, *Screen International*, 27 November 1976.

Part XI Losey en France

39 THE LOSEY CULT: A CHEVALIER ON TOUR

1 Raymond Lefevre, 'Joseph Losey', *Image et son*, February 1967, p. 47.
2 André Bazin, 'De la Politique des Auteurs'; J. Dudley Andrew, *The Major Film Theories*, London, 1976.
3 P. Mayersberg, int.
4 Ciment, pp. 145–6; B. Tavernier, 'Æ la recherche de Losey', *Positif*, No. 293/4, July–August 1985, pp. 28–9; B. Tavernier, int.; L. Chauvet, quoted, R. Lefevre, 'Joseph Losey', p. 47.
5 J. Serguine, 'Éducation du Spectateur ou l'École du Mac Mahon', *Cahiers*, No. 111, September 1960, pp. 39–45; Ciment, int.
6 P. Rissient, *Cahiers*, No. 111, pp. 26–30, 33; B. Tavernier, int.
7 Christian Ledieu, *Losey*, Paris, 1963, p. 29; R. Lefevre, 'Joseph Losey'.
8 J. Douchet, *Cahiers*, No. 117, March 1961, p. 47.
9 K. Reisz, int.; see also Gilles Jacob, 'Joseph Losey or the camera calls', p. 64; Gilles Jacob, *Le Cinéma Moderne*, Lyon, 1964, pp. 45–53.
10 Ciment, pp. 144–5; M. Ciment, 'Mémoire de Losey', pp. 30–31.
11 JL to L. Eisner, 4 May 1964; undated memo by JL, Item 10; Norman F. Salter to R. Fox, 28 September 1964.
12 M. Mourlet, 'Beauté de la Connaissance', *Cahiers*, No. 111, September 1960, pp. 34–8; Michel Mourlet, *Sur un art ignoré*, Paris, 1965, pp. 52–9, 71–7, 123–7, 166.
13 N. Priggen, int.; T. Courtenay, int.
14 B. Tavernier, int.; P. Rissient, int.
15 P. Rissient, *Cahiers*, No. 111, September 1960, p. 27; P. Rissient, *Losey*, pp. 100, 108, 110; P. Rissient, int.; B. Tavernier, int.

40 PRESIDENT OF THE JURY

1 P. Rissient, int.
2 G. Jacob, 'Accident', *Cinéma 67*, No. 119, September–October 1967.
3 J. Stanley to JL, 5 September 1967, 10 November 1967.
4 On Langlois in 1968: Peter Lennon, *Guardian*, 15 February 1968; Richard Roud, *A Passion for Films. Henri Langlois and the Cinémathèque Française*, London, 1983. JL to Maître G. Kiejman, 2 January 1978.
5 JL to M. Duras, 20 January 1969; Lee Langley, int.
6 *Positif*, No. 104, April 1969; B. Tavernier to JL, 2 July 1968.
7 B. Tavernier, int.; E. Kahane, int.
8 JL to E. Kahane, 24 March 1971; E. Kahane, int.

9 JL, int. by Gordon Gow, *F&F*, October 1971.

10 JL, int., *Film*, April 1973.

11 On the Cannes jury: PL, int.

12 P. Rissient, int.

13 JL to F. Malraux, 10 July 1968; F. Malraux to JL, 30 September 1968.

14 JL to E. Landau, 9 September 1974; E. Kahane, int.

15 On the rue du Bac: PL, diary, 16 November 1975. On the *Mr. Klein* auction room: PL, diary, 8 December 1975; PL, int.

41 COLLABORATION AND RESISTANCE: LOSEY'S FRENCH FILMS

1 Ciment, p. 351.

2 Ciment, p. 346; PL, diary, 26 October 1975.

3 Alexandre Trauner, *Décors de cinéma*, Paris, 1988, pp. 204–5; JL to PL, 22 January 1976; M. Castro, int.

4 JL to A. Delon, 20 July 1973; PL, diary, 17 July 1974.

5 JL to PL, 10 and 22 January 1976; R. Danon, int.

6 JL to PL, 10 and 22 January 1976.

7 T. Milne, 'Mr. Klein', *BMFB*, May 1977, p. 103.

8 PL, diary, 17 September 1976.

9 PL, diary, 21 February 1977; B. Tavernier, int.; M. Ciment, int.

10 JL to PL, 31 December 1975.

11 P. Kael, *When the Lights Go Down*, New York, 1980, first published in *New Yorker*, 6 February 1978.

12 R. Danon to JL, 15 March 1977; R. Danon, int.

13 M. Ciment, 'Mémoire de Losey', pp. 31–2; M. Ciment, letter, 3 March 1991.

14 C. Aillaud, int.

15 JL to P. Mahoney, 17 March 1983; E. Arnold, int.; JL to Francine Crescent, 19 January 1978.

16 J. Semprun, *L'Express*, 1–7 May 1978; F. Malraux, int.; M. Ciment, int.

17 Y. Rousset-Rouard, int.

18 PL, diary, 25 May 1977, 17 June 1977.

19 Y. Rousset-Rouard, int.; Yves Rousset-Rouard, *Profession Producteur*, Paris, 1979, p. 105; Ciment, p. 358–9; G. Fisher, int.

20 A. Trauner, *Décors de Cinéma*, p. 205; JL to C. Lastricati, 27 October 1967; JL to N. Stéphane, 17 January 1973; JL to Rosalyn Lastricati, 15 November 1978.

21 Y. Rousset-Rouard, *Profession Producteur*, p. 152; accounts: Trinacra to G. Beaume.

42 *DON GIOVANNI* AND *BORIS GODUNOV*

1 Roland Gelatt, 'Don Giovanni: Opera Into Film', *American Film*, April 1979.

2 JL to H. Moore, 21 November 1975, 5 December 1975; JL to PL, 11 January 1976.

3 JL, int., *Nouvel Observateur*, 19–25 November 1979; John Higgins, *The Times*, 18 September 1990; R. Gelatt, '*Don Giovanni*', p. 47; Pierre-Jean Rémy, *Don Giovanni Mozart-Losey*, Paris, 1979, p. 32.

4 France-Inter discussion, about 21 November 1979.

5 A. Trauner, *Décors de Cinéma*, p. 208.

6 R. Gelatt, '*Don Giovanni*', pp. 34, 47.

7 JL, int., *American Film*, November 1980; J. Reiss, int.; D. Toscan du Plantier, int.; M. Castro, int.

8 R. Gelatt, 'Don Giovanni', p. 47.

9 John Higgins, *The Times*, 18 September 1980.

10 D. Toscan du Plantier, *Le Monde*, 14 May 1985; N. Seydoux, int.

11 D. Toscan du Plantier, int., *Los Angeles Times*, 7 April 1980; N. Seydoux, int.

12 JL, int., *NYT*, 17 June 1980; Tom Sutcliffe, *Guardian*, 16 May 1960.

13 JL, int. by Susan Heller Anderson, *NYT*, 17 June 1980; J. Reiss, int.

14 JL to D. Rayfiel.

15 JL to Grazia Porazzini, 23 November 1979; JL to Kiri Te Kanawa, 19 September 1983.

Part XII The Bitter End

43   TAXES, QUARRELS AND TROUT

1 *Le Matin*, 3 October 1981.

2 R. Roud, 'Remembering Losey'; M. Lange, 'En attendant Losey', pp. 4–5; M. Ciment, 'Mémoire de Losey', pp. 31–2.

3 R. Overstreet, int.; B. Bray, int.; M. Castro, int.

4 PL, diary, January 1979.

5 JL to F. Sagan, 22 January 1979.

6 JL to Maître A. Dauman, 5 June 1979.

7 JL to M. Guy, 11 December 1981.

8 On the G. Beaume agency: press clipping, 10 December 1974 – source not known.

9 JL to G. Beaume, 8 June 1978; G. Beaume to JL, 13 June 1978.

10 H. Timms, letter, 30 October 1991.

11 L. Harbottle, int.; JL to L. Kaplan, 14 October 1980, 4 March 1981; A. Selepegno to L. Kaplan, 21 August 1980.

12 H. Bamberger to H. Timms, 23 June 1978; H. Bamberger, int.

13 See also Emmanuel Schlumberger to JL, 8 January 1982.

14 F. Raphael to JL, 7 June 1965; D. Bogarde, int.

15 B. Tavernier, int.; M. Lange, int.; Monique Lange, *Les Scénaristes français*, Paris, 1991.

16 N. Seydoux, int.

17 D. Toscan du Plantier, int.; M. Lange, 'En attendant Losey', pp. 4–5. M. Lange, int.

18 Henri Alekan, *Des Lumières et des ombres*, Paris, 1984, p. 291; H. Alekan, 'Miettes de souvenirs', pp. 15–19; Ciment, pp. 381–2.

19 I. Huppert to JL, undated, Item 281; D. Toscan du Plantier, int.; M. Lange, int.
20 M. Urman, int.; D. Toscan du Plantier, *Le Monde*, 14 May 1985.
21 Henri Alekan, 'Miettes de souvenirs', pp. 15–19.
22 PL, int.

44   FURTHER DISASTERS

1 JL to R. Bolt, 17 March 1975; F. Solinas to JL, 10 June 1978.
2 JL to P. Stone, 29 December 1978; JL to B. Bray, 21 August 1979; Francesca et Francesco Solinas, 'Histoire de la Battaglia', *Positif*, No. 293/4, July–August 1985, pp. 50–52; B. Bray, int.
3 G. Greene to JL, 15 April 1978; JL to G. Greene 13, 21 and 28 March 1978, 10 and 25 April 1978, 12 December 1978, 12 January 1979; G. Greene to JL, 15 April 1978, 13 and 16 February 1979; Quentin Falk, *Travels in Greeneland. The Cinema of Graham Greene*, London, 1984, pp. 178–89.
4 J. Mortimer, letter 16 September 1989; J. Pringle, int.
5 G. Greene, letter, 3 October 1989; J. Pringle, int.
6 J. Heyman to JL, 7 December 1979.
7 JL, int., *American Film*, November 1980; B. Tavernier, 'À la recherche de Losey', p. 29; J. Heyman, int. and letter, 22 October 1990; D. Rayfiel, int.; B. Tavernier, int.
8 M. Lange, 'En attendant Losey', pp. 4–7; D. Bogarde, int.
9 L. Archibald, *The Aquarian*, 22 June 1983; Jean H. O'Neill, letter, 26 September 1989; Jean H. O'Neill to Mary Losey Field, 10 March 1989.
10 J. Moat, int.
11 JL to D. Potter, 1 December 1982, 17 August 1983; V. Bacon, int.
12 G. Cabrera Infante, 'Joe Losey américain', pp. 13–14; JL to P. Syvertsen, 12 October 1983.
13 R. Dalton, int.; J. Loring, int.; PL, int.
14 J. Heyman, letter, 22 October 1990; J. Heyman, int.
15 JL to P. White, 1 October 1974; P. White to JL, 23 April 1974.
16 C. Challis, letter, 27 January 1993; G. Cabrera Infante, 'Joe Losey américain', p. 14; V. Redgrave, letters, 8 November 1991, 20 November 1992.
17 PL's responses to JL's report of notes from Georgina Hale, 22 February 1984; PL, int.; B. Bruce, int.; G. Hale, int.
18 G. Fisher, int.; J. Heyman, int. and letter, 20 December 1990; D. Slocombe, int.; C. Challis, int.
19 R. Durgnat, *A Mirror for England*, pp. 3, 230, quoting *BFMB*s, August 1954, April and July 1956, May 1957, March 1958, September 1959, November 1960, May 1963; R. Durgnat, 'Losey', *F & F*, May 1966, p. 28.
20 G. Jacob, 'Joseph Losey or the camera calls', pp. 62–7; P. Rissient, *Losey*, p. 148.
21 J. Baumbach, *Partisan Review*, 1974, No. 1, p. 88; P. French, *Observer*, 1 July 1984.
22 G. Jacob, 'Joseph Losey or the camera calls'; Jean Pierre Coursodon, 'Losey' in *American Directors*, pp. 205–7; D. Thompson, *A Biographical Dictionary of the Cinema*, p. 333.

23 T. Elsaesser, 'Joseph Losey: Time Lost and Found', pp. 171–5; R. Combs, *BMFB*, June 1985, p. 171.

45   HE'S JUST PASSED US

1 JL to A. Smith, 4 June 1982; A. Smith to JL, 1 July 1982.
2 A. Smith to JL, 24 March 1983; A. Smith to L. Harbottle, 5 July 1984.
3 Ints. with: N. Dunn, M. Williams, R. Dalton.
4 R. Dalton, int.
5 Ints. with: R. Dalton, PL, J. Loring, M. Williams.
6 Ints. with: R. Dalton, V. Bacon, J. Heyman.
7 D. Bogarde, int.

Part XIII The Films: A Viewing

46   THE FILMS IN PART III

1 Milne, p. 69.
2 Ciment, pp. 104–6.
3 Ciment, pp. 112–13.

47   THE FILMS IN PART IV

1 Ciment, pp. 139–40.
2 Richard Macdonald, int., *Isis*, 1 February 1964.
3 Ciment, p. 173.
4 D. Deutsch, int.
5 *Oxford Opinion*, int. with JL, typescript; *Cahiers*, No. 111, September 1960, pp. 11–12.
6 Philip French, *S & S*, Spring 1990.
7 Ciment, p. 200; Mary Losey Field, int.
8 For a detailed analysis of the prologue, see M. Riley and J. Palmer, *The Films of Joseph Losey*.
9 Hirsch, pp. 175–9; see JL, Int. in *Cinéma*, pp. 38–9.

48   THE FILMS IN PARTS V-X

1 Hirsch, p. 210; Losey production notes, Item 21.
2 Hirsch, pp. 189–94, 202.
3 M. Riley and J. Palmer, *The Films of Joseph Losey*.

49   THE FILMS IN PARTS XI AND XII

1 Ciment, pp. 353–4.
2 On Pierre's motivation, see Tom Milne, 'Mr. Klein', *BMFB*, May 1977, p. 103.

Acknowledgements

Joseph Losey's widow Patricia has most generously made available to me every kind of material – private documents, diaries, correspondence, videos of elusive Losey films, personal recollections and introductions. At the outset Patricia Losey granted me the indispensable permission to quote from Joe Losey's work now under her copyright. However, an 'authorized' biography is not necessarily, or desirably, or in this case (as ultimately became clear) an 'approved' one. The biographer has many debts but no creditors.

Obviously access to Joe Losey's inner family has been invaluable, and I am grateful for many discussions with his sister, Mary Losey Field, his sons, Gavrik and Joshua, and his former wives Luisa and Dorothy.

Among Joe's friends and colleagues, I am particularly indebted to Harold Pinter for his long recall and long-suffering kindness.

To the staff of the British Film Institute, which holds the Losey Collection, I am also much indebted. Janet Moat, Losey's archivist at the BFI, has been tirelessly helpful, informative and perceptive. My thanks also to Deac Rossell, David Meeker, Jacquie Morris and Julie Rigg, of the BFI, and to members of the Stills Department, for invaluable advice and assistance.

I am grateful to the Authors' Foundation of the Society of Authors for a most timely research grant. Joe's good friend John Loring generously subsidized research assistance by Karen Etcoff in New York – my gratitude to them both.

My wife Martha not only sustained me during a four-year labour with every kind of practical assistance, but finally brought the critical eye of a professional editor to the typescript. She sustained my spirit when the silencers became loudest.

I would like to thank the following for interviews, conversations and correspondence:

Charlotte Aillaud, Eve Arnold, Naim Atallah, Victoria Bacon, Henry Bamberger, Rosalyn Baxandall, Reginald Beck, George A. Bell, Bettina Berch, Pierre Blondeau, Sir Dirk Bogarde, John Bowen, Barbara Bray, Dorothy Bromiley (Phelan), Brenda Bruce, Freddy Buache, Gerry Burke, Jean Butler.

Simon Callow, Alex Cameron, Sir Raymond Carr, Marie Castro, Christopher Challis, Rosalind Chatto, Julie Christie, Michel Ciment, Leon Clore, Sidney Cole, David Cook, Dr Barrington Cooper, Peter Cotes, Theo Cowan, Tom Courtenay, Thomas Quinn Curtiss.

Richard Dalton, Paul Daneman, Raymond Danon, Basil Davidson, David Deutsch, Philippa Drummond, Nell Dunn, Frederick V. Field, Ann Firbank, Gerry Fisher, Stanley Forman, Edward Fox, Irma Fraad, Dr Lewis Fraad, Freddie Francis.

Sir John Gielgud, David Glassman, Graham Greene, Georgina Hale, Jemison Hammond, Rosine Handelmann, Laurence Harbottle, Richard Hartley, John Heyman, Foster Hirsch, Jim Hoberman, Lynn Horsford, Luisa Stuart Hyun (Louise Losey),

Glenda Jackson, Diane Jacob, Joseph Janni, Evan Jones, Eric Kahane, Leon Kaplan, Alexander Knox.

Monique Lange, Lee Langley, Michel Legrand, Carol Levi, Oscar Lewenstein, Viveca Lindfors, Ruth Lipton, John Loring, Sidney Lumet, Richard Macdonald, Florence Malraux, Robert M. Marr, Paul Mayersberg, Leo McKern, Alec McCowen, Daphne Mercer-Hadari, Paul Mills, Jeanne Moreau, John Mortimer, Nicholas Mosley, Victor Navasky, Dan Nissen, Loretta Nockels, Jean H. O'Neill, Richard Overstreet.

James W. Palmer, Celia Parker, Mauri Pasanen, Brian Phelan, Louise and Otto Plaschkes, Marsha Hunt Presnell, Norman Priggen, John Pringle, Gerald Rabkin, Frederic Raphael, David Rayfiel, Vanessa Redgrave, Janine Reiss, Betsy Blair Reisz, Karel Reisz, Robert Riley, Harold C. Ripley, Pierre Rissient, Pieter Rogers, Richard Roud, Yves Rousset-Rouard.

Jonathan Sa'adah, Alan Sapper, Nicolas Seydoux, Yola Sigerson, Douglas Slocombe, Anthony Smith, Peter Smith, Nicole Stéphane, Peter Stone.

Bertrand Tavernier, Virgil Thomson, Heather Timms, Bruno Tolusso, Ghigo Tolusso, Daniel Toscan du Plantier, Erkki Tuomioja, Vappu Tuomioja, Cleo Trumbo, Mark Urman, Bernard Vorhaus, Sam Wanamaker, David Warner, Chic Waterson, Moira Williams, Thomas Wiseman, Michael York.

Index